A Descriptive Catalogue of Recent Shells
by Lewis Weston Dillwyn

Address:
HardPress
8345 NW 66TH ST #2561
MIAMI FL 33166-2626
USA
Email: info@hardpress.net

ANNEX

Dillwy
QHP

A

DESCRIPTIVE CATALOGUE

OF

RECENT SHELLS,

ARRANGED ACCORDING TO

THE LINNÆAN METHOD;

WITH

PARTICULAR ATTENTION

TO THE

Synonymy.

BY

LEWIS WESTON DILLWYN, F.R.S. AND F.L.S.

HONORARY MEMBER OF THE GEOLOGICAL SOCIETY OF LONDON,
THE LINNÆAN SOCIETY OF PHILADELPHIA, &c.

IN TWO VOLUMES.

—

VOL. II.

London:

PRINTED FOR JOHN AND ARTHUR ARCH, CORNHILL.

—

1817.

J. M'Creery, Printer,
Black-Horse-court, Fleet Street, London.

DESCRIPTIVE CATALOGUE

OF

RECENT SHELLS.

Genus XXIV.

BUCCINUM:

SHELL UNIVALVE, SPIRAL AND GIBBOUS; APERTURE OVATE, ENDING IN A SHORT CANAL WHICH BENDS TO THE RIGHT, AND THE PILLAR-LIP FLATTENED.

Subdivisions.†

    * Inflated, rounded, thin, slightly transparent and brittle.

    ** With a short, exserted, reflected beak, and the outer lip unarmed outwardly.

    *** Resembling the last division, but the outer lip on the outside is spinous at the base.

    **** With the pillar-lip dilated and thickened.

    ***** With the pillar-lip appearing as if worn flat.

---

† The following indistinct species of Gmelin have been omitted. *B. asperum*, p. 3503. *B. canaliculatum*, p. 3505. *B. cancellatum*, p. 3489. *B. Chalys*, p. 3497. *B. cuspidatum*, p. 3505. *B. Digitellus*, p. 3504. *B. edentulum*, p. 3505. *B. fasciolatum*, p. 3504. *B. lividulum*, p. 3505. *B. Macloviense*, p. 3493. *B. nanum*, p. 3497. *B. nigropunctatum*, p. 3497. *B. niveum*, p. 3471. *B. obliquum*, p. 3504. *B. punctulatum*, p. 3503. *B. Pugio*, p. 3505. *B. squalidum*, p. 3487. *B. striatum*, p. 3477. *B. stromboides*, p. 3489. *B. trifasciatum*, p. 3489. *B. tuberculatum*, p. 3503. *B. verrucosum*, p. 3497.

****** Somewhat polished, and not enumerated
       in the former divisions.
******* Angulated, and not included in the for-
        mer divisions.
******** Turreted, subulate, and slightly polished.

———

* *Inflated, rounded, thin, slightly transparent, and
   brittle.*

OLEARIUM. 1. Shell roundish, transversely ribbed,
    with an elevated line in the interstices ; aper-
    ture without teeth.

Buccinum Olearium. *Linnæus Syst. Nat.* p. 1196. *Mar-
    tini,* iii. p. 403. t. 117. f. 1076 and 1077. *Schroeter Einl.*
    i. p. 307. *Gmelin,* p. 3469. *Schreibers Conch.* i. p. 138.
    Bruguiere Enc. Method. p. 243.
Rumphius, t. 27. f. D. *Petiver Gaz.* t. 99. f. 11, and *Amb.*
    t. 9. f. 7. *Gualter,* t. 44. f. T. *Seba,* iii. t. 69. two first
    figures. *Knorr,* v. t. 12. f. 1.
Inhabits the Indian Ocean. *Linnæus.* China, and the East
Indies. *Humphreys.*
Shell two, three, or four inches long, and more than two thirds
    as broad, pale brown or fawn-colour, and sometimes mark-
    ed with a few darker irregular spots or streaks ; there are on
    the body-whirl about eighteen broad ribs, with narrow inter-
    stices having an elevated line in the middle ; the spire con-
    sists of five small whirls marked with a distinct groove.

GALEA. 2. Shell ovate, transversely ribbed ; ribs
    convex, and becoming double near the edge ;
    aperture without teeth.

Buccinum Galea. *Linnæus Syst. Nat.* p. 1197. *Marti-
    ni,* iii. p. 393. t. 116. f. 1070. *Born Mus.* p. 239.
    Schroeter Einl. p. 308. t. 2. f. 1. *Gmelin,* p. 3469.
    Schreibers Conch. i. p. 138. *Bruguiere Enc. Method.* p.
    244.
Dolium Galea. *Lamarck Syst. des Anim.* p. 79.
Bonanni Rec. 3. f. 183, and *Kirch.* f. 179. *Lister Conch.*
    t. 898. f. 18. *Gualter,* t. 42. *Favanne,* t. 27. f. B 1.
Inhabits the Mediterranean. *Linnæus.* Coasts of Naples.
    Ulysses.

Shell eight or ten inches long, and almost equally broad, varying in colour from pale grey to ochraceous, ventricose, transversely ribbed, and the ribs near the edge divided by a longitudinal indenture. The number of ribs on the body-whirl is about twenty.

PERDIX. 3. Shell ovate, inflated, slightly grooved, and undulated with white; aperture without teeth.

Buccinum Perdix. *Linnæus Syst. Nat.* p. 1197. *Martini,* iii. p. 406. t. 117. f. 1078 to 1080. *Born Mus.* p. 240. *Schroeter Einl.* i. p. 309. *Gmelin,* p. 3470. *Schreibers Conch.* i. p. 138. *Bruguiere Enc. Method.* p. 245. *Dorset Cat.* p. 44. t. 15. f. 14. *Montagu Test.* p. 244. t. 8. f. 5. *Maton and Racket, in Lin. Trans.* viii. p. 133.

La Tesan. *Adanson Senegal,* p. 107. t. 7. f. 5.

Bonanni Rec. 3. f. 191, and *Kirch.* f. 189. *Lister Conch.* t. 984. f. 43. *Rumphius,* t. 27. f. C. *Petiver Gaz.* t. 153. f. 13, and *Amb.* t. 4. f. 11. *Gualter,* t. 51. f. F. *Argenville,* t. 17. f. A. *Knorr,* iii. t. 8. f. 1. *Favanne,* t. 27. f. A 1.

Variety. Shell rounder, and the spire more distinctly grooved at the sutures.

Buccinum maculosum. *Solander's MSS. Portland Cat.* p. 136. lot 3050.

Seba, iii. t. 68. f. 16.

Inhabits the coasts of America. *Linnæus.* Jamaica. *Lister.* Amboyna. *Rumphius.* Coast of Senegal. *Adanson.* West Indies, China, and South Seas, *Humphreys;* who also says the variety is from New Holland.

Shell from three to five inches long, and about three fourths as broad, thin, rather brittle, of a greyish, pale brown, or fawn colour, longitudinally undulated with white spots; the ribs formed by the shallow transverse grooves, of which there are about twenty on the body-whirl, are more or less flat, and the pillar is umbilicated. Dr. Solander, for his *B. maculosum,* has referred to Seba's fig. 16, and it appears to be only a variety of this species.

POMUM. 4. Shell ovate, with transverse convex ribs; pillar wrinkled, and the outer lip toothed and thickened.

Buccinum Pomum. *Linnæus Syst. Nat.* p. 1197. *Marti-ni*, ii. p. 58. t. 36. f. 370 and 371. *Born Mus.* p. 240. *Schroeter Einl.* i. p. 310. *Gmelin*, p. 3470. *Schreibers Conch.* i. p. 139. *Bruguiere Enc. Method.* p. 247. *Bonanni Rec.* and *Kirch.* 3. f. 22. *Lister Conch.* t. 792. f. 45. *Rumphius*, t. 23. f. 4, and t. 27. f. B. *Petiver Amb.* t. 12. f. 6. *Gualter*, t. 51. f. C. *Argenville*, t. 17. f. L. *Seba*, iii. t. 70. f. 3 and 4. *Knorr*, vi. t. 23. f. 2. *Favanne*, t. 27. f. G.
Inhabits the coasts of Spanish America. *Bonanni.* Amboyna. *Rumphius.* Java. *Linnæus.* China. *Humphreys.*
Shell from an inch·and a half to two inches and a half long, and about two thirds as broad, white mottled with yellowish or testaceous spots, or pale brownish yellow with white spots; there are about thirteen convex ribs on the body-whirl; the aperture is rather contracted, and the outer lip is thick, marginated, and strongly toothed.

SULCOSUM. 5. Shell ovate, with transverse flat ribs; pillar smooth, and the outer lip toothed and thickened.

Buccinum sulcosum. *Born Museum,* p. 241.
Buccinum fasciatum. *Bruguiere Enc. Meth.* p. 247.
Seba, iii. t. 68. f. 17. *Martini*, iii. p. 406. t. 118. f. 1081. *Favanne*, t. 27. f. B 2.
Inhabits the coasts of Coromandel. *Martini.* Chinese Seas. *Humphreys.*
Shell three inches and three quarters long, and hardly two thirds as broad, white, with four reddish or yellowish transverse bands, which gradually vanish before they approach the margin of the outer lip; there are twenty-one or twenty-two flat approximated ribs on the body-whirl.

DOLIUM. 6. Shell ovate, inflated, with remote semicylindrical ribs; beak rather prominent.

Buccinum Dolium. *Linnæus Syst. Nat.* p. 1197. *Martini*, iii. p. 397. t. 117. f. 1073 and 1074. *Born Mus.* p. 241. *Schroeter Einl.* i. p. 311. *Gmelin*, p. 3470. *Schreibers Conch.* i. p. 139. *Bruguiere Enc. Meth.* p. 246.
Variety A. White or pale brown, with reddish brown or orange spots on the ribs.
Bonanni Rec. 3. f. 16, 17, and 25, and *Kirch.* f. 16, 17,

and 28. *Lister Conch.* t. 899. f. 19. *Rumphius,* t. 27. f. A. *Petiver Amb.* t. 12. f. 5. *Gualter,* t. 39. f. E. *Argenville,* t. 17. f. C. *Seba,* iii. t. 68. f. 9 to 11, and t. 70. f. 1. *Knorr,* iii. t. 8. f. 4. *Favanne,* t. 27. f. C 1, and C 2.

Variety B. Pale brown, without any spots on the ribs.

Buccinum Allium. *Solander's MSS.*

Martini, iii. p. 396. t. 117. f. 1072, and t. 118. f. 1082. *Brookes's Intr.* t. 6. f. 82.

Inhabits the coasts of Sicily and Barbary. *Bonanni.* Amboyna. *Rumphius.* Philippine Islands. *Petiver.* Senegal. *Adanson.* Tranquebar. *Martini.* Very common about Tarentum. *Ulysses.*

Shell about three or four inches long, and three fourths as broad, with twelve or thirteen distant semicylindrical ribs on the body-whirl, and there are sometimes one or two faint elevated lines in the interstices; the shell is inflated and thin, and the pillar twisted.

CHINENSE. 7. Shell globose, inflated, ribbed and striated transversely; aperture spreading; outer lip toothed.

Buccinum Australe seu Chinense. *Chemnitz,* xi. p. 85. t. 188. f. 1804 and 1805.

Inhabits the coasts of Java and China. *Chemnitz.*

It appears, from the description of Chemnitz, that this species is nearly allied to *B. Dolium,* of which it seems doubtful whether it is more than a variety, as elevated striæ in the interstices of the ribs frequently occur in that species, and the outer lip, though it is said to be toothed, appears in the figure to be only crenated. The length is seven inches, the breadth five, and the height four inches; the shell is white, with the transverse ribs brown, and marked with longitudinal somewhat undulated bands of brown, and there is a row of brown spots at the inner margin of the outer lip.

CAUDATUM. 8. Shell ovate, with transverse rounded ribs, and the beak rather prominent.

Buccinum caudatum. *Gmelin,* p. 3471. *Schreibers Conch.* i. p. 150.

Buccinum, No. 13. *Schroeter Einl.* i. p. 359.

Knorr, v. t. 3. f. 4. *Martini,* iii. p. 408. t. 118. f. 1083 and 1084.

Inhabits ――――

Shell about an inch and three quarters long, and an inch broad, of a brownish or reddish colour, umbilicated, with a wide aperture, and the outer lip plaited and toothed within ; spire produced, and composed of six rather inflated whirls. Having been unable to ascertain this species, I have left it in the place which Gmelin has assigned to it, but it ought probably to be removed to the Murices, as some references which belong to *M. clandestinus* have been confounded with it. Schreibers has removed it to the third division of Buccinum, and in his Conchylien it stands next to *B. Glans.*

** *With a short exserted reflected Beak, and the outer Lip unarmed outwardly.*

ECHINOPHORUM. 9. Shell with four tuberculated belts, and the beak prominent.

Buccinum Echinophorum. *Linnæus Syst. Nat.* p. 1198, *Martini,* ii. p. 86. t. 41. f. 407 and 408. *Born Mus.* p. 242, and Vign. at 238. f. *a, b.* *Schroeter Einl.* i. p. 312. *Gmelin,* p. 3471. *Schreibers Conch.* i. p. 141.
Cassida Echinophora. *Bruguiere Enc. Meth.* p. 437.
Bonanni Rec. and *Kirch.* 3. f. 18 and 19. *Lister Conch.* t. 1003. f. 68. *Rumphius,* t. 37. f. 1. *Gualter,* t. 43. f. 3. *Argenville,* t. 17. f. P, and Zoom. t. 3. f. H. *Seba,* iii. t. 68. f. 18. *Knorr,* i. t. 17. f. 1. *Favanne,* t. 26. f. E 3, and t. 70, f. P 1.
Inhabits the Mediterranean. *Linnæus.* Common in the Bay of Naples. *Ulysses.*

Shell about three and a half inches long, and near two thirds as broad, of a pale brown or fawn-colour marked with transverse elevated striæ, and four strongly tuberculated belts ; the epidermis is often worn from the summits of the tubercles which then appear white, and the aperture is also white.

NODOSUM. 10. Shell with five rather acute belts, of which the upper are tuberculated ; beak prominent.

Buccinum strigosum. *Gmelin,* p. 3476.
Buccinum Echinophorum, Var. 3. *Gmelin,* p. 3472.
Cassidea carinata. *Bruguiere Enc. Meth.* p. 438.
Buccinum, No. 97. *Schroeter Einleitung,* i. p. 380.
Lister Conch. t. 1011, f. 71 f.

Inhabits ——

Shell an inch and a half long, and eleven lines broad, and differs from *B. Echinophorum* in being smaller, and in having five transverse belts, of which only the two upper are studded with tubercles. M. Bruguiere says it is found in a fossil state at Courtagnon in Champagne. Gmelin has three species under the name of *B. strigosum,* and as Bruguiere's appellation of *carinata* was before occupied, it became necessary to find another.

RUGOSUM. 11. Shell with crowded transverse elevated striæ, of which two of the upper are tuberculated and broader than the others; aperture toothed on both sides.

Buccinum rugosum. *Linnæus Mantissa,* p. 549.
Buccinum Tyrrhenum. *Chemnitz,* x. p. 192. t. 153. f. 1461 and 1462. *Gmelin,* p. 3478.
Buccinum Echinophorum, Var. *Martini,* ii. p. 88. *Gmelin,* p. 3472.
Buccinum ochroleucum. *Gmelin,* p. 3477.
Buccinum, No. 113. *Schroeter Einleitung,* i. p. 385.
Cassidea Thyrrena. *Bruguiere Enc. Meth.* p. 439.
Bonanni Rec. 3. f. 160, and *Kirch.* f. 162. *Lister Conch.* t. 1011. f. 71 e. *Gualter,* t. 43. f. 2. *Ginanni Op. Post.* ii. t. 5. f. 44, and t. 6. f. 45. *Favanne,* t. 26. f. E 1, and E 2.
Inhabits the shores of Tuscany. *Chemnitz.* Common in the Mediterranean, and particularly on the coasts of Italy and Sardinia. *Bruguiere.*
Shell about three inches and a quarter long, and two inches broad, white or yellowish with numerous crowded elevated transverse striæ, of which the fifth and sixth, reckoning downwards from the suture, are tuberculated and broader than the others; the aperture is white, and the outer lip strongly marginated.

ABBREVIATUM. 12. Shell with crowded transverse elevated equal striæ; whirls inflated, and decreasing gradually; base abbreviated; aperture toothed on both sides.

Buccinum abbreviatum. *Chemnitz,* x. p. 194. t. 153. f. 1463 and 1464. *Gmelin,* p. 3468.
Variety. Crenated at the sutures.

Chemnitz, x. p. 194. t. 153. f. 1465 and 1466.
Inhabits the Indian Ocean, and the Variety is from the coasts
of America. *Chemnitz.*
Shell an inch and a half long, and an inch broad, ovate, whitish, with a conical spire composed of six inflated whirls becoming gradually smaller. The pillar is wrinkled, and the
outer lip marginated and toothed within; the beak is very
short, and the base has somewhat of a truncated appearance.
The shell which I have followed Chemnitz and Gmelin in
placing as a variety, is only about half the size, of a reddish
or yellowish colour, with a crenated appearance at the sutures, and is most probably a perfectly distinct species.

PLICATUM. 13. Shell slightly plaited on the fore-
part, and somewhat decussated with striæ;
aperture toothed; beak recurved.

Buccinum plicatum. *Linnæus Syst. Nat.* p. 1198. *Mar-
tini,* ii. p. 68. t. 37. f. 379 and 380. *Schroeter Einl.* i.
p. 313. *Gmelin,* p. 3472. *Schreibers Conch.* i. p. 142.
Cassidea Crumena. *Bruguiere Enc. Meth.* p. 428.
Bonanni Rec. 3. f. 161. *Lister Conch.* t. 1002. f. 67.
Gualter, t. 40. f. C. *Favanne,* t. 26. f. I.
Inhabits the coasts of Jamaica. *Linnæus.* Ascension Island.
Lister.
Shell about two and a half inches long, and an inch and three-
quarters broad, of a cinereous or pale brownish flesh-colour; the upper part of the body-whirl is plaited, and the
lower part marked with decussated striæ. Solander for *B.
plicatum,* has quoted Martini, f. 381 and 382, which all
other Conchologists have considered to be *B. tuberosum* of
Linnæus, and the description in the Systema Naturæ is so
short, and the references so discordant, that *B. plicatum*
will probably always continue to be rather a doubtful
species.

CORNUTUM. 14. Shell turbinated, and armed with
three transverse rows of tubercles; inner lip
much dilated and rounded; aperture toothed;
beak recurved.

Buccinum cornutum, Var. β. *Schroeter Einl.* i. p. 315.
Gmelin, p. 3473. *Schreibers Conch.* i. p. 142.
Cassidea cornuta, Coquille adulte. *Bruguiere Enc. Meth.*
p. 435.

Cassis cornuta, Coquille vieille. *Lamarck Syst. des Anim.*
p. 80.

Cassis labiata. *Chemnitz*, xi. p. 71. t. 184 and 185.
Lister Conch. t. 1008. f. 71 *b. Rumphius*, t. 23. f. A,
Petiver Amb. t. 7. f. 10, and t. 11. f. 10. *Martini*, ii.
t. 35. f. 362. *Favanne*, t. 26. f. A 2.

Junior. Buccinum cornutum. *Linnæus Syst. Nat.* p. 1198.
Martini, ii. p. 33. t. 33. f. 348 and 349. *Born Mus.*
p. 243. *Schroeter Einl.* i. p. 314. *Gmelin*, p. 3472.
Schreibers Conch. i. p. 142.

Cassidea cornuta, Coquille jeune. *Bruguiere Enc. Methodique*, p. 434.

Cassis cornuta, Coquille jeune. *Lamarck Syst.* p. 80.
Bonanni Rec. and *Kirch.* 3. f. 155. *Lister Conch.* t.
1006. f. 70, and t. 1009. f. 71 *c. Rumphius*, t. 23. f. 1.
Petiver Gaz. t. 151. f. 9, and *Amb.* t. 7. f. 14. *Gualter*,
t. 40. f. D. *Seba*, iii. t. 73. f. 7, 8, 17, and 18. *Knorr*,
iii. t. 2. f. 1. *Favanne*, t. 26. f. A 1.

Inhabits Amboyna. *Rumphius.* Coasts of America. *Linnæus?*
East Indies and China. *Humphreys.*

This shell when at maturity, according to Born, is sometimes
more than a foot long, but the length is more commonly
about nine inches, and the extreme breadth near seven ; the
whole of the outside is of a dirty white, and indistinctly reticulated with shallow grooves which form obsolete nodules,
besides which there are also three transverse rows of large
distant pointed tubercles ; the pillar is purplish black with
white elevated plaits, and a large dilated flesh-coloured lip
which is rounded at the extremity ; the outer lip is dilated
and flesh-coloured, and has about eight distant broad white
teeth which are bordered with black. Young shells are
smaller, and are marked with reddish or chestnut-coloured
spots ; they have three smooth transverse convex bands,
which become obsolete when the shell arrives at maturity,
and in their place the rows of large tubercles are then
formed. Linnæus appears only to have known this species
in this early stage of its growth.

RUFUM. 15. Shell with decussated striæ, and nodulous transverse belts, between which is a
double line ; aperture toothed ; beak recurved.

Buccinum rufum. *Linnæus Syst. Nat.* p. 1198. *Martini*,
ii. p. 20. t. 32. f. 341, and t. 33. f. 346 and 347. *Born*

Mus. p.   . *Schroeter Einl.* p. 315. *Gmelin,* p. 3473.
Schreibers Conch. i. p. 143.
Cassidea rufa. *Bruguiere Enc. Methodique,* p. 433.
Bonanni Rec. 3. f. 328 and 329, and *Kirch.* f. 326 and
327. *Rumphius,* t. 23. f. B. *Petiver Amb.* t. 5. f. 5.
Gualter, t. 40. f. F. *Knorr,* ii. t. 9. f. 2, and iv. t. 1.
Seba, iii. t. 73. f. 3, 4, and 9. *Regenfuss,* i. t. 12. f. 69.
Favanne, t. 26. f. D 2.
Junior.   Buccinum pennatum. *Gmelin,* p. 3476.
Buccinum ventricosum. *Gmelin,* p. 3476.
Buccinum Pullum. *Born Mus.* p. 245.
Buccinum Rumphii. *Shaw's Nat. Misc.* xv. t. 596.
Buccinum, No. 2, and No. 94. *Schroeter Einl.* p. 357,
and p. 379.
Cassidea pennata. *Bruguiere Enc. Method.* p. 427.
Lister Conch. t. 1007. f. 71. *Rumphius,* t. 23. f. C. *Pe-*
tiver Amb. t. 10. f. 10. *Martini,* ii. p. 61. t. 36. f. 372
and 373.
Inhabits the shores of Amboyna. *Rumphius.* China and Tranquebar. *Regenfuss.* Madagascar. *Humphreys.*
Shell four or five inches long, and rather more than two thirds
as broad, with longitudinal striæ, and transverse double lines;
the colour on the back is white mottled with brown and
red, and the aperture and expanded lip are red; the teeth
on both sides the aperture are white.

TUBEROSUM.   16. Shell turbinated, and armed with
three transverse. rows of tubercles; inner
lip much dilated and triangular; aperture
toothed; beak recurved.

Buccinum tuberosum. *Linnæus Syst. Nat.* p. 1198.?
Martini, ii. p. 36. t. 38. f. 381 and 382. *Born Mus.*
p. 244. *Schroeter Einl.* i. p. 317. *Gmelin,* p. 3473.
Schreibers Conch. i. p. 143. *Shaw's Nat. Misc.* xii. t.
467.
Cassidea tuberosa. *Bruguiere Enc. Method.* p. 436.
Gualter, t. 41. *Seba,* iii. t. 73. f. 2. *Knorr,* iii. t. 10. f.
1 and 2. *Favanne,* t. 25. f. B 2.
Inhabits the American Ocean. *Linnæus.* West Indies. *Hum-*
phreys. Common on the shores of Guadaloupe, Martinique,
and St. Domingo. *Bruguiere.*
Shell commonly from five to seven inches long, and nearly
three-fourths as broad, of a greyish white or brown, with
longitudinal interrupted zic-zac streaks of purplish brown;
the pillar is blackish with white plaits; the expanded lips

are pale flesh-colour, and there are generally six black spots on the outer lip, as well as on the appendage which extends from the suture behind the projection of the inner lip to the reflected beak. It is nearly allied to *B. cornutum*, but differs in being smaller, and in having the pillar-lip triangular. Linnæus has described his *B. tuberosum* with only two rows of tubercles, and the present shell, I believe, has always three, but it answers to Gualter's t. 41. which is the only figure he has referred to.

FLAMMEUM. 17. Shell longitudinally plaited, transversely nodulous, and slightly coronated; plaits on the spire imbricated; aperture toothed; beak recurved.

Cassidea flammea. *Bruguiere Enc. Methodique*, p. 429. *Lister Conch.* t. 1004. f. 69, and 1005. f. 72. *Petiver Gaz.* t. 153. f. 1. *Seba*, iii. t. 73. f. 5, 6, 14, 15, and 16. *Martini*, ii. t. 34. f. 359. *Favanne*, t. 25. f. E.

Junior. Buccinum flammeum. *Linnæus Syst. Nat.* p. 1199. *Martini*, ii. p. 34. t. 34. f. 353 and 354. *Born Mus.* p. 245. *Schroeter Einl.* i. p. 318. *Gmelin*, p. 3473. *Schreibers Conch.* i. p. 144.

Bonanni Rec. 3. f. 156. *Rumphius*, t. 23. f. 2. *Seba*, iii. t. 73. f. 19 and 20. *Knorr*, iv. t. 4. f. 1.

Inhabits the coasts of Jamaica. *Lister*. West Indies. *Martini*.

Shell when at maturity about four and a half inches long, and near two thirds as broad, brown with transverse rows of white arcuated spots, which become fainter with age; the longitudinal plaits are thickened at the end of every whirl, and in adults there are from two to five transverse rows of tubercles on the body-whirl, but these are wanting in young shells, from one of which the description in the Systema Naturæ was obviously taken.

TESTICULUS. 18. Shell ovate, with elevated longitudinal and decussated striæ; aperture toothed; beak recurved.

Buccinum Testiculus. *Linnæus Syst. Nat.* p. 1199. *Martini*, ii. p. 64. t. 37. f. 375 and 376. *Born Mus.* p. 246. *Schroeter Einl.* i. p. 319. *Gmelin*, p. 3474. *Schreibers Conch.* i. p. 144.

Cassidea Testiculus. *Bruguiere Enc. Method.* p. 426. *Bonanni Rec.* 3. f. 162, and *Kirch.* f. 163. *Lister Conch.*

t. 1001. f. 66. *Rumphius*, t. 23. f. 3. *Petiver Gaz.* t. 152. f. 17. *Gualter*, t. 39. f. C. *Seba*, iii. t. 72. f. 17 to 21. *Knorr*, iii. t. 8. f. 2, and iv. t. 6. f. 1. *Favanne*, t. 26. f. D 3.

Junior. Martini, ii. t. 37. f. 377 and 378.

Inhabits the coasts of Jamaica. *Lister.* Isle of France, and shores of Coromandel. *D'Avila.* Madagascar and Guadaloupe. *Bruguiere.*

Shell about two inches and a half, or three inches long, and rather more than half as broad, of a whitish or pale yellowish red, generally marked with darker red spots, either placed irregularly or in transverse rows; the elevated longitudinal lines are placed thrice nearer to each other than those transverse ones which they cross over, and the side next the pillar is less strongly marked than the other; among the teeth of the pillar is a longitudinal groove, and the outer lip is marginated, and marked with dark brown spots.

DECUSSATUM. 19. Shell decussately striated, and covered with small square scales; aperture toothed; beak recurved.

Buccinum decussatum. *Linnæus Syst. Nat.* p. 1199. *Martini*, ii. p. 44. *Born Mus.* p. 246. *Schroeter Einl.* i. p. 320. *Gmelin*, p. 3474. *Schreibers Conch.* i. p. 144. Cassidea decussata. *Bruguiere Enc. Methodique*, p. 425.

Variety A. Olivaceous, with transverse rows of square yellowish spots.

Bonanni Rec. and *Kirch.* 3. f. 157. *Lister Conch.* t. 1000. f. 65. *Gualter*, t. 40. right hand fig. B. *Knorr*, ii. t. 10. f. 3 and 4. *Martini*, ii. t. 35. f. 360 and 361.

Variety B. Whitish, with longitudinal yellow stripes.

Gualter, t. 40. left hand fig. B. *Martini*, ii. t. 35. f. 367 and 368.

Inhabits the African Ocean. *Linnæus.* Coast of Portugal. *Bonanni.* Mediterranean. *Martini.* China. *Humphreys.*

Shell about an inch and three-quarters long, and one inch broad, marked with hollow transverse striæ and elevated longitudinal lines, which divide the surface into small square compartments; the raised belt-like appendages, of which there are six, are pointed at their upper extremities, and two of them are extended along the body-whirl on each side of the aperture. The Variety with longitudinal stripes differs only in colour and markings from that with square spots, and the inside of both is brownish red.

AREOLA. 20. Shell with the body-whirl smooth, and the spire elevated and rough with decussated striæ; aperture toothed; beak recurved; pillar wrinkled.

Buccinum Areola. *Linnæus Syst. Nat.* p. 1199. *Martini,* ii. p. 39. t. 34. f. 355 and 356. *Born Mus.* p. 247. *Schroeter Einl.* i. p. 321. *Gmelin,* p. 3475. *Schreibers Conch.* i. p. 145.

Cassidea Areola. *Bruguiere Enc. Methodique,* p. 423.

Bonanni Rec. and *Kirch.* 3. f. 154. *Lister Conch.* t. 1012. f. 76. *Rumphius,* t. 25. f. B, and 1. *Petiver Amb.* t. 2. f. 11. *Argenville,* t. 15. f. I. *Gualter,* t. 39. f. H. *Seba,* iii. t. 70. f. 7 to 9. *Klein Ost.* t. 6. f. 102. *Knorr,* iii. t. 8. f. 5. *Favanne,* t. 24. f. I.

Inhabits the Mediterranean and Java. *Linnæus.* Amboyna. *Rumphius.* East Indies. *D'Avila.* China. *Humphreys.*

Shell about two inches and a half long, and near one and a half broad, white with transverse rows of square yellow or orange spots; the body-whirl is polished and smooth, except at the base which is slightly grooved transversely.

STRIGATUM. 21. Shell sub-ventricose; body-whirl rather smooth, and the spire rough with decussated striæ; aperture toothed; beak recurved; pillar wrinkled.

Buccinum strigatum. *Gmelin,* p. 3477.

Buccinum cassideum strigatum. *Chemnitz,* x. p. 189. t. 153. f. 1457 and 1458.

Buccinum rugosum. *Gmelin,* p. 3476.

Buccinum Areola, Var. *Linnæus Mus. Lud. Ulr.* p. 605. *Schroeter Einl.* i. p. 321.

Buccinum, No. 99. *Schroeter Einl.* i. p. 320.

Cassidea Areola, Var. *Bruguiere Enc. Methodique,* p. 424.

Lister Conch. t. 1014. f. 78. *Rumphius,* t. 25. f. 2. *Petiver Amb.* t. 21. f. 11. *Argenville,* t. 15. f. D. *Martini,* ii. t. 34. f. 356 *A. Favanne,* t. 24. f. D.

Inhabits Amboyna. *Rumphius.* Mediterranean and Indian Seas. *Martini.*

Shell near three inches and a half long, and more than two and a quarter broad, white with longitudinal yellow stripes. The shell, besides the difference in its markings, is larger and more distended than B. *Areola.*

SABURON. 22. Shell with the whirls inflated, rounded, and marked with narrow transverse grooves; aperture toothed; beak recurved; pillar-lip strongly wrinkled transversely at the base.

Cassidea Saburon. *Bruguiere Enc. Meth.* p. 420.
Cassis geometrica. *Mus. Gevers.* p. 392. No. 1284.
Le Saburon. *Adanson Senegal,* p. 112. t. 7. f. 8.
Bonanni Rec. and *Kirch.* 3. f. 20.
Junior. Buccinum ignave. *Solander's MSS.*
  Lister Conch. t. 1012. f. 76. *Rumphius,* t. 25. f. C. *Petiver Amb.* t. 9. f. 6. *Gualter,* t. 39. f. G.
Inhabits the Mediterranean and the shore at Lisbon. *Bonanni.* Isle of Goree. *Adanson.*
Shell two inches and a quarter long, and nearly one inch and a half broad, rather thin, somewhat pellucid, of a dirty white colour with transverse rows of square yellow or ferruginous spots. It differs from *B. granulatum,* with which Martini has confounded it, in not having any longitudinal striæ except those which mark the growth of the shell, in the want of granules on the pillar, and in being destitute of any longitudinal belt-like appendages on the spire. When at maturity, it has a longitudinal belt extending along the pillar-lip from the suture to the base of the body-whirl, but this is wanting in young shells.

GRANULATUM. 23. Shell transversely grooved, and longitudinally striated; aperture toothed; beak recurved; pillar-lip granulated.

Buccinum granulatum. *Born Mus.* p. 248.
Cassidea granulosa. *Bruguiere Enc. Meth.* p. 421.
Buccinum Areola, Var. β. *Gmelin,* p. 3475.
Buccinum trifasciatum. *Gmelin,* p. 3477. *Schreibers Conch.* i. p. 155.
Buccinum, No. 111. *Schroeter Einl.* i. p. 384.
Bonanni Rec. and *Kirch.* 3. f. 158. *Lister Conch.* t. 1056. f. 9. *Martini,* ii. t. 32. f. 344 and 345, and t. 34. f. 350 to 352. *Favanne,* t. 25. f. A 4.
Inhabits the Mediterranean. *D'Avila.* West Indies. *Humphreys.*
Shell from one to four inches long, and the breadth rather exceeds two thirds of the length. It is nearly allied to *B. Areola,* from which, as well as *B. Saburon,* it may be at once

distinguished by the granules with which the lower part of the pillar-lip is covered, and the longitudinal striæ on the body-whirl will serve to distinguish it from *B. undulatum.*

UNDULATUM. 24. Shell thick, with transverse broad convex ribs ; aperture toothed ; beak recurved ; pillar-lip granulated.

Buccinum undulatum. *Gmelin,* p. 3475.
Buccinum porcatum. *Solander's MSS.*
Buccinum cassideum undulatum. *Chemnitz,* xi. p. 78. t. 186. f. 1794 and 1795.
Buccinum, No. 89. *Schroeter Einl.* i. p. 377.
Cassidea sulcosa. *Bruguiere Enc. Methodique,* p. 422.
Bonanni Rec. and *Kirch.* 3. f. 159. *Lister Conch.* t. 996. f. 61. *Petiver Gaz.* t. 152. f. 8. *Gualter,* t. 39. f. B.
Seba, iii. t. 68. f. 14 and 15. *Favanne,* t. 25. f. A 3.
Variety. With the ribs narrower, and the spire more acute.
Buccinum gibbum. *Gmelin,* p. 3476.
Buccinum, No. 92. *Schroeter Einleitung,* i. p. 378.
Lister Conch. t. 999. f. 64.
Inhabits Barbadoes. *Lister.* Mediterranean. *Chemnitz.*
Shell about three inches and a half long, and two and a half broad, of a brownish white colour with transverse rows of unequal yellowish brown spots ; the body-whirl and two first whirls of the spire have only broad transverse convex ribs, but the five upper whirls are marked with decussated striæ. Gmelin's *B. gibbum* is taken wholly from Lister's figure, and in Dr. Solander's MSS. the same figure is quoted for a variety of this species.

INFLATUM. 25. Shell ventricose, and slightly furrowed transversely ; aperture toothed ; beak recurved ; pillar-lip wrinkled and granulated.

Buccinum inflatum. *Shaw Nat. Misc.* xxii. t. 959.
Buccinum cassideum tessellatum. *Chemnitz,* xi. p. 76. t. 186. f. 1792 and 1793.
Inhabits the Indian and African Seas. ? *Shaw.*
Shell four and a half inches long, and three inches and a quarter broad, whitish, with transverse ferruginous bands more or less irregularly broken into oblong spots ; the spire is considerably produced, and the whirls rounded ; the whole shell is marked with obsolete, broad, slightly convex transverse ribs, and the upper whirls with decussated striæ. The mar-

gin of the outer lip is much narrower than in *B. undulatum,* and the expansion of the inner lip is confined to the base of the pillar, whereas in that species it extends over the upper angle of the aperture.

TESSELLATUM. 26. Shell slightly plaited longitudinally, and granulated above ; spire rather depressed ; aperture toothed ; beak recurved ; pillar-lip wrinkled and granulated.

Buccinum tessellatum. *Gmelin,* p. 3476. *Schreibers Conch.* i. p. 140.
Buccinum maculosum. *Gmelin,* p. 3476.
Buccinum Rumphii. *Gmelin,* p. 3491.
Buccinum granosum. *Solander's MSS. Portland Cat.* p. 55. No. 1293.
Buccinum, No. 1. *Schroeter Einl.* i. p. 356.
Cassis tuberculata. *Mus. Gevers.* p. 392. No. 1295.
Cassidea fasciata. *Bruguiere Enc. Meth.* p. 430.
Lister Conch. t. 997. f. 62. *Rumphius,* t. 25. f. 3. *Seba,* iii. t. 73. f. 1, 12, and 13. *Martini,* ii. p. 57. t. 36. f. 369, and t. 37. f. 374. *Favanne,* t. 26. f. B 1.
Junior. Buccinum Senegalicum. *Gmelin,* p. 3477.
Le Fasin. *Adanson Senegal,* p. 111. t. 7. f. 7.
Inhabits Amboyna. *Rumphius.* South Seas. *D'Avila.* Coast of Guinea. *Humphreys.*
Shell frequently four inches and a quarter long, and two and a half broad, thin, rather pellucid, of a pale brown colour, with four or five white transverse bands ornamented with large purplish brown spots ; the body-whirl is slightly plaited longitudinally, and armed with three transverse rows of granules on the upper part, and also on the spire.

BILINEATUM. 27. Shell smooth, with a double row of tubercles on the body-whirl ; aperture toothed ; beak recurved ; pillar-lip wrinkled and granulated.

Buccinum bilineatum. *Gmelin,* p. 3476. *Montagu Test.* p. 244. *Maton and Racket, in Lin. Trans.* viii. p. 134. *Dorset Cat.* p. 44. t. 17. f. 8.
Buccinum decussatum. *Pennant Brit. Zool. App.* t. 79. two lower figures.
Buccinum porcatum. *Pulteney's Dorset.* p. 41.

Lister Conch. t. 998. f. 63.?
Inhabits the shore at Weymouth. *Pennant.*
This species, of which I believe only a single specimen has
 been dredged up at Weymouth, is thus described, " Shell
 size of a large hazel nut, ovated, smooth, whitish, girdled
 with from two to four bands of brown spots; the upper part
 set round with two series of tubercles; upper volutions
 smooth; aperture large, sub-oval; outer lip slightly dentat-
 ed; inner lip rugose and granulated." Mr. Montagu says
 that foreign specimens exceed two inches in length. Dr.
 Pulteney considered this shell to be *B. porcatum* of Solander, and from the MSS. of the latter Naturalist, in Sir Joseph Banks's Library, it appears that he considered Lister's
 figure 63. t. 998. to be this species.

CICATRICOSUM. 28. Shell ovate, smooth, covered
 with hollow dots; spire elongated; aperture
 toothed; beak recurved; pillar-lip wrinkled
 and granulated.

Buccinum cicatricosum. *Gronovius Zooph.* p. 303. t. 19.
 f. 1 and 2. *Gmelin,* p. 3475.
Cassis cicatricosa. *Mus. Gevers.* p. 392.
Inhabits the Indian Ocean. *Gronovius.*
Gronovius says this shell is " white, and has a transverse row
 of small unequal contiguous granules; outer lip thickened,
 double, and armed with transverse equal teeth; pillar-lip
 wrinkled and granulated; spire elongated, and the whirls
 rounded."

RECURVIROSTRUM. 29. Shell slightly striated longitudinally; body-whirl inflated; spire rather
 prominent; outer lip toothed; beak recurved; pillar smooth.

Buccinum recurvirostrum. *Gmelin,* p. 3477. *Schreibers*
 Conch. i. p. 146.
Buccinum, No. 101. *Schroeter Einleitung,* i. p. 381.
Lister Conch. t. 1016. f. 75.
Inhabits Barbadoes. *Lister.*
Shell two inches and three quarters long, and one inch and
 three quarters broad, pale rufous, with obsolete crowded
 bands of whitish spots on the body-whirl, and the spire marbled with pale cinereous and brown; aperture with a thick
 coat of yellowish enamel on the pillar, and at the upper extremity of the inner lip.

CASSIS.   30.   Shell thickly striated transversely; beak recurved; inner lip membranaceous and united to the pillar; outer lip thin.

Buccinum Cassis. *Gmelin,* p. 3477. *Schreibers Conch.* i. p. 146.
Buccinum cassideum. *Chemnitz,* x. p. 188. t. 152. f. 1456.
Inhabits the Bay of Naples. *Chemnitz.*
Shell rather more than an inch and a half long, and one inch broad, of an uniform yellowish brown colour, except the pillar which is white, and marked with crowded transverse slightly elevated striæ. The term *Cassis* is so far from being well applied to this species, that from Chemnitz's figure it appears rather doubtful whether it should not be placed in another division, and it has not the general habit of a *Helmet.*

*** *Resembling the last Division, but the outer Lip on the outside is spinous at the base.*

ERINACEUS.   31.   Shell slightly plaited longitudinally, and crowned with papillæ; outer lip muricated at the base.

Buccinum Erinaceus. *Linnæus Syst. Nat.* p. 1199. *Martini,* ii. p. 48. t. 35. f. 363, and t. 38. f. 383 and 384. *Born Mus.* p. 248. *Schroeter Einl.* i. p. 322. *Gmelin,* p. 3478. *Schreibers Conch.* i. p. 148.
Buccinum ponderosum. *Gmelin,* p. 3477.
Buccinum, No. 100. *Schroeter Einl.* i. p. 381.
Cassidea Erinaceus. *Bruguiere Enc. Method.* p. 418.
Bonanni Rec. and *Kirch.* 3. f. 152 and 153. *Lister Conch.* t. 1015. f. 73, and t. 1016. f. 74. *Rumphius,* t. 25. f. D, and f. 7. *Petiver Amb.* t. 9. f. 9. *Gualter,* t. 39. f. D, and I. *Argenville,* t. 14. f. G. *Seba,* iii. t. 53. f. 8, 11, 12, 29, and 30. *Favanne,* t. 24. f. H 2.
Variety. With the outer lip unarmed.
Cassidea maculata. *Meuschen, Mus. Gevers.* p. 390. No. 1283.
Lister Conch. t. 1013. f. 77. *Seba,* iii. t. 53. f. 9 and 10. *Martini,* ii. t. 38. f. 385 and 386.
Inhabits the American Ocean, and coasts of Alexandria. *Linnæus.* Amboyna. *Rumphius.* Tranquebar. *Martini.* China. *Humphreys.*

Shell about an inch and a half long, and rather more than half
as broad, of a whitish, fawn, or flesh-colour, variously
marked with pale yellowish bands, or longitudinal streaks, or
reddish spots, and a row of dark brown spots on the outer
lip; the body-whirl is longitudinally plaited, particularly on
the side of the outer lip, and a transverse row of nodules is
formed on the shoulder, as also sometimes on the two first
whirls of the spire. Born and Bruguiere have followed Martini, and refer only to his figure 363 for this species, whereas Schroeter, Gmelin, and Schreibers have cited all the
figures above referred to, and the gradation appears to me to
be too regular to admit of any specific distinction. Dr. Solander, on the contrary, has not made any reference to Martini for *B. Erinaceus*, but quotes f. 363 for his *B. Meles*,
and the other four figures for another species which appears
in the Portland Catalogue under the name of *B. Pantheri-
na*. The lip is sometimes not muricated, but I much doubt
whether this difference, as well as that which often happens in
the nodules of the spire, may not be produced by age; and
Linnæus was probably of the same opinion, for he has referred to both D and I of Gualter's plate 39, and also to six
figures in Seba which comprise all these different appearances.

BIARMATUM. 32. Shell slightly plaited longitudinally, and crowned with papillæ; outer lip
with two rows of sharp spines.

Buccinum nodulosum. *Gmelin*, p. 3479. *Schreibers Conch.*
i. p. 140.
Buccinum, No. 107. *Schroeter Einleitung*, i. p. 383. t. 2.
f. 9.
Inhabits ——
Shell about an inch long and two thirds as broad, and is shorter
as well as thicker than *B. Erinaceus*, from which, according to Schroeter's description, it may be at once known by
its having the inside as well as outside of the outer lip muricated. In colour it resembles *B. Erinaceus*, and has a
dark spot at the termination of the Canal, as well as a row
of spots along the upper edge of the margin of the outer lip.
There are two *Buccina* in Gmelin with the name of *nodulosum*, and I have therefore changed that of the present species.

c 2

FIMBRIA.   33. Shell longitudinally plaited, crowned with papillæ, and the spire cancellated ; outer lip toothed within, and muricated at the base.

Buccinum Fimbria.   *Gmelin,* p. 3479.
Buccinum plicatum, Var. β.   *Gmelin,* p. 3472.
Buccinum Tulipa.   *Solander's MSS.*
Buccinum cassideum, plicis et nodis quasi crispatum. *Chemnitz,* x. p. 191. t. 153. f. 1459 and 1460.
Knorr, iii. t. 28. f. 1.   *Seba,* iii. t. 53. f. 1 and 2.   *Favanne,* t. 25. f. D 4.
Inhabits the East Indian Seas. *Chemnitz.*
Shell two inches and three-quarters long, and an inch and a half broad, white with longitudinal waved yellow stripes; the body-whirl is longitudinally plaited, and nodulous on the shoulder, and the whirls of the spire are cancellated and nodulous; the outer lip is toothed on the inside, besides which, it has a shorter row of exterior spines.

GLAUCUM.   34. Shell smooth and crowned with papillæ; spire cancellated, and the outer lip muricated at the base.

Buccinum glaucum.   *Linnæus Syst. Nat.* p. 1200.   *Martini,* ii. p. 23. t. 32. f. 342 and 343.   *Born Mus.* p. 249. *Schroeter Einleitung,* i. p. 323.   *Gmelin,* p. 3478. *Schreibers Conch.* i. p. 148.
Cassidea glauca.   *Bruguiere Enc. Method.* p. 419.
Lister Conch. t. 996. f. 60.   *Rumphius,* t. 25. f. A.   *Petiver Amb.* t. 7. f. 4, and t. 11. f. 18.   *Gualter,* t. 40. f. A.   *Seba,* iii. t. 71. f. 11 to 16.   *Knorr,* iii. t. 8. f. 3. *Favanne,* t. 25. f. D 3.
Inhabits the coasts of Amboyna. *Rumphius.* China. *Humphreys.*
Shell three or four inches long, and two thirds as broad, with the body-whirl smooth or only very obsoletely ribbed transversely, and the spire cancellated and coronated. Old shells are of a nearly uniform cinereous colour, but younger ones are marked with pale yellowish bands. The inside is brown or violet, and the outer margin of the outer lip is tinged with the same colour. The outer lip is toothed within, and armed with three or four spines at the base.

VIBEX.   35. Shell entirely smooth and level, and the outer lip muricated at the base.

Buccinum Vibex. *Linnæus Syst. Nat.* p. 1200. *Martini,* ii. p. 52. t. 35. f. 364 to 366. *Born Mus.* p. 249, and Vign. at p. 238, fig. *d. Schroeter Einl.* i. p. 324. *Gmelin,* p. 3479. *Schreibers Conch.* i. p. 149.

Cassidea Vibex. *Bruguiere Enc. Methodique,* p. 417.

Bonanni Rec. 3. f. 151. *Rumphius,* t. 25. f. E, and 9. *Petiver Amb.* t. 4. f. 9. *Gualter,* t. 39. f. F. *Seba,* iii. t. 53. f. 3, 4, 18 to 23, and 31. *Argenville,* t. 17. f. H. *Regenfuss,* i. t. 10. f. 40. *Favanne,* t. 25. f. H 1.

Inhabits the coasts of Alexandria, and Jamaica. *Linnæus.* Amboyna. *Rumphius.* Tranquebar and Frederick's Islands. *Regenfuss.*

Shell about two inches long, and very little more than half as broad, extremely smooth all over, and nearly white with longitudinal waved yellow bands. The outer lip has not any internal teeth, but is externally muricated towards the base, and the pillar is slightly plaited.

PAPILLOSUM. 36. Shell tuberculated all over, and the outer lip muricated at the base.

Buccinum papillosum. *Linnæus Syst. Nat.* p. 1200. *Martini,* iv. p. 63. t. 125. f. 1204 and 1205. *Born Mus.* p. 250. *Schroeter Einl.* i. p. 325. *Gmelin,* p. 3479. *Schreibers Conch.* i. p. 149. *Bruguiere Enc. Method.* p. 270.

Lister Conch. t. 969. f. 23. *Rumphius,* t. 29. f. M. *Petiver Amb.* t. 9. f. 16. *Gualter,* t. 44. f. G. *Argenville,* t. 9. f. I. *Seba,* iii. t. 49. f. 57 to 59. *Knorr,* ii. t. 27. f. 2. *Favanne,* t. 31. f. G 2.

Inhabits the Asiatic Ocean. *Linnæus.* Amboyna. *Rumphius.* Coasts of Tranquebar, Java, and the Moluccas. *Bruguiere.* Madagascar. *Humphreys.*

Shell about an inch and a half long, and half as broad, yellowish-red with a rose-coloured summit, and covered all over with rows of tubercles, by which, and its muricated lip, it may be readily known.

GLANS. 37. Shell smooth, with the summit of the spire longitudinally grooved, and the outer lip muricated.

Buccinum Glans. *Linnæus Syst. Nat.* p. 1200. *Martini,* iv. p. 60. t. 125. f. 1196 to 1198. *Born Mus.* p. 251. *Schroeter Einl.* i. p. 326. *Gmelin,* p. 3480. *Schreibers Conch.* i. p. 149. *Bruguiere Enc. Meth.* p. 269.

Lister Conch. t. 981. f. 40.　*Rumphius,* t. 29. f. P.　*Petiver Amb.* t. 13. f. 5.　*Seba,* iii. t. 39. f. 56, 57, and 60.　*Knorr,* iii. t. 5. f. 5.　*Regenfuss,* ii. t. 12. f. 55.　*Favanne,* t. 33. f. L.

Variety. Slightly tubercular at the sutures.

Martini, iv. t. 125. f. 1199 and 1200.

Inhabits the Asiatic Ocean. *Linnæus.* Amboyna. *Rumphius.*

Shell an inch and three-quarters long, and near an inch broad, white with a few large irregular brown spots, and marked throughout with parallel dark transverse lines, which on the body-whirl are near two lines apart; the inside is also marked with delicate pale striæ. Linnæus described the pillar-lip with two teeth, but I have never been able to find more than one, and this is placed near the upper angle of the aperture.

GIBBUM. 38. Shell sub-ventricose, smooth, slightly striated, and the outer lip muricated at the base.

Buccinum gibbum.　*Bruguiere in Enc. Method.* p. 267.

Buccinum tessulatum. *Gmelin,* p. 3479.　*Schreibers Conch.* i. p. 150.

Buccinum foliorum.　*Gmelin,* p. 3493.

Buccinum, No. 3.　*Schroeter Einleitung,* i. p. 357.

Bonanni Rec. 3. f. 63.　*Lister Conch.* t. 975. f. 30, and *Exer. Anatom.* t. 8. f. 7.　*Rumphius,* t. 29. f. Y.　*Petiver Amb.* t. 13. f. 25.　*Gualter,* t. 44. f. B.　*Martini,* ii. t. 38. f. 387 and 388.　*Favanne,* t. 33. f. S 2, and t. 77. f. A 7.

Inhabits the coasts of Naples. *Bonanni.* Among the leaves and branches of maritime shrubs at Amboyna. *Rumphius.* Coasts of Spain and Italy. *Bruguiere.*

Shell near an inch and a quarter long, and three-quarters of an inch broad, of a reddish or yellowish brown colour, variously speckled or striped with yellow or white, and always marked with a whitish band spotted with brown at the sutures; the upper part of the pillar-lip becomes abruptly rounded by the inflated contour of the whirl, and old shells have two or three sharp spines at the base of the outer lip. As there can be no doubt that Martini's figures 387 and 388, belong to *B. gibbum* of Bruguiere, I have retained this name, because it is more applicable than that of *tessulatum,* and *B. gibbum* of Gmelin is either a Variety of *B. undulatum,* or otherwise appears to be undeserving of notice.

**** *With the Pillar-lip dilated and thickened.*

ARCULARIA. 39. Shell plaited longitudinally and transversely striated, with the whirls papillary above; pillar-lip dilated and thickened.

Buccinum Arcularia. *Linnæus Syst. Nat.* p. 1200. *Martini*, ii. p. 89. t. 41. f. 409 to 412. *Born Mus.* p. 250. *Schroeter Einl.* i. p. 327. *Gmelin*, p. 3480. *Schreibers Conch.* i. p. 152. *Bruguiere in Enc. Meth.* p. 278.
Nassa Arcularia. *Lamarck Syst. des Anim.* p. 76.
Bonanni Rec. and *Kirch.* 3. f. 175, and f. 340. *Lister Conch.* t. 970. f. 24. *Rumphius*, t. 27. f. M. *Petiver Amb.* t. 12. f. 9. *Gualter*, t. 44. f. O, Q, and R. *Argenville*, t. 14. f. C. *Seba*, iii. t. 53. f. 32, 35, 37, 40, and 41. *Knorr*, vi. t. 22. f. 3. *Favanne*, t. 33. f. F 3.
Variety. With a broad band on the body-whirl.
Buccinum fasciolatum. *Gronovius Zooph.* t. 19. f. 7 and 8.
Inhabits the coasts of Amboyna. *Rumphius.* Java. *Linnæus.* Isle of France and Madagascar. *Bruguiere.* China. *Humphreys.*
Shell about fourteen lines long, white, and elegantly cancellated with longitudinal plaits and transverse striæ; the tubercles at the angulated shoulder of every whirl are crossed on their summits by a transverse groove; the outer lip is crenated on the margin and striated within, and the throat is marked with broad reddish brown, or flesh-coloured bands.

CORONATUM. 40. Shell striated at the base; whirls smooth and tuberculated at the sutures; outer lip spinous.

Buccinum coronatum. *Bruguiere in Enc. Meth.* p. 277.
Buccinum mutabile. *Schroeter Einl.* i. p. 329. t. 2. f. 4.
Seba, iii. t. 53. f. 28 and 39.
Inhabits the Sea at Foulepoint in the Island of Madagascar. *Bruguiere.*
Shell one inch long and seven lines broad, olivaceous or brown, striped irregularly towards the outer lip, and marked in the upper part of the body-whirl with a pale yellowish band which follows the course of the spire; the tubercles at the sutures are white. Bruguiere says it also differs from *B. Arcularia*, in not having the pillar-lip expanded over the

aperture, but I have a specimen which appears to be the same species in a more advanced stage of growth, and which has the lip equally expanded.

HEPATICUM. 41. Shell ribbed longitudinally, and papillary at the sutures; pillar-lip dilated and rugose.

Buccinum hepaticum. *Pulteney's Dorset Cat.* p. 44. t. 15. f. 13. *Montagu Test.* p. 243. t. 8. f. 1. *Maton and Racket, in Lin. Trans.* viii. p. 135.

Inhabits the coasts of Dorsetshire, rare. *Pulteney.*

Shell an inch long and five-eighths broad, of a dull brownish colour variegated with a few ferruginous spots, and there is sometimes a white band within the outer lip; spire sharp-pointed, with seven or eight slightly tumid and strongly ribbed whirls; pillar-lip folded back, and having one tooth-like ridge on the upper part.

PULLUS. 42. Shell gibbous, obliquely striated and cancellated; pillar-lip dilated and thickened.

Buccinum Pullus. *Linnæus Syst. Nat.* p. 1201. *Schroeter Einl.* i. p. 328. t. 2. f. 2, *a* and *b*. *Gmelin*, p. 3481. *Bruguiere Enc. Method.* p. 276.

Lister Conch. t. 970. f. 25. *Rumphius*, t. 27. f. N. *Petiver Amb.* t. 12. f. 10. *Gualter*, t. 44. f. N. *Adanson Senegal*, p. 117. t. 8. f. 11.

Inhabits the straights of Malacca. *Rumphius.* Coasts of Senegal. *Adanson.* Mediterranean. *Linnæus.*

Shell half, or sometimes near three-quarters of an inch long, and two thirds as broad, white, pale grey or fawn-colour, and sometimes slightly marked with spots or transverse bands of brown; the longitudinal elevated striæ are somewhat oblique, and are crossed by others transversely.

THERSITES. 43. Shell gibbous, with half the body-whirl and the whole spire longitudinally plaited; pillar-lip dilated and thickened.

Buccinum Thersites. *Bruguiere Enc. Meth.* p. 279.
Buccinum Arcularia, Var. *β. Gmelin*, p. 3480.
Buccinum, No. 4. *Schroeter Einl.* i. p. 357.
Lister Conch. t. 971. f. 26. *Seba*, iii. t. 53. f. 44 to 46. *Martini*, ii. t. 41. f. 413.

Inhabits the Asiatic Ocean. *Bruguiere.*

Shell three-quarters of an inch long, and two thirds as broad, pale olivaceous or cinereous, sometimes marked with a darker or brownish band, and the inside is of an uniform brown with the exception of a pale narrow transverse stripe; about the middle of the back of the body-whirl there is a singular gibbosity, the space between which and the outer lip is smooth, and the other half as well as the whole spire is longitudinally and somewhat obliquely plaited.

VERRUCOSUM. 44. Shell gibbous, and tuberculated; pillar-lip dilated and thickened.

Buccinum verrucosum. *Bruguiere Enc. Method.* p. 279.
Cassis perlata. *Meuschen Mus. Gevers.* p. 390, No. 1275.
Lister Conch. t. 972. f. 27.
Inhabits the East Indies. *Lister.* Ceylon, and among the rocks in the sea at Foulepoint in Madagascar. *Bruguiere.*
Shell nearly of the same size and appearance as *B. Thersites,* but has four rows of tubercles on the body-whirl, of which one is situated by the suture, two in the middle, and one at the base; the row at the base is said by Bruguiere to be sometimes wanting, and always smaller than the others.

GIBBOSULUM. 45. Shell gibbous, smooth; pillar-lip dilated and thickened.

Buccinum gibbosulum. *Linnæus Syst. Nat.* p. 1201.
Schroeter Einl. i. p. 329. t. 2. f. 3. *a* and *b.* *Gmelin,* p. 3481. *Schreibers Conch.* i. p. 153. *Bruguiere Enc. Meth.* p. 280.
Bonanni Rec. 3. f. 383. *Lister Conch.* t. 973. f. 28. *Knorr,* vi. t. 22. f. 6. *Martini,* ii. t. 41. f. 414 and 415.
Inhabits the Mediterranean. *Linnæus.* Asiatic Ocean. *Bruguiere.*
Shell rather more than half an inch long, and almost equally broad, with a short pointed spire, of which nearly the whole of one side is buried in the expanded margin of the aperture; the middle of the back of the body-whirl is gibbous, and on its side towards the outer lip is a large irregular indenture.

MUTABILE. 46. Shell oval-oblong, smooth, with the upper whirls longitudinally grooved; pillar two-plaited.

Buccinum mutabile. *Linnæus Syst. Nat.* p. 1201. *Born Mus.* p. 252. t. 9. f. 13.

Buccinum nitidulum. *Martini*, iv. p. 59. t. 125. f. 1194 and 1195.? *Schroeter Einl.* i. p. 347. *Gmelin*, p. 3497.

Buccinum Miran. *Bruguiere Enc. Method.* p. 268.

Le Miran. *Adanson Senegal*, p. 50. t. 4. f. 1.

Bonanni Rec. and *Kirch.* 3. f. 60. *Gualter*, t. 44. f. R. *Favanne*, t. 33. f. S 1.

Variety. With a spotted band at the sutures.

Chemnitz, xi. p. 89. t. 188. f. 1810 and 1811.

Inhabits the coasts of Senegal. *Adanson.*

Almost all authors have disagreed respecting *B. mutabile,* and Bruguiere with much probability has conjectured that Linnæus, supposing *B. gibbum* to differ from the present shell only in age, used the name to express such a remarkable change in the appearance. The shell is about thirteen lines long, and rather less than half as broad, white or of an agate-colour, and is said by Bruguiere to be uniform without any markings; Chemnitz has, however, figured a shell which agrees in all other respects with this species, but which resembles *B. gibbum* in being freckled, and in having a similar white spotted band at the suture; about one fourth of the body-whirl is transversely striated at the base, and of the nine whirls which compose the spire, the eight uppermost are finely grooved longitudinally. It would stand better near to *Buccinum Glans,* than in the place which Linnæus has assigned it among the shells of this division.

NERITEUM. 47. Shell convex, obtuse, depressed and smooth; pillar-lip obsoletely dilated, and thickened.

Buccinum neriteum. *Linnæus Syst. Nat.* p. 1201. *Born Mus.* p. 252. t. 10. f. 3 and 4. *Schroeter Einl.* i. p. 331. *Gmelin*, p. 3481. *Bruguiere Enc. Meth.* p. 284.

Trochus Vestiarius, Var. β. *Gmelin*, p. 3578.

Trochus, No. 40. *Schroeter Einl.* i. p. 259.

Trochus, No. 82. *Schreibers Conch.* i. p.

Trochus nanus. *Ulysses's Travels*, p. 468.

Plancus, t. 3. f. 3. *Bonanni Rec.* and *Kirch.* 3. f. 212. *Gualter*, t. 65. f. C and I. *Chemnitz*, v. t. 166. f. 1602, No. 1, 2, and 3. *Favanne*, t. 11. f. Q.

Inhabits the Adriatic. *Plancus.* Mediterranean. *Linnæus.* Isle of France, and the Moluccas. *Favanne.* Tarentum. *Ulysses.*

The greatest diameter is about half an inch, and when the shell is placed on its flattened side, its height, to the summit of the depressed spire, scarcely exceeds two lines; the colour is white, reddish, yellow, or olivaceous, either without markings, or with a purplish line along the sutures, or mottled somewhat in the same manner as *Trochus vestiarius*, with which however it possesses no other affinity; the form has more resemblance to a Nerite, but the aperture is emarginate at the base.

***** *Pillar-Lip appearing as if worn flat.*

HARPA. 48. Shell with longitudinal keeled mucronate remote ribs, and longitudinally striated; pillar smooth.

Buccinum Harpa. *Linnæus Syst. Nat.* p. 1201. *Born Mus.* p. 253. *Schroeter Einl.* i. p. 331. *Gmelin*, p. 3482. *Schreibers Conch.* i. p. 155. *Bruguiere Enc. Meth.* p. 249.

Variety A. Ribs broad, and mottled.
Buccinum Testudo. *Solander's MSS. Portland Cat.* p. 98. lot 2148.
Harpa ventricosa. *Lamarck Syst. des Anim.* p. 79.
Bonanni Rec. and *Kirch.* 3. f. 185. *Rumphius*, t. 32. f. K. *Petiver Amb.* t. 7. f. 8. *Knorr*, ii. t. 19. f. 1 and 2. *Regenfuss*, ii. t. 6. f. 51. *Martini*, iii. t. 119. f. 1090. *Favanne*, t. 28. f. A 3.

Variety B. Ribs narrow, and rose-coloured.
Buccinum Testudo, Var. *Solander's MSS.*
Martini, iii. t. 119. f. 1094.

Variety C. Ribs with transverse stripes in threes.
Buccinum Cithara. *Solander's MSS.*
Rumphius, t. 32. f. L. *Petiver Amb.* t. 2. f. 2. *Knorr*, i. t. 9. f. 3. *Martini*, iii. t. 119. f. 1091. *Favanne*, t. 28. f. A 1.

Variety D. Ribs with transverse stripes in pairs.
Buccinum Lyra. *Solander's MSS.*
Lister Conch. t. 994. f. 57: *Rumphius*, t. 32. f. M. *Petiver Amb.* t. 15. f. 10. *Klein Ost.* t. 6. f. 105. *Martini*, iii. t. 119. f. 1095 and 1097.

Variety E. Ribs with single transverse distant stripes.
Buccinum Barbiton. *Solander's MSS. Portland Cat.* p. 3. lot 10.
Martini, iii. t. 119. f. 1092. *Chemnitz*, x. t. 152. f. 1451.
Inhabits the coasts of Bengal. *Linnæus.* Isle of France. *Lis-*

ter. Amboyna. *Rumphius*. Coromandel. *Martini*. China. *Humphreys*.

Shell most commonly from two to three inches long, and nearly two thirds as broad, variously marbled, undulated, and banded with fawn-colour, reddish brown, rose-colour, and white, and marked with large irregular brown spots on the pillar. All its numerous varieties in size and colour approach each other by so many almost imperceptible gradations, that it is difficult to find any specific marks to distinguish them.

CANCELLATUM. 49. Shell with longitudinal, mucronate ribs, and their interstices transversely striated; pillar smooth.

Harpa cancellata. *Chemnitz*, x. t. 152. f. 1453. *Humphreys in Callone's Cat*. p. 18. No. 293.
Buccinum Harpa, Var. *Gmelin*, p. 3482. *Bruguière Enc. Meth*. p. 450.
Inhabits the coasts of Tranquebar. *Chemnitz*.
Shell about an inch and a half long, and near two thirds as broad, with fifteen or sixteen longitudinal ribs, and it differs principally from *B. Harpa* in having the spaces between the ribs transversely striated. Mr. Humphreys suspects it may be the young of Solander's *B. Testudo*.

COSTATUM. 50. Shell with crowded longitudinal mucronate ribs; pillar smooth.

Buccinum costatum. *Linnæus Syst. Nat*. p. 1202. *Martini*, iii. p. 418. t. 119. f. 1093. *Schroeter Einl*. i. p. 333. *Chemnitz*, x. p. 184. t. 152. f. 1452. *Gmelin*, p. 3482. *Schreibers Conch*. i. p. 156.
Buccinum Harpa, Var. E. *Bruguière Enc. Meth*. p. 250. *Argenville App*. t. 2. f. F. *Favanne*, t. 28. f. A 4.
Inhabits the Philippine Islands. *Humphreys*.
This shell is still more nearly allied to *B. Harpa* than the preceding, and in a specimen now before me the ribs on the side next the pillar-lip are much crowded, whilst those on the other half of the shell become abruptly twice as remote, and Linnæus has therefore probably erred in arranging it as a separate species.

PERSICUM. 51. Shell scabrous, with a crenulated outer lip and flat pillar.

Buccinum Persicum. *Linnæus Syst. Nat.* p. 1202. *Martini*, iii. p. 40. t. 69. f. 760. *Born Mus.* p. 254. *Schroeter Einl.* i. p. 334. *Gmelin*, p. 3482. *Schreibers Conch.* i. p. 157.

Purpura Persica. *Lamarck Syst. des Anim.* p. 77.

Lister Conch. t. 987. f. 46. *Rumphius*, t. 27. f. E. *Argenville*, t. 17. f. E. *Gualter*, t. 51. f. H and L. *Seba*, iii. t. 72. f. 10 and 11. *Knorr*, iii. t. 2. f. 5. *Favanne*, t. 27. f. D 2.

Variety. With transverse nodulous bands.

Buccinum Rudolphi. *Chemnitz*, x. p. 196. t. 154. f. 1467 and 1468.

Buccinum fornicatum. *Gmelin*, p. 3487.

Buccinum torvum. *Solander's MSS.*

Lister Conch. t. 987. f. 47. *Seba*, iii. t. 72. f. 12 to 16. *Knorr*, iv. t. 5. f. 4. *Favanne*, t. 27. f. D 3. *Kæmmerer Cab. Rudolst.* t. 9. f. 1.

Inhabits the Asiatic Ocean. *Linnæus.* Amboyna. *Rumphius.* Persian Gulf. *Martini.* China. *Humphreys.*

Shell about two or two and a half inches long, and sometimes rather more than two thirds as broad, brown with transverse elevated bands and lines, which are generally spotted with white. *B. Rudolphi* of Chemnitz has the spire more produced, the aperture not so wide, and the transverse bands are nodulous, but the intermediate gradations are so numerous that it appears almost impossible to draw any separating line; Gmelin has placed it as a variety of *B. Persicum,* and afterwards as a separate species under the name of *B. fornicatum.*

PATULUM. 52. Shell muricated; outer lip crenated without, and the pillar oblique and falcated.

Buccinum patulum. *Linnæus Syst. Nat.* p. 1202. *Martini*, iii. p. 38. t. 69. f. 758 and 759. *Schroeter Einl.* i. p. 335. *Gmelin*, p. 3483. *Schreibers Conch.* i. p. 157. *Bonanni Rec.* 3. f. 368, and *Kirch.* f. 361. *Lister Conch.* t. 989. f. 49. *Petiver Gaz.* t. 152. f. 3. *Gualter*, t. 51. f. E. *Adanson Senegal*, t. 7. f. 3. *Knorr*, vi. t. 24. f. 1. *Favanne*, t. 27. f. D 4.

Inhabits America. *Linnæus.* Barbadoes. *Lister.* Fort St. George in the East Indies. *Petiver.* Senegal. *Adanson.* West Indies. *Humphreys.*

Shell from two to four inches long, and more than half as broad, nearly black, with whitish transverse lines, and there are sometimes two broad paler bands towards the base; the whole surface is strongly muricated, and there are six or seven spinous elevated belts on the body-whirl.

MONODON.   53.   Shell transversely ribbed, and lon-
       gitudinally wrinkled; outer lip crenulated,
       and armed with a subulate tooth at its base.

Buccinum Monodon. *Solander's MSS. Portland Cat.* p.
     17. lot 372.   *Gmelin*, p. 3483. *Schreibers Conch.* i. p.
     147.
Buccinum Monoceros.   *Chemnitz*, x. p. 197. t. 154. f.
     1469 and 1470. *Bruguiere Enc. Meth.* p. 253.
Buccinum, No. 5.   *Schroeter Einleitung*, i. p. 357.
Pallas Spic. Zool. fasc. 10. p. 33. t. 3. f. 3 and 4. *Knorr*,
     iv. t. 30. f. 1.   *Regenfuss*, ii. t. 7. f. 2.   *Martini*, iii. t.
     69. f. 761.   *Favanne*, t. 27. f. D 1.   *Martyn Univ.
     Conch.* i. t. 10, and ii. t. 50.
Inhabits the coasts of Terra-del-fuego. *Solander.* Cape Horn
     and the Straits of Magellan. *Bruguiere.* Falkland Islands.
     Humphreys.
Shell two, or two inches and a half long, and nearly three-
fourths as broad, of a pale chestnut-colour, and the inside
white; the transverse ribs are made scaly by the crossings of
the longitudinal wrinkles. Bruguiere has described two shells
under the names of *B. Narval*, and *B. Unicorne*, which
have a subulate tooth like that of *B. Monodon*, and the difference is so small that it may be doubted whether they are
more than varieties; the former is said to be smooth, with
only obsolete transverse lines, and the other thick, smooth,
and somewhat angulated transversely.

HAUSTRUM.   54. Shell transversely ribbed, and the
       spire short; pillar oblique; outer lip crenu-
       lated and striated within.

Buccinum Haustrum.   *Martyn Univ. Conch.* i. t. 9.
Buccinum Hauritorium.   *Chemnitz*, x. p. 183. t. 152. f.
     1449 and 1450.
Buccinum Haustorium.   *Gmelin*, p. 3498. *Schreibers
     Conch.* i. p. 141.
Inhabits the coasts of New Zealand. *Martyn.*
Shell about two inches long, and nearly three fourths as broad,
of a dark chestnut colour, and the pillar and throat white;
it resembles *B. Monodon*, from which it may be distin-

guished by the smoothness of its transverse ribs, and by the want of any tooth on the outer lip.

CONCHOLEPAS. 55. Shell with the spire obliquely recurved; aperture very large; outer lip reflected; pillar-lip with two obsolete teeth at the base.

Buccinum Concholepas. *Bruguiere in Enc. Meth.* p. 252.
Patella Concholepas. *Portland Cat.* p. 186, lot 3964.
Patella Lepas. *Gmelin,* p. 3697.
Patella, No. 64. *Schroeter Einl.* ii. p. 466.
Le Grand Concholepas. *Favanne,* i. p. 543. t. 4. f. H 2.
Argenville, t. 2. f. D. *Humphreys's Conch.* t. 2. f. 7, and t. 5. f. 9. *Chemnitz,* x. Vign. 25, at p. 320. f. *A* and *B.*
Inhabits the straights of Magellan, and coasts of Peru. *Humphreys.*
Shell from two to four inches long, and three-fourths as broad at the base, rugged, of a chestnut-colour, and the inside white. In form it is a good deal like *Patella militaris,* but M. Bruguiere says, " J'ai eu occasion de voir plus de trente de ces coquilles chez mon ami, M. Dombey, qui les avoit ramassées lui-même sur les côtes du Pérou, et j'ai eu la satisfaction de trouver à chaque individu son opercule tendineux, ce qui, en etablissant le discernement éclairé de M. Dombey, suffit d'ailleurs pour démontrer que le *Concholépas* n'est point une Patelle, mais qu'il appartient incontestablement au Genre du Buccin. Ce qui est encore confirmé par l'échancrure de sa base, et par l'existence de sa spire, quoiqu'elle soit peu marquée, et d'une forme qui n'est point ordinaire."

HÆMASTOMA. 56. Shell sub-muricated; outer lip striated within; pillar flattish, and the throat and aperture fulvous.

Buccinum Hæmastoma. *Linnæus Syst. Nat.* p. 1202. *Born Mus.* p. 255. *Schroeter Einl.* i. p. 356. *Gmelin,* p. 3483. *Schreibers Conch.* i. p. 157. *Chemnitz,* xi. p. 80. t. 187. f. 1796 and 1797.
Lister Conch. t. 988. f. 48. *Gualter,* t. 51. f. A. *Martini,* iii. t. 101. f. 966.
Variety. Shell broader, and the tubercles nearly obsolete.
Martini, iii. t. 101. f. 964 and 965.
Inhabits the European Ocean. *Linnæus.* Mediterranean and Æthiopic Seas. *Gmelin.* Coasts of Turkey. *Humphreys.*

Shell from two to four inches long, and nearly three fourths as broad, coarse cinereous or greyish brown, transversely striated, and marked with about four elevated belts, of which the two upper are nodulous. The Linnæan *Murex Mancinella* belongs to the same natural family, and these two species ought not to be so far separated.

ARMIGERUM.   57.   Shell turbinated, and armed with about three transverse rows of large conical tubercles on the body-whirl; aperture white.

Buccinum Armigerum.   *Chemnitz,* xi. p. 82. t. 187. f. 1798 and 1799.
Inhabits the South Seas. *Chemnitz.*
Shell near three inches long, and more than half as broad, tapering almost equally both above and below in the manner of *Voluta Ceramica,* white, marked between the rows of conical spines on the body-whirl with transverse brown stripes; the spire is armed with a row of pointed tubercles on each whirl, and is white mottled with brown.

LUTEOSTOMUM.   58.   Shell with crowded transverse striæ, and four rows of tubercles on the body-whirl; aperture yellowish.

Buccinum luteostomum.   *Chemnitz,* xi. p. 83. t. 187. f. 1800 and 1801.
Inhabits the South Sea, and the coasts of China. *Chemnitz.*
Shell two inches long, and more than half as broad, of an ash-grey colour, and armed with tubercles, of which there are four rows on the body-whirl, and two on the spire.

LAMELLOSUM.   59.   Shell channelled transversely, and longitudinally ribbed; body-whirl lamellated towards the outer lip, and the inside bright blue.

Buccinum lamellosum.   *Gmelin,* p. 3498.   *Schreibers Conch.* i. p. 166.
Buccinum plicatum.   *Martyn Univ. Conch.* ii. t. 44.
Buccinum compositum.   *Chemnitz,* x. p. 179. Vign. 21. f. *A* and *B.*
Inhabits the coasts of New Zealand. *Martyn.*

Shell two inches and a half long, and half as broad, either wholly brown on the outside, or the two or three uppermost ribs of each whirl are greenish brown, and the inside is bright blue; it is deeply channelled transversely, and the longitudinal ribs are remote and narrow.

CRISPATUM. 60. Shell ribbed transversely, and longitudinally wrinkled with curled imbricated membranes; aperture ovate.

Buccinum crispatum. *Chemnitz,* xi. p. 84. t. 187. f. 1802 and 1803.

Inhabits King George's Sound, and the coasts of New Zealand. *Chemnitz.*

Shell an inch and a half, or two inches long, and about half as broad, of a brown, or brownish or reddish white colour, and the inside is white. Martyn, in his Universal Conchology, has given a good figure of this species, but the plate is without a number, so that it is impossible to make a reference, and on the other hand I have never seen a copy of the work, which contains any figure like *B. lamellosum,* and I have therefore quoted it wholly on the authority of Chemnitz.

LAPILLUS. 61. Shell ovate, acute, striated, without any protuberances; pillar flattish.

Buccinum Lapillus. *Linnæus Syst. Nat.* p. 1202. *Pennant Brit. Zool.* iv. p. 118. t. 72. f. 89. *Martini,* iii. p. 429. t. 121. f. 1111 and 1112, and iv. p. 22. t. 122. f. 1124, 1125, 1128, and 1129, and t. 123. f. 1136 and 1137. *Born Mus.* p. 255. *Schroeter Einl.* i. p. 337. *Gmelin,* p. 3484. *Schreibers Conch.* i. p. 158. *Bruguiere Enc. Meth.* p. 256. *Donovan,* i. t. 11. *Montagu Test.* p. 239. *Maton and Racket, in Lin. Trans.* viii. p. 135. *Dorset Cat.* p. 44. t. 15. f. 1, 2, 3, 4, 9, and 12.

Tritonium Lapillus. *Muller Zool. Dan. Prodr.* p. 244.

Purpuro-Buccinum. *Da Costa Brit. Conch.* p. 125. t. 7. f. 1, 2, 3, 4, 9, and 12.

Lister Anim. Ang. t. 3. f. 5 and 6, and *Conch.* t. 965. f. 18 and 19. *Petiver Gaz.* t. 18. f. 5. *Adanson Senegal,* t. 7. f. 4. *Knorr,* vi. t. 29. f. 4.

Inhabits the coasts of Britain. *Lister, &c.* Sweden. *Linnæus.* Iceland. *Olaffsen.* Denmark. *Muller.* France. *Reaumur.* Surinam. *Statius Muller.* Azores and Canaries. *Adanson.*

Shell one or two inches long, and about half as broad, more or
less striated transversely as well as longitudinally, and usually
of an uniform dirty white or yellowish colour, but sometimes
banded with brown or yellow; in some specimens the longitudinal striæ are membranaceous and wrinkled, and others
are almost smooth; aperture oval, and the outer lip slightly
toothed within.

FILOSUM. 62. Shell ovate, whitish, with red transverse striæ; spire rather prominent; aperture
oval; outer lip striated with red, and the pillar sub-umbilicated.

Buccinum filosum. *Gmelin*, p. 3486. *Schreibers Conch.*
i. p. 161.
Buccinum, No. 23. *Schroeter Einl.* i. p. 363.
Martini, iii. p. 433. t. 121. f. 1113 and 1114.
Inhabits ——
Shell an inch long, and near two thirds as broad, of a yellow-
ish white colour, with narrow transverse red striæ, and ap-
pears to be nearly allied to *B. Lapillus.*

SULCATUM. 63. Shell oblong-ovate, with the whirls
contiguous and transversely grooved; outer
lip crenulated, and striated within.

Buccinum sulcatum. *Bruguiere Enc. Method.* p. 255.
Variety A. Blackish, with white spots.
Buccinum sulcatum. *Born Mus.* p. 258. t. 10. f. 5 and 6.
Gmelin, p. 3491.
Buccinum, No. 159. *Schroeter Einl.* i. p. 397.
Lister Conch. t. 976. f. 31.
Variety B. White, with square black spots.
Buccinum, No. 47. *Schroeter Einl.* i. p. 369.
Lister Conch. t. 980. f. 39.
Variety C. Greyish, with longitudinal waved black stripes.
Buccinum pyramidalis. *Gmelin*, p. 3488. *Schreibers Conch.*
i. p. 159.
Buccinum, No. 66. *Schroeter Einl.* i. p. 369.
Martini, iv. p. 44. t. 124. f. 1170 and 1171.
Inhabits the coasts of Tranquebar. *Martini.*
Shell about an inch and a quarter long, and half as broad, and
Bruguiere says that all the above-mentioned varieties differ
only in colour. Messrs. Favanne have referred to Martini's

figures for their *Grand Point d'Hongrie*, vol. ii. p. 20 and
132, and t. 79. f. G, but it appears to be a different species (See *Trochus Iris*).

SERTUM. 64. Shell ovate-oblong, with transverse
crenulated striæ, and the pillar livid.

Buccinum sertum. *Bruguiere Enc. Method.* p. 262.
Buccinum coronatum. *Gmelin*, p. 3486. *Schreibers
Conch.* i. p. 162.
Buccinum, No. 24. *Schroeter Einl.* i. p. 363.
Murex rusticum. *Mus. Gevers.* p. 320, No. 722.
Lister Conch. t. 986. f. 45. *Klein Ost.* t. 4. f. 75. *Martini*, iii. p. 433. t. 121. f. 1115 and 1116.
Inhabits the coasts of Tranquebar. *Martini.*
Shell an inch and a half or two inches long, and rather more
than half as broad, reddish, fawn or chestnut-colour, with
a band on the body-whirl of paler chestnut, which Bruguiere
says appears as if composed of longitudinal spots, and one
of Martini's figures has large irregular dark chestnut spots ;
the transverse striæ are crossed by longitudinal impressed
lines, which render them crenulated. Bruguiere has used
the name of *coronatum* for another species to which it is
more applicable.

SMARAGDULUS. 65. Shell ovate, acute, and very
smooth; outer lip crenated, and striated
within; pillar slightly plaited.

Buccinum Smaragdulus. *Linnæus Syst. Nat.* p. 1203.
Born Mus. p. 256. *Gmelin*, p. 3484.
Buccinum rusticum. *Gmelin*, p. 3486. *Schreibers Conch.*
i. p. 162.
Buccinum Rullus. *Mus. Gevers.* p. 296, No. 515.
Murex sulcatus. *Gmelin*, p. 3549.
Buccinum, No. 21. *Schroeter Einl.* i. p. 362.
Lister Conch. t. 831. f. 55. *Gualter*, t. 43. f. X. *Adanson*, t. 9. f. 25. *Argenville*, t. 6. f. P. *Seba*, iii. t. 54.
f. 14 to 16. *Knorr*, iii. t. 14. f. 5. *Martini*, iii. t. 120.
f. 1104 and 1105.
Inhabits the coasts of Tranquebar. *Martini.*
Shell an inch and three-quarters long, and rather more than
half as broad, thick, somewhat ponderous, covered with a
greenish epidermis which generally leaves more or less of a
stain, and underneath which it is white marked with crowded
transverse brown stripes ; aperture beaked, and the pillar

has five or six plaits on its middle, so that the species might without impropriety be considered a Volute, but according to Adanson these plaits occur only on the shells of females.

VARIUM.　66. Shell ovate, coarse, with transverse elevated nodulous ribs; aperture ovate, and the pillar not plaited.

Buccinum varium. *Gmelin,* p. 3486. *Schreibers Conch.* i. p. 163.

Buccinum, No. 22. *Schroeter Einl.* i. p. 362.

Knorr, vi. t. 23. f. 3. *Martini,* iii. t. 121. f. 1106.

Inhabits ——

Shell an inch and a half long, and one inch broad, of a yellowish brown colour with darker bands, and a few distant longitudinal blue streaks on the spire. Martini's figure is copied from Knorr's, and it appears to be a very doubtful species.

TUBA.　67. Shell sub-fusiform, with the body-whirl nearly smooth, and thrice as long as the spire, which is cancellated.

Buccinum Tuba. *Gmelin,* p. 3484. *Schreibers Conch.* i. p. 171.

Buccinum, No. 6. *Schroeter Einl.* i. p. 358.

Martini, iii. p. 201. t. 94. f. 908.

Inhabits the East Indian Ocean. *Martini.*

Shell three inches and three-quarters long, and near one and a half broad, of a yellowish brown colour; the body-whirl is full thrice as long as the other five whirls put together, and gradually tapers from near its upper extremity to the canal in which it ends at the base.

PYRUM.　68. Shell pyriform, with a short spire, and the body-whirl ventricose; pillar smooth, and the aperture orange.

Buccinum Pyrum. *Gmelin,* p. 3484. *Schreibers Conch.* i. p. 162.

Buccinum, No. 7. *Schroeter Einl.* i. p. 358.

Martini, iii. p. 202. t. 94. f. 909 and 910.

Inhabits the Red Sea and coast of Coromandel. *Martini.*

Shell an inch and three-quarters long, and about thirteen lines broad, of a pale brownish colour with reddish streaks, and transversely striated at the base; spire depressed, and composed of five whirls.

PLUMBEUM. 69. Shell sub-globose, ponderous, glabrous, with a transverse groove near the
base of the body-whirl ; pillar lip very thick.

Buccinum plumbeum. *Chemnitz*, xi. p. 86. t. 188. f. 1806
and 1807.
Inhabits the coasts of California. *Chemnitz.*
Shell about an inch and a quarter or an inch and a half long, and
near four-fifths as broad, of a pale chestnut-colour ; towards
the base of the body-whirl there is a deep transverse groove,
and another more obsolete lower down, of which the former
ends in a projecting tooth on the outer lip ; the spire is
short and depressed, and the pillar-lip is thickly coated with
a vitreous substance.

SPADICEUM. 70. Shell pear-shaped, smooth, of a
chestnut-colour, with transverse undulated
white lines.

Buccinum spadiceum. *Gmelin*, p. 3485.
Buccinum Pyrum, Var. *Schreibers Conch.* i. p. 162.
Buccinum, No. 8. *Schroeter Einl.* i. p. 358.
Martini, iii. t. 94. f. 911.
Inhabits the West Indies. *Martini.*
Shell an inch and a half long, and about half as broad ; the
transverse stripes are described to be white, but in Martini's
figure they appear to be dark brown, and it is altogether an
uncertain species.

UMBILICATUM. 71. Shell oblong, sub-turbinated,
and slightly plaited longitudinally ; spire nodose ; aperture grooved within, and the pillar sub-umbilicated.

Buccinum umbilicatum. *Gmelin*, p. 3485. *Schreibers
Conch.* i. p. 150.
Buccinum, No. 10. *Schroeter Einl.* i. p. 358.
Martini, iii. p. 205. t. 94. f. 915 and 915 *a*, and Vign. 32,
at p. 198. f. 1.
Inhabits ———
Shell about two inches and a half long, and one and a half
broad, of a pale brownish yellow, transversely striated at the
base, and marked with irregular longitudinal wrinkled plaits ;
the spire is pyramidal, but not one third so long as the body-
whirl, and is nodulous at the sutures.

CANDIDUM.   72. Shell ovate, ventricose, ponderous, smooth, and white without any markings; spire short.

Buccinum candidum.   *Gmelin*, p. 3485.
Buccinum, No. 11.   *Schroeter Einl.* i. p. 359.
Martini, iii. p. 206. Vign. 31, at p. 191. f. 3.
Inhabits ———
Shell about two inches long, and rather more than half as broad, and white without any markings; it appears from Martini's description to resemble *Voluta gravis*, but has not any plaits on the pillar.

CRASSUM.   73. Shell sub-globose, ventricose, and glabrous; aperture oval, and the pillar thickened, with two callosities at the base.

Buccinum crassum.   *Gmelin*, p. 3485.   *Schreibers Conch.* i. p. 155.
Buccinum, No. 18.   *Schroeter Einl.* i. p. 361.
Martini, iii. p. 424. t. 120. f. 1099 and 1100.
Inhabits ———
Shell two inches long and one and a half broad, of a brownish white colour, and bears a considerable resemblance to *Voluta gibbosa*, but is without any plait on the pillar; the spire is small, and consists of five whirls.   Gmelin, at p. 3487, has another species under the name of *B. crassum*, which appears at most to be only a Variety of *Strombus Vexillum*.
B. Labyrinthus, as well as *B. marginatum*, which Gmelin has placed after this species, appear to be fossil shells.

ORBITA.   74. Shell ovate, thick, with transverse, distant, knotty, reflected ribs, and other smaller ones in the interstices; outer lip plaited.

Buccinum Orbita.   *Chemnitz*, x. p. 199. t. 154. f. 1471 and 1472.   *Gmelin*, p. 3490.   *Schreibers Conch.* i. p. 147.
Buccinum bicostatum.   *Bruguiere Enc. Meth.* p. 248.
Buccinum succinctum.   *Martyn Univ. Conch.* ii. t. 45.
Variety.   Smaller, and rather more oblong.
Buccinum lacunosum.   *Bruguiere Enc. Meth.* p. 258.
Buccinum striatum.   *Martyn Univ. Conch.* i. t. 7.

Buccinum Orbita lacunosa. *Chemnitz*, x. p. 200. t. 154. f. 1473.

Buccinum, No. 116. *Schreibers Conch.* i. p. 170.

Inhabits the coasts of New Zealand. *Martyn.*

Shell two or three inches long and two thirds as broad, white, and the inside somewhat pearly; the transverse ribs are striated, knotty, and reflected in the direction of the spire, and their interstices are made uneven by other smaller ones. This is undoubtedly *B. bicostatum* of Bruguiere, though his references are all erroneous; and it appears to me that his *B. lacunosum* is a Variety, or perhaps only a young shell of the same species.

SCALA. 75. Shell ovate, with dilated recurved transverse white belts, and their interstices longitudinally striated.

Variety A. With four belts.

Buccinum Scala. *Gmelin*, p. 3485. *Schreibers Conch.* i. p. 150.

Buccinum, No. 15. *Schroeter Einl.* i. p. 360. t. 2. f. 8.

Variety B. With three belts.

Buccinum cingulatum. *Linnæus Mantissa*, p. 549 and 550. *Gmelin*, p. 3506.

Buccinum Trochlea. *Bruguiere Enc. Method.* p. 248.

Buccinum, No. 16. *Schroeter Einl.* i. p. 360.

Murex planatus. *Mus. Gevers.* p. 320, No. 717.

Lister Conch. t. 1059. f. 2. *Petiver Gaz.* t. 101. f. 14. *Knorr*, iii. t. 7. f. 2. *D'Avila*, t. 8. f. V. *Martini*, iii. t. 118. f. 1089, *a* and *b*.

Variety C. With two belts.

Favanne, t. 34. f. E.

Inhabits the East Indian Ocean. *Martini.* Straights of Magellan, and Cape of Good Hope. *Bruguiere.* Maryland. *Lister?*

Shell most commonly about three-quarters of an inch long, and rather more than half as broad; there are four whirls, of which the upper parts are flattened like steps, and the lower parts have from two to four elevated broad belts, whose dilated margins are somewhat recurved; the belts are rather glabrous and white, and their interstices brown or grey, and striated longitudinally; the aperture is toothed on the outer lip, and strongly grooved within. The shell figured by Lister was probably a Fossil, and a fossil specimen with four belts, also from Maryland, is mentioned in the Portland Catalogue, p. 137, lot 3516. Chemnitz, in his eleventh volume, p. 91, has described a shell under the name of *B. annulatum*, which he considers to be nearly allied to this species, but his figure answers badly to the description, and appears to be entirely different.

****** *Somewhat polished, and not enumerated in the former Divisions.*

SPIRATUM. 76. Shell smooth, with the whirls separated by a deep canal; pillar abrupt and perforated.

Buccinum spiratum. *Linnæus Syst. Nat.* p. 1203. *Born Mus.* p. 256. *Martini*, iv. p. 14. t. 122. f. 1118. *Schroeter Einl.* i. p. 338, and *Inn. Bau Conch.* p. 37. t. 4. f. 3. *Gmelin*, p. 3487. *Schreibers Conch.* i. p. 160. *Bruguiere Enc. Meth.* p. 262.
Bonanni Rec. 3. f. 370, and *Kirch.* f. 362. *Lister Conch.* t. 983. f. 42 c. *Rumphius*, t. 49. f. D. *Petiver Gaz.* t. 101. f. 13, and *Amb.* t. 9. f. 21. *Argenville*, t. 17. f. N. *Seba*, iii. t. 73. f. 21, 22, 24, and 25. *Knorr*, ii. t. 6. f. 5, and iii. t. 3. f. 4. *Regenfuss*, i. t. 10. f. 41. *Favanne*, t. 33. f. E 1.
Variety. With the spire produced, and the whirls not separated by a canal.
Bonanni Rec. and *Kirch.* 3. f. 70. *Lister Conch.* t. 981. f. 41. *Rumphius*, t. 49. f. C. *Petiver Amb.* t. 9. f. 20. *Klein Ost.* t. 2. f. 46. *Seba*, iii. t. 73. f. 23 and 26. *Martini*, iv. t. 122. f. 1120 and 1121. *Favanne*, t. 33. f. E 2.
Inhabits the East Indies. *Petiver.* Coasts of Tranquebar. *Regenfuss.* Mediterranean. *Linnæus?* Coasts of China. *Martini.* Arabia Felix. *Humphreys.*
Shell an inch and a half, or two inches long, and about two thirds as broad, with the whirls more or less sunk into each other, and separated by a deep canal; the colour is white, with brown, yellow, or saffron-coloured spots, of which those on the upper parts of the whirls are largest. A shell differing in no other respect than having the whirls more produced, and only flattened at the suture, is not uncommon, and has been considered to be a Variety of this species by all authors except Martini.

ZEYLANICUM. 77. Shell smooth, with the whirls produced; pillar abrupt, and umbilicated; umbilicus large, and toothed at the margin.

Buccinum Zeylanicum. *Bruguiere Enc. Meth.* p. 264.
Buccinum glabratum, Var. *Schroeter Einl.* i. p. 341. *Gmelin,* p. 3489. *Schreibers Conch.* i. p. 161.
Lister Conch. t. 982. f. 42. *Gualter,* t. 51. f. B. *Klein Ost.* t. 2. f. 47. *Martini,* iv. t. 122. f. 1119.
Inhabits the coasts of Ceylon. *Bruguiere.*
Shell about two inches long, and rather more than half as broad, white with fawn or chestnut-coloured spots like those of *B. spiratum,* from the Variety of which species it differs in having the whirls more compleatly drawn out, and the umbilicus larger, with a tinge of violet, and toothed on the margin. It is much more allied to *B. spiratum,* than to *B. glabratum,* of which it has been considered a Variety.

GLABRATUM. 78. Shell umbilicated, highly polished, and the sutures obsolete; body-whirl channelled and produced at the base.

Buccinum glabratum. *Linnæus Syst. Nat.* p. 1203. *Martini,* iv. p. 10. t. 122. f. 1117. *Born Mus.* p. 257. *Schroeter Einl.* i. p. 341. *Gmelin,* p. 3489. *Schreibers Conch.* i. p. 161. *Bruguiere Enc. Meth.* p. 264.
Voluta butyracea. *Solander's MSS.*
Bonanni Rec. and *Kirch.* 3. f. 149. *Lister Conch.* t. 974. f. 29. *Gualter,* t. 43. f. T. *Argenville,* t. 9. f. G. *Knorr,* ii. t. 16. f. 4 and 5. *Favanne,* t. 31. f. F 1.
Inhabits the American Ocean, and coasts of Tranquebar. *Linnæus.*
Shell two or three inches long, and half as broad, of a pale orange colour mixed with white, and very glossy all over; the body-whirl has a double belt at the base, and a white vitreous mass is formed over the aperture, in both of which respects, as well as some others, this species bears a considerable affinity to the Olives.

TURGITUM. 79. Shell ovate, sub-umbilicated, smooth, with transverse rows of red spots, and the outer lip sinuated.

Buccinum turgitum. *Gmelin,* p. 3490. *Schreibers Conch.* i. p. 147.
Buccinum maculatum. *Martyn Univ. Conch.* ii. t. 49.

Buccinum adspersum.   *Bruguiére Enc. Meth.* p. 265.
Chemnitz, x. p. 201. t. 154. f. 1475 and 1476.
Inhabits the coasts of New Zealand. *Martyn.*
Shell two inches long, and near an inch and a quarter broad,
yellowish, with numerous transverse rows of small red spots,
and the inside pale orange ; there are six whirls, of which
the three lower are extremely level and smooth, and the
other upper ones are slightly plaited longitudinally.

SCUTULATUM.   80. Shell ovate, smooth, brown with
longitudinal veins ; whirls flattish, and the
beak obtuse.

Buccinum scutulatum. *Martyn Univ. Conch.* ii. t. 55.
Chemnitz, x. p. 179. Vign. 21. f. *C* and *D.   Gmelin,* p.
3498.   *Schreibers Conch.* i. p. 157.
Inhabits the coasts of New Zealand. *Martyn.*
Shell about two inches and a quarter long, and rather more
than half as broad, brown, with obsolete darker bands, and
longitudinal veins, and somewhat spotted with white.

TESTUDINEUM.   81. Shell oblong-ovate, smooth,
with transverse rows of crowded dark brown
spots, and somewhat produced at the base.

Buccinum testudineum. *Chemnitz*, x. p. 187. t. 152. f.
1454. *Gmelin,* p. 3498.   *Schreibers Conch.* i. p. 151.
Bruguiere Enc. Meth. p. 266.
Inhabits the coasts of New Zealand. *Chemnitz.*
Shell an inch and a half long, and half as broad, of a bluish
white colour, with numerous transverse rows of squarish
dark brown spots, which become confluent on the spire ; it
is narrower in proportion to its length than *B. turgitum,*
and has the inside of the outer lip spotted with dark brown.

CATARRACTA.   82. Shell ovate, rough with crowded
minute transverse grooves, and marked longitudinally with undulated decurrent stripes.

Buccinum Cataracta.   *Chemnitz*, x. p. 188. t. 152. f. 1455.
Gmelin, p. 3498.   *Schreibers Conch.* i. p. 146.
Inhabits the coasts of New Zealand. *Chemnitz.*   Cape of
Good Hope. *Humphreys.*
Shell an inch and a half long, and nearly half as broad, of a
bluish white colour, with the whirls tinged at both ends with
pale brown, and marked with longitudinal decurrent undu-
lated brown stripes.

LÆVISSIMUM. 83. Shell oblong-ovate, polished, obliquely truncated at the base, and the apex obtuse.

Buccinum lævissimum. *Gmelin*, p. 3494. *Schreibers Conch.* i. p. 151.
Buccinum flammeum. *Bruguiere Enc. Meth.* p. 266.
Buccinum, No. 66. *Schroeter Einl.* i. p. 372.
Martini, iv. p. 72. t. 127. f. 1215 and 1216.
Inhabits the East Indian Seas. *Martini.*
Shell about two inches long, and rather more than half as broad, yellowish or whitish, with eight or nine longitudinal stripes of brown on the body-whirl, and a smaller number on the spire; the pillar-lip is concave, and the aperture obliquely truncated at the base. Martini's figure is white with the apex blue, and Bruguiere suspects it was taken from a bleached specimen, which in this species he says are the most commonly met with. Knorr, v. t. 18. f. 3, and Martini, iv. t. 125. f. 1193, to which Gmelin has referred for his *B. obtusum*, may probably have been made from other damaged specimens of this species, but the figures are so bad, and the description is so obscure, that they are hardly worth notice.

CYANEUM. 84. Shell ovate-oblong, brittle, bluish, transversely striated, and the whirls tiled at the sutures; pillar with one plait.

Buccinum cyaneum. *Bruguiere Enc. Meth.* p. 266.
Buccinum glaciale novum. *Schreibers Conch.* i. p. 141.
Chemnitz, x. p. 182. t. 152. f. 1448.
Inhabits the shores of Greenland. *Chemnitz.*
Shell an inch and three-quarters long, and half as broad, of a pale blue colour slightly mottled with white; there are five whirls, which overlap each other so as to conceal the sutures, and the whole shell is marked with minute distant striæ.

LÆVE. 85. Shell ovate-oblong, smooth, brown with darker bands, and finely striated transversely; aperture oval, and ending in a canal.

Buccinum læve. *Gmelin*, p. 3488.
Buccinum, No. 35. *Schroeter Einl.* iii. p. 366.

Martini, iv. p. 86. t. 124. f, 1150.
Inhabits the East Indies. *Martini.*
Shell about sixteen lines long and ten broad, of a pale brown
  colour with darker bands, and minute transverse rather distant striæ ; the body-whirl is slightly ventricose, and more
  than twice as large as the spire, which consists of four or
  five whirls.

IGNEUM.   86. Shell oblong, narrow, glabrous, yel-
    lowish clouded with red, and the upper
    whirls longitudinally striated.

Buccinum igneum.   *Gmelin,* p. 3494.   *Schreibers Conch.*
  i. p. 151.
Buccinum, No. 68.   *Schroeter Einl.* i. p. 372.
Martini, iv. p. 72. t. 127. f. 1217.
Inhabits ——
Shell an inch and a half long, and about seven lines and a half
  broad, with seven whirls of a yellowish colour, and marked
  irregularly with undulated red streaks and spots.

PLUMATUM.   87.   Shell oblong, transversely striat-
    ed ; pillar-lip with a tooth at the upper end,
    and the outer-lip striated.

Buccinum plumatum. *Gmelin,* p. 3494. *Schreibers Conch.*
  i. p. 152.
Buccinum, No. 69.   *Schroeter Einl.* i. p. 373.
Murex accinctus.   *Born Mus.* p. 316.
Lister Conch. t. 822 b. f. 41. *Knorr,* iv. t. 21. f. 6. *Mar-*
  tini, iv. t. 127. f. 1218 to 1220.
Inhabits the coasts of Jamaica. *Lister.*   Curaçoa and Ascension Island. *Martini.*
Shell an inch and a half or two inches long, and less than half as
  broad, with eight slightly convex whirls, of a purplish or
  blackish brown or chestnut-colour, with transverse rows of
  darker spots, and a narrow reticulated white band which extends only to the second whirl.   Born's reference to Martini
  is erroneous, and this is certainly his *Murex accinctus.*

OCELLATUM.   88.   Shell smooth, black, with rows
    of white spots and dots ; body-whirl ventri-
    cose, and the spire rather prominent and no-
    dulous.

Buccinum ocellatum. *Gmelin*, p. 3488. *Schreibers Conch.*
i. p. 145.
Buccinum, No. 41. *Schroeter Einl.* i. p. 367.
Martini, iv. p. 39. t. 124. f. 1160 and 1161.
Inhabits the East Indian Seas. *Martini.*
Shell about an inch long, and two thirds as broad, black with
transverse rows of white spots, of which those on the upper
part of the body-whirl are broadest ; the inside is blue.

GLABERRIMUM. 89. Shell very smooth, minute,
with the base truncated.

Buccinum glaberrina. *Gmelin*, p. 3488.
Variety A. White, with a brown line at the sutures.
Buccinum, No. 49. *Schroeter Einl.* i. p. 369.
Martini, iv. t. 125. f. 1177.
Variety B. Uniform reddish brown.
Martini, iv. t. 125. f. 1180.
Variety C. Reddish brown, with a tessellated band at the two
lower sutures.
Buccinum, No. 50. *Schroeter Einl.* i. p. 369.
Martini, iv. t. 125. f. 1178.
Variety D. Reddish brown, with transverse dotted lines.
Buccinum, No. 51. *Schroeter Einl.* i. p. 369.
Martini, iv. t. 125. f. 1179.
Variety E. Reddish brown, with longitudinal dark undulated
stripes.
Buccinum, No. 52. *Schroeter Einl.* i. p. 369.
Martini, iv. t. 125. f. 1181.
Variety F. Reddish brown, with transverse rows of spots and
dots.
Buccinum, No. 53. *Schroeter Einl.* i. p. 369.
Martini, iv. t. 125. f. 1182.
Inhabits ———
Shell from five to seven lines long, and half as broad, white
or reddish brown, and variously marked with a darker co-
lour.

NUCLEUS. 90. Shell oblong-ovate, minute, transversely striated ; pillar obliquely grooved,
and the aperture truncated at the base.

Buccinum Nucleus. *Bruguiere Enc. Meth.* p. 254.
Buccinum strigosum. *Gmelin*, p. 3488. ?
Buccinum, No. 86. *Schroeter Einl.* i. p. 376, and No.
54. p. 369. ?

Lister Conch. t. 976. f. 32.   *Martini,* iv. t. 125. f. 1183. ?
Inhabits the coasts of New Zealand and Madagascar. *Bruguiere.*

Shell sometimes seven lines long, and four broad, but is usually smaller, and Bruguiere says it varies in colour from brown to black, with the inside paler, or sometimes white. Gmelin, who at p. 3476 and 3494 has two other species under the name of *B. strigosum,* in the present instance has referred to some figures of Martini's as varieties, which are obviously different, and his description is so extremely loose as to be undeserving of notice.

LINEATUM.   91.   Shell oblong, minute, and transversely striped; spire acute, and pyramidal; outer lip expanded.

Buccinum lineatum.   *Da Costa Brit. Conch.* p. 130. t. 8. f. 5.   *Donovan Brit. Shells,* i. t. 15.   *Montagu Test.* p. 245. *Maton and Racket, in Lin. Trans.* viii. p. 136. *Dorset Cat.* p. 45. t. 14. f. 5.
Martini, iv. t. 125. f. 1188 and 1189. ?
Inhabits the coasts of Cornwall, and the West Indies. *Da Costa.*   Dorsetshire. *Pulteney.*   Brittany. *Bruguiere.*

Shell a quarter of an inch long, and half as broad, smooth, sometimes white with only a few scattered transverse narrow brown or yellow bands, and not unfrequently regularly banded with equally broad alternate white and dark brown lines; aperture oval, and the outer lip considerably expanded. *B. lineatum* of Gmelin is a very indistinct species taken from Knorr, iii. t. 14. f. 4, and the same figure is quoted by Favanne (Vol. ii. p. 136), for a Trochus, which may probably be a variety of *T. zic-zac.*

EXILE.   92.   Shell oblong, minute, and cancellated; aperture expanded, spotted, and crenated.

Buccinum exile.   *Gmelin,* p. 3497.
Buccinum, No. 58.   *Schroeter Einl.* i. p. 370.
Martini, iv. t. 125. f. 1190 and 1191.
Inhabits ——

This shell appears, from Martini's figures, to be rather larger than *B. lineatum,* and of a brownish red colour with blackish cancellated lines. Martini has figured several small shells which he has confounded together in one description,' and

Gmelin has constituted his *B. nanum* (p. 3497), from one of these figures, with the appearance of which his specific character is notwithstanding at variance.

PRÆROSUM.   93.   Shell ovate, smooth, black, with the spire carious, and the pillar glabrous.

Buccinum prærosum. *Linnæus Syst. Nat.* p. 1203. *Chemnitz*, ix. part 2. p. 40. t. 121. f. 1035 and 1036.   *Gmelin*, p. 3489.   *Schreibers Conch.* i. p. 161.
Bulimus prærosus.   *Bruguiere Enc. Meth.* p. 361.
Inhabits Southern Europe.   In the aqueduct at Seville. *Linnæus.*
The shell figured by Chemnitz is about seven lines long, and four or five broad, and the spire terminates abruptly, and appears as if bitten off. It may perhaps be doubted whether the Linnæan *B. prærosum* was more than a damaged shell of *Helix palustris*, which Montagu says is sometimes almost black, and not uncommonly worn or decorticated about the apex.

COCHLIDIUM.   94.   Shell oblong, smooth, with the whirls flattened at the sutures ; aperture oval and effuse.

Buccinum Cochlidium.   *Chemnitz*, xi. p. 275. t. 209. f. 2053 and 2054.
Inhabits the Islands in the South Sea, and is a land shell. *Chemnitz.*
Shell about three inches long, and half as broad, white, without any markings.

AUSTRALE.   95.   Shell oblong, smooth, with narrow transverse variegated bands and veins ; aperture oval, and entire.

Buccinum australe.   *Gmelin*, p. 3490.
Buccinum Tritonis fluviatile.   *Chemnitz*, ix. part 2. p. 38. t. 120. f. 1033 and 1034.
Helix solida.   *Born Mus.* p. 393. t. 13. f. 18 and 19. *Gmelin*, p. 3651.
Helix, No. 197.   *Schroeter Einl.* ii. p. 231.
Trochus Phasianella.   *Brookes's Introd.* p. 163. t. 7. f. 96.
Le Faisan.   *Favanne Cat. Rais.* p. 11. t. 1. f. 46.
Inhabits fresh-water streams in New Zealand. *Chemnitz.*
This beautiful shell, generally known by the name of the

Pheasant, is about two or three inches long, and half as broad ; it is white, variously ornamented with narrow bands, composed of red lines and crescent-shaped spots, and other intermediate spots and lines of a pale olive or fawn-colour. Gmelin says it is an intermediate species between Buccinum, Bulla, and Helix ; and some of the Genus Bulimus, to which it probably belongs, have been placed in all these genera.

******* *Angulated, and not included in the former Divisions.*

UNDOSUM.   96.   Shell ovate, with transverse, elevated, glabrous striæ, and the body-whirl obtusely five-angled.

Buccinum undosum. *Linnæus Syst. Nat.* p. 1203. *Martini,* iv. p. 24. f. 122. f. 1126 and 1127. *Born Mus.* p. 258. *Schroeter Einl.* i. p. 342. *Gmelin,* p. 3490. *Schreibers Conch.* i. p. 163.
Lister Conch. t. 938. f. 33. *Rumphius,* t. 29. f. O. *Petiver Amb.* t. 13. f. 4. *Argenville,* t. 9. f. N. *Seba,* iii. t. 52. f. 26.
Inhabits the coasts of Amboyna. *Rumphius.* Island of Banda. *Seba.*
Shell nearly two inches long, and rather more than half as broad, white or yellowish, with elevated transverse brown striæ ; the body-whirl has five or six angles, and the pillar is slightly plaited at the base.

AFFINE.   97.   Shell ovate, with transverse, elevated, glabrous striæ, and the body-whirl rounded ; outer-lip swollen, and the pillar toothed.

Buccinum affine. *Gmelin,* p. 3490.
Buccinum constrictum. *Solander's MSS.*
Buccinum, No. 29. *Schroeter Einl.* i. p. 364.
Knorr, ii. t. 14. f. 4 and 5. *Martini,* iv. p. 32. t. 123. f. 1135.
Inhabits the Straights of Malacca. *Martini.* South Seas. *Solander.* Coasts of Madagascar. *Humphreys.*
This shell resembles the last in size and colour, but the body-whirl is not angulated, the outer lip is much swollen, and the pillar distinctly toothed.

FUMOSUM. 98. Shell nearly oval, plaited longitudinally, and transversely ribbed, with the interstices striated; aperture effuse.

Buccinum fumosum. *Solander's MSS.*
Buccinum strigosum. *Gmelin,* p. 3494. ?
Martini, iv. t. 123, upper figures 1145 and 1146, and Vign. 38, at p. 49. ?
Inhabits ――――
This shell is nearly allied to *B. undosum,* but differs in having the longitudinal angles or plaits smaller and more numerous, the aperture more effuse, and the interstices of the ribs striated. Of Martini's Vignette above referred to I cannot find that he has given any description, and it is probably at most only a variety of this species.

INDICUM. 99. Shell ovate, reticulated with transverse ribs and elevated longitudinal striæ; aperture effuse.

Buccinum Indicum. *Gmelin,* p. 3495.
Buccinum, No. 30. *Schroeter Einl.* i. p. 364.
Martini, iv. p. 32. t. 123. f. 1138 and 1139.
Inhabits the East Indies. *Martini.*
Shell an inch and a quarter long, and two thirds as broad, brown, with the inside white.

TRANQUEBARICUM. 100. Shell ovate, twelve-angled, and transversely striated; aperture toothed; with the outer lip orange, and the pillar perforated.

Buccinum Tranquebaricum. *Gmelin,* p. 3491.
Buccinum, No. 34. *Schroeter Einl.* i. p. 365.
Martini, iv. p. 35. t. 123. lower fig. 1146 to 1149.
Inhabits the coasts of Coromandel. *Martini.*
Shell from one to two inches long, and half as broad, of a pale brown colour, with darker stripes; the angles are somewhat nodulous, and their summits are yellow or orange, especially at the margin of the outer lip, which is slightly swollen.

VERSICOLOR. 101. Shell coarse, transversely striated, with two intermediate rows of black dots; spire with four concave whirls.

Buccinum versicolor. *Gmelin,* p. 3491.
Buccinum, No. 33. *Schroeter Einl.* i. p. 365.
Martini, iv. p. 34. t. 123, lower fig. 1145.
Inhabits the East Indian Seas. *Martini.*
Shell about an inch and three quarters long, and rather more
than two-thirds as broad, of a dirty brown colour with three
transverse rows of irregular large darker spots on the body-
whirl, besides the markings which are above described.

CRUENTATUM. 102. Shell ovate, white, minutely
striated transversely, and marked with squarish red spots ; whirls convex.

Buccinum cruentatum. *Gmelin,* p. 3491.
Buccinum, No. 32. *Schroeter Einl.* i. p. 365.
Martini, iv. p. 34. t. 123. f. 1143 and 1144.
Inhabits ——
Shell an inch and a half long, and one inch broad, white with
rather large scattered squarish red spots ; the upper parts of
the whirls are somewhat flattened, which gives an angular
projection to the outer lip.

BEZOAR. 103. Shell roundish, wrinkled, with the
whirls lamellated above; pillar perforated.

Buccinum Bezoar. *Linnæus Syst. Nat.* p. 1204. *Schroe-
ter Einl.* i. p. 343. *Gmelin,* p. 3491. *Schreibers Conch.*
i. p. 163.
Buccinum luxurians. *Solander's MSS. Portland Cat.*
p. 143. lot 3163.
Muriciformis trifoliatum. *Mus. Gevers.* p. 316. No. 683.
Murex rapiformis, Var. β. *Born Mus.* p. 307.
Argenville, t. 15. f. G. *D'Avila,* t. 11. f. E. *Martini,*
iii. t. 68. f. 754 and 755.
Inhabits the coasts of China. *D'Avila.*
Linnæus has described this species to be " as large as an apple,
coarse, with decussated wrinkles or striæ, and the base perforated; body-whirl on the upper part imbricated with prominent, undulated, numerous membranes ; spire angular
with straight sides, and the upper part flattish, and plaited
or toothed above." He adds that it is very nearly allied to
the Murices. Born for this species has referred to Martini,
f. 398 and 399, which is *B. calcaratum,* and has quoted f.
755 as a variety of his *Murex rapiformis.*

BULBOSUM. 104. Shell roundish, obliquely contracted at the base, with two transverse rows of
vaulted spines; whirls channelled at the sutures; aperture and umbilicus very wide.

Buccinum bulbosum. *Solander's MSS.*
Buccinum Bezoar, Var. *β*. *Kæmmerer Cab. Rudolstadt.*
Murex rapiformis. *Born Mus.* p. 307.
Murex Rapa. *Gmelin,* p. 3545.
Murex, No. 8. *Schroeter Einl.* i. p. 545.
Muriciformis Radix. *Mus. Gevers.* p. 316. No. 378.
Lister Conch. t. 894. f. 14. *Knorr,* v. t. 21. f. 2. *Martini,* iii. t. 68. f. 750 to 753.
Inhabits the coasts of Tranquebar. *Martini.*
Shell three or four inches long, and about three-fourths as
broad, of a pale yellowish brown colour, and armed on the
body-whirl with two transverse rows of vaulted spines, of
which the upper is largest, and is extended throughout the
spire. Several other of Gmelin's Murices have an equal
claim to be removed to this Genus, and it is only because
some authors have considered it to be a variety that I have
placed this species next to *Buccinum Bezoar.*

GLACIALE. 105. Shell ovate-oblong, smooth, and
somewhat striated; the body-whirl slightly
keeled.

Buccinum glaciale. *Linnæus Syst. Nat.* p. 1204. *Chemnitz,* x. p. 180. t. 152. f. 1446 and 1447. *Gmelin,* p.
3491. *Bruguiere Enc. Meth.* p. 259. *Donovan,* v. t.
154. *Maton and Racket, in Lin. Trans.* viii. p. 136.
Montagu Supp. p. 109.
Tritonium glaciale. *Muller Zool. Dan. Prod.* 2942. *Fabricius Fauna Groenl.* p. 397.
Inhabits the Northern Ocean. Coasts of Spitzbergen. *Linnæus.* Denmark. *Muller.* Greenland. *Fabricius.* Ork-
ney Islands. *Donovan.*
Shell two or three inches long, and about half as broad, pale
brown, transversely striated, and the upper part of the whirls
longitudinally plaited; there are about nine volutions, and the
carinated ridge is said by Muller, Fabricius, and Chemnitz,
as well as Linnæus, to be confined to the body-whirl, in
which respect it differs from *B. carinatum.*

CARINATUM. 106. Shell oblong-conical, transversely striated ; upper whirls with many oblique and obtuse angles, and the lower whirls with a single keel.

Buccinum carinatum. *Phipps's Voyage to the North Pole,* p. 197. t. 13. f. 2. *Gmelin,* p. 3493.

Inhabits the Arctic Ocean at Spitzbergen. *Phipps.*

Shell about two inches and a half long, and is probably the shell mentioned as a variety, in the description of *B. glaciale,* by Mr. Donovan, who says that the carinated ridge of that species is not always confined to the body-whirl, as described by Linnæus, for that in a specimen in Lord Tankerville's Cabinet it traverses the spire nearly to the apex.

LAMELLATUM. 107. Shell oblong, lamellated longitudinally, white with a purple inside, and the pillar-lip white.

Buccinum lamellatum. *Gmelin,* p. 3498.

Buccinum, No. 43. *Kæmmerer Cab. Rudolst.* p. 134. t. 9. f. 2.

Inhabits ——

Shell about an inch and a half long, and appears to be a badly constituted and rather uncertain species.

UNDATUM. 108. Shell oblong, coarse, with transverse striæ, and curved longitudinal plaits.

Buccinum undatum. *Linnæus Syst. Nat.* p. 1204. *Pennant Zool.* iv. p. 121. t. 73. f. 90. *Martini,* iv. p. 66. t. 126. f. 1206 to 1211. *Born Mus.* p. 259. *Schroeter Einl.* i. p. 344. *Gmelin,* p. 3492. - *Bruguiere Enc. Meth.* p. 258. *Schreibers Conch.* i. p. 164. *Montagu Test.* p. 237. *Donovan,* iii. t. 104. *Maton and Racket, in Lin. Trans.* viii. p. 137. *Dorset Cat.* p. 45. t. 17. f. 6. *Lamarck Syst. des Anim.* p. 77.

Buccinum vulgare. *Da Costa, Brit. Conch.* p. 122. t. 6. f. 6.

Tritonium undatum. *Muller Zool. Dan.* ii. p. 12. t. 50. *Fabricius Fauna Groenl.* p. 394.

Bonanni Rec. 3. f. 189, and *Kirch.* f. 191. *Lister Conch.* t. 962. f. 14, and *Anim. Ang.* t. 3. f. 2. *Seba,* iii. t. 39. f. 61, and 76 to 80. *Knorr,* iv. t. 19. f. 1. *Favanne,* t. 32. f. D.

Junior. Tritonium viridulum. *Fabricius Fauna Groenl.* p. 402.

Buccinum viridulum. *Gmelin,* p. 3493.

Variety B. With the longitudinal plaits obsolete.

Buccinum striatum. *Pennant Zool.* iv. p. 121. t. 74. f. 91. *Lister Conch.* t. 962. f. 15, and *Anim. Ang.* t. 3. f. 3.

Variety C. With the whirls reversed.

Buccinum Bornianum. *Chemnitz,* ix. t. 105. f. 892 and 893.

Born Mus. t. 9. f. 14 and 15.

Inhabits the sea on the Northern coasts of Europe, Britain. *Lister.* Sweden. *Linnæus.* Belgium. *Seba.* Denmark. *Muller.* Greenland. *Fabricius.* Iceland. *Olaffsen.*

Shell from two to five inches long, and more than half as broad, with seven or eight ventricose whirls, transversely striated, and longitudinally ribbed or plaited with rounded, curved, oblique ribs; the colour is dirty white, or pale, dark, or yellowish brown, and Seba has figured a blue variety, which he says is found in the Zuyder Zee, but is very rare. It appears to me very doubtful whether *Tritonium ciliatum* of Fabricius is more than an old shell of this species, and many of the young shells on our shores answer to the description of his *Tritonium viridulum.*

CILIATUM. 109. Shell turreted, patulous, somewhat beaked, angulated, longitudinally ciliated, and the pillar slightly plaited.

Buccinum ciliatum. *Gmelin,* p. 3492. Tritonium ciliatum. *Fabricius Fauna Groenl.* p. 401.

Inhabits the coasts of Greenland. *Fabricius.*

Fabricius says this shell is nearly allied to the foregoing, and is six inches long, whitish, with the inside glabrous, and coated externally with a ciliated plaited epidermis; it has but five whirls, which are less convex than in *B. undatum,* and strongly striated transversely, except at the summit of the spire, which is glabrous; aperture large, and the pillar with one plait.

SOLUTUM. 110. Shell ovate, with unequally distant longitudinal tubercles on the body-whirl; outer lip channelled, and somewhat detached.

Buccinum solutum. *Hermann Naturf.* p. 52. t. 2. f. 3 and 4. *Gmelin,* p. 3493.

Buccinum undatum, Var. C.   *Schreibers Conch.* i. p. 164.
Inhabits ——

This shell is said to be allied to *B. undatum*, but is described
with six unequal ribs, and the first and second whirl five times
as broad as the others; spire obtuse, and the colour whitish
mixed with yellow; it is a doubtful species, and has been
quoted by Schreibers for a variety of *B. undatum*.

PAPYRACEUM.   111.   Shell ovate-oblong, slender,
   obsoletely striated transversely, and the whirls
   depressed at their summits.

Variety A. Dirty white, with transverse brown stripes in pairs
at the base.
Buccinum papyraceum. *Bruguiere Enc. Meth.* p. 260.
Variety B. Yellowish, transversely striped with brown through-
out.
Buccinum Anglicum.   *Gmelin,* p. 3494.
Buccinum, No. 68.   *Schroeter Einl.* i. p. 371.
Lister Conch. t. 963. f. 17.   *Martini,* iv. p. 70. t. 126. f.
   1212.
Inhabits ——

This shell is described by Bruguiere to be twenty-two lines
long, and ten and a half broad, and that figured by Martini,
which seems to differ only in colour, is rather larger; the
surface is rather glossy and marked with slightly elevated
striæ, particularly on the upper, and on the base of the lower
whirls.   The name of *B. Anglicum,* which has been given
to the shell figured by Martini, is improper, as it is not a
native of these shores.

OTAHEITENSE.   112.   Shell oblong, transversely
   grooved, with the grooves wrinkled and dotted, and the sutures finely crenulated.

Buccinum Otaheitense.   *Chemnitz,* x. p. 202. t. 154. f.
   1477.   *Schreibers Conch.* i. p. 170.   *Bruguiere Enc.*
   Meth. p. 257.
Buccinum Tahitense.   *Gmelin,* p. 3498.
Inhabits the coasts of Otaheite. *Chemnitz.*
Shell an inch and a half or two inches long, and nearly half as
broad, of a bluish grey, with the ribs on the lower whirls
tinged with chestnut-colour, and the pillar-lip white and
shining; aperture ovate, and the outer lip somewhat plaited.

PORCATUM. 113. Shell ovate-oblong, sub-ventricose, thick, transversely ribbed, and the interstices striated; pillar convex.

Buccinum porcatum. *Gmelin*, p. 3494.
Buccinum Mexicanum. *Bruguiere Enc. Meth.* p. 260.
Buccinum, No. 64. *Schroeter Einl.* i. p. 372.
Martini, iv. p. 71. t. 126. f. 1213 and 1214.
Variety. With the transverse ribs obsolete.
Lister Conch. t. 963. f. 16.?
Inhabits the coasts of Mexico. *Bruguiere.*
Shell about two inches and a half long, and half as broad, generally brown, but sometimes of a bluish colour, and the inner margin of the lips nearly white; there are five whirls, and the summit is rather obtuse.

LYRATUM. 114. Shell fusiform, longitudinally ribbed, and the ribs transversely striated; apex blackish.

Buccinum lyratum. *Gmelin*, p. 3494.
Buccinum, No. 70. *Schroeter Einl.* i. p. 373.
Martini, iv. p. 74. t. 127. f. 1221 and 1222.
Inhabits ——
Shell an inch and a half long, and one third as broad, with about six whirls, of which the three lower are of a pale reddish brown, and the others nearly black.

PYROZONIAS. 115. Shell with transverse plaits, and undulated striæ, and the base and spire a little prominent; body-whirl with a double, and the other whirls with a single fulvous band.

Buccinum pyrozonias. *Gmelin*, p. 3488.
Buccinum, No. 12. *Schroeter Einl.* i. p. 359.
Martini, iii. t. 109. f. 1017.
Inhabits ——
Martini's figure is about thirteen lines long and nine broad, but I cannot find that he has any where described it, and the foregoing specific character of Gmelin's appears to have been taken in a very loose way from the figure only.

TEXTUM. 116. Shell turreted, longitudinally ribbed, and transversely striated; whirls flattish at the top.

Buccinum textum. *Gmelin*, p. 3493.
Buccinum, No. 61. *Schroeter Einl.* i. p. 371.
Martini, iv. p. 62. t. 125. f. 1201 and 1202.
Inhabits ——
This shell has been very imperfectly described, but appears from Martini's figure to be about an inch long, and two thirds as broad, and of a brownish white colour. Gmelin says the whirls are distant, and the figure represents the upper part to be flattened, and somewhat concave in the body-whirl.

CLATHRATUM. 117. Shell ovate, ventricose, longitudinally ribbed, decussated with transverse undulated striæ, and channelled at the sutures.

Buccinum clathratum. *Born Mus.* p. 261. t. 9. f. 17 and 18. *Schroeter Einl.* i. p. 397. *Gmelin*, p. 3471 and 3495. *Schreibers Conch.* i. p. 166. *Bruguiere Enc. Meth.* p. 275.
Bonanni Rec. and *Kirch.* 3. f. 62. *Petiver Gaz.* t. 56. f. 5.
Inhabits the East Indies. *Petiver.*
Shell about fifteen lines long, and eleven broad, with seven whirls, of a bluish or greenish white colour, and the body-whirl much more ventricose than the others; the outer lip is crenated and the inside grooved.

NIVEUM. 118. Shell sub-oval, with elevated cancellated striæ, and the body-whirl ventricose, and produced at the base. ⠀⠀⠀⠀⠀,

Buccinum niveum. *Gmelin*, p. 3495.
Buccinum, No. 26. *Schroeter Einl.* i. p. 363.
Martini, iv. p. 21. t. 122. f. 1122 and 1123.
Inhabits the coasts of Tranquebar. *Martini.*
This shell appears, from Martini's figure and description, to be very nearly allied to *B. clathratum*, but the transverse striæ are not undulated, and the base is produced into a longer beak. Gmelin at p. 3471 and 3504, has two other species under the name of *B. niveum*.

LIMA. 119. Shell oval, ventricose, acuminated, and cancellated with longitudinal ribs and transverse elevated striæ; aperture roundish and effuse.

Buccinum Lima. *Chemnitz,* xi. p. 87. t. 188. f. 1808 and 1809.

Inhabits the East Indian Seas. *Chemnitz.*

Shell about seventeen lines long and ten broad, white, with two pale tawny bands, and composed of eight whirls; it is narrower in proportion to its length than *B. clathratum,* nor are the transverse striæ undulated, and the beak at the base is much shorter than in *B. niveum;* it however appears to be very nearly allied to both these species, and Chemnitz says of this, what Bruguiere relates of the former, that it is found in a fossil state at Courtagnon in France.

RETICULATUM. 120. Shell ovate-oblong, transversely striated, and longitudinally ribbed; aperture toothed.

Buccinum reticulatum. *Linnæus Syst. Nat.* p. 1204. *Pennant Zool.* iv. p. 122. t. 72. f. 92. *Da Costa, Brit. Conch.* p. 131. t. 7. f. 10. *Born Mus.* p. 260. t. 9. f. 16. *Schroeter Einl.* i. p. 346. t. 2. f. 5. *Gmelin,* p. 3495. *Schreibers Conch.* i. p. 164. *Bruguiere Enc. Meth.* p. 273. *Donovan,* iii. t. 76. *Montagu Test.* p. 240. *Maton and Racket, in Lin. Trans.* viii. p. 137. *Dorset Cat.* p. 45. t. 15. f. 10.

Buccinum vulgatum. *Gmelin,* p. 3496. *Schreibers Conch.* i. p. 159.

Buccinum Nassula. *Ulysses's Travels,* p. 458.

Buccinum, No. 42. *Schroeter Einl.* i. p. 367.

Lister Conch. t. 966. f. 21. *Petiver Gaz.* t. 75. f. 4. *Gualter,* t. 44. f. C. *Seba,* iii. t. 49. f. 62 and 67. *Adanson Senegal,* t. 8. f. 9. *Martini,* iv. t. 124. f. 1162 to 1164.

Variety. With a yellowish band in the middle of the body-whirl.

Buccinum Tænia. *Gmelin,* p. 3490.

Buccinum, No. 152. *Schroeter Einleitung,* i. p. 395. *Knorr,* v. t. 10. f. 3.

Inhabits the Mediterranean. *Linnæus.* Coasts of Britain. *Lister,* &c. Islands of Teneriffe, and the Azores. *Adanson.* Coasts of France. *Reaumur.*

Shell an inch or an inch and a half long, and half as broad, with seven or eight whirls, longitudinally ribbed, and the transverse striæ rising into tubercles on the ribs; colour brown of various shades, and becoming white by exposure on the shore. *Buccinum Tænia* is taken wholly from Knorr, v. t. 10. f. 3, and Gmelin has also cited the same figure for the present species, of which it appears to be only a slight Variety.

AMBIGUUM.   121. Shell sub-pyramidal, with distant longitudinal ribs, and striated transversely; outer lip slightly denticulated.

Buccinum ambiguum.   *Pulteney's Dorset Cat.* p. 42. *Montagu Test.* p. 242. t. 9. f. 7.   *Maton and Racket, in Lin. Trans.* viii. p. 138. t. 4. f. 5.   *Dorset Cat.* p. 45. t. 18. f. 19.

Buccinum Pullus.   *Pennant Brit. Zool.* iv. p. 118. t. 72. f. 88.

Buccinum vulgatum, Var.   *Gmelin,* p. 3496.

Buccinum, No. 43.   *Schroeter Einl.* i. p. 368.

Gualter, t. 44. f. V.   *Martini,* iv. p. 42. t. 124. f. 1165 and 1166.?

Inhabits the coasts of Britain. *Pulteney,* &c.

Shell rather more than half an inch long, and three-eighths as broad, with six whirls, and differs from *B. reticulatum,* not only in being much broader in proportion to its length, but also in having the ribs more distant, and the aperture rounder.

MACULA.   122. Shell ovate, longitudinally ribbed and striated transversely; outer lip gibbous.

Buccinum Macula.   *Montagu Test.* p. 241. t. 8. f. 4. *Maton and Racket, in Lin. Trans.* viii. p. 138. t. 4. f. 4. *Dorset Cat.* p. 45. t. 15. f. 8.

Buccinum minutum.   *Pennant Zool.* iv. p. 122. t. 79.

Tritonium incrassatum.   *Muller in Acta Nidros.* iv. p. 96. t. 16. f. 25.

Murex incrassatus.   *Gmelin,* p. 3547.

Inhabits the coasts of Norway. *Muller.*   Britain. *Pennant,* &c.

Shell about half an inch long, and half as broad, and resembles the young of *B. reticulatum,* but may be at once distinguished by its gibbous outer lip, and by a small dark purplish spot at the outer edge of the canal.

STOLATUM.   123. Shell oblong-ovate, with one side of the body-whirl nearly smooth, and the rest plaited longitudinally and obsoletely striated transversely; aperture roundish.

Buccinum stolatum. *Gmelin*, p. 3496. *Schreibers Conch.*
i. p. 159.
Buccinum Miga. *Bruguiere Enc. Meth.* p. 274.?
Buccinum, No. 44. *Schroeter Einleitung*, i. p. 368.
Martini, iv. p. 43. t. 124. f. 1167 to 1169.
Inhabits the coasts of Tranquebar. *Martini.*
Shell three-quarters of an inch or an inch long, and more than
half as broad, with about seven whirls, of a brown or reddish colour with white bands, or white with brown or reddish bands; aperture toothed on both sides, and there is a
solitary tubercle near the upper angle of the inner lip; the
longitudinal plaits are distant, and excepting a broad rib by
the edge of the outer lip, the half of the body-whirl on that
side is nearly smooth, which circumstance is not noticed in
the description of *B. Miga*, though Bruguiere has referred
to Martini's figures of this species. A shell allied to *B.
stolatum* is figured in a Vignette in Clarke's Travels, and is
mentioned at page 276 of the second volume under the name
of *B. Galileum.*

CINCTUM. 124. Shell conical, closely ribbed longitudinally, and the interstices obsoletely striated transversely; apex acute; aperture oval.

Buccinum cinctum. *Montagu Test.* p. 246. t. 15. f. 1.
Maton and Racket, in Lin Trans. viii. p. 139. *Dorset
Cat.* p. 45. t. 14. f. 17.
Inhabits the Sea near Weymouth. *Mr. Bryer.*
Shell a quarter of an inch long, and hardly half as broad, with
six or seven whirls, white, and marked round the middle of
each whirl with a transverse narrow band of rufous brown.

MINIMUM. 125. Shell acuminated, minute, and reticulated with longitudinal elevated ribs and
transverse striæ; aperture oval.

Buccinum minimum. *Montagu Test.* p. 247. t. 8. f. 2.
Maton and Racket, in Lin. Trans. viii. p. 139.
Buccinum brunneum. *Donovan British Shells*, v. t. 179.
f. 2.
Inhabits the coasts of Devonshire. *Montagu.* Cornwall.
Donovan. Langland Bay near Swansea.
Shell about one fifth of an inch long, and not half so broad,
of a pale or dark chestnut-colour, without any spots or
markings.

PLICATULUM. 126. Shell oblong, longitudinally plaited and transversely striated, brownish white with darker brown bands, and the inside violet.

Buccinum plicatulum. *Gmelin*, p. 3496. *Schreibers Conch.* i. p. 161.
Buccinum, No. 39. *Schroeter Einl.* i. p. 367.
Gualter, t. 44. f. E. *Martini*, iv. t. 124. f. 1158 and 1159.
Favanne, t. 33. f. V 2.
Inhabits the East Indies. *Martini.*
Shell an inch long, and rather more than half as broad, and somewhat turreted, with the whirls rather flattish on the upper part.

PISCATORIUM. 127. Shell cancellated, and nodulous at the intersections on the lower whirls; pillar with one plait, and umbilicated; aperture effuse.

Buccinum Piscatorium. *Gmelin*, p. 3496.
Buccinum, No. 36. *Schroeter Einl.* i. p. 365.
Voluta echinata. *Solander's MSS.*
Martini, iv. p. 37. t. 124. f. 1151 and 1152.
Inhabits the East Indies. *Martini.*
Shell rather more than an inch long, and about two thirds as broad, of a pale brown colour; it appears from the description to possess considerable affinity with *Murex literatus* of Born, and both the species have a strong claim to be placed among the Volutes. Gmelin's *B. alatum* is probably a Variety, or is otherwise undeserving of notice.

MAURITII. 128. Shell ovate, with longitudinal ribs which form on the body-whirl four transverse rows of tubercles; outer lip six-toothed.

Buccinum Sti. Mauritii. *Gmelin*, p. 3496. *Schreibers Conch.* i. p. 165.
Buccinum, No. 37. *Schroeter Einl.* i. p. 365.
Martini, iv. p. 37. t. 124. f. 1153 and 1154.
Inhabits the Island of St. Maurice. *Martini.*
Shell about half an inch long, and half as broad, white, with dark brown tubercles, and the inside yellowish.

ARMILLATUM. 129. Shell ovate, with each whirl crowned by a row of tubercles; aperture large and toothless.

Buccinum armillatum. *Gmelin*, p. 3496.
Buccinum, No. 38. *Schroeter Einl.* i. p. 366.
Murex rugosus. *Born Mus.* p. 305. t. 11. f. 6 and 7.
Martini, iv. p. 38. t. 124. f. 1155 to 1157.
Inhabits ——
Shell rather more than an inch and a half long, and halfas broad, white, with reddish brown tubercles; pillar with a large umbilicus; aperture oval, and ending in a short beak.

NITIDULUM. 130. Shell ovate-oblong, polished, transversely striated and marked with articulated bands; outer lip toothed within.

Buccinum nitidulum. *Linnæus Syst. Nat.* p. 1205. *Bruguiere Enc. Meth.* p. 281.
Gualter, t. 52. f. C. *Adanson Senegal*, p. 135. t. 9. f. 27.
Junior. Buccinum lævigatum. *Linnæus Syst. Nat.* p. 1205.
Schroeter Einl. i. p. 348. *Gmelin*, p. 3497.
Inhabits the Mediterranean. *Linnæus.* Coasts of Goree. *Adanson.*
Linnæus says B. *nitidulum* ' varies in colour and bands, is often marked with a black belt, and that the pillar-lip is not replicated.' The shell which Martini, Schroeter, and Gmelin have considered to be this species, neither answers to the description or reference, and is much more like the Linnæan B. *mutabile.* Bruguiere says it is eleven lines long, and half as broad, generally brown, with white spots on the spire, and eight or ten dark brown bands, which are spotted with white on the body-whirl, but that the colour is sometimes greenish or livid, with white bands spotted with brown; he remarks that the longitudinal wrinkled striæ which Linnæus has mentioned are only markings occasioned by the growth of the shell, and describes it to be minutely striated transversely. Linnæus says his B. *lævigatum* differs from B. *nitidulum* in wanting a lip on the pillar, and teeth at the aperture, and Bruguiere has ascertained that it is only the same species, in an early stage of its growth.

VENTRICOSUM. 131. Shell ovate-oblong, brown striated with white, and somewhat plaited.
Buccinum ventricosum. *Gmelin*, p. 3498.

Martyn Univ. Conch. ii. t. 47.
Inhabits St. George's Bay. *Martyn.*
Having been unable to find any copy of Martyn's work which
  contains this species, I cannot make any addition to this
  short description of Gmelin's.

******** *Turreted, subulate, and slightly polished.*

MACULATUM.   132.   Shell   turreted,   sub-fusiform,
      with smooth undivided very entire whirls.

Buccinum maculatum. *Linnæus Syst. Nat.* p. 1205. *Mar-
    tini,* iv. p. 284. t. 153. f. 1440. *Born Mus.* p. 261.
    Schroeter Einl. i. p. 348. *Gmelin,* p. 3499. *Schreibers
    Conch.* i. p. 167.
Terebra maculata.   *Lamarck Syst. des Anim.* p. 78.
Bonanni Rec. 3. f. 317, and *Kirch.* f. 313. *Lister Conch.*
    t. 846. f. 74. *Rumphius,* t. 30. f. A. *Petiver Amb.* t.
    5. f. 4. *Gualter,* t. 56. f. I. *Argenville,* t. 11. f. A.
    Adanson Senegal, t. 4. f. 5. *Seba,* iii. t. 56. f. 6. *Knorr,*
    iii. t. 23. f. 2, and vi. t. 19. f. 6. *Favanne,* t. 39. f. A.
Variety. With the rows of spots nearly equally large.
Buccinum subulatum. *Martini,* iv. p. 288. t. 153. f. 1441.
    Schroeter Einl. i. p. 349. *Gmelin,* p. 3499. *Schrei-
    bers Conch.* i. p. 167.
Inhabits the coasts of Asia and Africa. *Linnæus.*   Amboyna.
Rumphius.   East Indies. *Seba.*   Senegal. *Adanson.* Ma-
    dagascar. *Humphreys.*
Shell from three to nine inches long, and about one fourth as
  broad, rather ponderous, with about fourteen whirls, of
  which the upper are divided more distinctly, by a transverse
  spiral line in the middle, than the lower ones; the colour
  varies from dirty white to a pale bluish brown somewhat
  mottled with white, and there are generally two rows of dark
  spots on each whirl, of which the upper row is generally
  much larger than the lower.   The shell which Martini has
  figured for the Linnæan *B. subulatum* appears to be only a
  variety of this species, and differs principally in having the
  rows of spots nearly equally large.

OCULATUM.   133.   Shell turreted, sub-fusiform, with
      the upper half of the whirls convex and ocel-
      lated.

Buccinum maculatum, Var. *Schroeter Einl.* i. p. 349. t. 2.

f. 6. *Gmelin*, p. 3499. *Schreibers Conch.* i. p. 167. *Rumphius*, t. 30. f. D. *Petiver Amb.* t. 2. f. 4. *Seba*, iii. p. 56. f. 11. *Martini*, iv. t. 153. f. 1442. *Favanne*, t. 40. f. Z.

Inhabits the East Indian Seas. *Martini.*

Shell from three to six inches long, and about one-fifth as broad, of a pale chestnut-colour, with a row of large white spots on the tumid upper part of the whirls ; the whirls are more numerous than in *B. maculatum*, and it is a perfectly distinct species. It is well figured in Martyn's Univ. Conchology, but I cannot ascertain the number of the plate.

SUBULATUM. 134. Shell turreted, subulate, smooth, with undivided very entire whirls.

Buccinum subulatum. *Linnæus Syst. Nat.* p. 1205. *Born Mus.* p. 262. t. 10. f. 9. *Shaw Nat. Misc.* xix. t. 799.

Buccinum subulatum, Var. *Gmelin,* p. 3499. *Schreibers Conch.* i. p. 167.

Buccinum dimidiatum. *Born Mus.* p. 266. *Kæmmerer Cab. Rudolst.* p. 153.

Buccinum, No. 167. *Schroeter Einl.* i. p. 400.

Bonanni Rec. and *Kirch.* 3. f. 118. *Lister Conch.* t. 842. f. 70. *Rumphius*, t. 30. f. B. *Petiver Amb.* t. 4. f. 2. *Gualter*, t. 56. f. B. *Argenville*, t. 11. f. X. *Seba*, iii. t. 56. f. 16, 23, 24, and 27. *Knorr*, i. t. 23. f. 4. *Martini*, iv. t. 154. f. 1443. *Favanne*, t. 40. f. D.

Variety. With the whirls divided by a transverse line.

Buccinum Taurinum. *Portland Cat.* p. 142. lot 3158.

Buccinum dimidiatum, Var. β. *Kæmmerer Cab. Rudolst.* p. 153.

Buccinum, No. 168. *Schroeter Einl.* i. p. 400.

Lister Conch. t. 841. f. 69. *Martini*, iv. t. 154. f. 1446. *Favanne*, t. 39. f. I.

Inhabits the Indian Ocean. *Linnæus.* Amboyna. *Rumphius.* China. *Humphreys.*

Shell from three to six inches long, and the breadth scarcely exceeds an eighth of the length ; the colour is white with a tinge of yellow, red or brown, and marked with three rows of squarish dark ferruginous spots on the body-whirl, and two on the other whirls, of which there are generally more than twenty ; the upper half of the whirls is slightly tumid, and a faint sunk line is observable in the lower whirls, which follows the course of the spire, and becomes stronger as it approaches the summit ; these markings being stronger than is usual may probably give the shell that angulated appearance which is represented in Lister's figure, and which in the Portland Catalogue is quoted for *B. Taurinum* of Solander, but among his MSS. in Sir Joseph Banks's Library, there is not any description under this name. The Linnæan references for *B. subulatum* and *B. dimidiatum* are much confused, and Seba, t. 56. f. 16. is quoted for both these species.

FELINUM. 135. **Shell turreted, pellucid, with all the whirls slightly emarginated on the back.**

Buccinum tigrinum. *Gmelin*, p. 3502.
Buccinum, No. 170. *Schroeter Einl.* i. p. 401.
Gualter, t. 56. f. G. *Seba*, iii. t. 56. f. A. *Martini*, iv. p. 297. t. 154. f. 1448.
Inhabits ———
Shell an inch and a half long, and full one-fourth as broad, white with two transverse rows of reddish spots on the body, and one on each other whirl. Gmelin, at p. 3475, has another species from Martyn with the name of *B. tigrinum*.

CRENULATUM. 136. **Shell turreted, with the whirls transversely divided, and crenated on their margins.**

Buccinum crenulatum. *Linnæus Syst. Nat.* p. 1205. *Martini*, iv. p. 294. t. 154. f. 1445. *Born Mus.* p. 264. *Schroeter Einl.* i. p. 350. *Gmelin*, p. 3500. *Schreibers Conch.* i. p. 167.
Buccinum varicosum. *Gmelin*, p. 3505.
Lister Conch. t. 846. f. 75. *Rumphius*, t. 30. f. E. *Gualter*, t. 57. f. L. *Seba*, iii. t. 56. f. 9 and 10. *Knorr*, i. t. 8. f. 7. *Favanne*, t. 40. upper fig. A 1.
Variety. Entirely white without any spots.
Buccinum candidum. *Born Mus.* p. 263. t. 10. f. 8.
Buccinum, No. 215. *Schroeter Einl.* i. p. 413.
Inhabits the African Ocean. *Linnæus.* Coasts of Amboyna. *Rumphius.* China. *Humphreys.*
Shell from three to five inches long, and nearly one-fifth as broad, of a pale or brownish flesh-colour, with a transverse row of ferruginous dots, which is double on the body-whirl, and there are also short ferruginous stripes between the crenellæ on the thickened upper margins of the whirls. Born's *B. candidum* is entirely white, and the margins of the whirls in the figure appear to be entire, but they are said in the description to be crenulated.

HECTICUM. 137. Shell turreted, with the whirls transversely divided, and the upper margin compressed and attenuated.

Buccinum hecticum. *Linnæus Syst. Nat.* p. 1206. *Chemnitz,* xi. p. 95. t. 188. f. 1817 and 1818.
Inhabits the African Ocean. *Linnæus.*
Linnæus has not given any description or further account of this species than what is contained in the above short specific character, and he has only referred to Gualter, t. 56. f. C, and Seba, iii. t. 56. f. 21, both of which are doubtful figures. Meuschen suspects it is the same as *B. dimidiatum,* and the shell figured by Chemnitz may probably be only a variety of that species. It is four and a half inches long, and about one sixth as broad ; the upper compressed part of the whirls is of a pale reddish brown, and the lower division is marked with large spots of the same colour on a white ground.

GEMINUM. 138. Shells turreted ; whirls transversely divided, with the lower divisions slightly striated, and the upper more protuberant.

Buccinum geminum. *Linnæus Mant.* p. 550. *Gmelin,* p. 3506.
Inhabits ——
Linnæus describes this species to be subulate, white, with the whirls transversely divided; the lower division is broadest, and is marked with transverse obsolete striæ, of which the uppermost are thickest, and the lower division is narrower, but more elevated and smooth.

PROXIMATUM. 139. Shell turreted, with the whirls transversely divided ; lower division slightly striated, and the upper filiform.

Buccinum proximatum. *Linnæus Mantissa,* p. 550. *Gmelin,* p. 3505.
Inhabits ——
No subsequent author has ascertained this species, and it is described by Linnæus to be subulate, glossy, with the lower division of the whirls broadest, and obsoletely striated transversely.

MONILE. 140. Shell turreted ; whirls transversely divided, with the lower division grooved, and the upper moniliform.

Buccinum monile.   *Linnæus Mant.* p. 550.   *Gmelin,* p. 3505.

Inhabits ——

Linnæus, who alone has noticed this species, says it is subulate, white or yellowish, with a transverse row of nodules on the upper division of the whirls, which is narrowest, and the lower division grooved longitudinally.   Linnæus, on the same page of the Mantissa, has given two almost exactly similar descriptions of this species.

VITTATUM.   141.   Shell turreted, and striated transversely ; whirls with a double crenulated suture on the upper margin.

Buccinum vittatum.   *Linnæus Syst. Nat.* p. 1206.   *Martini,* iv. p. 305. t. 155. f. 1461 to 1463.   *Born Mus.* p. 264.   *Schroeter Einl.* i. p. 352. t. 2. f. 7, and *Inn. Bau Conch.* p. 52. t. 3. f. 8.   *Gmelin,* p. 3500.   *Schreibers Conch.* i. p. 168.   *Chemnitz,* xi. p. 92. t. 188. f. 1814 and 1815.

Lister Conch. t. 977. f. 34.   *Petiver Gaz.* t. 98. f. 15.   *Klein Ost.* t. 7. f. 121.   *Favanne,* t. 40. f. C 2.

Variety. Elongated, and marked with decussated striæ.   *Chemnitz,* xi. p. 94. t. 188. f. 1816.

Inhabits the coasts of Ceylon. *Chemnitz.*

Shell generally about an inch and a half long, and more than one-third as broad, of a dull bluish white or pale livid colour, mixed sometimes with a tinge of red.   The variety figured by Chemnitz is marked with longitudinal as well as remote transverse striæ, and the breadth scarcely exceeds one-fourth of the length.

DIGITALE.   142.   Shell conical, sub-turreted, glabrous, with the aperture effuse at the base.

Buccinum digitale.   *Meuschen Mus. Gevers.* p. 296. No. 507.

Buccinum vittatum, Var.   *Gmelin,* p. 3500.   *Schreibers Conch.* i. p. 168.

Buccinum, No. 180.   *Schroeter Einl.* i. p. 404.

Lister Conch. t. 977. f. 33.   *Petiver Gaz.* t. 102. f. 15.   *Adanson Senegal,* t. 4. f. 1.   *Martini,* iv. p. 310. t. 155. f. 1468 and 1469, and t. 157. f. 1491.

Variety. Narrower, and somewhat elongated.

Buccinum, No. 181, and No. 182.   *Schroeter Einl.* i. p. 405.

Adanson Senegal, t. 4. f. 2. *Knorr*, v. t. 22. f. 5. *Martini*, iv. t. 155. t. 1470 and 1471. *Favanne*, t. 40. f. C 1.
Inhabits the coasts of Bombay. *Petiver.* Senegal. *Adanson.*
Coromandel. *Martini.*
Shell about an inch and a half long, and more than one-third
as broad, whitish or brown, with a paler or bluish band at
the upper margin of the whirls. I have followed Kæmmerer, Schroeter, and Martini, in arranging this as a separate
species, and it differs from *B. vittatum* in not having either
any transverse striæ or crenatures at the upper margin of the
whirls. *Martini*, f. 1491, probably differs only in being
coated with a green epidermis.

STRIGILATUM. 143. Shell turreted, with the whirls
transversely divided, and obliquely striated.

Buccinum strigilatum. *Linnæus Syst. Nat.* p. 1206. *Martini*, iv. p. 302. t. 155. f. 1456.? *Schroeter Einl.* i. p.
353. *Gmelin*, p. 3501.
Lister Conch. t. 845. f. 73. *Rumphius*, t. 30. f. H. *Knorr*,
vi. t. 22. f. 8 and 9.
Variety. Yellowish, and somewhat tesselated with red.
Buccinum commaculatum. *Gmelin*, p. 3502. *Schreibers
Conch.* i. p. 172.
Martini, iv. t. 154. f. 1452.
Inhabits the Asiatic Ocean. *Linnæus.*
Shell about two inches and three-quarters long, and scarcely
more than one-eighth as broad, whitish with yellow spots,
which at times form square somewhat tesselated patches, and
marked with a narrow elevated band both above and below
the suture. The Linnæan description is so short, and the
references so discordant, that this must always continue to be
rather a doubtful species, and Born, under the name of *B.
strigilatum*, has figured a very different shell. Martini's fig.
1452, from which Gmelin has constituted his *B. commaculatum*, is very indistinct, particularly about the aperture, and
is most probably only a variety of this species.

CONCINNUM. 144. Shell subulate, longitudinally
striated, with the whirls undivided, and
marked at the sutures with a spotted belt.

Buccinum strigilatum. *Born Mus.* p. 265. t. 10. f. 10.
Buccinum strigilatum, Var. β. *Gmelin*, p. 3501.
Buccinum, No. 173. *Schroeter Einl.* i. p. 401.
F 2

Murex strigilatus. *Gmelin*, p. 3564.
Gualter, t. 57. f. O. *Argenville*, t. 11. f. R.
Inhabits ———
Shell an inch and a quarter long, and about one fifth as broad,
with a white belt at the sutures, which is marked with large
brown spots. Two of the three figures which Linnæus has
cited for *B. strigilatum*, belong to the present species, and
are quite different from Rumphius, t. 30. f. H, which best
answers to his description.

DUPLICATUM. 145. Shell turreted, with the whirls
transversely divided, and longitudinally
grooved.

Buccinum duplicatum. *Linnæus Syst. Nat.* p. 1206. *Martini*, iv. p. 301. t. 155. f. 1455. *Born Mus.* p. 265.
Schroeter Einl. i. p. 354. *Gmelin,* p. 3501. *Schreibers Conch.* i. p. 169.
Bonanni Rec. and *Kirch.* 3. f. 110. *Lister Conch.* t. 837.
f. 64. *Gualter*, t. 57. f. N. *Knorr*, vi. t. 18. f. 6, and
t. 24. f. 5.
Inhabits the Indian Ocean. *Linnæus.* Coasts of Haynam.
Humphreys.
Shell about three inches long, and but little more than one-
sixth as broad, longitudinally grooved, and the whirls, at
about two-thirds of their length from the lower margin, divided by a transverse groove ; the colour is pale livid or steel
colour, and often nearly white towards the base.

CINEREUM. 146. Shell turreted, subulate, with the
upper part of the whirls longitudinally plaited.

Buccinum cinereum. *Born Mus.* p. 267. t. 10. f. 11 and
12. *Gmelin,* p. 3405. *Schreibers Conch.* i. p. 174.
Buccinum chalybeum. *Gmelin*, p. 3504.
Buccinum lividum. *Solander's MSS.*
Buccinum, No. 197. *Schroeter Einl.* i. p. 409.
Rumphius, t. 30 f. I.
Inhabits Amboyna. *Rumphius.*
Shell an inch and a quarter long, and about one-fifth as broad,
of a steel-grey colour, with a blackish obsolete transverse
band near the suture ; the upper plaited part of the whirls is
nearly white, and the summit of the spire brown.

SUCCINCTUM. 147. Shell turreted, subulate, minutely striated transversely, and the whirls transversely divided.

Buccinum succinctum. *Gmelin*, p. 3502. *Schreibers Conch.* i. p. 171.
Martini, iv. p. 298. t. 154. f. 1451.
Inhabits the East Indian Ocean. *Martini.*
Shell about two inches long, and one sixth as broad, white or straw-coloured, and Martini says the upper divisions of the whirls are so distinct as to have the appearance of a transverse belt.

ACUS. 148. Shell turreted, subulate; whirls transversely. divided, crenulated, and wrinkled; pillar twisted spirally.

Buccinum Acus. *Gmelin*, p. 3502. *Schreibers Conch.* i. p. 171.
Buccinum, No. 171. *Schroeter Einl.* i. p. 401.
Martini, iv. p. 297. t. 154. f. 1449.
Inhabits ———
Shell near an inch and a half long, and about one sixth as broad, whitish, with horizontal undulated lines.

LANCEATUM. 149. Shell turreted, smooth; whirls entire, and marked with testaceous longitudinal stripes.

Buccinum lanceatum. *Linnæus Syst. Nat.* p. 1206. *Martini*, iv. p. 297. t. 154. f. 1450. *Born Mus.* p. 266. *Schroeter Einl.* i. p. 354. *Gmelin*, p. 3501. *Schreibers Conch.* i. p. 169.
Rumphius, t. 30. f. G. *Petiver Amb.* t. 13. f. 20. *Gualter*, t. 56. f. D. *Argenville*, t. 11. f. Z. *Knorr*, vi. t. 24. f. 4.
Inhabits India. *Linnæus.* Coasts of Amboyna. *Rumphius.* China. *Humphreys.*
Shell most commonly about an inch and a half long, and one sixth as broad, white, with longitudinal fulvous or brown stripes; there are fifteen undivided whirls, of which the lower ones are level, and the upper longitudinally plaited.

DIMIDIATUM. 150. Shell turreted, with the whirls transversely divided, and smooth.

Buccinum dimidiatum. *Linnæus Syst. Nat.* p. 1206. *Martini*, iv. p. 292. t. 154. f. 1444. *Schroeter Einl.* i. p. 355. *Gmelin*, p. 3501. *Schreibers Conch.* i. p. 170.
Buccinum dimidiatum, Var. γ. *Born Mus.* p. 266. *Kæmmerer Cab. Rudolst.* p. 153.
Bonanni Rec. and *Kirch.* 3. f. 107. *Rumphius*, t. 30. f. C. *Gualter*, t. 57. f. 1 and M. *Seba*, iii. t. 56. f. 15 and 19. *Knorr*, i. t. 23. f. 5, and vi. t. 18. f. 5. *Favanne*, t. 40. f. Y.
Variety. Ferruginous, with obsolete white spots.
Buccinum ferrugineum. *Born Mus.* p. 263. t. 10. f. 7. *Lister Conch.* t. 843. f. 71.
Inhabits the African Ocean. *Linnæus.* Amboyna. *Rumphius.* China. *Humphreys.*
Shell about three inches long, and one sixth as broad, yellow or pale orange, with white longitudinal stripes on the lower part of the whirls, and the upper part plain and flat.

MURINUM. 151. Shell turreted ; whirls somewhat angular, with three muricated striæ.

Buccinum murinum. *Linnæus Syst. Nat.* p. 1206. *Schroeter Einl.* i. p. 356. *Gmelin*, p. 3502.
Gualter, t. 57. f. P.
Inhabits Africa. *Linnæus.*
Linnæus, to the above specific character and reference to Gualter, has only added that the shell is black, gibbous at the base, and the whirls often white at the base. Gualter's figure is about an inch and a half long, and one fifth as broad, and I cannot find that the species has been ascertained by any subsequent author.

PERTUSUM. 152. Shell turreted, subulate, with longitudinal elevated striæ, and transverse rows of excavated dots ; whirls transversely divided.

Buccinum pertusum. *Born Mus.* p. 267. t. 10. f. 13.
Buccinum duplicatum, Var. β. *Gmelin*, p. 3501.
Buccinum plicatum, Var. *Schreibers Conch.* i. p. 142.
Buccinum aciculatum. *Gmelin*, p. 3503. *Schreibers Conch.* i. p. 171.
Buccinum, No. 177, and No. 178. *Schroeter Einl.* i. p. 402 and 403.
Martini, iv. p. 303. t. 155. f. 1457.
Inhabits ——

Shell two inches and a half long, and one sixth as broad, white, or of a saffron-colour, with scattered fulvous dots. Martini's figure is far from good, and the lower whirls do not appear bipartite, but it is quoted by Born, and was obviously intended for the same species.

HASTATUM. 153. Shell turreted, sub-fusiform, with alternate brown and white bands, and longitudinal striæ.

Buccinum hastatum. *Gmelin*, p. 3502. *Schreibers Conch.* i. p. 171.
Buccinum, No. 176. *Schroeter Einl.* i. p. 402.
Martini, iv. p. 300. t. 154. f. 1453 and 1454.
Inhabits ——
Shell an inch long, and about one third as broad, and the first of Martini's two figures has the whirls slightly ventricose, and appears broader in proportion to its length than the other.

SINUATUM. 154. Shell turreted, with the whirls longitudinally plaited, and attenuated near the sutures ; aperture emarginate at both ends.

Buccinum sinuatum. *Born Mus.* p. 268.
Buccinum Phallus. *Gmelin*, p. 3503. *Schreibers Conch.* i. p. 172.
Buccinum, No. 179. *Schroeter Einl.* i. p. 403.
Martini, iv. p. 307. t. 155. f. 1464 and 1465.
Inhabits the East Indies, in rivers ?
Shell near two inches and a half long, and an inch and three-quarters broad, thin, semi-transparent, covered with a brown or black epidermis, under which it is blackish or reddish, and the ribs are sometimes white ; the spire has nine whirls, and the ribs are slightly curved.

BIFASCIATUM. 155. Shell subulate, glabrous, white, with two bands, and the whirls continuous.

Buccinum niveum. *Gmelin*, p. 3504. *Schreibers Conch.* i. p. 172.
Buccinum, No. 193. *Schroeter Einl.* i. p. 407.
Bonanni Rec. and *Kirch.* 3. f. 109. *Petiver Gaz.* t. 53. f. 5. *Klein Ost.* t. 7. f. 117.
Inhabits the East Indies. *Bonanni.*
Shell about two inches and a quarter long, and one fourth as

broad, and Bonanni says it is white and shining like alabaster, and twice encircled with a dark violet band which is shaped like a tongue, or the leaf of wheat, and terminates in a fine point; the whirls are continuous, and the sutures obsolete. Petiver calls it the 'Grass-girdled Indian Unicorn,' and says, 'it is a beautiful shell and rarely met with.' Gmelin, at p. 3471 and 3495, has two other species under the name of *B. niveum.*

RADIATUM. 156. Shell turreted, with transverse granulated striæ; whirls convex, and the first twice as large as the next.

Buccinum radiatum. *Gmelin,* p. 3504. *Schreibers Conch.* i. p. 173.
Buccinum, No. 199. *Schroeter Einl.* i. p. 409.
Murex strombiformis. *Ulysses's Travels,* p. 466.
Gualter, t. 52. f. D.
Inhabits the coast of Naples. *Ulysses.*
Shell about an inch long, and one third as broad, white, with two or three large reddish spots somewhat radiating from the base; the body-whirl is more ventricose, and twice as large as the next.

VIRGINEUM. 157. Shell turreted, with the whirls of the spire flattish, and the aperture large and oval.

Buccinum Virgineum. *Gmelin,* p. 3505. *Schreibers Conch.* i. p. 173.
Buccinum, No. 217. *Schroeter Einl.* i. p. 414.
Helix, No. 74. *Schroeter Einl.* ii. p. 198.
Lister Conch. t. 113. f. 7. *Petiver Gaz.* t. 104. f. 8.
Martini Berlin Mag. iv. p. 347. t. 10. f. 48.
Inhabits rivers in Virginia. *Lister.*
Shell about an inch long, and one third as broad, of a greenish yellow colour, with two transverse red bands.

ACICULA. 158. Shell subulate, minute, with the whirls well defined by an oblique suture; aperture oblong.

Buccinum Acicula. *Muller Verm.* ii. p. 150.
Buccinum terrestre. *Montagu Test.* p. 248. t. 8. f. 3.
Maton and Racket, in Lin. Trans. viii. p. 139.
Buccinum longiusculum. *Adams's Micros.* p. 639. t. 14. f. 26.

Bulimus Acicula. *Bruguiere Enc. Meth.* p. 311.
Helix octona. *Schroeter Fluss.* p. 350. t. 8. f. 6. *a* and *b*,
and *Einl.* ii. p. 162. *Gmelin*, p. 3653.
L'Aiguilette. *Geoffroy*, p. 59. t. 2.
Gualter, t. 6. f. B, B. *Walker's Minute Shells*, t. 2. f. 60.
Inhabits moss on old walls about Paris. *Geoffroy.* Moss, and
among the roots of grass on Barham Downs in Kent. *Montagu, &c.*
Shell hardly a quarter of an inch long, and about one-fifth as
broad, white, pellucid, and glabrous, and remarkable for the
uncommon obliquity of the sutures. Linnæus has erroneously
cited Gualter's excellent figure of this shell for *Helix octona*, which led Muller to quote that species for his *B. Acicula*, though he remarked that it did not accord with the Linnæan character. Schroeter has committed the same error,
and has been followed by Gmelin, who has very carelessly
arranged the true *Helix octona* as a variety of the same species. *B. Acicula* of Gmelin is a different shell, and the
species has been constituted entirely from Lister, t. 1055. f.
7, which is probably a variety of *Strombus ater*.

𝕲enus XXV.

STROMBUS:

SHELL UNIVALVE, SPIRAL, WITH THE OUTER LIP MUCH DILATED, AND THE APERTURE ENDING IN A PRODUCED CANAL WHICH BENDS TO THE LEFT.

Subdivisions.†

* With linear segments or claws at the margin of the outer lip.
** With the outer lip lobed.
*** With the outer lip very large.
**** Turreted, with a very long spire.

---

* *With linear Segments or Claws at the margin of the outer lip.*

FUSUS.   1.   Shell turreted, smooth; and slightly ventricose; aperture ending in an obliquely incurved beak; and the outer lip toothed.

Strombus Fusus. *Linnæus Syst. Nat.* p. 1207. *Martini,* iv. p. 332. t. 158. f. 1495 and 1496. *Born Mus.* p. 270. *Schroeter Einl.* i. p.   . *Kæmmerer Cab. Rudolst.* p. 154. *Gmelin,* p. 3506. *Schreibers Conch.* i. p. 175. *Shaw Nat. Misc.* xxiii. t. 499.   *Brookes's Introd.* p. 116. t. 7. f. 87.

---

† Gmelin's *S. affinis,* p. 3520, *S. lævis,* p. 3520, and *S. raninus,* p 3511, appear to me to be undeserving of notice.   *S. spinosus,* p. 3518, and *S. sinister,* p. 3524, are Fossils; and the former has been placed among the Volutæ by Lamarck.

Murex Fusus. *Forskael Descript. Anim.* p. 33.
Alata Fusus. *Mus. Geversianum*, p. 334. No. 821.
Fusus ventricosus. *Callone's Cat.* p. 35. No. 642.
Lister Conch. t. 854. f. 12. *Seba*, iii. t. 56. f. 1. *Knorr,*
v. t. 6. f. 1, and t. 7. f. 1. *Regenfuss*, ii. t. 7. f. 1.
Junior. With the outer lip not expanded.
Strombus, No. 10. *Schroeter Einl.* i. p. 455.
Seba, iii. t. 56. f. 3. *Favanne*, t. 24. f. B 1. *Martini*, iv.
t. 158. f. 1497.
Inhabits America. *Linnæus.* Frequent in the Red Sea, particularly about the Island of Ghorab. *Forskael.* Coasts of
Arabia Felix. *Humphreys.*
Shell about seven inches long, and two and a quarter broad,
with twelve whirls of a yellowish chestnut-colour, and the
inside white; the outer lip is thickened and toothed at its
lower margin, and the aperture terminates in a slightly incurved beak, which is about an inch long. In young shells, as
is commonly the case in this genus, the outer lip is not either
thickened or toothed at the margin.

UNICORNIS. 2. Shell turreted, smooth; aperture
ending in a long straight beak, and the outer
lip toothed.

Strombus Fusus, Var. *Schroeter Einl.* i. p. . *Kæmmerer Cab. Rudolst.* p. 154. *Gmelin*, p. 3506. *Schreibers Conch.* i. p. 175.
Murex Fusus. *Linnæus Mus. Reg. Ulr.* p. 638. *Syst. Nat.*
edit. 10. p. 752.
Alata lacinea. *Meuschen Mus. Gevers.* p. 334. No. 823.
Fusus longirostratus. *Callone's Cat.* p. 35. No. 643.
Rostellaria subulata. *Lamarck Hist. des Anim.* p. 81.
Bonanni Rec. and *Kirch.* 3. f. 121. *Lister Conch.* t. 916.
f. 9. *Argenville*, t. 10. f. D. *Seba*, iii. t. 56. f. 2. *Fa-
vanne*, t. 34. f. B 3. *Martini*, iv. p. 338. t. 159. f. 1500,
and Vign. at p. 344.
Junior. With the outer lip not expanded.
Strombus Clavus. *Linnæus Mant.* p. 549. *Martini*, iv. p.
342. t. 159. f. 1501 and 1502. *Schroeter Einl.* i. p. 424.
Gmelin, p. 3510. *Schreibers Conch.* i. p. 178.
Argenville, t. 10. f. A. *Favanne*, t. 34. f. B 2.
Inhabits the East Indies. *Chemnitz.* Coasts of Sumatra. *Humphreys.*
Shell, including the beak, about seven inches long, and one and
a half broad, of a yellowish or pale chestnut-colour; it differs

from *S. Fusus,* not only in being narrower, but also in having the beak quite straight, and twice as long ; the base of the body-whirl is striated transversely, and the upper whirls longitudinally.

FISSUS. 3. Shell turreted, smooth, with a longitudinal fissure extending from the aperture to the summit, of which the outer margin and also the outer lip are toothed.

Strombus Fusus fissus aculeatus. *Chemnitz,* xi. p. 141. t. 195 A. f. 1869.
Favanne, t. 79. f. Y.

Inhabits ———

Shell about two inches and a half long, and half as broad, of a yellowish white colour; in form it resembles *S. Fusus,* but has a longitudinal fissure similar to that of the Linnæan *S. Fissurella,* and from this species it differs in having the outer lip and the corresponding margin of the fissure strongly toothed.

PES-PELECANI. 4. Shell turreted, with the whirls tuberculated, and the outer lip expanded, four-clawed, and glabrous.

Strombus Pes-Pelecani. *Linnæus Syst. Nat.* p. 1207. *Pennant Zool.* iv. p. 122. t. 75. f. 94. *Martini,* iii. p. 144. t. 85. f. 848 to 850. *Born Mus.* p. 271. *Schroeter Einl.* i. p. 418. *Gmelin,* p. 3507. *Schreibers Conch.* i. p. 176. *Donovan,* i. p. 4. *Montagu Test.* p. 253. *Maton and Racket, in Lin. Trans.* viii. p. 141. *Dorset Cat.* p. 46. t. 15. f. 7.
Aporrhais quadrifidus. *Da Costa, Brit. Conch.* p. 136. t. 7. f. 7.
Tritonium Pes-Pelecani. *Muller Zool. Dan.* iii. p. 10. t. 87. f. 1 and 2.
Bonanni Rec. and *Kirch.* 3. f. 85 and 87. *Lister Conch.* t. 865. f. 20, t. 866. f. 21, and t. 1059. f. 3. *Petiver Gaz.* t. 79. f. 6, and t. 127. f. 11. *Gualter,* t. 53. f. A, B, and C. *Argenville,* t. 14. f. M. *Klein Ost.* t. 2. f. 41 and 42. *Seba,* iii. t. 62. f. 17. *Knorr,* iii. t. 7. f. 4. *Favanne,* t. 21. f. D 1, and D 2.

Inhabits the Atlantic and Mediterranean Seas, and the coasts of Europe, America, and Norway. *Linnæus.* Portugal. *Bonanni.* Great Britain. *Pennant, &c.*

Shell about an inch and a half or two inches long, and in full

grown shells, which have the outer lip most expanded, the
breadth about equals the length, but young shells are without this expansion, and might be mistaken for a Murex; the
outer lip is divided into four strong projecting segments or
claws, of which the upper one is attached to the spire, and
they are all more or less channelled; it is of a pale brownish
flesh-colour.

CHIRAGRA. 5. Shell ovate, and tuberculated; outer
lip with six strong curved diverging claws,
and the throat striated.

Strombus Chiragra. *Linnæus Syst. Nat.* p. 1207. *Martini,* iii. p. 149. t. 86. f. 853 and 854, and t. 87. f. 856
and 857. *Born Mus.* p. 271. *Schroeter Einl.* i. p.
419. *Gmelin,* p. 3507. *Schreibers Conch.* i. p. 176.
Bonanni Rec. 3. f. 312 and 313. *Lister Conch.* t. 870. f.
24. *Rumphius,* t. 35. f. A. *Petiver Amb.* t. 14. f. 3.
Gualter, t. 35. f. B. *Knorr,* i. t. 27. f. 1. *Favanne,*
t. 21. f. C 2.
Junior. With the outer lip not expanded.
Lister Conch. t. 875. f. 31. *Rumphius,* t. 35. f. B and C,
and t. 37. f. 1. *Petiver Amb.* t. 14. f. 1 and 2. *Gualter,* t. 26. f. B, and t. 35. f. A. *Seba,* iii. t. 62. f. 34, t.
73. f. 29, and t. 83. f. 1 and 2. *Martini,* iii. t. 85. f.
851 and 852, and t. 92. f. 895 to 901. *Favanne,* t. 21.
f. C 1, C 3, and C 4.
Inhabits the coasts of Banda. *Rumphius.* Isle of France.
Martini. China. *Humphreys.*
Shell about five inches long, and three inches and a half broad,
without measuring the claws, which vary in full grown specimens from two to three inches, and of which the two upper
and two lower are curved in contrary directions; the colour
is white mottled with brown, and the lips are rose-coloured.

SCORPIUS. 6. Shell ovate, tuberculated, and transversely striated; aperture with seven nodulous claws, of which that at the base is longest, and the throat striated.

Strombus Scorpius. *Linnæus Syst. Nat.* p. 1208. *Martini,* iii. p. 159. t. 88. f. 860. *Born Mus.* p. 272. *Schroeter Einl.* i. p. 421. *Gmelin,* p. 3508. *Schreibers Conch.*
i. p. 176.
Bonanni Rec. 3. f. 314 and 315. *Lister Conch.* t. 867. f.
22. *Rumphius,* t. 36. f. K. *Petiver Amb.* t. 3. f. 2.

Gualter, t. 36. f. C.   *Argenville*, t. 14. f. B.   *Knorr*,
ii. t. 3. f. 1.   *Favanne*, t. 22. f. B.
Inhabits the coasts of Amboyna. *Rumphius*.   China. *Humphreys.*
Shell about two inches and a half long, and an inch and three-
quarters broad without including the spines, of which the one
at the base measures about an inch; it is white, mottled
with brownish flesh-colour, and the lips are violet, with
white elevated striæ.

LAMBIS.   **7. Shell ovate, and tuberculated; aperture with seven claws, and the throat and
lips smooth.**

Strombus Lambis.   *Linnæus Syst. Nat.* p. 1208.   *Martini*, iii. p. 154.   *Born Mus.* p. 273.   *Schroeter Einl.*
i. p. 422, and *Inn. Bau Conch.* t. 2. f. 1.   *Gmelin*, p.
3508.   *Schreibers Conch.* i. p. 177.   *Shaw Nat. Misc.*
xxiii. t. 1011.   *Brookes's Introd.* p. 115. t. 7. f. 86.
Pterocera Lambis.   *Lamarck Syst. des Anim.* p. 81.
Variety A.   With the claws nearly straight.
Lister Conch. t. 866. f. 21.   *Rumphius*, t. 35. f. H.   *Gual-
ter*, t. 36. f. A and B.   *Argenville*, t. 14. f. E.   *Knorr*,
i. t. 28. f. 1.   *Martini*, iii. t. 87. f. 858 and 859.
Variety B.   With the claws longer and more slender, and the
throat orange.
Strombus Scorpio.   *Chemnitz*, x. p. 224. t. 158. f. 1508
and 1509.
Strombus, No. 58.   *Schroeter Einl.* i. p. 469. t. 2. f. 15,
and t. 7. f. 1.
Knorr, v. t. 4. f. 3.
Variety C.   With the claws strong and somewhat hooked.
Strombus Camelus.   *Chemnitz*, x. p. 205. t. 155. f. 1478.
Rumphius, t. 55. f. F.
Junior.   With the lip partially or not expanded.
Rumphius, t. 35. f. D, and t. 36. f. G.   *Petiver Amb.* t.
14. f. 8, and 14.   *Gualter*, t. 30. f. A.   *Seba*, iii. t. 61.
f. 9 and 10.   *Knorr*, ii. t. 27. f. 4, and iii. t. 7. f. 1.
Martini, iii. t. 90. f. 884, t. 91. f. 888 and 889, and t.
92. f. 902 and 903.
Inhabits the coasts of Asia. *Linnæus*.   Amboyna. *Rumphius*.
Banda, and Frederick's Islands. *Regenfuss*.   Red Sea. *For-
skael*.   Batavia. *Martini*.   China, Madagascar, and the
South Sea. *Humphreys*.
Shell commonly about four inches long, and two inches and a
half broad, without including the claws, and the Variety C

is sometimes considerably larger; the two claws at the upper end are generally longer and straighter than the others, and those on the sides in the Variety C are remarkably bent back, and hooked; the colour is greyish white, slightly mottled with brown, and the throat more or less orange coloured.

TRUNCATUS. 8. Shell ovate, with the spire very obtuse and knotty; aperture with eight straightish claws, of which one is smaller.

Strombus truncatus. *Humphreys in Portland Cat.* p. 133. lot 2967.

D'Avila, i. p. 190, and p. 566. t. 14. *Favanne*, t. 21. f. E 3.

Junior. With the outer lip partially or not expanded.
Strombus Bryonia. *Gmelin*, p. 3520.
Strombus radix Bryoniæ. *Chemnitz*, x. p. 227. t. 159. f. 1512 to 1515.
Strombus, No. 9. *Schroeter Einl.* i. p. 454.
Murex Gigas. *Gmelin*, p. 3557. ?
Murex, No. 240. *Schroeter Einl.* i. p. 629. ?
Lister Conch. t. 882. f. 4, and t. 931. *Seba*, iii. t. 63. f. 3. *D'Avila*, t. 12 and 13. *Martini*, iii. t. 93. f. 904 and 905. *Favanne*, t. 21. f. E 1, and E 2.

Inhabits the East Indies, and coasts of China. *Humphreys.*
I never saw a full grown shell of this species, which is said by Mr. Humphreys to be very rare; it is sometimes a foot long, and eight inches broad, of a pale yellowish brown colour mottled with white; the four uppermost whirls are depressed, so as to give the spire a truncated appearance, and each of the other whirls has a row of strong rounded knobs. It appears to me that *S. Bryonia* is the shell in its earliest stage of growth, and that *Murex Gigas* may be the same at a more advanced age with the lip partially expanded.

MILLEPEDA. 9. Shell rather flattish and gibbous on the back; aperture with about ten inflected claws, and the throat striated.

Strombus Millepeda. *Linnæus Syst. Nat.* p. 1208. *Martini*, iii. p. 161. t. 88. f. 861 and 862. *Born Mus.* p. 274. *Schroeter Einl.* i. p. 423. *Gmelin*, p. 3509. *Schreibers Conch.* i. p. 177.

Variety A.   With ten distinct claws.
Bonanni *Rec.* 3. f. 311, and *Kirch.* f. 315. *Lister Conch.*
t. 868, and t. 869. f. 23.   *Rumphius*, t. 36. f. I.   *Pe-
tiver Amb*. t. 14. f. 7.   *Argenville*, t. 15. f. B.   *Fa-
vanne*, t. 22. f. A 6.
Variety B.   With nine claws connected by an extension of
the outer lip.
Strombus novem dactylis instructus.   *Chemnitz*, x. p. 207.
t. 155. f. 1479 and 1480.
Variety C.   With twelve claws.
Strombus multipes.   *Chemnitz*, x. p. 216. t. 157. f. 1494
and 1495.
Inhabits the Asiatic Ocean. *Linnæus.*   Coasts of Coromandel,
China, and Ceylon. *Humphreys.*
Shell about four inches long, and two inches and a quarter
broad, of a greyish white colour slightly mottled with brown;
the throat is saffron-coloured, and both the lips are purplish,
covered with numerous white elevated irregular striæ.

** *With the outer lip lobed.*

LENTIGINOSUS.   10. Shell ovate, warty, coronated,
and the base obtuse; outer lip thickened,
and three lobed above.
Strombus lentiginosus. *Linnæus Syst. Nat.* p. 1208. *Martini*, iii. p. 120. t. 80. f. 825 to 828.   *Born Mus*. p. 275.
Schroeter Einl. i. p. 425. *Gmelin*, p. 3510. *Schreibers
Conch.* i. p. 179.
Bonanni *Rec.* and *Kirch.* 3. f. 300.   *Lister Conch.* t. 861.
f. 18.   *Rumphius*, t. 37. f. Q.   *Petiver Amb.* t. 14. f.
10. *Gualter*, t. 32. f. A. *Argenville*, t. 15. f. C. *Seba*,
iii. t. 62. f. 11 and 30. *Knorr*, iii. t. 13. f. 2. *Favanne*,
t. 20. f. B 1.
Junior.   With the outer lip thin, and not expanded.
Martini, iii. t. 91. f. 892.
Inhabits the coasts of Amboyna. *Rumphius.*   China. *Humphreys.*
Shell about two inches and a half long, and an inch and a half
broad, white, marbled with brown, and the outer lip spotted;
the throat is pale orange; on the body-whirl there are irregular transverse rows of small rounded knobs, and the spire
is coronated with larger tubercles.

PAPILIO. **11.** Shell transversely striated with transverse nodulous bands, and the spire coronated; outer lip growing to the spire, and sinuated at the base; throat smooth and white.

Strombus Papilio. *Chemnitz*, x. p. 226. t. 158. f. 1510 and 1511.

Strombus, No. 31. *Schroeter Einl.* i. p. 462.

Strombus lentiginosus, Var. β. *Gmelin*, p. 3510.

Seba, iii. t. 52. f. 17 and 18. *Knorr*, iii. t. 26. f. 2 and 3.

Inhabits the East Indian Seas. *Chemnitz.*

This is rather an uncertain species, with which I am unacquainted, and Chemnitz says it is nearly allied to *S. lentiginosus*, but considers it distinct; he describes it to be an inch and three-quarters long, and an inch and a quarter broad, of a yellowish white colour, prettily variegated with brown, and the gutter at the base brown on its inside; of the above figures which he has quoted, the inside of Knorr's is painted dark red, and Seba's is said to be of a saffron colour.

FASCIATUS. **12.** Shell somewhat grooved transversely, and the body-whirl and spire coronated; outer lip sinuated only at the base, and the throat white.

Strombus fasciatus. *Gmelin*, p. 3510.

Strombus lentiginosus, Var. *Schroeter Einl.* i. p. 426. *Schreibers Conch.* i. p. 179.

Alata torosa. *Meuschen Mus. Gevers.* p. 344, No. 902.

Alata Lentigo rosacea. *Martini*, iii. p. 127. t. 82. f. 833 and 834.

Le Kalan. *Adanson Senegal*, p. 137. t. 9. f. 30.

Lister Conch. t. 860. f. 17. *Klein Ost.* t. 6. f. 107. *Seba*, iii. t. 62. f. 6 to 8. *Favanne*, t. 20. f. B 2.

Junior. With the outer lip not expanded.

Lister Conch. t. 883. f. 5. *Knorr*, v. t. 16. f. 4.? *Martini*, iii. t. 91. f. 893.

Inhabits the coasts of Jamaica. *Lister.* Goree. *Adanson.*

Shell about three inches long and two broad, variegated and banded with yellowish brown and rose-colour, and a broad rose-coloured band traverses the row of large pointed tubercles towards the summit of the body-whirl, and on the spire; besides this row of large tubercles, there are also on the body-whirl two indistinct rows of smaller nodules, and the outer lip is much more expanded and acute than in *S. lentiginosus.*

POLYFASCIATUS. 13. Shell sub-ovate, with dark transverse interrupted stripes, and a row of pointed tubercles on each whirl; outer lip sinuated only towards the base, and the throat reddish yellow.

Strombus polyfasciatus. *Chemnitz,* x. p. 209. t. 155. f. 1483 and 1484.

Strombus fasciatus, Var. *Gmelin,* p. 3511.

Alata flavigula. *Meuschen Mus. Gevers.* p. 338, No. 848.

Junior. With the outer lip less expanded.

Strombus fasciatus. *Born Mus.* p. 278.

Strombus lentiginosus, Var. *a. Gmelin,* p. 3510. *Schreibers Conch.* i. p. 179.

Alata sagittis lineata. *Martini,* iii. p. 97. t. 78. f. 800 to 802.

Seba, iii. t. 61. f. 7.

Inhabits the Red Sea. *D'Avila.* Coasts of Bengal. *Favanne.*

Shell about two inches long, and fourteen lines broad, white, with seven or eight dark brown transverse, more or less interrupted, narrow stripes; the throat is yellowish or orange, and the transverse stripes on the outside appear through the substance of the shell, which is rather thin and pellucid.

GALLUS. 14. Shell transversely ribbed, and nodulous, and the spire short; outer lip lobed near the base, and produced and angulated above; throat smooth.

Strombus Gallus. *Linnæus Syst. Nat.* p. 1209. *Martini,* iii. p. 139. t. 84. f. 841 and 842, and t. 85. f. 846. *Born Mus.* p. 275. *Schroeter Einl.* i. p. 427. *Gmelin,* p. 3511. *Schreibers Conch.* i. p. 179.

Bonanni Rec. 3. f. 309 and 310, and *Kirch.* f. 310 and 311. *Lister Conch.* t. 874. f. 30. *Rumphius,* t. 37. f. 5. *Gualter,* t. 32. f. M. *Seba,* iii. t. 62. f. 1. *Knorr,* iv. t. 12. f. 1. *Favanne,* t. 21. f. A 1.

Variety. With three large pointed knobs on the body-whirl.

Strombus tricornis. *Humphreys in Port. Cat.* p. 5, lot 50.

Strombus, No. 4, and No. 5. *Schroeter Einl.* i. p. 452.

Pugil tricornis. *Martini,* iii. p. 140. t. 84. f. 843 to 845, and t. 85. f. 847.

Lister Conch. t. 873. f. 29.

Junior. With the outer lip not expanded.

Lister Conch. t. 891. f. 11. *Martini,* iii. t. 91. f. 891.

Inhabits the coasts of Jamaica and Barbadoes. *Lister.* Red Sea. *Forskael.* Martinique. *Humphreys.*

Shell sometimes three inches and a half long, and two inches and three-quarters broad, but is commonly smaller ; the colour is white, or reddish clouded with grey, and the spire is only about half as long as the body-whirl ; the outer lip is large and effuse, and at the upper end is drawn out into a projecting more or less linear beak, which is often elevated above the summit of the spire. The Variety has only three very large pointed protuberances on the shoulder of the body-whirl, and is less distinctly ridged transversely.

LACINIATUS. 15. Shell transversely ribbed and no-dulous ; outer lip sinuated near the base, and much scolloped towards its upper angle ; throat smooth.

Strombus laciniatus. *Chemnitz,* x. p. 223. t. 158. f. 1506 and 1507.

Strombus sinuatus. *Humphreys in Portland Cat.* p. 189, lot 4022.

Strombus Gallus, Var. *Gmelin,* p. 3512.

Strombus, No. 50. *Schroeter Einl.* i. p. 466.

Strombus, No. 39. *Schreibers Conch.* i. p. 189.

Seba, iii. t. 62. f. 3. *Spengler Selt. Conch.* t. 3. f. A. *Favanne,* t. 22. f. A 2.

Inhabits the East Indian Seas. *Chemnitz.*

Shell four inches and a half long, and two inches and three-quarters broad, and has the spire more than half as long as the body-whirl ; the outer lip is curved inwards in the middle, and is sinuated near the base, and plaited and laciniated towards the upper extremity ; the colour is white clouded with yellowish spots, and the throat brown.

AURIS DIANÆ. 16. Shell transversely ribbed and nodulous, and the spire rather short ; outer lip pointed above, and the beak at the base of the aperture recurved.

Strombus Auris Dianæ. *Linnæus Syst. Nat.* p. 1209. *Martini,* iii. p. 135. t. 84. f. 838 and 839. *Born Mus.* p. 276. *Schroeter Einl.* i. p. 428. *Gmelin,* p. 3512. *Schreibers Conch.* i. p. 180.

G 2

Lister Conch. t. 872. f. 28. *Rumphius,* t. 37. f. R. *Gualter,* t. 32. f. D and H. *Argenville,* t. 14. f. O. *Klein Ost.* t. 6. f. 106. *Seba,* iii. t. 61. f. 3 to 6, and t. 62. f. 13. *Knorr,* ii. t. 15. f. 1 and 2. *Favanne,* t. 21. f. A 5, and A 6.

Variety B. With the pillar-lip and aperture blackish.

Strombus Auris Dianæ adusta. *Chemnitz,* x. p. 211. t. 156. f. 1487 and 1488.

Strombus, No. 16. *Schroeter Einl.* i. p. 456.

Lister Conch. t. 872. f. 27.

Variety C. With a large chestnut spot on the pillar-lip, and the throat ribbed.

Strombus Auris Dianæ Zelandiæ Novæ. *Chemnitz,* x. p. 210. t. 156. f. 1485 and 1486.

Strombus Auris Dianæ, Var. *Portland Cat.* p. 29. lot 679. *Gmelin,* p. 3512. *Schreibers Conch.* i. p. 180. *Shaw Nat. Misc.* xxii. t. 926.

Alata Aratrum. *Martyn Univ. Conch.* i. t. 1.

Junior. With the outer lip not expanded.

Seba, iii. t. 61. f. 1 and 2, and t. 62. f. 16. *Martini,* iii. t. 84. f. 840.

Inhabits the Asiatic Ocean. *Linnæus.* Coasts of Amboyna. *Rumphius.* China. *Humphreys.*

Shell two and a half or three inches long, and more than half as broad, with transverse ribs, and one, two, or three rows of tubercles on the body-whirl; the spire is nearly as long as the body-whirl; the colour is white variously mottled with brown or fawn-colour, and the throat is more or less tinged with orange. I am far from certain that the variety C is not a distinct species.

PUGILIS. 17. Shell ponderous, transversely spinous, and the base and spire striated; outer lip rounded at the projecting summit, and three-lobed below.

Strombus Pugilis. *Linnæus Syst. Nat.* p. 1209. *Martini,* iii. p. 124. t. 81. f. 830 and 831. *Born Mus.* p. 277. *Schroeter Einl.* i. p. 429. *Gmelin,* p. 3512. *Schreibers Conch.* i. p. 181. *Lamarck Syst. des Anim.* p. 80. *Brookes's Introd.* p. 115. t. 7. f. 85.

Bonanni Rec. 3. f. 299, and *Kirch.* f. 301. *Gualter,* t. 32. f. B. *Argenville,* t. 15. f. A. *Knorr,* i. t. 9. f. 1, and iii. t. 16. f. 1. *Favanne,* t. 21. f. B 2.

Variety B. Brown, with a band of white spots, and the spire unarmed.

Strombus alatus. *Gmelin*, p. 3513.

Strombus Pugilis, Var. *Kæmmerer Cab. Rudolst.* p. 158.
  Schreibers Conch. i. p. 181.

Strombus, No. 8. *Schroeter Einl.* i. p. 454. t. 2. f. 14.
Martini, iii. t. 91. f. 894.

Variety C. With the spines double and distorted.

Strombus Pugil duplicatus. *Chemnitz*, x. p. 215. t. 156. f.
  1493.

Junior. With the outer lip unexpanded.

  Rumphius, t. 36. f. No. 6. *Gualter*, t. 31. f. G. *Seba*, iii.
    t. 62. f. 44 and 50. *Knorr*, vi. t. 29. f. 6 and 7. *Marti-
    ni*, iii. t. 90. f. 882.

Inhabits the coasts of Jamaica. *Linnæus.* West Indies. *Martini.* Florida. *Humphreys.*

Shell about three inches and a half long, and two inches broad,
  generally of a pale red or flesh-colour, but is sometimes
  brown, and marked with a paler band round the row of spines
  on the shoulder of the body-whirl ; the row of spines on the
  spire becomes gradually smaller, and the upper whirls are
  strongly striated transversely ; the aperture is highly polished,
  and the outer lip is more or less distinctly striated within.
  Dr. Leach, in the Zoological Miscellany, t. 22, under the
  name of *Strombus Sloanii*, has figured what may probably
  be a Lusus of *S. Pugilis*, with elevated compressed quadrate
  processes round the upper end of the body-whirl, and at all
  events it appears by the aperture that the shell has not sufficiently attained its full growth to warrant its introduction as
  a species, till some more mature specimen has been found.

MARGINATUS. 18. Shell transversely keeled, slightly
  nodulous, and both the lips pointed above;
  outer lip with a broad thickened margin.

Strombus marginatus. *Linnæus Syst. Nat.* p. 1209.

Inhabits the coasts of China. *Humphreys.*

Schroeter, Gmelin and Chemnitz appear to have mistaken
  S. accinctus for this species, and a shell which Mr. Humphreys sold me, under the name of *S. marginatus*, answers
  better to the Linnæan definition ; it is about sixteen lines
  long, and half as broad, and the spire occupies about two-
  fifths of the length ; the colour is pale chestnut or fawn-
  colour, slightly mottled with white, and the margin of the
  aperture and the pillar are white ; both the lips are much
  thickened, and their pointed terminations are attached to the
  second whirl of the spire ; the aperture is yellow.

LUHUANUS. 19. Shell smooth, with the summit of the outer lip a little prominent ; aperture reddish, with a black margin to the pillar-lip.

Strombus Luhuanus. *Linnæus Syst. Nat.* p. 1209. *Martini*, iii. p. 91. t. 77. f. 789 to 791. *Born Mus.* p. 277. *Schroeter Einl.* i. p. 432, and *Inn. Bau Conch.* p. 26. t. 3. f. 3. *Gmelin*, p. 3513. *Schreibers Conch.* i. p. 182.

Lister Conch. t. 850. f. 5, and t. 851. f. 6. *Rumphius*, t. 37. f. S. *Petiver Gaz.* t. 98. f. 10, and *Amb.* t. 14. f. 12. *Gualter*, t. 31. f. H and I. *Seba*, iii. t. 61. f. 11, 12, and 20; and t. 62. f. 31 and 32. *Knorr*, v. t. 16. f. 5. *Favanne*, t. 20. f. D 1.

Variety. Without any black margin to the pillar-lip, and the outer lip white at the edge.

Strombus Luhuanus, Var. *Chemnitz*, x. p. 218. t. 157. f. 1499 and 1500.

Favanne, t. 20. f. D 2.

Junior. With the outer lip imperfect.

Lister Conch. t. 849. f. 4. *Seba*, iii. t. 61. f. 14.

Inhabits the Asiatic Ocean. *Linnæus.* Coasts of Amboyna. *Rumphius.* Isle of France. *Lister.* China. *Humphreys.*

Shell about two inches long, and half as broad, with the first whirl of the spire slightly gibbous, and the upper whirls somewhat ribbed longitudinally; the colour is white, with spots and interrupted transverse bands of yellowish brown. The shells which belong to the above-mentioned variety are generally rather larger, and more rounded on the shoulder of the body-whirl, than those which have the broad black stripe along the margin of the pillar-lip.

GIBBERULUS. 20. Shell smooth, slightly striated at the base, and towards the margin of the outer lip ; spire distorted, and the whirls gibbous.

Strombus gibberulus. *Linnæus Syst. Nat.* p. 1210. *Martini*, iii. p. 95. t. 77. f. 792 to 798. *Born Mus.* p. 278. *Schroeter Einl.* i. p. 433. *Gmelin*, p. 3514. *Schreibers Conch.* i. p. 182.

Bonanni Rec. and *Kirch.* 3. f. 150. *Lister Conch.* t. 847. f. 1. *Rumphius*, t. 37. f. V. *Petiver Amb.* t. 14. f. 13. *Gualter*, t. 31. f. N. *Argenville*, t. 14. f. N. *Seba*, iii. t. 61. f. 17 to 19, and 51 to 53; and t. 62. f. 48 to 49. *Knorr*, ii. t. 14. f. 3, and iii. t. 13. f. 4.

Junior. With the outer lip imperfect.

Seba, iii. t. 61. f. 21. *Martini,* iii. t. 88. f. 863 and 864.

Inhabits the Asiatic Ocean. *Linnæus.* Coasts of Amboyna. *Rumphius.* Isle of France. *Martini.* China. *Humphreys.* Shell an inch and three-quarters or two inches long, and about half as broad, and the body-whirl is flattened on the side next the pillar; the first whirl of the spire is very gibbous, and, has obtained for the shell its common name of the *Powter,* from a fancied resemblance to the breast of the bird so called; it is white variegated with brown, or sometimes tinged with flesh-colour, and within the aperture is a broad red or purplish stripe running parallel to the margin of the outer lip.

ONISCUS. 21. Shell ob-ovate, with about three transverse nodulous belts; pillar-lip granulated, and the outer lip thickened and toothed within.

Strombus Oniscus. *Linnæus Syst. Nat.* p. 1210. *Born Mus.* p. 279. *Schroeter Einl.* i. p. 434, and *Inn. Bau Conch.* p. 12. t. 4. f. 8. *Gmelin,* p. 3514. *Schreibers Conch.* i. p. 182. *Chemnitz,* xi. p. 143. t. 195 A. f. 1872 and 1873.

Cypræa conoidea. *Scopoli del Ins.* ii. p. 78. t. 24. f. 3. *Gmelin,* p. 3414.

Cassidea Oniscus. *Bruguiere Enc. Meth.* p. 432.

Cassis. *Martini,* ii. p. 42. t. 34. f. 357 and 358.

Lister Conch. t. 791. f. 44. *Petiver Gaz.* t. 48. f. 16. *Gualter,* t. 22. f. I. *Seba,* iii. t. 55. f. 23. *Knorr,* iv. t. 12. f. 4, and vi. t. 15. f. 6. *Favanne,* t. 26. f. K.

Inhabits the West Indian Seas. *Martini.*

This shell has been arranged among the Helmets by Lister, Martini, and Bruguiere, among the Cones by Gualter, with the Strombi by Linnæus, and as a Cypræa by Scopoli. It is sometimes sixteen lines long and ten broad, but is generally smaller; the colour is most commonly white with irregular brown spots, but I have a specimen which is dark brown with only a few white markings; the spire is depressed, and terminated by a small pointed summit; the aperture is long and narrow; in young shells the outer lip is thin, and the transverse nodulous belts of the body-whirl give its margin a sinuated appearance.

*** *With the outer Lip very large.*

GIGAS. **22. Shell crowned with large conical spines, and slightly ribbed transversely; outer lip much expanded.**

Strombus Gigas. *Linnæus Syst Nat.* p. 1210. *Martini,* iii. p. 117. t. 80. f. 824. *Born Mus.* p. 281. *Schroeter Einl.* i. p. 436, and *Inn. Bau Conch.* p. 20. t. 1. f. 1. *Gmelin,* p. 3515. *Schreibers Conch.* i. p. 184.

Bonanni Rec. and *Kirch.* 3. f. 304. *Gualter,* t. 33. f. A, and t. 34. *Favanne,* t. 20. f. C 1.

Junior. With the outer lip not expanded.

Strombus Lucifer. *Linnæus Syst. Nat.* p. 1210. *Martini,* iii. p. 173. t. 90. f. 878, 879, 881, 885, and 886. *Born Mus.* p. 280. *Schroeter Einl.* i. p. 435. *Gmelin,* p. 3515. *Schreibers Conch.* i. p. 184.

Bonanni Rec. 3. f. 303, and *Kirch.* f. 305. *Lister Conch.* t. 886. f. 7, t. 887. f. 8, and t. 888. f. 9. *Gualter,* t. 54. f. M. *Argenville,* t. 14. f. I. *Klein Ost.* t. 4. f. 85. *Seba,* iii. t. 62. f. 38 to 40. *Knorr,* ii. t. 29. f. 1.

Inhabits the southern coasts of America. *Linnæus.*   Jamaica. *Lister.*   West India Islands. *Martini.*

This well known shell is sometimes a foot long, and ten inches broad, and on account of its large rich rose-coloured aperture is often used as an ornament for mantel-pieces.   Some shells are much more distinctly ribbed than others.

LATISSIMUS. **23.   Shell somewhat nodose, and the spire unarmed; outer lip rounded and very large.**

Strombus latissimus. *Linnæus Syst. Nat.* p. 1211. *Martini,* iii. p. 126. t. 82. f. 832, and t. 83. f. 835. *Schroeter Einl.* i. p. 437. *Gmelin,* p. 3516. *Schreibers Conch.* i. p. 185. *Shaw Nat. Misc.* xiii. t. 519.

Strombus latus. *Gmelin,* p. 3520. ?

Strombus Goliath. *Chemnitz,* xi. p. 147. t. 195 B.

Lister Conch. t. 856. f. 12 c, and t. 862. f. 18 a. *Rumphius,* t. 36. f. L. *Petiver Amb.* t. 14. f. 9. *Seba,* iii. t. 63. f. 1 and 2.

Junior. With the outer lip unexpanded.

Seba, iii. t. 83. f. 12 to 14. *Martini,* iii. t. 89. f. 874.

Inhabits the Asiatic Ocean. *Linnæus.*   Coasts of Amboyna. *Rumphius.*   Pulo Condore and China. *Humphreys.*

Shell about a foot long, and nine inches broad, thick, ponderous, and is nearly allied to S. *Gigas,* from which it differs in being unarmed with conical spines ; the outer lip expands so much upwards that it generally extends beyond the summit of the spire. Young shells are more or less mottled with brown and white, but they afterwards become of nearly an uniform brownish white, and the aperture is tinged with rose-colour ; some specimens are more distinctly ribbed transversely than others, and the S. *Goliath* of Chemnitz differs principally from the shell figured by Martini in having attained to greater maturity.

ACCIPITER. **24.** Shell with the body-whirl crowned with spinous tubercles, and their interstices striated; first whirl of the spire ribbed, and the next striated transversely.

Strombus costatus. *Gmelin,* p. 3520. *Schreibers Conch.* i. p. 188.
Strombus, No. 2. *Schroeter Einl.* i. p. 450.
Ala accipitrina. *Martini,* iii. p. 121. t. 81. f. 829.
Lister Conch. t. 863. f. 18 b. *Favanne,* t. 20. f. C 2.?
Junior. With the outer lip unexpanded.
Martini, iii. t. 91. f. 887.
Inhabits the Asiatic Ocean. *Martini.*
Shell about five inches and a half long, and five inches broad, and is very nearly allied to S. *Gigas.* D'Avila, in his Cat. Syst. p. 185. No. 325, says, " Elle en diffère d'ailleurs en ce que la bouche n'en est point couleur de Rose, que les tubercules du premier orbe en sont plus gros à proportion, et que ceux des autres orbes sont presque entièrement cachés sous les extrémités des orbes précédens ; l'aîle est fort épaisse dans l'une, et papyracée dans l'autre." Gmelin's name of *costatus* had been before used by Da Costa for a different species.

EPIDROMIS. **25.** Shell with the spire somewhat plaited longitudinally, and transversely nodulous ; outer lip rounded and short.

Strombus Epidromis. *Linnæus Syst. Nat.* p. 1211. *Martini,* iii. p. 112. t. 79. f. 821. *Born Mus.* p. 281. *Schroeter Einl.* i. p. 439. *Gmelin,* p. 3516. *Schreibers Conch.* i. p. 185.
Lister Conch. t. 853. f. 10. *Rumphius,* t. 36. f. M. Pe-

tiver Amb. t. 14. f. 18. *Seba,* iii. t. 62. f. 21, 22, and 26. *Knorr,* vi. t. 33. f. 2. *Favanne,* t. 20. f. A 6.
Inhabits the Asiatic Ocean. *Linnæus.* Red Sea. *Forskael.* Coasts of Amboyna. *Rumphius.* China. *Humphreys.*
Shell about three inches long, and rather less than half as broad, and is white, or fawn-coloured mottled with white; the aperture is white, and the outer lip nearly semi-circular, and not expanded at its upper end.

MINIMUS. 26. Shell somewhat plaited longitudinally, and transversely nodulous; aperture sinuated at both ends, and the outer lip slightly reflected at the margin.

Strombus minimus. *Linnæus Mantissa,* p. 549. *Schroeter Einl.* i. p. 439. t. 2. f. 11. *Chemnitz,* x. p. 214. t. 156. f. 1491 and 1492. *Gmelin,* p. 3516.
Rumphius, t. 36. f. P. *Petiver Amb.* t. 14. f. 16. *Gualter,* t. 31. f. L.
Inhabits the East Indies. *Linnæus.* Coasts of Amboyna. *Rumphius.* Tranquebar. *Chemnitz.*
Shell an inch and a quarter or an inch and a half long, and rather more than half as broad; it is white, and has four or five broad pale yellowish brown transverse bands, with darker longitudinal stripes; the shoulder of the body-whirl, as well as the spire, is slightly plaited and nodulous, in which respect it differs from *S. Epidromis,* and it may be at once distinguished from *S. Urceus* by its much more expanded outer lip.

CANARIUM. 27. Shell thick, ventricose, and somewhat heart-shaped; aperture polished, and the outer lip short, thick, and obtuse at the margin.

Strombus Canarium. *Linnæus Syst. Nat.* p. 1211. *Martini,* iii. p. 108. t. 79. f. 818. *Born Mus.* p. 282. *Schroeter Einl.* i. p. 440. *Gmelin,* p. 3517. *Schreibers Conch.* i. p. 185.
Bonanni Rec. and *Kirch.* 3. f. 146. *Lister Conch.* t. 853. f. 9. *Rumphius,* t. 36. f. N. *Petiver Gaz.* t. 98. f. 11, and *Amb.* t. 14. f. 17. *Gualter,* t. 32. f. N. *Argenville,* t. 14. f. Q. *Klein Ost.* t. 4. f. 73. *Seba,* iii. t. 62. f. 23 to 25, 28 and 29. *Knorr,* i. t. 18. f. 5. *Favanne,* t. 20. f. A 5.

Variety. With the spire more produced, and the whirls rounded.

Bonanni Rec. 3. f. 147. *Gualter,* t. 32. f. L. *Martini,* iii. t. 79. f. 817.

Inhabits the Asiatic Ocean. *Linnæus.* Red Sea. *Forskael.* Coasts of Amboyna. *Rumphius.* Fort St. George in the East Indies. *Petiver.* China. *Humphreys.*

Shell about an inch and a half long, and an inch and a quarter broad, of a pale brownish or yellowish white, and often marked with crowded zic-zac longitudinal stripes; the pillar and outer lip are generally more or less covered with a glassy coating; the variety is sometimes two inches and a half long, and is considerably narrower in proportion to its length.

VITTATUS. 28. Shell longitudinally plaited, and the spire produced with an elevated belt at the sutures; outer lip rounded, short and sinuated at both ends.

Strombus vittatus. *Linnæus Syst. Nat.* p. 1211. *Martini,* iii. p. 110. t. 79. f. 819, 820, 822 and 823. *Born Mus.* p. 282. *Schroeter Einl.* i. p. 141. *Gmelin,* p. 3517. *Schreibers Conch.* i. p. 185.

Strombus vittatus Mari Rubri. *Chemnitz,* x. p. 217. t. 157. f. 1496. ?

Lister Conch. t. 852. f. 8. *Rumphius,* t. 36. f. O. *Petiver Gaz.* t. 98. f. 12, and *Amb.* t. 7. f. 9. *Argenville,* t. 9. f. F. *Seba,* iii. t. 62. f. 18 to 20. *Knorr,* iii. t. 20. f. 2.

Variety. Turreted, with the spire more produced, and the body-whirl shorter.

Strombus vittatus angustior. *Chemnitz,* x. p. 208. t. 155. f. 1481 and 1482.

Lister Conch. t. 885. f. 12 b. *Favanne,* t. 20. f. A 8.

Inhabits the coasts of Asia. *Linnæus.* Amboyna. *Rumphius.* China, and the Red Sea. *Chemnitz.*

Shell about three inches long, and an inch and a quarter broad, of a pale brownish or fawn-colour with white dotted bands, or white with brown bands; the body-whirl is transversely grooved at the base; the spire is more produced, and an elevated belt at the sutures at once distinguishes this species from *S. Epidromis*; the aperture is white, and the outer lip striated at a short distance from the inner margin.

SULCATUS.   29.   Shell transversely grooved, with
the spire produced and channelled at the sutures; outer lip rounded, short, and sinuated.

Strombus sulcatus. *Chemnitz*, xi. p. 142. t. 195 A. f. 1870
and 1871.
Lister Conch. t. 852. f. 8.   *Favanne*, t. 20. f. A 4.
Inhabits the coasts of China. *Chemnitz.*
Shell about three inches long, and but little more than an inch
broad, of a yellowish brown colour ; the base of the body-
whirl and the spire are transversely grooved, without any longitudinal plaits, and it also differs from *S. vittatus* in having
a hollow channel instead of an elevated band at the sutures.

ACCINCTUS.   30.   Shell smooth, slightly grooved at
the base; outer lip thick, wrinkled within,
and extended over two whirls of the spire.

Strombus accinctus.   *Linnæus Syst. Nat.* p. 1212.   *Born
Mus.* p. 283.
Strombus succinctus. *Martini*, iii. p. 105. t. 79. f. 815
and 816.   *Schroeter Einl.* i. p. 442.   *Gmelin*, p. 3518.
Schreibers Conch. i. p. 186.
Strombus marginatus.   *Schroeter Einl.* i. p. 431. t. 2. f. 10.
Chemnitz, x. p. 212. t. 156. f. 1489 and 1490.   *Gmelin*,
p. 3513.
Lister Conch. t. 859. f. 16.   *Rumphius*, t. 37. f. X.   *Pe-
tiver Amb.* t. 14. f. 19.   *Gualter*, t. 33. f. B. *Argenville*,
t. 10. f. C.   *Seba*, iii. t. 61. f. 15.
Junior. With the outer lip not expanded.
Petiver Gaz. t. 98. f. 13.   *Martini*, iii. t. 89. f. 877.   *Born
Mus.* t. 10. f. 14 and 15.
Inhabits the coasts of Amboyna. *Rumphius.* Batavia. *Petiver.*
Tranquebar. *Chemnitz.*   China. *Humphreys.*
Shell commonly about an inch and three-quarters long, and half
as broad, of a pale yellowish or reddish brown colour, with
four or five white pencilled transverse bands ; the outer lip
is produced, and attached to the two lower whirls of the
spire, and has a rounded protuberance in a line with the
shoulder of the body-whirl ; the pillar-lip is somewhat gibbous, and is also extended over the spire ; the upper whirls
have a small crenated ridge near the suture.

FISSURELLA.   31. Shell turreted, with longitudinal
ribs, and a marginated fissure extending from
the aperture to the summit.

Strombus Fissurella. *Linnæus Syst. Nat.* p. 1212. *Martini*, iv. p. 337. t. 158. f. 1498 and 1499. *Schroeter Einl.* i. p. 444. *Gmelin*, p. 3518. *Schreibers, Conch.* i. p. 187. *Petiver Gaz.* t. 73. f. 7 and 8. *Argenville*, t. 29. line 2. f. 6. *Favanne*, t. 79. f. &.

Inhabits the East Indies. *Petiver*.

Shell about an inch and a quarter long, and half an inch broad, of a dirty white colour, and has a marginated cleft extending from the aperture to the summit; the outer lip is but little expanded. It is frequently met with in a fossil state, and Petiver asserts that some live shells have been brought from the East Indies.

URCEUS. 32. **Shell with nodulous plaits on the body-whirl and spire; outer lip double, bilobed, short, thickened above and striped within.**

Strombus Urceus. *Linnæus Syst. Nat.* p. 1212. *Martini*, iii. p. 100. t. 78. f. 803 to 809. *Born Mus.* p. 283. *Schroeter Einl.* i. p. 445. *Gmelin*, p. 3518. *Schreibers Conch.* i. p. 187.

Bonanni Rec. and *Kirch.* 3. f. 144. *Lister Conch.* t. 857. f. 13. *Rumphius*, t. 37. f. T. *Petiver Gaz.* t. 98. f. 14, and *Amb.* t. 14. f. 21. *Gualter*, t. 32. f. E. and G. *Seba*, iii. t. 60. f. 28 and 29. *Knorr*, iii. t. 13. f. 5.

Junior. With the outer lip imperfect.

Martini, iii. t. 88. f. 870.?

Inhabits the Asiatic Ocean. *Linnæus*. Coasts of Amboyna. *Rumphius*. Straights of Malacca, Sincapour, Java, and Luzone. *Petiver*. Isle of France. *Martini*. China. *Humphreys*.

Shell an inch and a half, or sometimes two inches and a half long, and about half as broad, and varies much in its colour and markings; the pillar-lip is thickened, and slightly plaited at both the extremities; the outer lip has a prominent ridge running parallel to the margin, and is but a little expanded.

ERYTHRINUS. 33. **Shell with nodulous plaits on the body-whirl and spire; outer lip striated on both sides, and the inner lip white and reflected.**

Strombus Erythrinus. *Chemnitz*, xi. p. 146. t. 195. A. f. 1874 and 1875.

Inhabits the Red Sea. *Chemnitz*.

Shell about thirteen lines long and six broad, shining, reddish, and tinged with rose-colour; the inner lip is thickened, somewhat reflected, white, and striated at the base.

SAMAR.   34. Shell smooth, thin, and plaited on the shoulder of the body-whirl; outer lip very little expanded and three-toothed; base and pillar dark brown or violet.

Strombus Samar.   *Chemnitz*, x. p. 221. t. 157. f. 1503.
Strombus tridentatus.   *Gmelin*, p. 3519.
Strombus Urceus, Var.   *Schreibers Conch.* i. p. 187.
Strombus, No. 1.   *Schroeter Einl.* i. p. 450.
Alata Samar. *Mus. Geversianum*, p. 336, No. 841. *Martini*, iii. p. 102. t. 78. f. 810 to 814.
Lister Conch. t. 858. f. 14.   *Rumphius*, t. 37. f. Y.   *Petiver Amb.* t. 14. f. 15.   *Gualter*, t. 33. f. C and D. *Seba*, iii. t. 61. f. 34.
Inhabits the coasts of Amboyna. *Rumphius*.   East Indian Seas. *Martini, &c.*
Shell about an inch and a half long, and about half as broad, of a yellowish white colour, spotted and clouded with brown or orange; the spire is produced, and consists of eight or nine slightly ventricose whirls; the outer lip is but very little expanded, and is thickened at its upper extremity.

DENTATUS.   35. Shell longitudinally plaited, with the outer lip slightly expanded and toothed, and sinuated near the base; inner lip thickened and reflected; throat striped.

Strombus dentatus.   *Linnæus Syst. Nat.* p. 1213. *Schroeter Einl.* i. p. 446. t. 2. f. 12.? *Chemnitz*, x. p. 220. t. 157. f. 1501 and 1502.   *Gmelin*, p. 3519.   *Schreibers Conch.* i. p. 188.
Seba, iii. t. 61. f. 25.
Inhabits the coasts of the Isle of France. *Chemnitz*.
Shell about an inch and a half long, and nearly half as broad, variegated with different shades of brown and white dots and stripes; it resembles *S. Urceus*, but is said to be more obliquely plaited, and to have the whirls more distant and the aperture more distinctly striated.

VEXILLUM.   36. Shell ovate, rather solid, and the spire small; outer lip slightly expanded, somewhat thickened, and toothed on the inside; pillar flat and glabrous.

Strombus Vexillum. *Chemnitz*, x. p. 222. t. 157. f. 1504 and 1505. *Gmelin*, p. 3520.

Strombus, No. 11. *Kæmmerer Cab. Rudolst*. p. 119. t. 7. f. 2 and 3.

Inhabits the East Indian Seas. *Chemnitz*.

Shell about ten lines long and six broad, with alternate reddish and ochraceous transverse bands ; the outer lip is finely toothed on its inside, and the throat is white ; it is said to be a very scarce shell, and judging from Chemnitz's figure, it has not much the appearance of a Strombus.

NORWEGICUS. 37. Shell oblong, smooth, with six produced rounded whirls ; aperture ovate, patulous, and the outer lip slightly expanded.

Strombus Norwegicus. *Chemnitz*, x. p. 218. t. 157. f. 1497 and 1498. *Gmelin*, p. 3520.

Seba, iii. t. 52. f. 9. ?

Inhabits the coasts of Norway. *Chemnitz*.

Shell about three inches long, and an inch and a half broad, of a dirty white colour ; the canal at the base of the aperture is slightly recurved, and the throat is said to be ivory-white ; the figure which Chemnitz has given has more the appearance of being a Murex allied to *M. antiquus*, than of any Strombus with which I am acquainted.

**** *Turreted, with a very long spire.*

TUBERCULATUS. 38. Shell ovate-oblong, with transverse rows of dark tubercles, and transversely striated ; outer lip thickened.

Strombus tuberculatus. *Linnæus Syst. Nat.* p. 1213. *Born Mus.* p. 284. t. 10. f. 16 and 17. *Schroeter Einl.* i. p. 477. *Gmelin*, p. 3521.

Turbo rostratus. *Martini*, iv. p. 327. t. 157. f. 1490.

Cerithium Morus. *Bruguiere Enc. Meth.* p. 500.

Murex tuberculatus. *Gmelin*, p. 3564.

Murex sordidus. *Gmelin*, p. 3561.

Murex, No. 171. *Schroeter Einl.* i. p. 598, and No. 45. p. 560.

Seba, iii. t. 55, fig. in the right hand corner of the group, No. 21.

Inhabits the Mediterranean. *Linnæus*.

Shell most commonly about fourteen lines long, and eight

broad, with seven coarse whirls of a brown colour, and
armed with transverse rows of oval convex smooth black
tubercles; the aperture is ovate, and the beak very short and
recurved.

PALUSTRIS.   39. Shell with longitudinal plaits, and
distant transverse grooves; outer lip channelled above, and crenulated.

Strombus palustris. *Linnæus Syst. Nat.* p. 1213. *Mar-
tini*, iv. p. 312. t. 156. f. 1472. *Schroeter Einl.* i. p.
448. *Gmelin*, p. 3521.
Strombus agnatus. *Gmelin*, p. 3528.
Strombus, No. 30. *Schroeter Einl.* i. p. 462.
Cerithium palustre. *Bruguiere Enc. Meth.* p. 486.
Lister Conch. t. 837. f. 63. *Rumphius*, t. 30. f. Q. *Petiver Amb.* t. 13. f. 13. *Seba*, iii. t. 50. f. 13, 14, 17,
18, and 19. *Knorr*, iii. t. 18. f. 1. *Martini Berl.
Mag.* iv. t. 9. f. 40. *Favanne*, t. 40. f. A 1.
Inhabits marshes by the sea-side in the East Indies. *Rumphius.* China. *Humphreys.*
Shell about four inches long, and an inch and a half broad, of
a greenish or yellowish brown colour with darker transverse
bands; it is somewhat pyramidal, and has sixteen or eighteen whirls; the body-whirl is slightly, and the spire distinctly, but not very strongly, plaited longitudinally; the
outer lip is thick, crenulated at the margin, and marked on
the inside with dark or reddish brown bands, and at the upper extremity is separated from the body-whirl by a straight
canal.

ATER.   40. Shell subulate, black, with smooth contiguous whirls; aperture ending in a canal
at both ends.

Strombus ater. *Linnæus Syst. Nat.* p. 1213. *Schroeter
Fluss.* p. 371, and *Einleitung*, i. p. 449. *Chemnitz*, ix.
part 2. p. 191. t. 135. f. 1227. *Gmelin*, p. 3521.
Strombus dealbatus. *Gmelin*, p. 3523.
Strombus, No. 32. *Schroeter Einl.* i. p. 462.
Nerita atra. *Muller Verm.* ii. p. 188.
Cerithium atrum. *Bruguiere Enc. Meth.* p. 485.
Lister Conch. t. 115. f. 10. *Rumphius*, t. 30. f. R. *Pe-
tiver Amb.* t. 13. f. 16. *Seba*, iii. t. 56. f. 13 and 14.
Martini Berl. Mag. iv. t. 9. f. 41. *Favanne*, t. 61. f.
H 11.

Junior. With the outer lip unexpanded.

Nerita lineata. *Muller Verm.* ii. p. 189. *Schroeter Fluss.* p. 339.

Strombus lineatus. *Gmelin*, p. 3521.

Helix, No. 77. *Schroeter Einl.* ii. p. 199.

Bulimus terebralis. *Bruguiere Enc. Meth.* p. 328.

Lister Conch. t. 116. f. 11.? *Martini Berlin Mag.* iv. t. 10. f. 50.

Inhabits marshes in the Island of Amboyna. *Rumphius.*

Shell about three inches long, and three-quarters of an inch broad, of an uniform brownish black or dark chestnut-colour, and the inside white; it consists of thirteen or fourteen flattish contiguous whirls, which are very finely striated transversely; the aperture is oval, and terminates at both ends in a somewhat similar canal. Bruguiere for this species has quoted Gualter, t. 6. f. E and F, and Muller has cited these figures, though with a mark of doubt, for his *Nerita punctata*, which has been twice described by Gmelin, under the names of *Strombus punctatus* and *Buccinum fluviatile*, and is too uncertain a species to be retained; for the same shell Lister, t. 976. f. 36. is referred to, and this has been also quoted by Gmelin, as well as by several other authors, for a Variety of *Buccinum strigilatum*. The shells which form the present turreted division of the Strombi belong to the same natural family as *Murex Aluco*, and *M. fuscatus*, and I am unable to discover any separating line.

AURITUS. 41. Shell turreted, and transversely striated, with a row of remote tubercles, and a brown band on each whirl; outer lip contracted at the upper extremity.

Strombus auritus. *Gmelin*, p. 3522.

Strombus Tympanorum, &c. *Chemnitz*, ix. part 2. p. 192. t. 186. f. 1265 and 1266.

Nerita aurita. *Muller Verm.* ii. p. 192.

Murex fasciatus. *Gmelin*, p. 3561.

Murex scolopaceus, Var. β. *Gmelin*, p. 3548.

Murex, No. 48. *Schroeter Einl.* i. p. 561.

Bulimus auritus. *Bruguiere Enc. Meth.* p. 331.

Lister Conch. t. 121. f. 16. *Seba*, iii. t. 50. f. 38. *Knorr*, iii. t. 26. f. 5. *Martini Berl. Mag.* iv. t. 10. f. 55.

Inhabits rivers in Guinea. *Chemnitz.*

Shell about twenty lines long, and seven broad, with nine whirls, of which the uppermost is rarely entire; the colour is a rich brown, with three transverse white bands on the body-whirl, and one on the other whirls, and on each a row of remote tubercles; the throat and pillar are white. Muller says it is sometimes yellow with brown bands, and has described one Variety with a white line at the sutures in addition to these markings. Bruguiere, from its thickness, doubts whether Lister and Chemnitz have been correct in considering it a fresh-water shell.

LIVIDUS. 42. Shell turreted, sub-angulated, and armed with spinous nodules; outer lip separated on the anterior side.

Strombus lividus. *Linnæus Syst. Nat.* p. 1213. *Chemnitz,* ix. part 2. p. 193. t. 136. f. 1269 and 1270. *Gmelin,* p. 3523.

Inhabits ——

Linnæus, in the Mus. Reg. Lud. Ulricæ, has described this shell to be " turreted, sub-angulated, with smooth pointed nodules, and a single row of straight conical acute spines in the middle of the whirls; colour livid with ferruginous spots; aperture oblong, and not contracted at the base." Muller says that his *Buccidum torridum* differs only in not having any ferruginous spots, but Bruguiere in his description of *Bulimus muricatus,* in the Encyclopédie Methodique, p. 330, expresses a contrary opinion. Bruguiere there alludes to a description of the Linnæan shell under the name of *Cerithe livide,* but I cannot find any such species in his Genus Cerithium. The shell which Chemnitz has figured is about an inch and a half long, and near an inch broad.

COSTATUS. 43. Shell turreted, with the whirls longitudinally ribbed, and a spiral line on each; aperture sub-orbicular, and the outer lip expanded.

Strombus costatus. *Donovan,* iii. t. 94. *Montagu Test.* p. 255. *Maton and Racket, in Lin. Trans.* viii. p. 142. *Dorset Cat.* p. 46. t. 14. f. 14.

Strombus costatus et transversim striatus. *Schroeter Fluss.* p. 373. t. 8. f. 14. ?

Strombiformis costatus. *Da Costa Brit. Conch.* p. 118. t. 8. f. 14.

Helix acicula. *Gmelin,* p. 3668. ?

Helix, No. 261. *Schroeter Einl.* ii. p. 251. ?

Variety.   Without a transverse line on the whirls.
   Strombus turboformis. *Montagu Supp.* p. 110, and p. 169.
      t. 30. f. 7.
Inhabits the coasts of the West of England. *Da Costa,* &c.
   Plentiful in Bantry Bay. *L. W. D.*
Shell about half an inch long, and one third as broad, of a
   dark brown or chestnut-colour ; it has ten or eleven rounded
   whirls, which are longitudinally ribbed, and furnished with
   a transverse line which becomes double at the base of the
   body-whirl ; the aperture is roundish, and ends in a short
   canal, which has not any inclination to the left, or any other
   mark to distinguish it from that of many of the turreted
   Murices.   *Strombus costatus* of Schroeter, which is *Helix
   Acicula* of Gmelin, appears to me to possess much more
   affinity with the present species than with any of the Helices,
   and is said to be a native of the coast of Coromandel.

𝔊𝔢𝔫𝔲𝔰 XXVI.

MUREX:

SHELL UNIVALVE, SPIRAL, ROUGH, WITH MEMBRA-
NACEOUS RIDGES; APERTURE ENDING IN A
STRAIGHT, OR SLIGHTLY ASCENDING CANAL.

Subdivisions.†

* Spinous, with a produced beak.
** Foliated, and the beak short.
*** With thick protuberant rounded varices.‡
**** Somewhat spinous, and without a beak.
***** Unarmed, with a long straight subulate
beak.
****** Turreted and subulate, with a very short
beak.

* *Spinous, with a produced Beak.*

HAUSTELLUM. 1. Shell sub-ovate, with three thick
varices, and intermediate smaller ribs; beak
long, subulate, and straight.

---

† Besides some others, which I have elsewhere mentioned, the following of Gmelin's species appear to me to be undeserving of notice: *M. Afer*, p. 3558. *M. affinis*, p. 3532. *M. cancellatus*, p. 3548. *M. candidus*, p. 3528. *M. eburneus*, p. 3564. *M. fusiformis*, p. 3549. *M. granularis*, p. 3557. *M. lignosus*, p. 3557. *M. Lima*, p. 3541. *M. Loco*, p. 3538. *M. niveus*, p. 3545. *M. nodulosus*, p. 3562. *M. semilunaris*, p. 3549. *M. trigonus*, p. 3549. *M. undulatus*, p. 3559. The following are fossils: *M. campanicus*, p. 3558. *M. costatus*, p. 3543. *M. fossilis*, p. 3555. *M. lævigatus*, p. 3555, and *M. triacanthus*, p. 3527.

‡ Varices are defined in Burrows's Elements of Conchology to be ' longitudinal gibbous sutures formed in the growth of the shell at certain proportionable distances on the whirls.' See tab. 1. f. *p*.

Murex Haustellum. *Linnæus Syst. Nat.* p. 1213. *Martini*, iii. p. 379. t. 115. f. 1066. *Born Mus.* p. 287. *Schroeter Einl.* i. p. 475, and *Inn. Bau Conch.* p. 31. t. 3. f. 1. *Gmelin*, p. 3524. *Schreibers Conch.* i. t. 191. *Lamarck Syst. des Anim.* p. 81. *Shaw Nat. Misc.* xii. p. 488. *Brookes's Introd.* p. 119. t. 7. f. 88. *Bonanni Rec.* 3. f. 268, and *Kirch.* f. 270. *Lister Conch.* t. 903. f. 23. *Rumphius*, t. 26. f. F. *Petiver Amb.* t. 4. f. 8. *Argenville*, t. 16. f. B. *Gualter*, t. 30. f. E. *Klein Ost.* t. 4. f. 81. *Seba*, iii. t. 78. f. 5 and 6. *Knorr*, i. t. 12. f. 2 and 3. *Favanne*, t. 38. f. B 2.
Junior ? With the beak not at all, or but little produced. *Martini*, iii. t. 115. f. 1067 and 1068. *Favanne*, t. 38. f. B 3.
Inhabits the Asiatic Ocean. *Linnæus.* Red Sea. *Bonanni.* Coasts of Amboyna. *Rumphius.* America. *D'Avila.* Coromandel. *Martini.* China. *Humphreys.*
Shell three or four inches long, of which the subulate beak at the base occupies the larger half, and the breadth is about an inch and a half or two inches; it is longitudinally divided by three strong elevated ribs, covered by a membrane which overlaps a part of their interstices, and the interstices are furnished with less elevated tuberculated ribs; it is of a flesh-colour transversely marked with blackish brown, and the inner margin of the aperture is rose-colour.

MOTACILLA. 2. Shell triangular, with three thick somewhat spinous varices, and transversely grooved; beak rather long, subulate, and slightly ascending.

Murex Motacilla. *Chemnitz*, x. p. 268. t. 163. f. 1563. *Gmelin*, p. 3530. *Schreibers Conch.* i. p. 198.
Inhabits the East Indian Seas. *Chemnitz.*
Shell about two inches and a half long, of which the beak occupies about an inch, and the breadth is an inch and a quarter; it has three strong varices armed with remote pointed tubercles, and both the varices and their interstices are transversely grooved; the colour is whitish with reddish brown spots and bands, and the throat is white.

SCOLOPAX 3. Shell sub-ovate, with three spinous varices, and darker transverse ribs; beak very long, straight, and armed with similar long spines.

Murex Tribulus maximus. *Chemnitz,* xi. p. 101. t. 189. f. 1819 and 1820.

Murex Tribulus, Var. *Martini,* iii. p. 363. t. 113. f. 1052. Inhabits the Red Sea. *Chemnitz.*

This shell differs from *M. Tribulus,* in being larger and stronger, and in the want of cancellated striæ; the three rows of spines are thicker, and their interstices marked with reddish slightly elevated ribs, and obsolete transverse striæ; the colour, except the ribs, is whitish, or tinged with reddish brown.

TRIBULUS. 4. Shell sub-ovate, with three spinous varices, and cancellated striæ; beak very long, straight, and armed with similar long spines.

Murex Tribulus. *Linnæus Syst. Nat.* p. 1214. *Martini,* iii. p. 368. t. 113. f. 1053 to 1056. *Born Mus.* p. 288. *Schroeter Einl.* i. p. 476. *Gmelin,* p. 3525. *Schreibers Conch.* i. p. 192.

Variety A. With the spines rather remote, and shorter than the beak.

Purpura Tribulus. *Mus. Geversianum,* p. 310, No. 624. *Bonanni Rec.* 3. f. 269, and *Kirch* f. 271. *Lister Conch.* t. 902. f. 22. *Rumphius,* t. 26. f. G. *Petiver Gaz.* t. 101. f. 16, and *Amb.* t. 6. f. 8. *Gualter,* t. 31. f. A. *Seba,* iii. t. 78. f. 4. *Knorr,* i. t. 11. f. 3 and 4. *Favanne,* t. 38. f. A 1.

Variety B. With crowded spines, and some as long as the beak.

Murex Tribulus duplicatus. *Chemnitz,* xi. p. 103. t. 189. f. 1821, and t. 190. f. 1822.

Purpura Histrix. *Mus. Geversianum,* p. 308, No. 621. *Rumphius,* t. 26. f. No. 3. *Gualter,* t. 31. f. B. *Argenville,* t. 16. f. A. *Seba,* iii. t. 78. f. 1 to 3. *Knorr,* v. t. 27. f. 1. *Regenfuss,* ii. t. 11. f. 46. *Favanne,* t. 38. f. A 2. *Shaw Nat. Misc.* xii. t. 460.

Inhabits the Asiatic Ocean, and coasts of Java. *Linnæus.* Amboyna. *Rumphius.*

Shell three or four inches long, of which the subulate beak occupies the larger half, and the breadth of the body-whirl is about an inch or an inch and a quarter; the spines issue from three varices, and are extended to the extremity of the beak; the interstices are slightly nodulous, transversely ribbed, and marked with cancellated striæ; the colour is dirty or pale brownish white.

CORNUTUS. 5. Shell roundish, with seven varices, and subulate oblique spines; beak long, subulate, straight, and irregularly spinous.

Murex cornutus. *Linnæus Syst. Nat.* p. 1214. *Martini*, iii. p. 372. t. 114. f. 1057. *Born Mus.* p. 289. *Schroeter Einl.* i. p. 478. *Gmelin*, p. 3525. *Schreibers Conch.* i. p. 192.

Le Bolin. *Adanson Senegal,* p. 127. t. 8. f. 20.

Bonanni Rec. 3, f. 283, and *Kirch.* f. 284. *Lister Conch.* t. 901. f. 21. *Rumphius*, t. 26. f. No. 5. *Gualter*, t. 30. f. D. *Seba*, iii. t. 78. f. 7 to 9. *Knorr,* vi. t. 17. f. 1. *Favanne*, t. 38. f. E 2.

Inhabits the coasts of Africa. *Linnæus.* Amboyna. *Rumphius.* Magdalen Islands. *Adanson.* Guinea. *Humphreys.* Shell five or six inches long, of which the beak occupies nearly half, and the breadth is about three inches; it is transversely striated, and has seven varices armed with strong oblique spines, which form two transverse rows, and are pointed in opposite directions; the colour is brownish white. Adanson, under the name of *Le Sirat* (p. 125. t. 8. f. 19), has described a Murex, of which he says the animal is exactly the same, and the shell nearly allied to *Le Bolin,* which is the present species. From this description of *Le Sirat,* Gmelin has derived both his *M. Senegalensis,* p. 3536, and his *M. costatus,* p. 3549. Martini (vol. iii. p. 369.) expresses a doubt whether *Le Sirat* is not the same as Argenville's t. 16. f. G, from which Gmelin, at p. 3528, has constituted one of the two species to which he has given the name of *M. candidus.*

BRANDARIS. 6. Shell sub-ovate, with seven varices, and subulate straight spines; beak rather short, and obliquely spinous.

Murex Brandaris. *Linnæus Syst. Nat.* p. 1214. *Martini*, iii. p. 376. t. 114. f. 1058 to 1061, and t. 115. f. 1062 to 1065. *Born Mus.* p. 289. *Schroeter Einl.* i. p. 479. *Chemnitz*, x. p. 276. t. 164. f. 1571. *Gmelin,* p. 3526. *Schreibers Conch.* i. p. 192. *Shaw Nat. Misc.* xv. t. 631.

Bonanni Rec. 3. f. 281 and 282, and *Kirch.* f. 282 and 283. *Lister Conch.* t. 900. f. 20. *Rumphius,* t. 26. f. No. 4. *Petiver Gaz.* t. 68. f. 12. *Gualter*, t. 30. f. F. *Argenville Zoom.* t. 4. f. C. *Knorr,* ii. t. 18. f. 1, and t. 22.

f. 4 and 5. *Regenfuss*, i. t. 6. f. 67. *Favanne*, t. 38. f. E 1.

Variety. Distorted with the beak strangely bent backwards. *Chemnitz*, xi. p. 293. t. 211. f. 2094 and 2095.

Inhabits the Mediterranean. *Linnæus, &c.* Adriatic. *Ginanni.* Coasts of Africa. *Bonanni.* Guinea. *Regenfuss.*

Shell about three inches and a quarter long, of which the beak occupies nearly one third, and the breadth is about an inch and three-quarters; it is transversely striated, and has six or seven varices, with two or three rows of large straight spines, which are more or less pointed in opposite directions; the varices extend from the body-whirl successively further down the beak, and each terminates in a large spine, of which an obliquely spiral row is thus formed.

TRUNCULUS. 7. Shell ovate, rough, with seven spinous varices and transverse striæ; beak short and perforated.

Murex Trunculus. *Linnæus Syst. Nat.* p. 1215. *Martini,* iii. p. 341. t. 109. f. 1018 to 1020. *Born Mus.* p. 290. *Schroeter Einl.* i. p. 480. *Gmelin,* p. 3526. *Schreibers Conch.* i. p. 193.

Bonanni Rec. 3. f. 271 and 272, and *Kirch.* f. 273 and 274. *Lister Conch.* t. 947. f. 42, and t. 952. f. 1. *Gualter,* t. 31. f. C. *Klein Ost.* t. 6. f. 104. *Seba,* iii. t. 52. f. 15 and 16. *Knorr,* iii. t. 13. f. 1, and v. t. 19. f. 6. *Favanne,* t. 38. f. D 2.

Variety. With the whirls reversed.

Murex Trunculus perversus. *Chemnitz,* ix. part 1. p. 64. t. 105. f. 897 and 898.

Inhabits the Mediterranean. *Linnæus.* Red Sea. *Bonanni.* Coasts of Jamaica, and other West India Islands. *Martini.* Bay of Naples. *Ulysses.*

Shell about three inches long, and an inch and three-quarters broad, most commonly of a brownish white colour, and sometimes variegated with brown or reddish transverse bands; the varices terminate abruptly on the beak, which is broad, and rarely more than three-quarters of an inch long. This is supposed to be the species from which the Ancients chiefly extracted the beautiful purple known by the name of the Tyrian dye.

ROSARIUM. 8. Shell ovate, with seven furbelowed spinous varices, and the beak short and perforated.

Murex Rosarium. *Chemnitz,* x. p. 245. t. 161. f. 1528 and 1529.

Murex Trunculus, Var. *Gmelin,* p. 3527.

Murex, No. 18. *Schreibers Conch.* i. p. 196.

Inhabits ——

Chemnitz has described this shell to be two inches and seven lines long, and an inch and three-quarters broad, of a whitish colour, with seven reddish foliated varices. It may possibly be nothing more than an extraordinary Variety of *M. Trunculus,* but Schreibers, as well as Chemnitz, considered it to be distinct, and the former has arranged it among the foliated species in the second division.

POMUM. 9. Shell ovate, nodose, with about seven oblique varices, and transverse ribs; beak broad.

Murex Pomum. *Gmelin,* p. 3527.

Murex, No. 20. *Schroeter Einl.* i. p. 549.

Purpura pomiformis. *Martini,* iii. p. 342. t. 109. f. 1021 to 1023, and t. 110. f. 1024 and 1025.

Le Cofar. *Adanson Senegal,* p. 131. t. 9. f. 22.

Lister Conch. t. 944. f. 39 *a.* *Knorr,* iii. t. 9. f. 1.

Inhabits the Mediterranean. *D'Avila.* Coasts of Senegal, and common on the shores of the Magdalen Islands. *Adanson.*

Shell about three inches long, and an inch and three-quarters broad, and is described by Gmelin with from three to seven varices, but Adanson says it has from seven to nine; it is thick and ponderous, and nearly allied to *M. Trunculus;* the colour is yellowish brown, or sometimes dark brown verging on black, and paler on the varices and transverse ribs.

MILIARIS. 10. Shell ovate, scabrous, with seven crenulated, sub-foliated, oblique varices; outer lip double and toothed; beak short, ascending, and narrow.

Murex miliaris. *Gmelin,* p. 3536.

Murex Purpura scabra. *Chemnitz,* x. p. 246. t. 161. f. 1532 to 1535.

Murex Brandaris, Var. *Gmelin,* p. 3526.

Murex, No. 12. *Kæmmerer Cab. Rudolst.* p. 107. t. 9. f. 4.

Murex, No. 19. *Schroeter Einl.* i. p. 549.
Valentyn Abh. t. 2. f. 14 to 18. *Knorr,* iii. t. 29. f. 5.
Martini, iii. Vign. at p. 303. f. 1 to 5.
Inhabits the coasts of the Nicobar Islands. *Chemnitz.*
Shell about one inch and eight lines long, and thirteen lines
broad, and has the varices crenulated, and somewhat foliated
on their margins; the colour is yellowish brown variegated
with white, and clouded or transversely striped with dark
brown; the aperture and throat are white and shining. The
colour is however variable, according to Gmelin, who has
very strangely placed this species in the third division after
M. Anus, and also as a variety of *M. Brandaris.*

MELANOMATHOS. 11. Shell sub-fusiform, transversely striated, with eight spinous varices,
and the spines black; spire nodulous and
prickly; beak short and acute.

Murex melanomathos. *Gmelin,* p. 3527. *Schreibers Conch.*
i. p. 220.
Murex radix. *Gmelin,* p. 3527.?
Murex, No. 16, and No. 17. *Schroeter Einl.* i. p. 548.
Argenville Zoom. t. 11. f. K. ? *D'Avila,* t. 15. f. H. *Martini,* iii. t. 108. f. 1015.
Inhabits the East Indian Seas. *Chemnitz.*
Shell about an inch and a quarter long, and three-quarters of
an inch broad, white, with about eight blackish varices,
which are armed with crowded spines of the same colour.
Argenville's figure, from which the *M. Radix* of Gmelin
has been wholly constituted, appears to differ only in having
the varices somewhat furbelowed as well as spinous.

** *Foliated, and the Beak short.*

RAMOSUS. 12. Shell transversely grooved, and somewhat nodulous, with three foliated and
branched varices; beak short and truncated.

Murex ramosus. *Linnæus Syst. Nat.* p. 1215. *Martini,*
iii. p. 310. t. 102. to t. 106, and f. 960 to 997. *Born
Mus.* p. 292. *Schroeter Einl.* i. p. 481. *Gmelin,* p.
3528. *Schreibers Conch.* i. p. 193. *Shaw Nat. Misc.*
xiii. t. 527. *Burrows's Elements,* p. 163. t. 18. f. 4.

Bonanni Rec. 3. f. 275 and 276, and *Kirch.* f. 280 and 281. *Lister Conch.* t. 946. f. 41. *Rumphius,* t. 26. f. A, C, and No. 1. *Petiver Amb.* t. 12. f. 1 and 2. *Gualter,* t. 37. f. D, G, H, I, and L, and t. 38. *Klein Ost.* t. 4. f. 82. *Argenville,* t. 16. f. C, E, and H. *Seba,* iii. t. 77. f. 1 to 4, and 7 to 12. *Knorr,* i. t. 25. f. 1 and 2, and t. 26. f. 1 and 2 ; ii. t. 7. f. 4 and 5 ; iii. t. 9. f. 3 ; and v. t. 11. f. 1. *Regenfuss,* i. t. 7. f. 6. *Favanne,* t. 36. f. E 1, E 2, F, H 1, H 2, I 1 and I 2.

Variety B. White, tinged with red, and the foliated varices dark purple.

Murex versicolor. *Gmelin,* p. 3530. *Schreibers Conch.* i. p. 195.

Murex, No. 163. *Schroeter Einl.* i. p. 596.

Knorr, v. t. 4. f. 1.

Variety C. Of an uniform blackish brown colour.

Murex Monachus Capucinus. *Chemnitz,* xi. p. 123. t. 192. f. 1849 and 1850.

Inhabits the Persian Gulf, and the coasts of Jamaica. *Linnæus.* Amboyna. *Rumphius.* West Indies. *Regenfuss.* East and West Indies, China, Madagascar, and Martinique. *Humphreys.*

This shell varies very much in its size and colour, and in the size of its foliated varices ; the length is from two to five inches, and the breadth is about half the length ; the colour is sometimes almost uniformly white, pale yellow, brownish, pale red, or nearly black, and sometimes the whirls are of one, and the foliations of another of these colours, or marked with dark transverse lines in pairs.

FOLIATUS. 13. Shell transversely ribbed, with three membranaceous foliated varices, and the aperture one-toothed.

Murex foliatus. *Gmelin,* p. 3529.

Murex Purpura alata. *Chemnitz,* x. p. 250. t. 161. f. 1538 and 1539.

Purpura foliata. *Martyn Univ. Conch.* ii. t. 66.

Inhabits King George's Sound. *Martyn.* Coasts of New Zealand. *Humphreys.*

Shell about two inches and a quarter long, and an inch and a half broad, of a greenish white colour more or less tinged with red ; the aperture has a broad erect tooth on the margin of the outer lip near the base, and the beak is closed.

LINGUA. 14. Shell sub-triangular, ovate, with three membranaceous varices, and somewhat nodulous; whirls contracted at the sutures, and the beak closed.

Murex Lingua vervecina. *Chemnitz,* x. p. 251. t. 161. f. 1540 and 1541.

Murex decussatus. *Gmelin,* p. 3527. *Schreibers Conch.* i. p. 196.

Le Jatou. *Adanson Senegal,* p. 129. t. 9. f. 21.

Inhabits the coasts of Goree. *Adanson.*

Shell an inch and a half long, and about half as broad, either wholly white, or sometimes the interstices between the membranaceous varices are dark brown; the aperture is elliptical, and the outer lip double. Gmelin, under the name of *M. decussatus,* appears, both by his description and references, to have strangely confounded this shell with some varieties of *M. Erinaceus.*

TRIPTERUS. 15. Shell long, narrow, and sub-triangular, with three membranaceous, sub-foliaceous varices; and the beak rather long.

Murex tripterus. *Born Mus.* p. 291. t. 10. f. 18 and 19. *Gmelin,* p. 3530. *Schreibers Conch.* i. p. 212.

Murex ramosus, Var. ε. *Gmelin,* p. 3528.

Murex, No. 172. *Schroeter Einl.* i. p. 598.

D'*Avila,* t. 16. f. K. *Martini,* iii. t. 110. f. 1031 and 1032, and t. 111. f. 1033.

Inhabits the coasts of Batavia. *Born.* Java. *Humphreys.*

Shell two inches or two inches and a half long, and the body-whirl, without including the varices, is not more than one third as broad; it has seven produced whirls, which are transversely wrinkled, and between the varices granulated; colour whitish.

TRIQUETER. 16. Shell long, narrow, and sub-triangular, with reticulated ribs, and three membranaceous varices; beak rather long, and closed.

Murex triqueter. *Born Mus.* p. 291. t. 11. f. 1 and 2. *Gmelin,* p. 3530. *Schreibers Conch.* i. p. 212.

Murex ramosus, Var. ζ. *Gmelin,* p. 3259.

Murex, No. 173. *Schroeter Einl.* i. p. 599.

D'Avila, t. 16. f. L. *Martini*, iii. t. 111. f. 1038.
Inhabits the coasts of Tranquebar. *Martini*. China. *Humphreys*.

Born describes this species to be rather more than two inches long, and only three-quarters of an inch broad, and white, marked with scattered red spots; the spire pyramidal, with six whirls; aperture elliptical, and the outer lip crenated.

SCORPIO.   17.   Shell contracted, with four foliated varices, and the spire capitate; beak rather long and truncated.

Murex Scorpio. *Linnæus Syst. Nat.* p. 1215. *Martini*, iii. p. 327. t. 106. f. 998 to 1003. *Born Mus.* p. 293. *Schroeter Einl.* i. p. 483. *Gmelin*, p. 3529. *Schreibers Conch.* i. p. 194.
Rumphius, t. 26. f. D. *Petiver Amb.* t. 11. f. 14. *Gualter*, t. 37. f. M. *Argenville*, t. 16. f. D. *Seba*, iii. t. 77. f. 13 to 16. *Knorr*, ii. t. 11. f. 4 and 5. *Da Costa Elements*, t. 5. f. 6. *Spengler Selt. Conch.* t. 2. f. C. *Favanne*, t. 36. f. G 3 and G 4.

Inhabits the Asiatic Ocean. *Linnæus*. Coasts of Amboyna. *Rumphius*. China. *Humphreys*.

Shell about two inches long, and three-quarters of an inch broad, without including the branched foliated processes which issue from the varices; the first whirl of the spire is turgid and contracted at its base, and the summit is rather obtuse; the colour is generally brown, more or less variegated with white, and the shell is sometimes entirely white. A variety with the whirls reversed is mentioned in the Portland Catalogue, lot 3991, and is said to be extremely rare. Its common name is the *Skeleton Shell*.

SAXATILIS.   18.   Shell transversely grooved, and somewhat nodulous, with five foliated varices, and the beak short.

Murex saxatilis. *Linnæus Syst. Nat.* p. 1215. *Martini*, iii. p. 328. t. 107. f. 1004 to 1010, and t. 108. f. 1011 and 1012. *Born Mus.* p. 294. *Schroeter Einl.* i. p. 484. *Gmelin*, p. 3529. *Schreibers Conch.* i. p. 194. *Shaw Nat. Misc.* xx. t. 927.
Murex fasciatus. *Gmelin*, p. 3528. ?
Murex, No. 116. *Schroeter Einl.* i. p. 582, and No. 169. p. 597.

Rumphius, t. 26. No. 2. *Klein Ost.* t. 6. f. 109. *Seba*, iii.
t. 77. f. 5 and 6. *Knorr*, iii. t. 9. f. 2, vi. t. 19. f. 1,
and t. 40. f. 6? and 7. *Regenfuss*, i. t. 1. f. 6, and t. 9.
f. 26. *Favanne*, t. 36. f. K.

Variety. With six foliated varices?

Murex diaphanus. *Gmelin*, p. 3529. *Schreibers Conch.*
i. p. 195.

Murex, No. 114. *Schroeter Einl.* i. p. 582.

Argenville, t. 16. f. F.

Junior. Murex striatus. *Gmelin*, p. 3530. *Schreibers Conch.*
i. p. 195.

Murex, No. 18. *Schroeter Einl.* i. p. 548.

Martini, iii. t. 109. f. 1016.

Inhabits the Asiatic Ocean. *Linnæus.* Coasts of Amboyna.
Rumphius. Guinea. *Martini.* Mediterranean about Tarentum. *Regenfuss.* Philippine Islands. *Humphreys.*

This shell varies in size and colour as much as *M. ramosus*,
which it resembles, but may be at once distinguished by the
greater number of its varices. Argenville has described his
t. 16. f. F, which is *M. diaphanus* of Gmelin, with six foliated varices, but there is not any appearance of more than
five in the figure, and according to Martini, Gmelin's *M.
striatus* is the young of this species. It appears by Knorr's
description, that f. 6 and 7 of his tab. 40 were intended for the
same shell, and the latter Gmelin himself has referred to for
M. saxatilis, though with the other he has constituted a separate species under the name of *M. fasciatus.* To increase
the confusion, Gmelin has also given the name of *fasciatus*
to another Murex at p. 3561.

ERINACEUS. 19. Shell angulated, with several
strong somewhat foliated varices, and scaly,
and transversely ribbed; aperture oval, and
the beak closed.

Murex Erinaceus. *Linnæus Syst. Nat.* p. 1216. *Pennant
Zool.* iv. p. 123. t. 76. f. 95. *Born Mus.* p. 294. t. 11.
f. 3 and 4. *Schroeter Einl.* i. p. 485. *Gmelin*, p. 3530.
Donovan, i. t. 35. *Montagu Test.* p. 259. *Maton and
Racket, in Lin. Trans.* viii. p. 142. *Dorset Cat.* p. 46.
t. 14. f. 7.

Murex Cichoreum. *Gmelin*, p. 3530. ?

Buccinum porcatum. *Da Costa, Brit. Conch.* p. 133. t. 8.
f. 7.

Gualter, t. 49. H. *Seba*, iii. t. 49. f. 78 and 79. *Martini*, iii. t. 110. f. 1026 to 1028.

Inhabits the Mediterranean. *Linnæus.* Coasts of Great Britain. *Pennant, &c.*

Shell an inch and a quarter or an inch and a half long, and about half as broad, thick, and of a brownish white or pale yellowish brown colour; it has six or seven slightly foliated varices, which are alternately stronger, and crossed by alternately large and small transverse ribs; these ribs, on close examination, appear to be formed of small imbricated scales, which make the whole surface very rough.

SACELLUM. 20. **Shell with the upper part of the whirls flattish, and muricated transverse ribs below; outer lip crenated.**

Murex Sacellum. *Chemnitz,* x. p. 267. t. 163. f. 1561 and 1562. *Gmelin,* p. 3530.

Inhabits the coasts of the Nicobar Islands. *Chemnitz.*

Chemnitz has described this shell to be twenty-one lines long, and sixteen broad, of a yellowish colour, spotted with brown on the spines of the ribs; the upper part of the whirls is flattish, and bordered by an imbricated margin, below which they are armed with transverse muricated ribs; these ribs may probably be formed of scales, somewhat in the same manner as those of *M. Erinaceus,* but it can hardly be said that the shell is foliated, nor has it any varices, and its claim to be placed in this division appears to be very doubtful.

*** *With thick protuberant rounded Varices.*

RANA. 21. **Shell with two opposite varices, and remote transverse spinous belts, and granulated striæ; whirls rounded, and the spines short; aperture ovate.**

Murex Rana. *Linnæus Syst. Nat.* p. 1216. *Born Mus.* p. 295. *Martini,* iv. p. 106. t. 133. f. 1268 and to 1271. *Schroeter Einl.* i. p. 486, and *Inn. Bau Conch.* p. 42. t. 4. f. 9. *Gmelin,* p. 3531. *Schreibers Conch.* i. p. 198. *Burrows's Elements,* p. 164. t. 18. f. 3.

Rana aspera. *Humphreys, Callone's Cat.* p. 33. No. 613. *Bonanni Rec.* 3. f. 182, and *Kirch.* f. 183. *Lister Conch.* t. 995. f. 58. *Rumphius,* t. 24. f. G. *Petiver Gaz.* t. 100. f. 12. *Gualter,* t. 49. f. L. *Seba,* iii. t. 60. f. 13, and 15 to 18. *Knorr,* ii. t. 13. f. 6 and 7. *Regenfuss,* i. t. 6. f. 64. *Favanne,* t. 32. f. B 3.

Inhabits the Asiatic Ocean. *Linnæus.* Coasts of Amboyna. *Rumphius.* Borneo and Bencoolen. *Petiver.* Tranquebar. *Regenfuss.* China. *Humphreys.*

Shell about two inches and a quarter long, and one inch and a quarter broad, of a whitish, greyish, or pale brown colour mottled with darker shades, and the transverse ribs are often blackish; the body-whirl has three transverse ribs, of which the uppermost is armed throughout with short strong spines, and the others are less regularly so, except at their intersections with the varices.

SPINOSUS. 22. Shell with two opposite varices, and remote transverse spinous belts; whirls flattened, and the spines on the varices very long; aperture ovate.

Murex Rana, Var. *Linnæus Syst. Nat.* p. 1216. *Born Mus.* p. 296. *Schroeter Einl.* i. p. 487. *Gmelin,* p. 3531. *Schreibers Conch.* i. p. 198.

Buccinum bufonium muricatum. *Martini,* iv. p. 110. t. 133. f. 1274 to 1276.

Rana aculeata. *Humphreys, Callone's Cat.* p. 33. No. 615. *Lister Conch.* t. 949. f. 44. *Seba,* iii. t. 60. f. 19. *Knorr,* iii. t. 7. f. 5. *Favanne,* t. 32. f. B 2.

Inhabits the coasts of Tranquebar. *Linnæus.*

This shell differs from *M. Rana,* in having the whirls much flatter and less granulated, and in having the spines on the varices very long; these characters are so constant, and give the shell such an entirely different appearance, that I have followed Martini and Humphreys in arranging it as a separate species. I have not Perry's Conchology to refer to, but if I recollect right, it is his *Biplex spinosus.*

CRASSUS. 23. Shell with two opposite very thick varices, and remote transverse obsolete belts, and granulated striæ; whirls rather flattened; aperture ovate, and the outer lip very thick.

Murex Rana, Var. *Martini,* iv. p. 106. t. 133. f. 1272 and 1273.

Rana crassa. *Humphreys, Callone's Cat.* p. 33. No. 614. *Favanne,* t. 32. f. B 4.

Inhabits the coasts of Madagascar. *Humphreys.*

Shell about two inches long, and an inch and a quarter broad, and is much thicker than either of the preceding species;

the outer lip is remarkably thick and strongly toothed, and the inner lip wrinkled; the transverse ribs are rather obsolete, and more or less nodulous, and the colour is yellowish brown.

GYRINUS. 24. Shell with two opposite varices, and tuberculated transverse belts; aperture orbicular.

Murex Gyrinus. *Linnæus Syst. Nat.* p. 1216. *Born Mus.* p. 296. *Martini,* iv. p. 79. t. 127. f. 1224 to 1227, and t. 128. f. 1229, 1230, 1234, and 1235. *Schroeter Einl.* i. p. 488. *Gmelin,* p. 3531. *Schreibers Conch.* i. p. 199.

Lister Conch. t. 939. f. 34. *Gualter,* t. 49. f. E. *Argenville,* t. 9. f. P. *Seba,* iii. t. 60. f. 21 to 27. *Knorr,* vi. t. 25. f. 5 and 6. *Favanne,* t. 32. f. B 5.

Variety. Smaller, and more attenuated at the base.

Murex Gyrinus. *Montagu Supp.* p. 170.'

Martini, iv. t. 128. f. 1231 and 1232. *Chemnitz,* xi. p. 129. t. 193. f. 1860 and 1861.

Inhabits the Mediterranean. *Linnæus.* And also the Atlantic, American, and Indian Shores. *Gmelin.* The variety has been found on the coasts of Scotland, near Dunbar. *Montagu.*

Shell most commonly about an inch and a quarter long, and three quarters of an inch broad, with narrow strongly tuberculated transverse belts, and very fine intermediate striæ; the colour is white, or greyish, or brownish white, sometimes marked with chestnut bands, or with two of the transverse belts chestnut, and the others milk white.

BUFONIUS. 25. Shell nodulous, with two opposite varices, and transverse granulated striæ; outer lip furrowed, and ending in a channel above.

Murex bufonius. *Gmelin,* p. 3534. *Chemnitz,* xi. p. 120. t. 192. f. 1843 to 1846.

Murex, No. 33. *Schroeter Einl.* i. p. 554.

Murex Gyrinus, Var. β. *Schreibers Conch.* i. p. 199.

Argenville, t. 9. f. R. *Seba,* iii. t. 60. f. 14 and 20. *Martini,* iv. t. 129. f. 1240 and 1241. *Favanne,* t. 32. f. B 1.

Inhabits the South Sea. *Chemnitz.*

Shell about two inches long, and an inch and a quarter broad,

or sometimes almost twice as large ; it somewhat resembles
M. Gyrinus, but has the surface more irregularly uneven,
and the outer lip is separated at its upper extremity by a narrow canal, nearly similar to that which forms the short beak
at the base ; the outer lip is furrowed, and the inner wrinkled ; the colour is white marbled with chestnut.

ARGUS.   26.   Shell with two sub-alternate varices,
and rounded nodules on transverse ribs; outer
lip double, and strongly toothed within.

Murex Argus. *Chemnitz*, x. p. 240. t. 160. f. 1522. *Gme-
lin*, p. 3547.   *Schreibers Conch.* i. p. 205.
Murex Olearium.   *Born Mus.* p. 297.   *Mus. Leskianum.*
p. 252.
Murex, No. 32.   *Schroeter Einl.* i. p. 554.
Rumphius, t. 49. f. B.   *Petiver Amb.* t. 6. f. 6.   *Klein
Ost.* t. 7. f. 128.   *Knorr*, v. t. 3. f. 3.   *Martini*, iv. t.
127. f. 1223.
Variety. With crimson somewhat ocellated bands.
Murex, No. 35.   *Schroeter Einl.* i. p. 555.
Murex, No. 41.   *Schreibers Conch.* i. p. 205.
Argus oculato nodosus, cingulis alternis bifidis nodosis fasci-
atus.   *Martini*, iv. p. 98. t. 131. f. 1255 and 1256.
Inhabits the coasts of Amboyna. *Rumphius.* Tranquebar.
Martini.   Mediterranean.   *Chemnitz.*
Shell generally about an inch and three-quarters long, but Martini has figured an oriental specimen which is more than twice
as large ; it is of a pale brown or whitish colour, finely striated transversely all over, and marked with slightly elevated
darker bands, ocellated with rounded distant nodules ; the
whole aperture is white, and the outer lip double and strongly toothed. With the variety I am entirely unacquainted, and
Martini's figure has much the appearance of a distinct species.

LAMPAS.   27.   Shell granulated, with two opposite
varices, and strong transverse nodulous belts;
aperture toothed on both lips, and the inside
orange.

Murex Lampas. *Linnæus Syst. Nat.* p. 1216. *Born Mus.*
p. 296. *Martini*, iv. p. 84.   *Schroeter Einl.* i. p. 459.
Gmelin, p. 3532.   *Schreibers Conch.* i. p. 199.
Variety, Bubo.   Shell about four inches long.

Lister Conch. t. 1023. f. 88. *Rumphius,* t. 28. f. C. *Petiver Amb.* t. 12. f. 16. *Gualter,* t. 50. f. D. *Martini,* iv. t. 129. f. 1238 and 1239.
Variety, Rubeta. Shell about fourteen inches long.
Bonanni Rec. and *Kirch.* 3. f. 103. *Rumphius,* t. 28. f. D. *Petiver Amb.* t. 12. f. 17. *Argenville,* t. 9. f. D. *Klein Ost.* t. 3. f. 59. *Knorr,* ii. t. 28. f. 1. *Martini,* iv. t. 128. f. 1236 and 1237. *Shaw Nat. Misc.* xxiii. t. 1020.

Inhabits the Mediterranean. *Linnæus.* Coasts of Tambucco in the East Indies. *Rumphius.* Madagascar. *Humphreys.*

This shell is said to vary from four to fourteen inches in length, and is about half as broad, and granulated all over; the spire is conical, and longer than the body-whirl; the aperture ovate, and the beak rather oblique; colour pale reddish brown, or chestnut with darker shades.

OLEARIUM. 28. Shell with two sub-alternate varices, and numerous tubercles; back on the hind part unarmed and striated; aperture toothless.

Murex Olearium. *Linnæus Syst. Nat.* p. 1216. *Schroeter, Inn. Bau Conch.* p. 43. *Gmelin,* p. 3532.
Murex Olearium, Var. *α. Schroeter Einl.* i. p. 492. t. 3. f. 1.? *Schreibers Conch.* i. p. 200.
Bonanni Rec. and *Kirch.* 3. f. 105. *Lister Conch.* t. 937. f. 32. *Gualter,* t. 49. f. G. *Knorr,* iii. t. 9. f. 5. *Martini,* iv. t. 130. f. 1242 and 1243.

Inhabits the coasts of Southern Europe, Africa, and the Mediterranean. *Linnæus.*

Linnæus's description of this species is short, and his references extremely discordant, so that it is almost impossible to ascertain his meaning. In the opinion of Schroeter and Schreibers, *M. Olearium, M. Lotorium* and *M. Pileare* all belong to the same species, and for the former they have referred to the above-mentioned figures of Martini; but Born considered *M. Argus* to be the Linnæan *M. Olearium,* and in his work on the Structure of Shells, Schroeter has quoted Martini's fig. 1243 for this species, though in the Einleitung it is cited for *M. Pileare.* The larger of Martini's two figures is four inches long, and an inch and a half broad, and the transverse ribs are represented much finer and more granulated than in *M. Lotorium.*

PARTHENOPUS. 29. Shell sub-fusiform, with transverse angulated ribs, and a solitary varix; beak produced.

Murex Parthenopus. *Ulysses's Travels,* p. 462. t. 7. f. 4.
Murex costatus. *Born Mus.* p. 297. *Museum Leskianum.* p. 251.
Murex Lotorium, Var. β. *Gmelin,* p. 3534.
Murex, No. 34. *Schroeter Einl.* i. p. 555.
Seba, iii. t. 57. f. 30. *Knorr,* v. t. 21. f. 1. *Martini,* iv. t. 131. f. 1252 and 1253.

Inhabits the Bay of Naples. *Ulysses.*

Born has described this shell to be five inches long, and half as broad, with a strong solitary varix opposite the outer lip; whirls about eight, with thick convex angulated ribs, and their interstices marked with elevated double striæ; aperture ovate-oblong, plaited, and attenuating into an umbilicated, channelled, recurved beak; colour pale yellow. In the Museum Leskianum the colour is said to be sub-ferrugineous, and the throat somewhat like opal. Gmelin has erroneously quoted Chemnitz's figure of *M. lyratus,* jointly with those of this species, for a variety of *M. Lotorium,* and has described two other shells under the name of *M. costatus,* of which one is probably nothing more than a variety of *M. cornutus,* and the other is a fossil. The name was also before given by Pennant to another species.

LYRATUS. 30. Shell with strong distant transverse ribs, and the beak flexuous and ascending; lips thick, and the outer double and sinuated.

Murex lyratus. *Gmelin,* p. 3531. *Shaw Nat. Misc.* xxi. t. 923.
Murex Glomus cereus. *Chemnitz,* x. p. 281. t. 169. f. 1634.
Murex, No. 145. *Schreibers Conch.* i. p. 230.
Buccinum lyratum. *Martyn Univ. Conch.* ii. t. 43.

Inhabits King George's Bay in New Zealand. *Martyn.* Coasts of Australasia. *Shaw.*

Shell about five inches long, and three broad, of a dull brownish yellow colour, and the throat violet; it has about seven strong much elevated ribs on the body-whirl, and both the inner and outer lips are thick.

FEMORALE. 31. Shell with two varices and transverse nodulous ribs; body-whirl triangular, and the aperture oblong.

Murex Femorale. *Linnæus Syst. Nat.* p. 1217. *Martini,* iii. p. 355. t. 111. f. 1039. *Born Mus.* p. 298. *Schroeter Einl.* i. p. 494. *Gmelin,* p. 3533. *Schreibers Conch.* i. p. 201. *Shaw Nat. Misc.* xv. p. 599.

Bonanni Rec. 3. f. 290, and *Kirch.* f. 291. *Lister Conch.* t. 941. f. 37. *Gualter,* t. 50. f. C. *Seba,* iii. t. 63. f. 7 to 10. *Knorr,* iv. t. 16. f. 1.

Variety. With the upper angle of the lip less prominent.

Murex Tripus. *Chemnitz,* xi. p. 128. t. 193. f. 1858 and 1859. *?*

Argenville, t. 10. f. B. *Knorr,* vi. t. 26. f. 2. *Regenfuss,* t. 2. f. 21. *Favanne,* t. 34. f. A 3.

Inhabits the Asiatic Ocean. *Linnæus.* Coasts of Jamaica. *Lister.* Guinea. *Martini.* Ceylon. *Humphreys.*

Shell varying from three to six inches in length, and the breadth is about half the length; it has several rather remote strong nodulous transverse ribs, and the interstices are transversely grooved; the upper angle of the outer lip is prominent, and forms a triangle with the opposite varix and the base; it has about eight whirls, of which the upper ends are flattened; the aperture is attenuated, and ends in rather a long ascending beak at the base; the colour is pale or yellowish brown, and the transverse ribs become white towards the aperture.

CUTACEUS. 32. Shell with one nodulous varix on the body-whirl; whirls rather smooth and flattish above, and transversely ribbed below; outer lip toothed on the margin, and grooved within.

Murex cutaceus. *Linnæus Syst. Nat.* p. 1217. *Martini,* iii. p. 410. t. 118. f. 1085 and 1086. *Born Mus.* p. 299. *Schroeter Einl.* i. p. 495, and *Inn. Bau Conch.* p. 36. t. 5. f. 5. *Gmelin,* p. 3533. *Schreibers Conch.* i. p. 202. *Lister Conch.* t. 942. f. 38. *Seba,* iii. t. 52. f. 10 and 11.

Variety. With the spire much produced.

Murex cutaceus maximus elongatus. *Chemnitz,* x. p. 266. t. 163. f. 1559 and 1560.

Inhabits the coasts of Barbary, the West Indies, and Coromandel. *Martini.* Guinea. *Chemnitz.*

Shell commonly about two inches and a half long, and an inch and a half broad, but Chemnitz has described the Variety to be three inches and a quarter long, and two inches and two lines broad; the whirls are flattened above, and ribbed below, somewhat in the same manner as *Buccinum Scala;*

the base is umbilicated, the throat white, and the outer lip double, toothed on the margin, and grooved on the inside; the colour is brownish or reddish white, sometimes marked with distant undulated darker lines. In the Variety, the spire appears as if it had been pulled out while the whirls were soft, in the same manner as often so remarkably occurs in *Buccinum spiratum.*

LOTORIUM. 33. Shell angulated, with decussated varices, and longitudinally tuberculated; beak flexuous, and the aperture toothed.

Murex Lotorium. *Linnæus Syst. Nat.* p. 1217. *Schroeter Inn. Bau Conch.* p. 43. *Kæmmerer Cab. Rudolst.* p. 132. *Gmelin,* p. 3533.
Murex Olearium, Var. β. *Schroeter Einl.* i. p. 493. t. 3. f. 2. *Schreibers Conch.* i. p. 201.
Lister Conch. t. 934. f. 29. *Rumphius,* t. 26. f. B. *Argenville,* t. 10. f. M. *Martini,* iv. t. 130. f. 1246 to 1249. *Favanne,* t. 34. f. G 4.
Inhabits the coasts of Jamaica. *Lister.* Amboyna. *Rumphius.*
All Conchologists appear to have entertained doubts respecting the Linnæan *M. Olearium, M. Lotorium,* and *M. Pileare,* and the references given for these species in the Systema Naturæ are so confused, that it may be doubted whether it will ever be possible satisfactorily to ascertain them. Martini, in vol. iv. p. 89, has referred to his figures 1242, 1243, 1246, 1247, 1248 and 1249, for one species, and of these the two former have been considered by Schroeter to belong to *M. Olearium,* and the others to *M. Lotorium.* In the Museum Leskianum, fig. 1245 and 1246 are quoted for *M. Lotorium,* and 1247 and 1248 for *M. Pileare,* and Born's opinion is adopted, that *M. Argus* is the Linnæan *M. Olearium.* Schroeter's references in the Einleitung to his own figures are obviously erroneous, and in his work on the Structure of Shells, he has considerably varied the synonyms. Of Martini's figures which I have above quoted, two are about two inches long and half as broad, and the others are almost twice as large; they differ from those of *M. Olearium* in having the transverse ribs much broader.

PILEARE. 34. Shell with decussated varices, and somewhat nodulous wrinkles; aperture toothed, and the beak slightly bent upwards.

Murex Pileare. *Linnæus Syst. Nat.* p. 1217. *Schroeter Inn. Bau Conch.* p. 43.? *Gmelin,* p. 3534. *Chemnitz,* xi. p. 115. t. 191. f. 1837 and 1838.?

Murex Olearium, Var. γ. *Schroeter Einl.* i. p. 493. t. 3. f. 3. *Schreibers Conch.* i. p. 201.

Murex Lotorium, Var. β. *Kæmmerer Cab. Rudolst.* p. 132.

Rumphius, t. 49. f. I. *Gualter,* t. 49. f. A. *Seba,* iii. t. 52. f. 1 and 2. *Martini,* iv. t. 130. f. 1243, and t. 131. f. 1250 and 1251.

Inhabits the Mediterranean. *Linnæus.*

The above figures have been referred to by Schroeter in the Einleitung, and by Gmelin, but the former has altered the references in his ' Innern Bau Conch.'; and there are hardly any two authors who at all agree respecting this species. Martini has only quoted Seba, iii. t. 57. f. 29, and I know not how to distinguish the figures which he has given from *M. Lotorium,* of which I think there can hardly be any doubt that *M. Pileare* is nothing more than a Variety. Chemnitz, under the name of *M. Spengleri,* has figured a shell which differs in having a concatenated row of granules in the middle of the larger transverse ribs.

CANDISATUS. **35.** Shell turreted, with alternate varices, and granulated all over; aperture channelled above, and the outer lip toothed within; beak short and straight.

Murex candisatus. *Chemnitz,* x. p. 254. t. 162. f. 1544 and 1545.

Murex conditus. *Gmelin,* p. 3565. *Schreibers Conch.* i. p. 229.

Inhabits ——

This shell is said by Chemnitz to be three inches long and one inch broad, but the breadth appears by the figures to be two fifths of the length; the colour is whitish, spotted with yellowish brown, especially on the varices, and the whirls are granulated all over in transverse rows; the aperture is somewhat channelled at the upper as well as lower extremity, and has the outer lip double; the throat and inside are striated.

MACULOSUS. **36.** Shell turreted, elongated, with alternate varices, and decussated striæ forming granules at their intersections; beak short and slightly ascending.

Murex maculosus. *Gmelin*, p. 3548. *Schreibers Conch.* i. p. 197, and p. 206.

Murex maculatus. *Chemnitz*, x. p. 260. t. 162. f. 1552 and 1553.

Murex, No. 37. *Schroeter Einl.* i. p. 556.

Buccinum vitiliginosum. *Mus. Geversianum*, p. 304, No. 585.

Buccinum maculosum. *Martini*, iv. p. 100. t. 132. f. 1257 and 1258.

Bonanni Rec. and *Kirch.* 3. f. 48. *Lister Conch.* t. 1022. f. 86. *Rumphius*, t. 49. f. G. *Petiver Amb.* t. 8. f. 15. *Seba*, iii. t. 51. f. 20 and 21. *Favanne*, t. 33. f, X 2.

Inhabits the coasts of Amboyna. *Rumphius.* Isle of France. *Chemnitz.*

Shell two or three inches long, and hardly two fifths as broad, of a pale yellowish brown colour somewhat banded, and the varices spotted with brown; the aperture is white, with its outer lip toothed, and the pillar slightly striated.

SPENGLERI. 37. **Shell angulated, with decussated varices, and a concatenated row of granules in the interstices of the transverse ribs; beak short and straight.**

Murex Spengleri. *Chemnitz*, xi. p. 117. t. 191. f. 1839 and 1840.

Inhabits the coasts of New South Wales. *Chemnitz.*

Chemnitz has described this shell to be three inches and three-quarters long, and two inches broad, of a yellowish colour, and the throat white; it has narrow decussated grooves like those of *M. Pileare*, and there is a row of concatenated granules in the interstices of the transverse ribs.

PYRUM. 38. **Shell ovate, with two varices, and transversely ribbed, and nodulous; beak long and flexuous.**

Murex Pyrum. *Linnæus Syst. Nat.* p. 1218. *Martini*, iii. p. 359. t. 67. f. 745 and 746. ? and t. 112. f. 1040 to 1044. *Born Mus.* p. 299. *Schroeter Einl.* i. p. 497. *Gmelin*, p. 3534. *Schreibers Conch.* i. p. 202.

Gualter, t. 37. f. F. *Argenville*, t. 10. f. O. *Knorr*, ii. t. 7. f. 2 and 3. *Regenfuss*, i. t. 6 f. 60. *Favanne*, t. 34. f. A 2.

Variety B. Shell smaller, and the beak subulate.

Regenfuss, i. t. 5. f. 50.   *Martini*, iii. t. 112. f. 1048 and
  1049.
Variety C.   With the body-whirl gibbous, the lips thick, and
  the beak recurved.
Murex, No. 26.   *Schroeter Einl.* i. p. 552.
Murex, No. 39.   *Schreibers Conch.* i. p. 205.
Martini, iii. t. 112. f. 1050 and 1051.
Inhabits the coasts of Coromandel. *Martini.*
Shell commonly about four inches long, and an inch and three-
  quarters broad, of a pale reddish or brownish yellow, and
  the throat is of the same colour ; the aperture is ovate, and
  ends in an attenuated flexuous beak, which is commonly
  about half as long, and in the Variety B, is nearly of the
  same length as the body-whirl ; the Variety C. is probably
  a distinct species.

CLAVATOR.   39. Shell ovate, with two varices, lon-
    gitudinally plaited and striated, and trans-
    versely ribbed ; beak long and flexuous.

Murex Clavator.   *Chemnitz*, xi. p. 110. t. 190. f. 1825 and
  1826.
Inhabits the coasts of Ceylon. *Chemnitz.*
Shell about two inches and a quarter long, and rather more
  than one inch broad, of a whitish colour with the plaits yellowish ; the interstices of the longitudinal plaits are striated
  in the same direction, and crossed transversely with stronger
  nodulous ribs ; the aperture is oval, ending in a long flexuous
  beak, and the outer lip double and toothed.

CAUDATUS.   40. Shell with two varices on the body-
    whirl, and transverse granulated ribs ; whirls
    excavated at the sutures, and the beak subu-
    late.

Murex caudatus.   *Gmelin*, p. 3535.
Murex, No. 25.   *Schroeter Einl.* i. p. 551.
Murex, No. 38.   *Schreibers Conch.* i. p. 205.
Lister Conch. t. 893. f. 13.   *Martini*, iii. t. 112. f. 1045
  to 1047.
Inhabits the coasts of Coromandel. *Martini.*
Shell about two inches long and half as broad, and the beak
  occupies rather more than half the length of the body-whirl,
  which on the shoulder is slightly nodulous transversely ; the
  ribs are granulated, and the interstices transversely striated ;
  the whirls are rounded, and deeply excavated at the sutures ;
  colour white, shaded with pale reddish or yellowish brown.

Rubecula. 41. Shell obtuse, with two alternate varices, and granulated transverse ribs with their interstices decussated; aperture toothed, and the throat grooved.

Murex Rubecula. *Linnæus Syst Nat.* p. 1218. *Martini,* iv. p. 104. t. 132. f. 1259 to 1267. *Born Mus.* p. 300. *Schroeter Einl.* i. p. 498. *Gmelin,* p. 3535. *Schreibers Conch.* i. p. 202.

Gualter, t. 49. f. F and I. *Argenville,* t. 9. f. K. *Seba,* iii. t. 49. f. 1 to 6. *Knorr,* i. t. 13. f. 3 and 4, and iii. t. 5. f. 2 and 3. *Favanne,* t. 34. f. G 1, and G 3.

Variety. Shell brown with white bands.

Murex varicosus. *Chemnitz,* x. p. 256. t. 162. f. 1546 and 1547.? *Mus. Leskianum,* p. 265.

Murex, No. 37. *Schreibers Conch.* i. p. 204.?

Inhabits the Red Sea. *Forskael.* Coasts of Africa. *Martini.* Bay of Naples. *Ulysses.*

Shell an inch and a quarter, or an inch and a half long, and about half as broad; it is somewhat plaited longitudinally, and transversely ribbed, with the ribs prettily granulated, and their interstices striated in both directions; the colour is reddish or yellowish orange, with the interstices somewhat violet, or ornamented with white transverse bands, and the varices spotted with white; from this richness of its livery it has received its common name of the ' *Footman.*' I have now before me a reddish brown Variety with white bands, but I rather doubt whether it is the same as the *M. varicosus* of Chemnitz, which has been blended with the synonyms of this species by Gmelin.

Scrobilator. 42. Shell with two nearly opposite furrowed varices, and rather smooth; aperture toothed, and ending at both ends in a channel.

Murex Scrobilator. *Linnæus Syst. Nat.* p. 1218. *Chemnitz,* x. p. 262. t. 163. f. 1556 and 1557.

Murex Scrobiculator. *Schroeter Einl.* i. p. 499. *Gmelin,* p. 3535.

Lister Conch. t. 943. f. 39. *Gualter,* t. 49. f. B. *Adanson Senegal,* t. 8. f. 13. *Favanne,* t. 32. f. E.

Inhabits the Mediterranean. *Linnæus.* Coasts of Senegal. *Adanson.* Naples. *Chemnitz.*

Shell about three inches long, and rather more than half as

broad; it has two nearly opposite varices, which are deeply furrowed or indented transversely, and the remainder of the surface is nearly level and smooth; the colour is yellowish brown, with a few darker transverse bands, and the throat is tinged with violet.

RETICULARIS. 43. Shell with two nearly opposite varices and decussated ribs, which are tuberculated at their intersections; aperture nearly toothless, and the beak slightly ascending.

Murex reticularis. *Linnæus Syst. Nat.* p. 1218. *Martini,* iv. p. 81. t. 128. f. 1228. *Born Mus.* p. 301. t. 11. f. 5. *Schroeter Einl.* i. t. 501. *Gmelin,* p. 3535. *Schreibers Conch.* i. p. 203.
Bonanni Rec. and *Kirch.* 3. f. 193. *Lister Conch.* t. 935. f. 30. *Petiver Gaz.* t. 153. f. 6. *Gualter,* t. 49. f. M, and t. 50. f. A.
Inhabits the coasts of Carolina. *Linnæus.* Bombay. *Petiver.* Mediterranean. *D'Avila.* Barbadoes. *Martini.* Naples and Sicily. *Humphreys.*
Shell sometimes four or even six inches long, but is generally much shorter, and about half as broad; the transverse ribs are stronger than the longitudinal ones, and small tubercles in regular rows are formed at their intersections; the colour is pale brownish or purplish white, and the tubercles of a darker brownish purple.

ANUS. 44. Shell nodulous, with very unequally gibbous whirls, and decussated ribs, which are tuberculated at their intersections; aperture surrounded by a thin dilated membrane, and the beak short and erect.

Murex Anus. *Linnæus Syst. Nat.* p. 1218. *Martini,* ii. p. 84. t. 41. f. 403 and 404. *Born Mus.* p. 301. *Schroeter Einl.* i. p. 501, and *Inn. Bau Conch.* p. 10. t. 4. f. 7. *Gmelin,* p. 3536. *Schreibers Conch.* i. p. 203.
Cassida Anus. *Mus. Geversianum,* p. 386, No. 1257.
Bonanni Rec. 3. f. 279 and 280, and *Kirch.* f. 278 and 279. *Lister Conch.* t. 833. f. 57. *Rumphius,* t. 24. f. F. *Petiver Gaz.* t. 99. f. 10, and *Amb.* t. 6. f. 4. *Argenville,* t. 9. f. H. *Gualter,* t. 37. f. B and E. *Seba,*

iii. t. 60. f. 4, 6, and 7.   *Knorr,* iii. t. 3. f. 5.   *Da Costa Elements,* t. 4. f. 2.   *Favanne,* t. 31. f. H 1.
Inhabits the Asiatic Ocean. *Linnæus.*   Coasts of Amboyna. *Rumphius.*   Pulo Condore and Siam. *Petiver.*   Mediterranean. *Martini.*

This strangely distorted shell is about an inch and three-quarters long, and more than an inch broad, and is variously carbuncled all over; the whirls are partially and alternately gibbous, and have irregular decussated ribs, which form nodules or granules of different sizes at their intersections; the aperture is surrounded by a widely expanded thin membrane over both its lips, and ends in a short recurved beak; the colour is white with indistinct bands of pale reddish yellow.

MULUS.   45. Shell with the whirls unequally gibbous, and decussated ribs slightly tuberculated at their intersections; aperture surrounded by a thin dilated membrane, and the beak produced and ascending.

Murex Anus, Var. β.   *Gmelin,* p. 3536. *Schreibers Conch.* i. p. 203.
Murex, No. 3.   *Schroeter Einl.* i. p. 543.
Cassida penita.   *Mus. Geversianum,* p. 388, No. 1258.
Gualter, t. 31. f. D.   *Seba,* iii. t. 60. f. 5.   *Martini,* ii. p. 85. t. 41. f. 405 and 406.   *Favanne,* t. 31. f. H 2.
Inhabits the coasts of Hitoe. *Rumphius.*
Shell about two inches, or two inches and a quarter long, and half as broad; it differs from *M. Anus* in being much less nodulous, in having the tubercles at the intersections of the ribs more regular, and the beak more produced and only slightly ascending; the colour is brownish white, and the membrane which expands round the aperture has a purplish tinge.

**** *Somewhat spinous, and without a Beak.*

RICINUS.   46. Shell sub-ovate, with subulate spines, and the aperture toothed on both sides.

Murex Ricinus.   *Linnæus Syst. Nat.* p. 1219.   *Born Mus.* p. 302.   *Museum Leskianum,* p. 255.
Murex neritoideus.   *Martini,* iii. p. 282. t. 101. f. 972 and

973, and t. 102. f. 976 and 977. *Schroeter Einl.* p. 502.
Gmelin, p. 3537.
Bonanni Rec. and *Kirch.* 3. f. 173. *Lister Conch.* t. 804.
f. 12 and 13. *Rumphius,* t. 24. f. E.? *Gualter,* t. 28.
f. N.? *Klein Ost.* t. 1. f. 30. *Seba,* iii. t. 60. f. 41 and
48. *Knorr,* i. t. 25. f. 5 and 6. *Favanne,* t. 24. f. A 1.
Inhabits the Asiatic Ocean *Linnæus.* Coasts of Coromandel.
Martini. China. *Humphreys.*
Shell varying in length from eleven to fifteen lines, and in
breadth from eight to thirteen; the throat in the larger shells
is most commonly yellowish, and in the smaller is more or
less tinged with violet; the outer surface is white or brownish white with transverse rows of remote pointed black tubercles, sometimes forming subulate spines, particularly on
the margin of the outer lip, though they are rarely so long as
is represented in the figures of Rumphius and Gualter, to
which Linnæus has referred. Martyn in his Universal
Conchology has figured the shell, but I cannot ascertain the
number of his plate. In Callone's Catalogue, p. 31, the
yellow-throated Variety is said to be the *Buccinum Ricinus,*
and the violet-coloured one the *Buccinum amethystinum*
of Solander, but among Solander's MSS. in Sir Joseph
Banks's library there is not any description under either of
these names.

NODUS. 47. Shell obovate, attenuated at both ends,
and armed with conical spines; outer lip
toothed, and the pillar smooth and flesh-
coloured.

Murex Nodus. *Linnæus Syst. Nat.* p. 1219. *Born Mus.*
p. 303. *Gmelin,* p. 3537. *Museum Leskianum,* p.
255.
Murex Hippocastanum, Var. β. *Gmelin,* p. 3539.
Murex, No. 10. *Schroeter Einl.* i. p. 545.
Buccinum strumosum. *Solander's MSS.*
Lister Conch. t. 991. f. 53. *Petiver Gaz.* t. 19. f. 10.
Seba, iii. t. 60. f. 11. *Martini,* iii. t. 100. f. 956 to 958.
Inhabits the coasts of Jamaica. *Lister.*
Shell about an inch and and a half or two inches long, and almost equally broad; the body-whirl is armed with one or
two transverse rows of large conical spines, and is much
larger than the spire; the aperture is ovate, and the inside
of the outer lip toothed; colour greyish white variegated
with black.

NERITOIDEUS. 48. Shell ovate, with about four rows of pointed nodules ; outer lip angulated, and the pillar obliquely flattened, with two black spots.

Murex neritoideus. *Linnæus Syst. Nat.* p. 1219. *Born Mus.* p. 303.
Murex Fucus. *Gmelin,* p. 3538.
Murex, No. 11. *Schroeter Einl.* i. p. 546.
Murex Moega. *Martini,* iii. p. 270. t. 100. f. 959 and 960.
Buccinum neritoideum. *Solander's MSS.*
Bonanni Rec. and *Kirch.* 3. f. 174. *Lister Conch.* t. 990. f. 50. *Gualter,* t. 66. f. B B.
Variety. Less nodulous and the outer lip plaited within.
Buccinum insignitum. *Solander's MSS. Callone's Cat.* p. 30, No. 537.
Martini, iii. t. 100. f. 961 and 962.
Inhabits the coasts of Guinea. *Martini.* South Sea. *Solander.*
Shell an inch or an inch and a half long, and about equally broad, with the spire small and not produced ; the colour is white or reddish. Some of the Linnæan references answer to Martini, f. 972 and 973, and others to the present shell with which the description best agrees, and the name was most probably derived from the circumstance of its having been placed by Gualter among the Nerites. *Murex neritoideus* of Chemnitz, x. p. 280. t. 165. f. 1577 and 1578, and of Gmelin, p. 3559, is quite different, and has more the appearance of a Buccinum.

HYSTRIX. 49. Shell sub-ovate, with acute crowded spines ; aperture toothless and repand.

Murex Hystrix. *Linnæus Syst. Nat.* p. 1219. *Martini,* iii. p. 285. t. 101. f. 974 and 975. *Schroeter Einl.* i. p. 505. *Gmelin,* p. 3538. *Schreibers Conch.* i. p. 216.
Argenville, t. 14. f. A. *Gualter,* t. 28. f. R, and t. 44. f. S. *Seba,* iii. t. 52. f. 30 and 31, and t. 60. f. 33, 43, and 47. *Regenfuss,* i. t. 3. f. 32.
Inhabits the East Indian Seas. *Martini.* Coasts of the Friendly Islands. *Humphreys.*
Shell about an inch and a quarter long, and an inch broad, with transverse granulated striæ, and crowded strong spines ; the colour is greyish or yellowish white, and the throat

tinged with violet; the aperture is not toothed, by which it may be at once distinguished from *M. Ricinus.*

MANCINELLA. 50. Shell ovate, and transversely striated with four rows of obsolete spines; outer lip toothless, and the pillar umbilicated, flattish, and striated.

Murex Mancinella. *Linnæus Syst. Nat.* p. 1219. *Martini,* iii. p. 275. t. 101. f. 967 and 968. *Born Mus.* p. 304. t. 9. f. 19 and 20. *Schroeter Einl.* i. p. 506. *Gmelin,* p. 3538. *Schreibers Conch.* i. p. 216.

Buccinum Mancinella. *Solander's MSS. Callone's Cat.* p. 30, No. 540.

Lister Conch. t. 957. f. 9. *Rumphius,* t. 24. f. No. 5. *Seba,* iii. t. 60. f. 45. *Knorr,* iii. t. 29. f. 6.

Inhabits the coasts of Amboyna. *Rumphius.* Madagascar. *Humphreys.*

Shell about an inch and three-quarters long, and an inch and a quarter broad; the colour is brownish or yellowish white, with the tubercles black or red, and the pillar-lip yellow with tawny striæ. It is too nearly allied to *Buccinum hæmastomum* to be placed in a different Genus, and most of the shells in this division agree better with the Linnæan definition of Buccinum than with that of Murex. I rather doubt whether the *Murex Pyrum* of Chemnitz, xi. p. 122. t. 192. f. 1847 and 1848, is more than a Variety of this species.

HIPPOCASTANUM. 51. Shell ovate, transversely striated, and armed with four rows of conical spines; pillar-lip sinuated, and the throat striated.

Murex Hippocastanum. *Linnæus Syst. Nat.* p. 1219. *Martini,* iii. p. 261. t. 99. f. 945 and 946. *Kæmmerer Cab. Rudolst.* p. 140. *Schroeter Einl.* i. p. 507. *Gmelin,* p. 3539. *Schreibers Conch.* i. p. 216.

Rumphius, t. 24. f. C. *Petiver Amb.* t. 4. f. 12. *Argenville,* t. 14. f. L. *Klein Ost.* t. 7. f. 112. *Knorr,* ii. t. 2. f. 3. *Seba,* iii. t. 52. f. 22, 23, 27, and 28. *Regenfuss,* i. t. 2. f. 18. *Favanne,* t. 24. f. C 1.

Variety. With the spines on the margin of the outer lip curled.

Buccinum crispatum. *Solander's MSS.*

Inhabits the coasts of the Island of Banda. *Linnæus.* Batavia. *Martini.*

Shell about an inch and a half long, and an inch and a quarter broad, and the spire somewhat produced and turbinated; the colour is white, variegated with black, and the throat smooth, and whitish with brownish yellow stripes. The shell, which Born has described under the name of *M. Hippocastanum,* probably belongs to *M. calcaratus.*

NODATUS. 52. Shell ovate, transversely striated, with four rows of compressed spines; outer lip toothed within, and the pillar slightly plaited.

Murex nodatus. *Gmelin,* p. 3586.
Buccinum nodatum. *Martyn Univ. Conch.* ii. t. 51.
Inhabits the coasts of New Holland. *Martyn.*
Shell about an inch and a half long, and rather more than half as broad, of a greyish white colour; it has four rows of compressed nodulous spines on the body-whirl, and one on the spire; the outer lip is rather strongly toothed at a small distance from the inner margin, and the pillar has two somewhat obsolete plaits.

LACERUS. 53. Shell ovate, transversely striated, and armed with two rows of compressed spines; spire muricated, and the whirls keeled.

Murex lacerus. *Born Mus.* p. 308.
Murex Africanus. *Martini,* iii. p. 266. t. 100. f. 951 to 953.
Murex Hippocastanum, Var. *Schroeter Einl.* iii. p. 508.
Buccinum tardum. *Solander's MSS.*
Lister Conch. t. 958. f. 11. *Petiver Gaz.* t. 101. f. 18.
Klein Ost. t. 3. f. 58. *Seba,* iii. t. 60. f. 30.
Inhabits the coasts of Guinea. *Martini.*
Shell an inch and a half, or two inches long, and about two-thirds as broad, and has two transverse rows of rather remote compressed pointed nodules on the body-whirl; the spire consists of five carinated whirls, and the keel is somewhat spinous; it is sometimes of a pale rose-colour, and sometimes brownish. Gmelin, with a mark of doubt, has referred to Martini's figures of this species for *M. Hippocastanum.*

PLICATUS. 54. Shell ovate, ventricose, with the whirls longitudinally plaited and striated transversely; throat violet.

Murex plicatus. *Martini*, iii. p. 268. t. 100. f. 954 and 955.

Murex Hippocastanum, Var. γ. *Gmelin*, p. 3539.

Murex, No. 9. *Schroeter Einl.* i. p. 545.

Murex, No. 110. *Schreibers Conch.* i. p. 220.

Buccinum crinitum. *Solander's MSS.*

Inhabits the East Indian Seas. *Martini.*

Shell about an inch long, and three-quarters of an inch broad, of a whitish colour, and the throat violet; the body-whirl is more or less keeled towards the middle, and the transverse striæ are somewhat membranaceous; the base is umbilicated.

MORBOSUS. 55. Shell ovate, transversely striated, and the whirls nodulous; aperture roundish, and the throat yellowish.

Murex Hippocastanum, Var. η. *Gmelin*, p. 3539.

Murex, No. 14. *Schroeter Einl.* i. p. 547.

Buccinum nodulosum. *Gmelin*, p. 3496.

Buccinum morbosum. *Solander's MSS.*

Lister Conch. t. 990 b. f. 51, and t. 991. f. 52. *Martini*, iv. t. 123. f. 1140.

Inhabits the coasts of the West India Islands. *Martini.*

Shell about an inch and a half long, and rather more than an inch broad, thick, and of a greyish white colour variegated with brown; the body-whirl is more or less transversely nodulous on the shoulder, and the lower whirls of the spire are tumid and somewhat lamellated.

SENTICOSUS. 56. Shell turreted, longitudinally ribbed, and cancellated with transverse acute striæ; pillar obliquely plaited.

Murex senticosus. *Linnæus Syst. Nat.* p. 1220. *Martini*, iv. p. 309. t. 155. f. 1466 and 1467. *Born Mus.* p. 306. *Schroeter Einl.* i. p. 508. *Gmelin*, p. 3539. *Schreibers Conch.* i. p. 217. *Chemnitz*, xi. p. 132. t. 193. f. 1864 to 1866.

Murex cancellatus. *Museum Leskianum.* p. 266. f. 6.

Buccinum senticosum. *Museum Geversianum.* p. 296. *Bruguiere Enc. Meth.* p. 272.

Bonanni Rec. and *Kirch.* 3. f. 35. *Rumphius*, t. 29. f. N. *Petiver Amb.* t. 9. f. 17. *Gualter*, t. 51. f. G and I. *Argenville*, t. 9. f. O. *Seba*, iii. t. 49. f. 45 to 48. *Knorr*, iv. t. 23. f. 4 and 5. *Favanne*, t. 31. f. L.

Inhabits the Adriatic. *Bonanni.* Coasts of Amboyna. *Rum-*

phius.   China, Nicobar Islands, and the South Sea. *Chem-nitz.*
Shell about an inch and a half long, and nearly half as broad,
of a brownish or yellowish white colour, sometimes marked
with darker transverse bands ; the longitudinal ribs are made
prickly by the crossings of the transverse elevated striæ ; the
pillar has two oblique plaits, on which account this species
has been removed to the Volutæ by Dr. Solander.

MELONGENA.   57.   Shell with three or four trans-
    verse rows of strong spines on the body-whirl,
    and the spire longitudinally plaited ; aperture
    oblong and toothless.

Murex Melongena.   *Linnæus Syst. Nat.* p. 1220. *Marti-*
    ni, ii. p. 74. t. 39. f. 389 to 392, and t. 40. f. 394. *Born*
    Mus. p. 306.   *Schroeter Einl.* i. p. 510.   *Chemnitz,* x.
    p. 271. t. 164. f. 1568.   *Schreibers Conch.* i. p. 217.
Buccinum Melongena.   *Solander's MSS.*
Bonanni Rec. and *Kirch.* 3. f. 186 and 296. *Lister Conch.*
    t. 904. f. 24.   *Rumphius,* t. 24. f. No. 2.   *Petiver Amb.*
    t. 21. f. 9.   *Gualter,* t. 26. f. F.   *Argenville,* t. 15. f. H.
    Knorr, i. t. 17. f. 5.   *Seba,* iii. t. 72. f. 3 to 9.   *Regen-*
    fuss, i. t. 5. f. 49.   *Favanne,* t. 24. f. E 2.
Variety.   With the spines obsolete.
    Rumphius, t. 24. f. No. 3.   *Petiver Amb.* t. 21. f. 8.
    Knorr, ii. t. 10. f. 1.   *Martini,* ii. t. 39. f. 393, and t.
    40. f. 395 to 397.
Inhabits America.   *Linnæus.*   Eastern Ocean.   *Bonanni.*
    Coasts of Jamaica. *Lister.*   Amboyna. *Rumphius.*
Shell from three to five inches long, and half or sometimes two
    thirds as broad, and rather ponderous ; the body-whirl is very
    large in proportion to the spire, and is generally armed with
    three or four transverse rows of strong conical spines ; the co-
    lour is glaucous or reddish brown, with white transverse
    bands.

CALCARATUS.   58. Shell somewhat turbinated, with
    transverse striæ, and spinous belts ; aperture
    sub-ovate.

Murex Hippocastanum.   *Born Mus.* p. 309.
Murex Melongena, Var. β. to ζ.   *Gmelin,* p. 3540.
Buccinum calcaratum.   *Solander's MSS.*
Variety A. With four transverse spinous belts.

Murex, No. 1. *Schroeter Einl.* i. p. 543.
Rumphius, t. 23. f. D. *Petiver Amb.* t. 8. f. 11. *Gualter*, t. 31. f. F. *Seba*, iii. t. 49. f. 80 and 82. *Martini*, ii, t. 40. f. 398 and 399.
Variety B. With three transverse spinous belts.
Murex, No. 165. *Schroeter Einl.* i. p. 596.
Rumphius, t. 24. f. No. 4. *Knorr*, vi. t. 24. f. 2, and t. 35. f. 3. *Martini*, ii. t. 40. f. 400 and 401.
Variety C. With two transverse spinous belts.
Murex, No. 164. *Schroeter Einl.* i. p. 596.
Knorr, v. t. 4. f. 2.
Variety D. With one tuberculated belt.
Murex, No. 2. *Schroeter Einl.* i. p. 543.
Murex, No. 107. *Schreibers Conch.* i. p. 220.
Seba, iii. t. 60. f. 9. *Martini*, ii. t. 40. f. 402.
Inhabits the coasts of Amboyna. *Rumphius.* China. *Humphreys.*
Shell an inch and a half, or two inches long, and the breadth is about three-fourths of the length ; the form is rather pear-shaped, and it differs from *M. Melongena* in its more produced spire, narrower base, and more ovate aperture ; the colour is brownish white, or brown with white transverse stripes or bands.

CONSUL. 59. Shell thick, ventricose, transversely striated and nodulous ; aperture wide, ovate, with the outer lip sinuated and plaited within.

Murex Consul. *Chemnitz*, x. p. 236. t. 160. f. 1516 and 1517. *Gmelin*, p. 3540. *Schreibers Conch.* i. p. 221.
Inhabits the East Indian Seas. *Chemnitz.*
Shell about three inches long, and almost equally broad, with a transverse row of large nodules on the shoulder of the body-whirl, and the spire nodulous ; the colour is whitish, and the throat yellow.

STRAMINEUS. 60. Shell ovate, grooved and striated transversely, with the upper ends of the whirls flattish and nodulous ; aperture wide, and the lips thickened.

Murex stramineus. *Gmelin*, p. 3542.
Murex Pes Struthio-cameli. *Chemnitz*, x. p. 238. t. 160. f. 1520 and 1521.

Murex, No. 194. *Schroeter Einl.* i. p. 608.
Buccinum papillosum. *Martyn Univ. Conch.* ii. t. 54.
Un Buccin extrémement rare. *Favanne Cat. Rais.* p. 187.
t. 4. No. 889.
Spengler Naturf. xvii. p. 24. t. 2. f. A and B. *Favanne,*
t. 79. f. S.
Inhabits the coasts of New Zealand. *Martyn.*
Shell sometimes three inches and a quarter long, and two inches
broad, but the specimen which Favanne described measured
only two inches and three quarters by twenty-one lines, and
the *Buccinum Vermis* of Martyn is hardly half so large; it
is of a dull straw colour with thick white lips, and the throat
brown and shining.

AUSTRALIS. 61. **Shell ovate, longitudinally striated with four plaits on the body, and three on
the next whirl; whirls channelled, and the
outer lip undulated.**

Murex australis. *Gmelin,* p. 3542.
Murex, No. 195. *Schroeter Einl.* i. p. 609.
Buccinum Vermis. *Martyn Univ. Conch.* ii. t. 53. ? ?
Spengler Naturf. xvii. t. 2. f. C and D.
Inhabits the South Sea. *Spengler.*
Shell an inch and a half long, and one inch broad, of a straw-
colour, with the pillar yellow and the lip white; it is said to
be nearly allied to *M. stramineus,* and may possibly be the
Buccinum Vermis of Martyn, but I have not Spengler's
work to compare the figures.

***** *Unarmed, with a long straight subulate Beak.*

CARIOSUS. 62. **Shell without a beak, somewhat
plaited, ovate, acuminated, and the summit
carious.**

Murex cariosus. *Linnæus Syst. Nat.* p. 1220. *Gmelin,*
p. 3541.
Inhabits the Aqueduct at Seville. Alstrœmer. *Linnæus.*
Linnæus has described this species to be " as large as a bean,
ovate, oblong, acuminated, cinereous, somewhat diaphanous,
and longitudinally but obsoletely grooved; apex carious;
base emarginate." No other author has noticed it.

CLAVATULUS. **63.** Shell somewhat turreted and coronated, with decussated striæ and prickly nodules ; whirls excavated at the sutures ; outer lip with a notch at the summit, and the beak short and obtuse.

Murex Turris coronata. *Chemnitz,* xi. p. 114. t. 190. f. 1831 and 1832.
Murex Babylonius, Var. β. *Gmelin,* p. 3541.
Murex, No. 215. *Schroeter Einl.* i. p. 620.
Clavatula coronata. *Lamarck Syst. des Anim.* p. 84.
Turris Babylonica coronata. *Martini,* iv. p. 176. Vign. 39. at p. 143. f. *C.*
Variety ?. With curled laminæ at the sutures.
Murex Taxus. *Chemnitz,* x. p. 259. t. 162. f. 1550 and 1551.
Murex Babylonius, Var. ζ. *Gmelin,* p. 3541.
Murex, No. 174. *Schreibers Conch.* i. p. 236.
Inhabits the coasts of Guinea. *Chemnitz.*
Shell about an inch and a half long, and nearly half as broad, of a pale chestnut-colour ; the beak is short and obtuse, and the pillar somewhat umbilicated ; the outer lip is sinuated by a notch at its upper extremity. *M. Taxus,* as it is figured by Chemnitz, has rather the appearance of being an accidentally distorted shell, and though it answers to many of the leading characters of this species, may possibly belong to some other.

GIBBOSUS. **64.** Shell turreted, with longitudinal plaits and transverse grooves, and an elevated belt at the sutures ; outer lip with a notch at the summit, and the beak short and obtuse.

Murex gibbosus. *Born Mus.* p. 321. t. 11. f. 12 and 13. *Gmelin,* p. 3564. *Schreibers Conch.* i. p. 211. *Chemnitz,* xi. p. 112. t. 190. f. 1829, 1830, 1833, and 1834.
Murex alatus. *Gmelin,* p. 3562. *Schreibers Conch.* i. p. 233.
Murex, No. 170, and No. 230. *Schroeter Einl.* i. p. 598, and p. 626.
Martini, iv. p. 344. t. 159. f. 1503 and 1504.
Inhabits the Red Sea. *Chemnitz.*
Shell two or three inches long, and about one-third as broad, of a white or whitish brown colour, sometimes marked with

pale red spots; the outer lip has a notch at its upper extremity, and a sinuosity near the base, which gives it some slight resemblance to the expanded wing of a Strombus; the pillar is somewhat umbilicated, and the beak short.

VIRGINEUS. 65. Shell turreted, with longitudinal grooves and transverse granulated striæ; whirls with a belt at the sutures; outer lip sinuated, and the beak short.

Murex Turris virginea. *Chemnitz,* xi. p. 99, and p. 115. t. 190. f. 1835 and 1836.

Inhabits the coasts of Guinea. *Chemnitz.*

Chemnitz has not described the size, but by his figures this shell appears to be near an inch long, and about half as broad, with alternately bluish grey and white transverse bands, and a few rows of darker spots.

BABYLONIUS. 66. Shell turreted, with acute spotted belts, and a notch in the outer lip; beak long and straight.

Murex Babylonius. *Linnæus Syst. Nat.* p. 1220. *Martini,* iv. p. 167. t. 143. f. 1331 and 1332. *Born Mus.* p. 309. *Schroeter Einl.* i. p. 512, and *Inn. Bau Conch.* p. 44. t. 2. f. 8. *Gmelin,* p. 3541. *Schreibers Conch.* i. p. 222. *Brookes's Introd.* p. 120. t. 7. f. 91.

Pleurotoma Babylonica. *Lamarck Syst. des Anim.* p. 84. *Lister Conch.* t. 917. f. 11. *Rumphius,* t. 29. f. L. *Petiver Amb.* t. 4. f. 7. *Gualter,* t. 52. f. N. *Argenville,* t. 9. f. M. *Knorr,* iv. t. 13. f. 2. *Seba,* iii. t. 79. several figures. *Regenfuss,* i. t. 1. f. 9. *Favanne,* t. 33. f. D.

Inhabits Asia. *Linnæus.* Coasts of Amboyna. *Rumphius.* China. *Humphreys.*

Shell sometimes three inches long, and about ten lines broad, but is generally smaller; it has about twelve whirls, which are ribbed and striated transversely, and keeled near the middle; the colour is greyish white, elegantly marked with purplish brown spots and dots in transverse rows.

JAVANUS. 67. Shell turreted, transversely striated, with a tuberculated keel on each whirl, and a notch at the summit of the outer lip.

Murex Javanus. *Linnæus Syst. Nat.* p. 1121. *Born Mus.*

p. 309. *Museum Leskianum*. p. 257. *Burrow's Elements*, p. 165. t. 18. f. 5.

Murex Turris. *Gmelin*, p. 3543.

Murex Babylonius, Var. δ. *Gmelin*, p. 3541. *Schreibers Conch*. i. p. 222.

Murex, No. 213. *Schroeter Einl.* i. p. 619.

Knorr, vi. t. 27. f. 3. *Martini*, iv. t. 143. f. 1334 and 1335. *Favanne*, t. 33. f. C 5.

Variety. With transverse granulated striæ.

Murex Turris australis. *Chemnitz*, xi. p. 111. t. 190. f. 1827 and 1828.

Inhabits the coasts of Java. *Linnæus*. Tranquebar. *Martini*. Amboyna and China. *Humphreys*. South Sea. *Chemnitz*.

Shell two or three inches long, and rather more than one-third as broad, of a greyish or yellowish white colour; the spire consists of nine whirls, with a transverse nodulous keel in the middle of each. Chemnitz says that Muller considered this shell to be the *M. lignarius* of Linnæus. *M. Turris* has been constituted from Bonanni, fig. 79, which is quoted by Born for *M. Javanus*, and certainly belongs either to this species or to *M. Babylonius*.

TORNATUS. 68. Shell turreted, smooth, with the whirls concave above and convex below; outer lip with a notch at the summit, and the beak rather long.

Murex Javanus. *Martini*, iv. p. 172. t. 143. f. 1336 to 1338. *Schroeter Einl.* i. p. 513. *Gmelin*, p. 3541. *Schreibers Conch*. i. p. 122.

Turris albida. *Humphreys, Call. Cat.* p. 34. No. 632.

Lister Conch. t. 915. f. 8.

Variety. With remote transverse ribs.

Murex Babylonius, Var. γ. *Gmelin*, p. 3541.

Murex, No. 214. *Schroeter Einl.* i. p. 620.

Turris candida. *Humphreys, Callone's Cat.* p. 34. No. 631.

Buccinum sinuatum. *Martyn Univ. Conch.* t. 94. right hand fig.

Argenville Zoom. t. 4. f. B. *Martini*, iv. Vign. 39 at p. 143. f. B.

Inhabits the coasts of Tranquebar; and the variety, Martinique. *Humphreys*. Straights of Magellan. *Martyn*.

Shell about three inches long, and rather more than an inch broad, white without spots, and sometimes variegated with pale yellowish stripes; the body-whirl is slightly grooved at the base, and the other parts are generally smooth, but in the above-mentioned variety the whole surface is slightly ribbed transversely.

DUBIUS.   69.   Shell with longitudinal plaits, and transverse ribs; spire rather prominent; aperture ovate, and the outer lip crenulated.

Murex asper.   *Gmelin*, p. 3543.
Murex, No. 229.   *Schroeter Einl.* i. p. 625.
Murex, No. 157.   *Schreibers Conch.* i. p. 233.
Turricula striis exasperata.   *Martini*, iv. p. 232. t. 150. f. 1396 and 1397.
Inhabits ——
This shell, as it is figured by Martini, appears to be an inch and a quarter long, and half as broad, and it is said to be of a reddish colour, with five or six whirls, and the ribs acute. It possesses no affinity with the Linnæan *M. asper*, and would perhaps stand better near *Buccinum plicatulum*.

FENESTRATUS.   70.   Shell oblong, angulated, nodulous, with decussated striæ, and the interstices excavated; outer lip double and toothed, and the beak rather short and straight.

Murex fenestratus.   *Chemnitz*, x. p. 249. t. 161. f. 1536 and 1537.
Murex Colus, Var. *i*.   *Gmelin*, p. 3543.
Murex, No. 20.   *Schreibers Conch.* i. p. 197.
Le Buccin à cul de Dé.   *Favanne Cat. Rais.* p. 188. No. 891.
Favanne, t. 35. f. C 1.
Inhabits ——
Shell about sixteen lines long, and three-quarters of an inch broad, of a whitish colour with reddish spots, and the throat white. I am unacquainted with the shell, but it does not appear by the figure to possess much affinity with *M. Colus*, of which Gmelin has placed it as a variety.

COLUS.   71.   Shell turreted, with a nodulous keel, and transverse ribs, and plaited longitudinally; outer lip crenulated, and the beak long and straight.

Murex Colus. *Linnæus Syst. Nat.* p. 1221. *Martini*, iv. p. 180. *Born Mus.* p. 310. *Schroeter Einl.* i. p. 514, and *Inn. Bau Conch.* p. 43. t. 2. f. 6. *Gmelin*, p. 3543. *Schreibers Conch.* i. p. 223. *Brookes's Introd.* p. 119. t. 7. f. 89.

Fusus longicauda. *Lamarck Syst. des Anim.* p. 82.

Variety A. Shell whitish.

Murex candidus. *Gmelin*, p. 3556. *Schreibers Conch.* i. p. 232.

Murex undatus. *Gmelin*, p. 3556.

Murex longissimus. *Gmelin*, p. 5556.

Lister Conch. t. 917. f. 10, and t. 918. f. 11 a. *Rumphius*, t. 29. f. F. *Petiver Amb.* t. 6. f. 5. *Gualter*, t. 52. f. L. *Klein Ost.* t. 4. f. 78. *Argenville*, t. 9. f. B. *Seba*, iii. t. 79. largest fig. *Knorr*, iii. t. 5. f. 1. *Da Costa Elements*, t. 4. f. 4. *Martini*, iv. t. 144. f. 1339 and 1342, and t. 145. f. 1343 and 1344.

Variety B. Shell white spotted with brown.

Murex Colus Nicobaricus. *Chemnitz*, x. p. 241. t. 160. f. 1523.

La Quenouille tigrée. *Favanne Cat. Rais.* No. 976. *Favanne*, t. 33. f. A 5.

Variety C. Shell brown.

Murex ansatus. *Gmelin*, p. 3556. *Schreibers Conch.* i. p. 232.

Murex Colus, Var. β. *Gmelin*, p. 3543.

Murex, No. 217. *Schroeter Einl.* i. p. 621.

Regenfuss, i. t. 12. f. 62. *Martini*, iv. t. 144. f. 1340.

Variety D. Shell distorted with two beaks.

Murex Colus monstrosus. *Chemnitz*, xi. p. 291. t. 211. f. 2088 and 2089.

Inhabits the Indies. *Linnæus.* Coasts of Amboyna. *Rumphius.* Red Sea. *Forskael.* Cape of Good Hope, China, and Ceylon. *Humphreys.*

Shell varying from three to nine inches in length, of which the beak occupies about a third, and the body-whirl is about one third as broad; of the above-mentioned varieties *M. ansatus* differs most, not only in colour, but also in being proportionably broader, and it appears by his reference to Regenfuss, that Linnæus considered this to be a variety. Born has quoted all the above-mentioned figures of Martini's for *M. Colus*, and I cannot discover any specific difference to distinguish them.

STRIATULUS. 72. Shell sub-fusiform, with the whirls rounded and striated transversely; outer lip crenulated, and the beak long and straight.

Murex striatulus. *Gmelin,* p. 3557. *Schreibers Conch.* i. p. 233.

Murex, No. 227. *Schroeter Einl.* i. p. 624.

Fusus tenerrimus leviter striatus. *Martini,* iv. p. 190. t. 146. f. 1351 and 1352.

Inhabits ——

This shell appears by Martini's figure to be about three and a half inches long, and thirteen lines broad, and covered with a brown epidermis, beneath which the colour is whitish ; it much resembles *M. Colus,* but has not any longitudinal plaits or nodules.

LANCEA. 73. Shell subulate, with longitudinal plaits and transverse ribs ; aperture ovate, ribbed within, and toothed on the margin ; pillar two-plaited, and the beak long and straight.

Murex Lancea. *Gmelin,* p. 3556. *Schreibers Conch.* i. p. 232.

Murex angustus. *Gmelin,* p. 3556. ?

Murex, No. 221. *Schroeter Einl.* i. p. 622 ; and No. 222. p. 623. ?

Valentyn Abh. t. 1. f. 6. ? *Martini,* iv. t. 145. f. 1347.

Inhabits the coasts of Amboyna. *Martini.*

Shell about two and a half inches long, of which the beak occupies more than a third, and the breadth is only one fifth of the length ; it appears to differ from *M. Colus* principally in being narrower ; the spire has eleven or twelve whirls ; the colour is white, with the interstices of the ribs and the aperture violet, and young shells are said to be reddish. *M. angustus* has been entirely constituted from Valentyn's figure, and is hardly worth any notice.

VERSICOLOR. 74. Shell turreted, and transversely ribbed, with the upper whirls plaited longitudinally ; beak straight and rather long.

Murex versicolor. *Gmelin,* p. 3556. *Schreibers Conch.* i. p. 218.

Murex, No. 223. *Schroeter Einl.* i. p. 623.

Knorr, iii. t. 14. f. 1. *Martini,* iv. p. 188. t. 146. f. 1348.

Inhabits the East Indian Seas. *Martini.*

Martini's figure is about six inches and a half long, and two inches broad, and is painted of a bluish colour with brown longitudinal streaks ; it differs from *M. Colus* in its want of any transverse keel or longitudinal plaits on the body-whirl. Gmelin's other *M. versicolor* at p. 3530, is only a Variety of *M. ramosus.*

VERRUCOSUS. 75. Shell turreted, umbilicated, with transverse ribs which are more elevated in the middle, and the whirls coronated; beak rather short and incurved.

Murex verrucosus. *Gmelin,* p. 3557. *Schreibers Conch.* i. p. 218.
Murex, No. 224. *Schroeter Einl.* i. p. 624.
Bonanni Rec. and *Kirch.* 3. f. 88.? *Martini,* iv. p. 189. t. 146. f. 1349 and 1350.
Junior. Shell much smaller:
Martini, iv. t. 144. f. 1341.
Inhabits the Red Sea. *Bonanni.*
Shell about four inches long, and one inch and a half broad, and Martini's figures are painted pale brown with darker small brown spots, except on the pillar-lip which is white; it is transversely ribbed, and longitudinally plaited or nodulous on the shoulder of the whirls; Martini says the beak is flexuous and incurved, and he has quoted Bonanni's figure, which in other respects more resembles a Variety of *M. Colus.* Kæmmerer has referred to Martini, fig. 1341, for a separate species, to which I cannot discover that it has any claim.

MORIO. 76. Shell ventricose, transversely striated, blackish with a white band, and the whirls somewhat nodulous; aperture dilated, and the pillar wrinkled.

Murex Morio. *Linnæus Syst. Nat.* p. 1221. *Martini,* iv. p. 139. t. 139. f. 1300 and 1301. *Born Mus.* p. 310. *Schroeter Einl.* i. p. 515, and *Inn. Bau Conch.* p. 45. t. 2. f. 4. *Gmelin,* p. 3544. *Schreibers Conch.* i. p. 223.
Bonanni Rec. 3. f. 357, and *Kirch.* f. 350. *Lister Conch.* t. 988. f. 22. *Adanson Senegal,* t. 9. f. 31. *Seba,* iii. t. 80, many figures. *Knorr,* i. t. 20. f. 1. *Regenfuss,* i. t. 11. f. 61. *Favanne,* t. 35. f. B 1.
Variety. Of a greyish colour, and beak dark brown at the base.
Murex, No. 202. *Schroeter Einl.* i. p. 614.
Martini, iv. p. 152. t. 140. f. 1302 and 1303.

Junior? Shell much smaller.
  Knorr, ii. t. 6. f. 2.
Inhabits the coasts of Africa. *Linnæus.* Jamaica. *Lister.*
  Islands of Goree and Magdalen. *Adanson.* Curaçoa. *S.*
  Muller.
Shell from four to seven inches long, and nearly half as broad,
  of a blackish brown colour, with a narrow white transverse
  band, and sometimes another yellowish band above it, or
  with two white bands close to each other ; it is striated
  transversely and slightly plaited longitudinally ; the whirls are
  transversely keeled, with their upper parts flattish, and somewhat nodulous on the shoulder ; Adanson says it is covered
  with a very thick, velvety, tough epidermis.

COCHLIDIUM.   77. Shell fusiform, brown, trans-
      versely grooved, and the whirls flattish above;
      aperture dilated, and the throat yellowish.

Murex Cochlidium. *Linnæus Syst. Nat.* p. 1221. *Chem-*
  nitz, x. p. 273. t. 164. f. 1569. *Gmelin*, p. 3544.
Argenville, t. 9. f. A. *Seba*, iii. t. 52. f. 6, and t. 57. f.
  27 and 28. *Favanne*, t. 35. f. B 3.
Inhabits the East Indian Seas. *Chemnitz.*
Chemnitz's specimen measured four inches and two lines in
  length, by one inch and three-quarters, and the spire is about
  as long as the body-whirl ; it has seven whirls, which are
  much flattened at their upper ends, and the upper whirls are
  plaited longitudinally ; the colour is brown. Born considered Gmelin's *M. Tuba* to be this species, and Meuschen
  has arranged *M. Cochlidium*, as well as most of the other
  shells belonging to this division, among the Buccina.

TUBA.   78. Shell fusiform, longitudinally plaited,
      and transversely striped ; whirls subcylindri-
      cal below, and the shoulders nodulous ; beak
      attenuated, and rather long.

Murex Tuba. *Gmelin*, p. 3554. *Schreibers Conch.* i. p.
  218.
Murex Cochlidium. *Born Mus.* p. 311. *Mus. Lesk.* p.
  258.
Murex, No. 212. *Schroeter Einl.* i. p. 619.
Martini, iv. p. 171. t. 143. f. 1333.
Inhabits the coasts of China. *Martini.*
Chemnitz's figure measures five and a half inches in length, and
  is about half as broad, of a brownish white colour, with pale

reddish brown regular transverse narrow stripes; the colour of the spire is rather darker, and it consists of eight whirls, which are somewhat keeled transversely; Kæmmerer has quoted Martini's figure for a Variety of *M. Ternatanus*, but it has a very different appearance.

SPIRILLUS. 79. Shell with the body-whirl ventricose, and the spire flattish, with its summit mucronated; beak sub-cylindrical, and very long.

Murex Spirillus. *Linnæus Syst. Nat.* p. 1221. *Born Mus.* p. 312. *Schroeter Einl.* i. p. 517. t. 3. f. 4. *Gmelin,* p. 3544. *Schreibers Conch.* i. p. 122.
Knorr, vi. t. 24. f. 3. *Spengler Selt. Conch.* t. 3. f. E. *Martini,* iii. t. 115. f. 1069.
Inhabits the coasts of Tranquebar. *Spengler.* Malabar. *Martini.*
Shell about two and a half inches long, of which the beak occupies the larger half, and the body-whirl at its transverse keel is an inch and a half broad; the upper part of the body-whirl, as well as the spire, is flattish, and the latter ends in an obtuse knob; the pillar is separated from the beak by a strong oblique plait, and in the length of the beak, as well as general appearance, this shell somewhat resembles *M. Haustellum;* the colour is dirty white with a few scattered small yellow spots.

CANALICULATUS. 80. Shell with the body-whirl ventricose, and the spire pyramidal; whirls grooved at the suture; aperture dilated, and the beak long.

Murex canaliculatus. *Linnæus Syst. Nat.* p. 1222. *Martini,* iii. p. 29. t. 67. f. 742 and 743. *Born Mus.* p. 312. *Schroeter Einl.* i. p. 518. *Gmelin,* p. 3544. *Schreibers Conch.* i. p. 224.
Lister Conch. t. 878. f. 2. *Gualter,* t. 47. f. A. *Ellis's Corallines,* t. 33. f. b. ?
Junior? Shell brown, or white variegated with brown.
Martini, iii. t. 66. f. 738 to 740.
Larva. In a long string of Ovaries.
Murex canaliculatus, Var. Granum. *Linnæus Syst. Nat.* p. 1222.
Murex Granum. *Schroeter Einl.* i. p. 519. *Gmelin,* p. 3545. *Schreibers Conch.* i. p. 225.

Lister Conch. t. 881. f. 3 a.　*Ellis's Corallines,* t. 33. f. a.
Inhabits the coasts of Canada. *Linnæus.* Virginia. *Ellis.*
Full grown shells are five or six inches long, and about half as
　broad, thick, heavy, and nearly of an uniform reddish brown
　colour; it has eight slightly striated whirls separated by a
　channelled suture, and they are transversely divided by a
　somewhat coronated keel, the part above which is flattish;
　the beak is formed by the gradual contraction of the aper-
　ture, and is about two inches long. Linnæus considered *M.*
　Granum to be the Larva of this species, which Ellis's figure
　resembles more than *M. perversus,* although the whirls ap-
　pear to be reversed, and this may probably have arisen from
　an error of the engraver.

CARICA.　81. Shell with the body-whirl ventricose,
　　and armed on the shoulder with large com-
　　pressed nodules; aperture dilated, and the
　　beak very long.

Murex Carica.　*Gmelin,* p. 3545.
Murex, No. 7.　*Schroeter Einl.* i. p. 544.
Lister Conch. t. 880. f. 3 b.　*Gualter,* t. 47. f. B.　*Knorr,*
　i. t. 30. f. 1, and vi. t. 27. f. 1.　*Martini,* iii. t. 67. f.
　744, and t. 69. f. 756 and 757.
Inhabits ——
Shell varying from five to eight inches in length, and is three-
　fifths as broad, of a yellowish or dirty white colour, some-
　times marked with greyish bands and irregular transverse
　stripes; it is thick and heavy, and has the shoulder of the
　body-whirl armed with large strong compressed nodules,
　which become gradually smaller on the spire; the outer lip
　is sharply angulated near its upper end by a transverse keel
　on the body-whirl.

FICUS.　82. Shell pyriform, umbilicated, smooth,
　　transversely striped, and the body-whirl co-
　　ronated; outer lip thick and toothed, and the
　　inside striped.

Murex Ficus.　*Chemnitz,* x. p. 269. t. 163. f. 1564 and
　1565, and xi. p. 125. t. 193. f. 1853.　*Gmelin,* p. 3545.
Murex canaliculatus, Var. b.　*Schreibers Conch.* i. p. 224.
Murex, No. 4, and No. 5.　*Schroeter Einl.* i. p. 544.
Gualter, t. 26. f. N.　*Martini,* iii. t. 66. f. 741.
Inhabits the Red Sea. *Chemnitz.*
Shell about an inch and a half, or two inches long, and the

breadth is about two thirds of the length; the spire is rather depressed, and the suture of the first whirl is grooved; the colour is greyish or dusky yellow, with dark narrow transverse stripes on the inside, as well as the out.

CLANDESTINUS. 83. Shell sub-ovate, with six rounded whirls with transverse and minute longitudinal striæ; aperture toothed, with the outer lip double, and ending in a straight somewhat produced beak.

Murex clandestinus. *Chemnitz*, xi. p. 127. t. 193. f. 1856 and 1857.
Buccinum caudatum, Var. β. *Gmelin*, p. 3471.
Buccinum, No. 14. *Schroeter Einl.* i. p. 359.
Lister Conch. t. 940. f. 36. *Klein Ost.* t. 3. f. 61. *Knorr*, vi. t. 29. f. 5.
Inhabits ——
This shell appears by the figures to be about twenty-two lines long and thirteen broad, of a straw colour, with reddish brown transverse ribs, and minute longitudinal striæ; the outer lip is double and toothed, and the inner lip crenulated; it a good deal resembles *Buccinum caudatum*, but the aperture ends in rather a narrow beak nearly half an inch long.

ARUANUS. 84. Shell ventricose, with the spire conical and coronated with spines; aperture dilated, with the beak long, and the pillar flexuous.

Murex Aruanus. *Linnæus Syst. Nat.* p. 1222. *Martini*, iv. p. 191, Vign. 39, at p. 143. f. D. *Schroeter Einl.* i. p. 520. *Gmelin*, p. 3546. *Schreibers Conch.* i. p. 225.
Bonanni Rec. and *Kirch.* 3. f. 101. *Rumphius*, t. 28. f. A. *Favanne*, t. 35. f. M.
Inhabits the coasts of New Guinea and China. *Linnæus*. Island of Aru. *Rumphius*.
Chemnitz's figure is five and a quarter inches long, of which the beak occupies two inches, and the breadth is about two inches; it is coarse and ponderous, and slightly ribbed transversely; Linnæus says that the pillar is flexuous like the letter S, and that the colour is a remarkable brownish blue, or sometimes whitish flesh-colour. *M. Aruanus* of Born appears to be *M. fornicatus*.

PERVERSUS. 85. Shell with the body-whirl ventricose, and the whirls reversed, and slightly coronated; aperture dilated, and the beak long.

Murex perversus. *Linnæus Syst. Nat.* p. 1222. *Born Mus.* p. 313. t. 11. f. 8 and 9. *Schroeter Einl.* i. p. 521. *Chemnitz*, ix. part 1. p. 67. t. 106. f. 900 to 903, and t. 107. f. 904 to 907. *Gmelin*, p. 3546.
Bonanni Kirch. 3. f. 402. *Lister Conch.* t. 907. f. 27, and t. 908. f. 28. *Argenville*, t. 15. f. F. *Seba*, iii. t. 68. f. 21 and 22. *Regenfuss*, ii. t. 3. f. 25. *Favanne*, t. 23. f. H 2.
Inhabits the American Seas. *Linnæus.* Bay of Campechy. *Lister.* Coasts of Jamaica, Mexico, and the West Indies. *Martini.* Florida and North America. *Humphreys.*
Shell five or six inches long, and about half as broad, and those which are proportionably narrower have not attained to maturity; Born has described his specimen to be one foot long and six inches broad; the body-whirl is transversely keeled, and nodulous on the shoulder, and the spire slightly coronated; the colour is whitish, more or less striped longitudinally with red.

ANTIQUUS. 86. Shell oblong, ventricose, with rounded whirls and slightly decussated striæ; aperture dilated, with a short beak, and the throat yellowish.

Murex antiquus. *Linnæus Syst. Nat.* p. 1222. *Martini*, iv. p. 127. t. 138. f. 1292 and 1294. *Schroeter Einl.* i. p. 522. *Gmelin*, p. 3546. *Schreibers Conch.* i. p. 127. *Maton and Racket, in Lin. Trans.* viii. p. 145. *Dorset Cat.* p. 47. t. 17. f. 4.
Murex despectus. *Pennant Zool.* iv. p. 124. t. 78. f. 98. *Born Mus.* p. 314. *Pulteney's Dorset*, p. 43. *Donovan*, i. t. 31. *Montagu Test.* p. 256.
Buccinum magnum. *Da Costa Brit. Conch.* p. 120. t. 6. f. 4.
Tritonium antiquum. *Muller Zool. Dan.* iii. p. 64. t. 118. f. 1 to 3. *Fabricius Fauna Grænl.* p. 397.
Bonanni Rec. 3. f. 190, and *Kirch.* f. 192. *Lister Conch.* t. 913. f. 4, and *Anim. Ang.* t. 3. f. 1.
Variety. With the whirls reversed.
Murex contrarius. *Linnæus Mantissa*, p. 551. *Chemnitz,*

ix. part 1. p. 58. t. 105. f. 894 and 895. *Gmelin*, p. 3564.

Murex antiquus, Var. *γ*. *Gmelin*, p. 3546.

Murex, No. 66. *Schroeter Einl.* i. p. 566.

Lister Conch. t. 950. f. 44 b. *Regenfuss*, ii. t. 4. f. 36. *Martini n. Mannigf.* iv. t. 2. f. 14. *Favanne*, t. 32. f. N, t. 79. lower f. F, and t. 80. f. R.

Inhabits the European Seas, and the coasts of Norway and Sweden. *Linnæus*. Britain. *Lister*, &c. Denmark. *Muller*.

Shell four or five inches long, and about half as broad, with seven or eight ventricose whirls; though thick and heavy, it is semitransparent, and is thickly set with transverse striæ which are crossed by other slighter longitudinal ones; the colour is reddish or dirty white, and the inside yellow. Rutty says that the tail of the inhabitant is more fat and tender than a lobster's. Linnæus has described his *M. contrarius* ' striis geminatis,' but I think there can be no doubt that it is only a Variety of this species.

MAGELLANICUS. 87. Shell oblong, ventricose, with decussated striæ, and a callosity on the upper angle of the aperture; beak short and dilated; throat white.

Murex Magellanicus. *Chemnitz*, x. p. 275. t. 164. f. 1570.

Murex Magellanicus, Var. *β*. *Gmelin*, p. 3548.

Murex, No. 142. *Schreibers Conch.* i. p. 230.

Inhabits the Straights of Magellan. *Chemnitz*.

Chemnitz has described this shell to be three inches and a quarter long, and one inch and three-quarters broad, and it is nearly allied to *M. antiquus*; it differs in having the whirls rather less rounded, the striæ more distinctly cancellated, and a tooth-like callosity on the upper angle of the aperture. It is entirely different from *M. Magellanicus*, Var. *α* of Gmelin, which is *M. lamellosus*.

FORNICATUS. 88. Shell ovate-oblong, with the whirls somewhat ventricose, striated longitudinally, and transversely carinated; aperture dilated, and the beak short.

Murex fornicatus. *Gmelin*, p. 3547.

Murex Aruanus. *Born Mus.* p. 313.

Murex despectus. *Schroeter Einl.* i. p. 523. t. 3. f. 5.

Murex carinatus. *Pennant Zool.* iv. p. 123. t. 77. f. 96,

and Frontispiece? *Donovan,* iv. t. 109. ? *Maton and Racket, in Lin. Trans.* viii. p. 147. ?

Murex antiquus, Var.   *Montagu Test.* p. 257.

Tritonium fornicatum.   *Fabricius Fauna Groenl.* p. 399.

Martini, iv. t. 138. f. 1295.

Inhabits the coasts of Greenland. *Fabricius.*

This shell is of the same size, and is nearly allied to *M. despectus,* but it has the upper part of its whirls more gradually sloped downwards, and has three or four transverse ridges on its body-whirl, and two on the whirls of the spire; some of the longitudinal striæ or plaits, which Fabricius mentions, are also observable in Pennant's Frontispiece, and I have hardly any doubt that *M. carinatus* belongs to the same species. Born appears to have mistaken this shell for the Linnæan *M. Aruanus.*

DESPECTUS.   89. Shell oblong, with the whirls ventricose, and two transverse elevated lines; aperture dilated, and the beak short.

Murex despectus.   *Linnæus Iter W. Goth.* p. 200. t. 5. f. 8, and *Syst. Nat.* p. 1222. *Gmelin,* p. 3547. *Donovan,* v. t. 180.

Inhabits the Northern Ocean, and coasts of Iceland. *Linnæus.*

The shell which Mr. Donovan has figured, and which he says answers to the Linnæan description and figure, is two inches and a quarter long, and an inch and a quarter broad. Mr. Donovan adds, " at the first glance this shell appears to be an intermediate kind, between Lister's shell and the *Murex carinatus* of Pennant and ourselves : indeed the principal difference we perceive between the true *M. despectus* and Lister's shell, is that the former has the whirls of the spire rather more ventricose, and distinctly marked with two slightly elevated spiral lines; from *Murex carinatus* it differs principally in the very prominent angulations of the *anfractibus,* where the ridges appear, and more particularly in the strong depression between the upper ridge and the suture of the whirls."

SUBANTIQUATUS.   90. Shell oblong, with eight whirls, angulated, and strongly keeled transversely; aperture dilated, and the beak short.

Murex subantiquatus.   *Maton and Racket, in Lin. Trans.* viii. p. 147.

Murex antiquus. *Pennant,* iv. p. 124. *Donovan,* iv. t. 119. *Montagu Test.* p. 257.
Martini, iv. t. 138. f. 1293 and 1296.
Inhabits the coasts of Britain. *Pennant,* &c.
Martini has arranged the Linnæan *M. antiquus* and *M. despectus* as varieties, and considered the shells which have since been called *M. subantiquatus* by English authors, and *M. fornicatus* by Fabricius, to be the same species. Montagu, at p. 112 of his Supplement, expresses considerable doubt on this subject, and I confess my inability to find any permanent character by which the present shell can be distinguished from either *M. fornicatus* or *M. despectus*; it is said to have the upper part of the whirls more flattened, and separated by a strongly keeled angulated ridge, with another transverse rib below, and these sometimes crossed by angular longitudinal ridges; but, as Mr. Montagu observes, it is with the others 'closely connected by every shade of gradation.'

TRITONIS. 91. Shell ventricose, with alternate varices, and the whirls crenulated at the sutures; inner lip grooved, and the beak short.

Murex Tritonis. *Linnæus Syst. Nat.* p. 1222. *Martini,* iv. p. 113. t. 134. f. 1277, and t. 135. f. 1282 and 1283. *Born Mus.* p. 315. *Schroeter Einl.* i. p. 525. *Gmelin,* p. 3549. *Schreibers Conch.* i. p. 226. *Shaw Nat. Misc.* xii. t. 479.
Bonanni Rec. and *Kirch.* 3. f. 188. *Lister Conch.* t. 959. f. 12. *Rumphius,* t. 28. f. B, and No. 1. *Petiver Gaz.* t. 151. f. 5, and *Amb.* t. 12. f. 15. *Gualter,* t. 48. f. A. *Klein Ost.* t. 7. f. 127. *Knorr,* ii. t. 16. f. 2 and 3, and v. t. 5. f. 1. *Regenfuss,* ii. t. 5. f. 46. *Favanne,* t. 31. f. G 1, and G 2.
Inhabits the Archipelago, and coasts of America. *Linnæus.* Barbadoes. *Lister.* Amboyna. *Rumphius.* Bay of Naples. *Ulysses.*
Shell a foot, or sometimes sixteen inches long, and about half as broad, with smooth, slightly elevated, broad transverse ribs, and the interstices striated; the colour is whitish, ornamented on the ribs with parallel curved reddish brown spots, which are shaded off towards each other, and have some resemblance to the feathers of a bird; the pillar-lip is grooved, and striped with dark brown. I have left this shell in the place which Linnæus has assigned it, but the varices indicate

its greater affinity with the shells of the preceding division. It is used by the Africans and by many of the Eastern nations as a military horn. Martini's fig. 1280 and 1281 differ in being much smaller, and have been erroneously quoted by Born for his *M. accinctus,* which is *Buccinum plumatum.*

NEREI. 92. Shell ventricose, with alternate varices, decussated striæ, and transverse nodulous ribs; inner lip with a transverse callosity, and the beak short.

Murex Tritonis, Var. *β. Schroeter Einl.* i. p. 525. *Gmelin,* p. 3550.
Murex Tritonium australe. *Chemnitz,* xi. p. 134. t. 194. f. 1867 and 1868.
Lister Conch. t. 960. f. 13. *Martini,* iv. p. 118. t. 136. f. 1284.
Inhabits the South Sea. *Chemnitz.*
Shell generally seven or eight inches long, and about half as broad, spotted with brownish white and chestnut; it has numerous narrow unequal transverse ribs, of which the broadest are nodulous, and fine decussated striæ. It is said to be used by the natives of some of the South Sea Islands for musical purposes.

VULPINUS. 93. Shell oblong, ventricose, glabrous, with the base and margins of the whirls striated; beak short and twisted.

Murex vulpinus. *Born Mus.* p. 317. t. 11. f. 10 and 11. *Gmelin,* p. 3558. *Schreibers Conch.* i. p. 221.
Murex, No. 274. *Schroeter Einl.* i. p. 640.
Inhabits ———
According to Born, this shell is seventeen lines long, and seven broad, oblong, with eight smooth somewhat rounded whirls, which are transversely striated near the sutures; body-whirl convex, and ending in a short beak, which is bent outwards; aperture smooth; colour reddish brown.

PUSIO. 94. Shell ventricose, oblong, smooth, with the spire striated and the whirls rounded; aperture toothless, and the beak short.

Murex Pusio. *Linnæus Syst. Nat.* p. 1223.? *Martini,* iv. p. 202. t. 147. f. 1357. *Born Mus.* p. 315. *Schroeter*

Einl. i. p. 526. *Gmelin*, p. 3550. *Schreibers Conch.* i. p. 226.

Buccinum Nifat. *Bruguiere Enc. Meth.* p. 282.

Le Nifat. *Adanson Senegal*, p. 52. t. 4. f. 3.

Lister Conch. t. 914. f. 7. *Rumphius*, t. 49. f. E. *Petiver Amb.* t. 7. f. 1. *Gualter*, t. 52. f. I. *Favanne*, t. 53. f. I.

Inhabits the Mediterranean. *Linnæus.*

Linnæus with the above character has quoted *Bonanni*, f. 40, and *Gualter*, t. 52. f. I, and has added the following description, "Shell as large as a hazel-nut, of a bluish colour, with grey longitudinal waved bands, and the suture simple." The present shell, which all authors, except Bruguiere, have considered to be the same, is about an inch and three-quarters long and three-quarters of an inch broad, with transverse rows of square spots corresponding with the above-mentioned figure of Gualter's, but Bonanni's has longitudinal stripes.

TULIPA. 95. Shell sub-fusiform, smooth, and striated by the sutures; aperture striated within, and the pillar two-plaited.

Murex Tulipa. *Linnæus Syst. Nat.* p. 1223. *Martini*, iv. p. 123. t. 136. f. 1286 and 1287, and t. 137. f. 1288 to 1291. *Born Mus.* p. 317. *Schroeter Einl.* i. p. 527. *Gmelin*, p. 3550. *Schreibers Conch.* i. p. 227. *Shaw Nat. Misc.* xxiv. t. 1022. *Brookes's Introd.* p. 120. t. 7. f. 90.

Fasciolaria Tulipa. *Lamarck Syst. des Anim.* p. 83.

Bonanni Rec. and *Kirch.* 3. f. 187. *Lister Conch.* t. 910. f. 1, and t. 911. f. 2. *Rumphius*, t. 49. f. H. *Gualter*, t. 46. f. A and E. *Argenville*, t. 10. f. K. *Seba*, iii. t. 71. f. 23 to 31. *Knorr*, v. t. 18. f. 5, and vi. t. 29. f. 1. *Regenfuss*, i. t. 9. f. 35, and ii. t. 8. f. 16. *Favanne*, t. 34. f. L.

Inhabits the coasts of Campeachy and Jamaica. *Lister.* Guinea. *Regenfuss.* Antilles and Caribbee Islands. *Martini.* Florida. *Humphreys.*

Shell about six inches long, and two and a half broad, of a dirty white colour mottled with purplish brown, or vice versa, and marked with blackish transverse lines; the whirls are smooth, except a narrow space below the sutures, which is marked with somewhat decussated striæ; the inside is finely striated, and the pillar has two oblique plaits; the beak is produced, and about an inch long.

CLATHRATUS. 96. Shell oblong, with longitudinal somewhat membranaceous ribs, and a short beak.

Murex clathratus. *Linnæus Syst. Nat.* p. 1223. *Schroeter Einl.* i. p. 528. *Gmelin*, p. 3551.
Buccinum muricinum. *Gmelin*, p. 3503.
Buccinum, No. 1381. *Gronovius Zooph.* p. 308.
Lister Conch. t. 926. f. 19. *Klein Ost.* t. 3. f. 67.
Inhabits the coasts of Iceland. *Linnæus.*
Linnæus says this shell resembles *Turbo Clathrus* both in size and habit, but has a short beak, and numerous erect compressed longitudinal plaits; he has quoted Klein, whose figure is a copy from Lister, t. 926, and the latter has been referred to by Gronovius for a *Buccinum*, with the following description, 'Shell reticulated and wrinkled; spire incurved, and the aperture crenated; pillar wrinkled, and the outer lip thick.'

LAMELLOSUS. 97. Shell ventricose, turreted, with parallel membranaceous longitudinal ribs, and transverse striæ; beak short, and the throat purplish.

Murex lamellosus. *Gmelin*, p. 3536.
Murex Magellanicus, Var. *α*. *Gmelin*, p. 3548. *Schreibers Conch.* i. p. 206.
Murex plicatus. *Portland Cat.* p. 104. No. 2284.
Murex, No. 38. *Schroeter Einl.* i. p. 557.
Muriciformis granatus. *Mus. Gevers.* p. 316. No. 685.
Buccinum Geversianum. *Pallas Spic. Zool.* x. p. 33. t. 3. f. 1 and 2.
Buccinum laciniatum. *Martyn Univ. Conch.* ii. t. 42.
Buccinum Harpa, Var. *β*. *Gmelin*, p. 3482.
Knorr, iv. t. 30. f. 2. *D'Avila*, t. 10. f. B and d. *Favanne*, t. 37. f. H 1.
Variety. Shell smaller, and rather narrower in proportion to the length.
Murex foliaceus minor. *Chemnitz*, xi. p. 108. t. 190. f. 1823 and 1824.
Favanne, t. 79. f. I.
Inhabits the Straights of Magellan. *D'Avila.* Coasts of the Falkland Islands. *Martyn.*
Shell three or four inches long, and two-thirds as broad, of a pale greyish or whitish colour, and the inside violet or pur-

plish brown; it has six or seven whirls with longitudinal membranaceous acute ribs, which at their upper ends are pointed and raised above the sutures; the aperture is ovate, and ends in a short beak; the ribs appear to be formed very much in the same manner as the varices of *M. ramosus* and its congeners.

SCALA. 98. Shell umbilicated, with longitudinal plaits and transverse striæ; whirls transversely angular, and the upper parts flattish; aperture sub-cordiform, and the beak very short.

Murex Scala. *Gmelin*, p. 3551.
Murex, No. 29. *Schroeter Einl.* i. p. 552.
Buccinum scalare. *Gmelin*, 3495.
Buccinum, No. 28. *Schroeter Einl.* i. p. 364.
Trochus Turris. *Meuschen Mus. Gevers.* p. 288. No. 448.
Pallas Spicel. x. p. 33. t. 3. f. 7 and 8. *Martini*, iv. p. 27, and Vign. 37. at p. 1. f. *a*, *b*, and *c*.
Inhabits the East Indies. *Chemnitz.*
This shell appears by Martini's figure to be about sixteen lines long, and eleven broad, with a large funnel-shaped umbilicus, and the base remarkably truncated; the whirls are somewhat cylindrical, and flattened at the upper end; it is of a brownish yellow colour with a paler band, and is said to be an extremely rare and valuable species.

FISCELLUM. 99. Shell ovate, longitudinally plaited, and transversely ribbed; outer lip double and toothed; beak short, and the throat violet.

Murex Fiscellum. *Chemnitz*, x. p. 242. t. 160. f. 1524 and 1525. *Gmelin*, p. 3552. *Schreibers Conch.* i. p. 229. *Martyn Univ. Conch.* t. 93. ?
Inhabits the coasts of Pulo Condore. *Chemnitz.*
Shell about an inch and a half long, and one inch broad, of a brownish white colour with darker longitudinal stripes, and the inside violet; the beak is very short and straight. Chemnitz says his specimen came from England, with the name of "Purple-mouthed Basket Buccinum from Pulo Condore;" and it is well figured in Martyn's Universal Conchology, but I am unable with sufficient certainty to determine the number of the plate.

UNDATUS. 100. Shell ovate, longitudinally plaited and transversely grooved; outer lip double and toothed; beak short, and the throat white.

Murex undatus. *Chemnitz,* xi. p. 124. t. 192. f. 1851 and 1852.

Inhabits the coasts of Tranquebar. *Chemnitz.*

This shell, by Chemnitz's figure, appears to be about fourteen lines long, and half as broad, of a blackish brown colour, and the throat white; it has angulated transverse ribs or plaits, crossed by transverse grooves, and is somewhat nodulous. *M. undatus* of Gmelin appears to be nothing more than a trifling variety of *M. Colus,* and the present shell is probably not distinct from *M. Fiscellum.*

VIRGATUS. 101. Shell ovate, longitudinally plaited and nodulous, and transversely ribbed; aperture ovate, and the outer lip toothed within.

Murex plicatus. *Gmelin,* p. 3551.
Murex, No. 31. *Schroeter Einl.* i. p. 554.
Buccinum virgatum. *Solander's MSS.*
Martini, iv. p. 33. t. 123. f. 1441 and 1442.

Inhabits the East Indian Seas. *Martini.*

Shell about three-quarters of an inch long, and half an inch broad, and possesses considerable affinity with *Buccinum Tranquebaricum;* the colour is purplish brown, with about four whitish ribs on the body-whirl, and of these the two upper are sharply nodulous, and spotted with orange; the inside is white, and appears striped from the transparency of the shell; the beak is very short and straight.

CORONA. 102. Shell ovate, with the whirls flat above, and crowned with undulated membranaceous scales; beak short.

Murex Corona. *Gmelin,* p. 3552. *Schreibers Conch.* i. p. 196.
Murex Corona Mexicana. *Chemnitz,* x. p. 243. t. 161. f. 1526 and 1527.
D'Avila, t. 9. f. A.

Inhabits the Gulf of Mexico. *D'Avila.*

Shell about two inches and a half long, and the breadth is nearly three-fifths of the length; the whirls become flattish towards their upper margins, and crowned with undulated some-what imbricated membranaceous scales, of which those on the body-whirl are incurvated; the colour is yellowish, with two or three broad transverse more or less interrupted brown bands. I never saw a specimen of this species, which is said to be very rare, but it appears from the description to bear some resemblance to *Buccinum Bezoar*.

DOLARIUM. 103. Shell with remote, much elevated, obtuse, narrow belts, and smaller longitudinal plaits; whirls flat above, and the summit obtuse.

Murex Dolarium. *Linnæus Syst. Nat.* p. 1223. *Born Mus.* p. 318. *Gmelin,* p. 3552.
Bonanni Rec. and *Kirch.* 3. f. 347. *Knorr,* ii. t. 24. f. 5, and v. t. 3. f. 5. *Martini,* iii. t. 118. f. 1087 and 1088.
Inhabits the coasts of Portugal. *Bonanni.*
Shell commonly about an inch and a half long, and an inch and a quarter broad, of an uniform greyish brown or horn-colour; it has about six narrow much elevated obtuse belts on the body-whirl, and two on the first whirl of the spire; the other whirls are much depressed, and form a rounded knob; the pillar is umbilicated, and the beak short. Some of its synonyms have been confounded by almost every author with those of *M. cutaceus*, from which it differs in having the whirls more produced, and in the want of a longitudinal varix.

CORNEUS. 104. Shell oblong, with eight convex whirls, striated transversely, and slightly wrinkled longitudinally; beak rather long, and somewhat ascending.

Murex corneus. *Linnæus Syst. Nat.* p. 1224. *Pennant Zool.* iv. p. 124. t. 76. f. 99. *Schroeter Einl.* i. p. 530. *Gmelin,* p. 3552. *Donovan,* ii. t. 38. *Montagu Test.* p. 258. *Maton and Racket, in Lin. Trans.* viii. p. 147. *Dorset Cat.* p. 47. t. 17. f. 5.
Murex Islandicus. *Gmelin,* p. 3555. *Schreibers Conch.* i. p. 232.
Murex, No. 206. *Schroeter Einl.* i. p. 616.
Fusus Islandicus. *Martini,* iv. p. 159. t. 141. f. 1312 and 1313.
Buccinum gracile. *Da Costa Brit. Conch.* p. 124. t. 6. f. 5.

Lister Anim. Ang. t. 3. f. 4, and *Conch.* t. 913. f. 5.
Inhabits the coasts of Southern Europe. *Linnæus.* Great
   Britain. *Lister, &c.* Iceland. *Martini.*
Shell about three inches long and an inch and a quarter broad,
   white, and covered when alive with a brown epidermis ;
   it has eight or nine rounded whirls, and the spire terminates in a small rounded knob; the aperture is oblong-
   oval, with rather a long, curved, and slightly ascending beak.

LINEATUS.   105. Shell oblong, whitish, with red-
    dish transverse striæ; beak short and straight.

Murex lineatus. *Chemnitz,* x. p. 278. t. 164. f. 1572. *Gme-*
   lin, p. 3559. *Schreibers Conch.* i. p. 230.
Buccinum Linea. *Martyn Univ. Conch.* ii. t. 48.
Inhabits the coasts of New Zealand. *Martyn.*
Shell about an inch and three-quarters long, and half as broad,
   of a whitish colour, prettily marked with brownish red transverse striæ.

LIGNARIUS.   106. Shell oblong, coarse, with ob-
    tusely nodulous whirls; aperture toothless,
    and the beak rather short and straight.
Murex lignarius. *Linnæus Syst. Nat.* p. 1224. *Gmelin,*
   p. 3552.
Bonanni Rec. 3. f. 32. *Seba,* iii. t. 52. f. 4.
Inhabits the coasts of Southern Europe. *Linnæus.*
Linnæus, in addition to the above character, has described this
   shell to be about as long as a finger, smooth, coarse, and
   armed with a single row of obtuse protuberances. He has
   quoted the above figures from Bonanni and Seba, and to
   these Gmelin has added Knorr, vi. t. 26. f. 5, which is more
   like *Voluta turrita.* Born has described *M. Nassa* under
   this name, and according to Muller, *M. lignarius* is the shell
   which Born considered to be the Linnæan *M. Javanus.*
   Ulysses says it corresponds tolerably well with Gualter, t. 52.
   f. 3, which Linnæus with a mark of doubt has quoted, and
   this appears to be different from either of the above-mentioned shells, so that the *M. lignarius* must be regarded as
   a doubtful species.

NASSA. 107. Shell ovate, transversely striated, with
    a row of nodulous plaits on the shoulder of
    the whirls, and a transverse belt below; beak
    short, and the pillar plaited.

Murex Nassa   *Gmelin*, p. 3551.   *Schreibers Conch.* i. p. 231.
Murex lignarius.   *Born Mus.* p. 313.
Murex, No. 30.   *Schroeter Einl.* i. p. 553.
Voluta fuscata.   *Gmelin*, p. 3465.
Voluta, No. 228.   *Schroeter Einl.* i. p. 302. t. 1. f. 15.
Buccinum piceum.   *Solander's MSS.*
Lister Conch. t. 828. f. 50.   *Knorr*, vi. t. 20. f. 7. *Spengler Selt. Conch.* t. 3. f. F.   *Regenfuss*, i. t. 7. f. 1.
Inhabits the coasts of Barbadoes. *Lister.*   St. Croix and Guinea. *Martini.*
Shell sometimes nearly three inches long, and rather more than half as broad, but is frequently not more than half so large; the colour is brown or sometimes blackish, with the summits of the nodules generally whitish, and a white transverse elevated belt about half way between the shoulder and the base of the body-whirl; the aperture is white, and the beak rather short; the pillar has four distinct plaits.

AMPLUSTRE.   108.   Shell with dark blue, yellow, and white bands, and a transverse tuberculated keel near the margins of the whirls; beak short and straight, and the pillar three plaited.

Murex Amplustre.   *Chemnitz*, xi. p. 119. t. 191. f. 1841 and 1842.
Murex Argus, Var. γ.   *Gmelin*, p. 3548.
Buccinum Amplustre.   *Martyn Univ. Conch.* i. t. 3.
Inhabits the North West Coasts of America. *Humphreys.* Friendly Islands. *Martyn.*
Shell about two inches and a quarter long, and one inch and a quarter broad, prettily variegated with transverse stripes of different colours; the beak is short, straight, and obtuse; the inside is white, with the inner margin of the outer lip striped with blue.   In the Portland Catalogue it is called the *American Flag*, and is said to be extremely scarce.

TRAPEZIUM.   109.   Shell oblong, obtusely angulated, and the whirls slightly nodulous; aperture toothed, and the beak rather short and straight.

Murex Trapezium.   *Linnæus Syst. Nat.* p. 1224.   *Born Mus.* p. 319.   *Martini*, iv. p. 136. t. 139. f. 1298 and

1299.   *Schroeter Einl.* i. p. 531.   *Gmelin,* p. 3552. *Schreibers Conch.* i. p. 228.   *Shaw's Nat. Misc.* xvii. t. 690.

Bonanni Rec. 3. f. 287, and *Kirch.* f. 288.   *Lister Conch.* t. 931. f. 26.   *Rumphius,* t. 49. f. K.   *Gualter,* t. 46. f. B.   *Argenville,* t. 10. f. F.   *Knorr,* iv. t. 20. f. 1. *Favanne,* t. 35. f. B 5.

Variety. Less angulated transversely, and the whirls more produced.

Murex Trapezium.   *Schroeter Inn. Bau Conch.* p. 45. t. 2. f. 10.

Gualter, t. 52. f. T.   *Argenville,* t. 10. f. H.   *Knorr,* ii. t. 15. f. 3.? *Martini,* iv. t. 140. f. 1310 and 1311.   *Favanne,* t. 34. f. H.

Inhabits the Red Sea. *Bonanni.*   Coasts of Amboyna. *Rumphius.*   China. *Humphreys.*

Shell from four to six inches long, and the breadth at the shoulder of the body-whirl is nearly two thirds of the length, but the Variety is proportionably narrower; it has eight transversely angulated whirls, and a row on each of obtuse nodulous longitudinal plaits; the aperture is striated transversely, and the pillar has three oblique plaits; the colour is brown, brownish-white, or yellowish-brown marked transversely with blackish lines in pairs. A specimen of a most extraordinary size, eighteen inches long, is mentioned in the Portland Catalogue, p. 189, lot 4015.

POLYGONUS.   110. Shell somewhat ventricose, with transverse striæ, and strong longitudinal obtuse plaits in the middle of the whirls; aperture oval, and the beak very short.

Murex polygonus.   *Gmelin,* p. 3555.   *Schreibers Conch.* i. p. 217.   *Ulysses's Travels,* p. 465.

Murex gibbulus.   *Gmelin,* p. 3557.   *Schreibers Conch.* i. p. 238.

Murex, No. 205, and No. 271.   *Schroeter Einl.* i. p. 615, and p. 639.

Murex Trapezium, Var.   *Mus. Leskianum,* p. 262.

Fusus brevis truncatus polygonus.   *Martini,* iv. p. 156. t. 140. f. 1306 to 1309, and t. 141. f. 1314 to 1316.

Inhabits the coasts of the Isle of France. *Martini.* Bay of Naples. *Ulysses.*

Of the seven figures which Martini has given of this species, the four first have the double lines, and greatly resemble *M. Trapezium;* the others also appear to be nearly allied to fig. 1310 and 1311, which all authors have quoted for that species; it is said to be about three and a half inches long, sometimes umbilicated, and sometimes not, and to have the pillar with or without plaits.

PUGILINUS. 111. Shell solid and nearly smooth, with a nodulous keel on the shoulder which has a transverse groove ; sutures channelled, and the beak rather long.

Murex pugilinus. *Born Mus.* p. 314. *Mus. Lesk.* p. 259.
Murex Vespertilio. *Gmelin*, p. 3553. *Schreibers Conch.* i. p. 218.
Fusus crassus carnarius. *Martini*, iv. p. 162. t. 142. f. 1323, 1324, 1326, and 1327.
Lister Conch. t. 884. f. 6 a. *Favanne*, t. 23. f. L 2.
Variety. With a second row of nodules about half way between the shoulder and the base.
Murex carnarius. *Chemnitz*, x. p. 270. t. 164. f. 1566 and 1567.
Lister Conch. t. 885. f. 6 b.
Inhabits the coasts of Tranquebar, and the Molucca Islands. *Martini.* Carolina. *Humphreys.* Nicobar Islands. *Chemnitz.*

Shell about four and a half inches long, and two and three-quarters broad, of a uniform reddish brown, or sometimes flesh-colour ; the whirls are slightly wrinkled in both directions, and the upper part above the shoulder is rather concave ; in my specimen a narrow groove traverses the keel, and is most conspicuous in the interstices of the compressed nodules ; the inside is of a pale flesh-colour. *M. carnarius* has been placed as a Variety by Gmelin, but is probably a distinct species.

SCOLYMUS. 112. Shell thin, diaphanous, ventricose, transversely striated, and nodulous on the shoulders of the whirls ; pillar straight, with three plaits on the inner lip.

Murex Scolymus. *Gmelin*, p. 3553. *Schreibers Conch.* i. p. 218.
Murex, No. 210. *Schroeter Einl.* i. p. 618.
Fusus carnarius legitimus. *Martini*, iv. p. 164. t. 142. f. 1325.
Inhabits ——

Shell about four and a half inches long, and two inches broad, of a whitish brown colour ; the two extremities of the body-whirl are strongly striated transversely, and the shoulder nodulous ; the spire consists of nine or ten whirls, and the nodules are said to be hollow.

TERNATANUS.   113. Shell oblong, with longitudinal plaits, and transverse striæ ; aperture oblong, and the beak long and straight.

Murex Ternatanus.   *Gmelin*, p. 3554.   *Schreibers Conch.* i. p. 231.
Murex, No. 203.   *Schroeter Einl.* i. p. 614.
Buccinum Tarnatanum.   *Mus. Gevers.* p. 300, No. 555.
Fusus brevis Ternatanus.   *Martini*, iv. p. 153. t. 140. f. 1304 and 1305.
Lister Conch. t. 892. f. 12. *Valentyn Abh.* t. 1. f. 2. *Seba*, iii. t. 52. f. 5.   *Knorr*, vi. t. 15. f. 4, and t. 26. f. 1. *Favanne*, t. 35. f. B 6.
Inhabits the Islands of Ternate. *Valentyn*.
Shell about two and a quarter inches long, and eleven lines broad, and is sometimes larger; the colour is pale brownish yellow, with darker transverse lines in pairs, like those of *M. Trapezium*, and to that species, as well as *M. polygonus*, it appears to be nearly allied.

COLUMBARIUM.   114. Shell with longitudinal plaits, and   transverse   somewhat   spinous   keels ; beak   short   and   straight,   and   the   throat violet.

Murex Columbarium.   *Chemnitz*, x. p. 284. t. 169. f. 1637 and 1638.   *Gmelin*, p. 3559.   *Schreibers Conch.* i. p. 204.
Buccinum spinosum.   *Martyn Univ. Conch.* i. t. 4.
Inhabits the coasts of the Friendly Islands. *Martyn*.   Pulo Condore. *Chemnitz*.
Shell about an inch and a half long, and eleven lines broad ; the colour is brown, with a white cylindrical belt in the middle of the whirls, which is bounded on both sides by an elevated margin, and sharply angulated by the longitudinal plaits ; the body-whirl towards the base has also one or two transverse articulated stripes, and the base is whitish ; the beak is short, straight, and rather broad.

INFUNDIBULUM. 115. Shell oblong, with longitudinal plaits, and elevated transverse brown striæ; umbilicus funnel-shaped, and the pillar two-plaited.

Murex Infundibulum. *Gmelin*, p. 3554.
Murex, No. 204. *Schroeter Einl.* i. p. 615.
Buccinum sulcatum. *Mus. Gevers.* p. 304, No. 576.
Fusus Ananas dictus. *Martini,* iv. p. 154, and Vign. 39, at p. 143. f. *A.*
Bonanni Rec. and *Kirch.* 3. f. 104. *Lister Conch.* t. 921. f. 14. *Seba,* iii. t. 50. f. 54.
Inhabits the West Indies. *Martini.*
Shell about three and a half inches long, and one third as broad; it consists of nine whirls of a pale brown colour, with darker brown elevated transverse stripes, and the aperture is yellowish white; the umbilicus is very large and funnel-shaped, and there are two small plaits on the pillar.

SYRACUSANUS. 116. Shell oblong, turreted, with a transverse tuberculated keel, longitudinally plaited and striated transversely; throat striated and the beak short.

Murex Syracusanus. *Linnæus Syst. Nat.* p. 1224. *Schroeter Einl.* i. p. 533. *Chemnitz,* x. p. 253. t. 162. f. 1542 and 1543. *Museum Leskianum,* p. 262. *Gmelin,* p. 3554. *Ulysses's Travels,* p. 465.
Murex asperrimus. *Gmelin,* p. 3559.
Bonanni Rec. and *Kirch.* 3. f. 80. *Favanne,* t. 34. f. D 2. *Kæmmerer Cab. Rudolst.* p. 137. t. 9. f. 7.
Inhabits the Mediterranean. *Linnæus.* Coasts of Syracuse. *Bonanni.* Sicilian Seas. *Chemnitz.* Near Naples and Tarentum. *Ulysses.*
Shell about two inches long, and one inch broad, with alternate white and yellow, or white and brownish transverse bands; the outer lip is crenulated, the pillar glabrous, and the throat striated. Ulysses mentions a Variety with white and crimson bands.

PERRON. 117. Shell fusiform, sub-turreted, with the whirls flat above, and transversely keeled and marginated; beak straight and rather long.

Murex Perron.   *Chemnitz*, x. p. 278. t. 164. f. 1573 **and** 1574.   *Gmelin*, p. 3559.   *Schreibers Conch.* i. p. 238. *D'Avila*, t. 5. f. L.
Inhabits the South Sea.   *Chemnitz*.
Shell near an inch and a half long, and about seven lines **broad**, of a brownish white colour ; it has about seven whirls, of which a small space at the upper end is flat, and separated by a projecting marginated keel.

CRATICULATUS,   118. Shell oblong, with rounded, longitudinally plaited and reticulated whirls; aperture toothed, and the pillar plaited ; beak rather short.

Murex craticulatus.   *Linnæus Syst. Nat.* p. 1224.   *Martini*, iv. p. 224. t. 149. f. 1382 and 1383.   *Born Mus.* p. 319.   *Gmelin*, p. 3554.
Voluta craticulata.   *Gmelin*, p. 3554.   *Schreibers Conch.* i. p. 121.
Voluta, No. 159.   *Schroeter Einl.* i. p. 284.
Lister Conch. t. 919. f. 13, and t. 967. f. 22.   *Seba*, iii. t. 50. f. 55 and 56, and t. 51. f. 31 and 32.   *Knorr*, ii. t. 3. f. 6.
Inhabits the Mediterranean.   *Linnæus*.
Shell about two inches long, and three-quarters of an inch broad, with rounded longitudinal chestnut-coloured plaits, crossed by transverse elevated striæ, and the interstices white ; the outer lip is crenulated, and striated within ; the pillar has three or four oblique plaits.

PARDALIS.   119. Shell sub-ventricose, with longitudinal plaits and transverse striæ ; summit obtuse, and the beak long.

Murex Pardalis.   *Gmelin*, p. 3557.   *Schreibers Conch.* i. p. 233.
Murex, No. 228.   *Schroeter Einl.* i. p. 624.
Knorr, ii. t. 3. f. 4.   *Martini*, iv. p. 226. t. 149. f. 1384.
Inhabits ――
This shell appears by the figures to be nearly two inches long, and about ten lines broad, of a whitish colour, with rows of violet spots on the longitudinal plaits. It is not by any means a well defined species.

PRISMATICUS. 120. Shell ovate, with longitudinal plaits, and transverse iridescent elevated belts ; beak short and straight, and the outer lip slightly toothed.

Murex prismaticus. *Chemnitz,* x. p. 282. t. 169. f. 1635 and 1636. *Gmelin,* p. 3559.

Buccinum prismaticum. *Martyn Univ. Conch.* i. t. 2.

Buccinum Iris. *Portland Cat.* p. 14, lot 301.

Inhabits the coasts of the Friendly Islands. *Martyn.* Pulo Condore. *Chemnitz.*

Shell nearly two inches long, and three-quarters of an inch broad, and is remarkable for its elevated glittering bluish green transverse belts ; the inside is yellowish, and the outer lip marked at the margin with bright blue striæ ; Mr. Humphreys in the Portland Catalogue says, ' the epidermis of this singular species when wet is of various colours, and it is exceeding scarce.' He has called it the *Buccinum Iris* of Solander, but among Solander's MSS. in Sir Joseph Banks's Library, there is not any species described under this name.

MAROCCENSIS. 121. Shell fusiform, narrow, with longitudinal nodulous plaits, and thickish transverse striæ ; whirls reversed, and the beak short and straight.

Murex Maroccensis. *Gmelin,* p. 3558.

Murex, No. 287. *Schroeter Einl.* i. p. 644.

Fusus Maroccanus. *Chemnitz,* ix. part 1. p. 62. t. 105. f. 896.

Buccinum Scævolum. *Mus. Gevers.* p. 304, No. 575.

Martini, n. Mannigfalt, iv. p. 422. t. 2. f. 17 to 19. *Schroeter Inn. Bau Conch.* p. 47. t. 4. f. 6. *Favanne,* t. 33. f. A 6.

Inhabits the coasts of Morocco. *Martini.*

Shell about an inch and a quarter long, and one fourth as broad, of an uniform pale yellowish brown colour ; it has eight reversed whirls, and the aperture at the upper end is rounded, and terminates below in a short straight beak.

HARPA. 122. Shell ventricose, with longitudinal transversely striated ribs ; spire prominent, and the whirls flattened above.

Murex Harpa. *Gmelin,* p. 3554.

Murex, No. 211. *Schroeter Einl.* i. p. 618.

Martini, iv. p. 166. t. 142. f. 1328 to 1330.
Inhabits ——
This species appears by the figures to vary from an inch to an
inch and a half long, and the breadth is about two-fifths of
the length; the two larger figures are yellowish brown, and
the smaller one (which also differs in having a more slender
form, and is probably a different species) is white, with two
or three narrow brown transverse stripes.

BAMFFIUS.   123. Shell ventricose, white, with acute
    longitudinal ribs; beak rather short, and
    slightly ascending.

Murex Bamffius. *Donovan,* v. t. 169. f. 1.   *Maton and
    Racket, in Lin. Trans.* viii. p. 149.   *Montagu Supp.* p.
    117.
Inhabits the coasts of Scotland. *Donovan.*   England. *Mont-
    agu.*
Shell half an inch, or sometimes an inch long, and half as
broad, with six ventricose whirls; Mr. Montagu says, that
younger shells are most commonly rufous brown, and the
ribs, which are seldom less than twenty in number, frequently white, but as they advance in age they become wholly
white.

GRACILIS.   124. Shell turreted, with longitudinal
    ribs interrupted by a flat space at the sutures,
    and transversely striated; beak rather long,
    and slightly ascending.

Murex gracilis. *Montagu Test.* p. 267, and p. 586. t. 15.
    f. 5.   *Maton and Racket, in Lin. Trans.* viii. p. 143.
    Dorset Cat. p. 46. t. 14. f. 18.
Murex marginatus.   *Donovan,* v. t. 169. f. 2.
Inhabits the coasts of Great Britain. *Montagu, &c.*
Shell about seven-eighths of an inch long, and two-eighths
broad, of a yellowish white colour, with a white band round
the middle of the body-whirl; it has nine or ten whirls, and
eleven or twelve longitudinal ribs; the ribs are not continuous, but are separated by a flat space at the upper extremity
of each whirl, and the transverse striæ are there continued
uninterruptedly in a spiral direction up the shell.

ATTENUATUS.   125. Shell fusiform, with the whirls
    sub-continuous, and nine equidistant strong
    ribs; beak rather long and straightish.

Murex attenuatus. *Montagu Test.* p. 266. t. 9. f. 6. *Maton and Racket, in Lin. Trans.* viii. p. 143.

Inhabits the coasts of the West of England. *Montagu.*

Shell half an inch long, and one eighth broad, with eight whirls regularly tapering to a very fine point, and destitute of striæ; colour yellowish white without markings, and it may be distinguished from *M. costatus* by its more slender form, larger size, and more numerous whirls.

NEBULA. 126. Shell turreted, with longitudinal ribs, and minutely reticulated; beak short and slightly bent.

Murex Nebula. *Montagu Test.* p. 267. t. 15. f. 6. *Maton and Racket, in Lin. Trans.* viii. p. 143. *Dorset Cat.* p. 46. t. 14. f. 16.

Inhabits the coasts of the West of England, and South Wales. *Montagu.*

Shell about half an inch long, and one third as broad, with eight whirls, and about the same number of longitudinal ribs; the colour is generally yellowish white, or sometimes pale reddish or purplish brown, and Mr. Montagu says it has been found of a blush-colour, with the decussated striæ white.

COSTATUS. 127. Shell oblong, with eight longitudinal elevated ribs; beak very short and obsolete.

Murex costatus. *Pennant Zool.* iv. p. 125. t. 79. *Donovan,* iii. t. 91. *Montagu Test.* p. 265. *Maton and Racket, in Lin. Trans.* viii. p. 144. *Dorset Cat.* p. 46. t. 14. f. 4.

Murex truncatus. *Gmelin,* p. 3547. ?

Buccinum costatum. *Da Costa Brit. Conch.* p. 128. t. 8. f. 4.

Muller, in Acta Nidros. iv. p. 97. t. 16. f. 26. ?

Inhabits the coasts of Great Britain. *Pennant, &c.* Norway. *Muller.*

Shell about three tenths of an inch long, and one eighth of an inch broad, with about six whirls, and eight equi-distant longitudinal ribs; the colour is whitish, pale purplish, or brown, generally more or less distinctly marked with transverse chestnut stripes, and I have sometimes found it in Langlan Bay near Swansea, with the ribs white, and their interstices purplish brown.

PROXIMUS.   128. Shell oblong, with about eleven longitudinal elevated ribs; beak short, and rather spreading at the end.

Murex proximus. *Montagu Supp.* p. 118. t. 30. f. 8.

Inhabits the coasts of Scotland near Dunbar. *Montagu.*

Shell nearly half an inch long, and Mr. Montagu says it might be readily mistaken for an extraordinary growth of *M. costatus*, " but it differs materially from that species in the number of ribs, being possessed of eleven on the body or lower volution, whereas the *costatus* never has more than eight, and usually only seven, and those are broader."

SEPTANGULARIS.   129. Shell oblong, with seven longitudinal continuous ribs scarcely interrupted by the sutures; beak short.

Murex septangularis. *Montagu Test.* p. 268. t. 9. f. 5. *Maton and Racket, in Lin. Trans.* viii. p. 144.

Murex septem-angulatus. *Donovan,* v. t. 179. f. 4.

Inhabits the coasts of the West of England. *Montagu, &c.*

Shell about five-eighths of an inch long, and two-eighths broad, of a pale purplish brown colour, and sometimes white at the sutures; it has seven or eight whirls, with seven longitudinal continuous ribs, which give the shell an heptagonal appearance.

TURRICULA.   130. Shell turreted, with longitudinal plaits and transverse striæ; whirls sub-cylindrical, and flattened at their upper extremities; beak short and broad.

Murex Turricula. *Montagu Test.* p. 262. t. 9. f. 1. *Maton and Racket, in Lin. Trans.* viii. p. 144. *Dorset Cat.* p. 47. t. 14. f. 15.

Murex angulatus. *Donovan,* v. t. 156.

Common on the coasts of England, Wales, and Ireland.

Shell about three-quarters of an inch long, and rather more than one third as broad, of a brownish white colour, and somewhat glossy; the whirls rise almost perpendicularly from each other, and form an abrupt slope at their summits; the aperture is oblong and narrow, and ends in a broad beak.

RUFUS.   131. Shell turreted, with longitudinal plaits and transverse striæ, and the whirls convex; beak short.

Murex rufus. *Montagu Test.* p. 263. *Maton and Racket, in Lin. Trans.* viii. p. 145.
Not uncommon on the coasts of England and South Wales.
Shell about three-tenths of an inch long, and one eighth broad, of an uniform reddish brown or chestnut-colour; it differs from *M. Turricula* in colour, in its smaller size, and in the form of the whirls, which are much more rounded, and not flattened at their summits.

SINUOSUS. 132. Shell turreted, with seven longitudinal ribs, and minute transverse striæ; beak very short, and the upper angle of the outer lip channelled.

Murex sinuosus. *Montagu Test.* p. 264. t. 9. f. 8. *Maton and Racket, in Lin. Trans.* viii. p. 145.
Found on the shore near Weymouth, by Mr. Bryer. *Montagu.*
Shell three-quarters of an inch long, and one third as broad, white, with six whirls, and seven much elevated longitudinal ribs; Mr. Montagu says, " it is at once distinguished from all other British shells, by the singular sinus or gutter in the upper part of the outer lip."

LINEARIS. 133. Shell with about nine longitudinal ribs crossed by elevated striæ, and the whirls convex; outer lip crenated, and the beak slightly produced.

Murex linearis. *Montagu Test.* p. 261. t. 9. f. 4. *Maton and Racket, in Lin. Trans.* viii. p. 148.
Murex elegans. *Donovan,* v. t. 179. f. 3.
Inhabits the coasts of the West of England. *Montagu.*
Shell about a quarter of an inch long, and half as broad, of a pale yellowish brown colour, with ochreous transverse bands, and the summit purplish; it has seven or eight rounded whirls, and nine or ten longitudinal strong ribs, which are crossed by elevated transverse striæ; the aperture is oval, and ends in a slightly produced straight beak.

PURPUREUS. 134. Shell oblong, acuminated, with convex cancellated whirls; pillar striated, and somewhat tuberculated; outer lip crenated.

Murex purpureus. *Montagu Test.* p. 260. t. 9. f. 3. *Maton and Racket, in Lin. Trans.* viii. p. 148.

Inhabits the coast of Devonshire. *Montagu.*
Shell about five-eighths of an inch long, and a quarter of an
inch broad, of a dark purple colour, sometimes marked with
a few irregular spots or blotches of white; Mr. Montagu
says it has nine or ten rounded whirls, tapering to an extremely fine sharp point, and furnished with nineteen or
twenty somewhat oblique longitudinal ribs, which are crossed
by numerous sharp ridges, rising into angles upon the ribs,
and making the shell very rough.

MURICATUS.   135. Shell oblong, and very rough;
    whirls ventricose, with longitudinal ribs and
    transverse elevated striæ, forming tubercles
    at their intersections; beak long and narrow.

Murex muricatus.   *Montagu Test.* p. 262. t. 9. f. 2. *Maton and Racket, in Lin. Trans.* viii. p. 149.
Inhabits the coasts of Devonshire. *Montagu.*
Shell about half an inch long, and half as broad, generally covered with an orange red epidermis, or extraneous matter, beneath which it is white tinged with flesh-colour, or sometimes
with green; Mr. Montagu says the aperture is oval, terminating in a long slender canal, which in length together rather exceed the rest of the shell; outer lip sharp, and den-
tated on the edge; margin within crenulated, and the pillar-
lip smooth.

MINUTISSIMUS.   136. Shell minute, with five spirally
    striated whirls, and remote ribs; beak closed.

Murex minutissimus.   *Adams in Lin. Trans.* iii. p. 65.
    Montagu Test. p. 273.   *Maton and Racket, in Lin.
    Trans.* viii. p. 149.
Inhabits the sea-sand on the coast of Pembrokeshire. *Adams.*
The only addition which Mr. Adams has made to the above
short character, is, that the shell is thin and pellucid.

ARENOSUS.   137. Shell minute, with decussated
    ribs on the lower whirls, and the three upper
    ones smooth; body-whirl very large, and the
    beak acute; aperture oval, and the outer lip
    toothed without.

Murex arenosus.   *Gmelin,* p. 3558.
Murex, No. 286.   *Schroeter Einl.* i. p. 644.
Spengler, n. Schr. Daen. i. p. 373. t. 2. f. 8.

Inhabits sand on the shores of India. *Spengler.*
No addition has been made by Gmelin to the above specific character, and I have not a copy of Spengler's work to refer to.

SCRIPTUS. 138. Shell fusiform, smooth, of a pale colour, with longitudinal brown striæ; inner lip toothed.

Murex scriptus. *Linnæus Syst. Nat.* p. 1225. *Gmelin,* p. 3554.
Inhabits the Mediterranean. *Linnæus.*
Linnæus says that this shell is larger than a grain of barley, smooth all over, of a whitish colour, with flexuous longitudinal dark stripes, and has only a very small beak.

***** *Turreted and subulate, with a very short Beak.*

OBELISCUS. 139. Shell ventricose, with four granulated ribs, of which the uppermost is tuberculated; pillar with one tooth, and the beak ascending.

Murex Sinensis. *Gmelin,* p. 3542. *Schreibers Conch.* i. p. 237.
Murex, No. 44. *Schroeter Einl.* i. p. 560.
Strombus acanthinus. *Mus. Geversianum,* p. 292. No. 483.
Cerithium Obeliscus. *Bruguiere Enc. Meth.* p. 472.
Obeliscus Chinensis. *Martini,* iv. p. 325. t. 157. f. 1489.
Lister Conch. t. 1018. f. 80. *Petiver Gaz.* t. 152. f. 4.
Gualter, t. 56. f. M. *Argenville,* t. 11. f. F. *Seba,* iii. t. 50. f. 26 and 27, and t. 51. f. 26. *Favanne,* t. 39. f. C 6.
Inhabits the coasts of Jamaica and Barbadoes. *Lister.* St. Domingo and Guadaloupe. *Bruguiere.*
Shell about two inches long, and the breadth is nearly two-fifths of the length; the colour is brownish, variegated with darker and white spots, and transverse brown lines; the body-whirl has four narrow transverse ribs, of which the lowermost is only minutely granulated, and the uppermost strongly tuberculated; the pillar is very short and reflected. Gmelin's name of *Sinensis* is improper for a West India species, and has been derived from some fanciful resemblance which it has been supposed to bear to a Chinese clock-tower.

VERTAGUS. 140. Shell ventricose, with the upper halves of the whirls longitudinally plaited ; pillar with one plait, and the beak ascending.

Murex Vertagus. *Linnæus Syst. Nat.* p. 1225. *Martini,* iv. p. 319. t. 156. f. 1479, and t. 157. f. 1480. *Born Mus.* p. 320. *Schroeter Einl.* i. p. 534. *Gmelin,* p. 3560. *Schreibers Conch.* i. p. 234. *Burrow's Elements,* p. 165. t. 18. f. 6.
Cerithium Vertagus. *Bruguiere Enc. Meth.* p. 473.
Bonanni Rec. and *Kirch.* 3. f. 84. *Lister Conch.* t. 1020. f. 83. *Rumphius,* t. 30. f. K. *Petiver Gaz.* t. 56. f. 4, and *Amb.* t. 13. f. 14. *Gualter,* t. 57. f. D. *Argenville,* t. 11. f. P. *Klein Ost.* t. 7. f. 118. *Seba,* iii. t. 50. f. 42, and t. 51. f. 24, 33, and 34. *Knorr,* vi. t. 40. f. 4 and 5. *Favanne,* t. 39. f. C 16.

Variety. Less ventricose, and ornamented with yellowish or reddish transverse stripes.
Murex Vertagus, Var. β. *Schroeter Einl.* i. p. 534. *Gmelin,* p. 3560. *Schreibers Conch.* i. p. 234.
Cerithium fasciatum. *Bruguiere Enc. Meth.* p. 474.
Turbo fasciatus. *Martini,* iv. p. 321. t. 157. f. 1481 and 1482.
Lister Conch. t. 1021. f. 85 b. *Gualter,* t. 57. f. F and H. *Seba,* iii. t. 50. f. 43 and 44. *Knorr,* iii. t. 20. f. 3, and v. t. 15. f. 6. *Favanne,* t. 39. f. C 15.
Inhabits the coasts of Amboyna. *Rumphius.* East Indies. *Martini.*

Shell about two inches and a half long, and rather more than one-fourth as broad, of a milk-white colour, when deprived of the brown epidermis with which live shells are covered ; the upper half of the lower whirls is strongly plaited longitudinally, and the uppermost whirls are almost wholly covered by the plaits ; the beak is rather long, and bent so as to form a right angle with the body-whirl. The variety is narrower, and has the plaits less prominent, and crossed by more distinct transverse striæ.

PLICATULUS. 141. Shell ventricose, transversely striated, with nodulous longitudinal plaits on the whirls, and the aperture oval.

Murex plicatulus. *Gmelin,* p. 3561. *Schreibers Conch.* i. p. 206.
Murex, No. 43. *Schroeter Einl.* i. p. 559.
Buccinum fuscatum. *Bruguiere Enc. Meth.* p. 282. ?

Martini, iv. t. 157. f. 1488.

Inhabits the East Indies. *Martini*.

This shell, as figured by Martini, appears to be sixteen lines long, and half as broad, of a yellowish colour with white longitudinal plaits, which extend from the upper end nearly to the base of the whirls, except in the body-whirl, of which they occupy a much smaller proportion; it has eight whirls, and a solitary ridge on the pillar.

ALUCO. 142. Shell with a transverse tuberculated line on the lower whirls, and the upper ones transversely striated; beak ascending.

Murex Aluco. *Linnæus Syst. Nat.* p. 1225. *Brookes's Introd.* p. 120. t. 7. f. 92.

Murex Aluco, Var. β. *Schroeter Einl.* i. p. 537. *Gmelin*, p. 3560. *Schreibers Conch.* i. p. 235.

Murex coronatus. *Born Mus.* p. 322.

Turbo muricatus, &c. *Martini*, iv. p. 317. t. 156. f. 1478.

Cerithium Aluco. *Bruguiere Enc. Meth.* p. 476.

Lister Conch. t. 1017. f. 79. *Rumphius*, t. 30. f. N. *Petiver Gaz.* t. 153. f. 2, and *Amb.* t. 13. f. 23. *Argenville*, t. 11. f. H. *Gualter*, t. 57. f. A. *Seba*, iii. t. 50. f. 37 and 39, and t. 51. f. 22, 23, 25, and 27. *Knorr*, iii. t. 16. f. 5. *Favanne*, t. 39. f. C 10.

Inhabits the Mediterranean. *Linnæus?* Coasts of Amboyna. *Rumphius.* Coromandel. *Bruguiere.*

Shell commonly about two inches long, and three quarters of an inch broad, of a greyish white colour, dotted and spotted with brown; the whirls have a transverse elevated line or keel, armed with remote compressed tubercles, which become gradually smaller, and disappear towards the summit of the spire; the pillar-lip has a strong rib near its upper extremity.

TUBEROSUS. 143. Shell transversely striated, with large pointed nodules on the lower part of the whirls; outer lip strongly grooved, and the beak slightly recurved.

Murex Aluco, Var. *Linnæus Mus. Lud. Ulr.* p. 643.

Murex Aluco. *Born Mus.* p. 321. *Martini*, iv. p. 314. t. 156. f. 1473 and 1474. *Schroeter Einl.* i. p. 534. *Gmelin*, p. 3560. *Schreibers Conch.* i. p. 235.

Cerithium nodulosum. *Bruguiere Enc. Meth.* p. 478. *Lamarck Syst. des Anim.* p. 85.

Lister Conch. t. 1025. f. 87. *Rumphius*, t. 30. f. O. *Pe-

tiver Amb. t. 7. f. 12.   *Gualter*, t. 57. f. G.   Seba, iii.
t. 50. f. 16.   Knorr, i. t. 16. f. 4.   *Favanne*, t. 39.
f. C 5.

Inhabits the coasts of Amboyna. *Rumphius.* Red Sea. *Forskael.*

Shell four or five inches long, and about an inch and three-
quarters broad, of a whitish colour, with dark brown transverse striæ, which are generally alternately larger; the base
of the body-whirl is transversely ribbed and tuberculated, and
all the lower whirls have a row on their under halves of large
compressed somewhat pointed nodules; the inside of the outer lip is strongly grooved, and the beak is much straighter
than in *M. Aluco.*

ADANSONI.   144.   Shell ventricose, with transverse
striæ, and a row of conical tubercles on each
whirl; outer lip crenulated.

Murex Aluco, Var. ζ.   *Gmelin,* p. 3561.
Cerithium Adánsoni.   *Bruguiere Enc. Meth.* p. 479.
Le Cerite.   *Adanson Senegal,* p. 155. t. 10. f. 2.
Gualter, t. 57. f. B.   Seba, iii. t. 50. f. 15.
Inhabits the River Gambia, opposite the factory. *Adanson.*
Shell about two inches long, and half as broad; it is shorter and
proportionably broader than the preceding species, to which
it is very nearly allied, and with which it appears to have
been frequently confounded; Adanson says that young shells
are white, and that with age they become spotted with brown.

CLAVA.   145.   Shell turreted, solid, with transverse striæ, and nodulous plaits above; outer lip double, and the beak ascending.

Murex clava.   *Chemnitz,* x. 256, and Vign. 22. at p. 233.
f. *A* and *B*.   *Gmelin,* p. 3565.
Clava maculata.   *Martyn Univ. Conch.* ii. t. 57.
Cerithium clava.   *Bruguiere Enc. Meth.* p. 479.
Inhabits the coasts of Pulo Condore. *Martyn.*
This shell appears by the figures to be about six inches long,
and an inch and three-quarters broad, and is of a yellowish
colour somewhat tesselated with darker spots; it has fifteen
or sixteen flattish whirls, with six or seven transverse narrow
grooves in each, and plaited at their upper extremities; the
outer lip is not crenulated on the margin, but within it is
slightly grooved, and the beak is bent upwards.

UNCINATUS. 146. Shell turreted, transversely striated, with the four lower whirls armed in the middle with hooked spines, the fifth and sixth ribbed, and the others glabrous.

Murex uncinatus. *Gmelin,* p. 3542.
Murex, No. 198. *Schroeter Einl.* i. p. 611.
Strombus Tympanorum. *Schroeter Fluss.* p. 379. t. 8. f. 15.
Inhabits ——
Shell about an inch and a quarter long, and half as broad, of a pale blackish brown colour; it is said to have nine whirls, an oval aperture, and the outer lip toothed, but it appears to be rather a doubtful species.

ATRATUS. 147. Shell turreted, with transverse striæ, and two tuberculated belts on each whirl; outer lip striated within, and the beak nearly straight.

Murex atratus. *Born Mus.* p. 329. t. 11. f. 17 and 18.
Gmelin, p. 3564. *Schreibers Conch.* i. p. 211.
Murex, No. 177. *Schroeter Einl.* i. p. 601.
Cerithium atratum. *Bruguiere Enc. Meth.* p. 480.
Inhabits ——
Shell about an inch and a quarter long, and half an inch broad, of a blackish colour; it has, according to Bruguiere, fifteen whirls, on each of which there is a granulated belt in the middle, and another near the upper extremity; Born has described the beak to be straight, and the pillar one-plaited; but Bruguiere says the former is slightly curved, and that the latter is not plaited at all. I have a shell with alternate varices, which in other respects answers the description, but is of a reddish brown colour,

ALUCOIDES. 148. Shell transversely striated with spinous plaits below the middle of the whirls, and crenulated above; beak slightly recurved.

Murex alucoides. *Olivi Adriat.* p. 153.
Murex Aluco, Var. δ, and ε. *Gmelin,* p. 3561.
Murex, No. 196. *Schroeter Einl.* i. p. 496.
Strombus thymelus. *Mus. Gevers.* p. 292. No. 482.
Cerithium vulgatum. *Bruguiere Enc. Meth.* p. 481.
Le Goumier. *Adanson Senegal,* p. 156. t. 10. f. 3.
Bonanni Rec. and *Kirch.* 3. f. 82. *Lister Conch.* t. 1019.

f. 82.   *Gualter,* t. 56. f. L.   *Seba,* iii. t. 50. f. 23. *Fa-vanne,* t. 39. f. C 1.

Inhabits the Mediterranean. *Lister.*   Adriatic. *Olivi.* Coasts of Teneriffe and Fayal. *Adanson.*   Provence and Langue-doc. *Bruguiere.*

Shell an inch and a half, or sometimes two inches and a half long, and the breadth is about one-third of the length; the lower parts of the whirls have longitudinal plaits ending in a sharp point in the middle, and the upper part is slightly con-cave, and crenulated by the suture; the colour is pale grey-ish brown, marbled with darker brown and white.

EBENINUS. 149.   Shell turreted, angulated, and transversely striated, with the whirls nodu-lous in the middle; outer lip sinuous.

Murex Aluco, Var. γ. *Schroeter Einl.* i. p. 537. *Gmelin,* p. 3560.
Murex Cochlear ebeninum. *Chemnitz,* x. p. 257. t. 162. f. 1548 and 1549.
Strombus aculeatus, Var. β.   *Gmelin,* p. 1523.
Clava Herculea. *Martyn Univ. Conch.* i. t. 13.
Cerithium ebeninum. *Bruguiere Enc. Meth.* p. 490.
Spengler Naturf. ix. p. 145. t. 5. f. 3.   *Favanne,* t. 79. f. lower N.

Inhabits the coasts of the Friendly Islands. *Martyn.* New Holland, and New South Wales. *Humphreys.* New Zea-land. *Bruguiere.*

Shell about three inches and a quarter long, and one inch and a quarter broad, with the lower whirls of a deep black, and the upper ones of a brownish black colour; it has broad lon-gitudinal plaits on the lower half of the whirls, which termi-nate in a row of pointed nodules at the middle; the aperture is whitish, with a yellowish brown border on the extended margin of the outer lip.

FUSCATUS. 150.   Shell turreted, with two trans-verse ribs on each whirl, of which one in the middle is granulated, and the other at the suture strongly muricated.

Murex fuscatus. *Linnæus Syst. Nat.* p. 1225. *Born Mus.* p. 323.   *Schroeter Einl.* i. p. 538.   *Gmelin,* p. 3562.
Murex scolopaceus. *Gmelin,* p. 3548.   *Schreibers Conch.* i. p. 214.

Strombus aculeatus. *Gmelin*, p. 3523.

Strombus Tympanorum aculeatus. *Chemnitz*, ix. part 2. p. 193. t. 136. f. 1267 and 1268.

Cerithium muricatum. *Bruguiere Enc. Meth.* p. 490.

Lister Conch. t. 121. f. 17. *Argenville*, t. 14. f. &. *Klein Ost.* t. 2. f. 39. *Knorr*, iii. t. 26. f. 4. *Favanne*, t. 39. f. C 19.

Inhabits the Mediterranean. *Linnæus?* Rivers of Senegal within reach of the tide. *Bruguiere.*

Shell about two inches long and ten lines broad, coated with a dark purplish brown epidermis; a small granulated stripe attends the middle of the whirls, and at the sutures, which are obsolete, there is a row of strong tubercles pointing to the apex; the base is flattish, marked with concentric grooves, and on the body-whirl near its margin there is a double row of small aculeated nodules; the outer lip is much expanded, and nearly horizontal. The description is so short, and the two figures referred to in the Systema Naturæ are so unlike each other, that it is impossible to come to any certain decision, but the present shell is most probably the Linnæan *M. fuscatus.* If, however, the expanded lip is the mark, by which the turreted Strombi are to be distinguished from the present division of Murices, this shell should be placed with the former; but in all respects they approach each other so closely, that it is almost impossible to draw any separating line.

TORULOSUS. 151. Shell turreted, very finely striated transversely, with a convex belt at the sutures of the lower, and longitudinal nodulous plaits on the upper whirls.

Murex torulosus. *Linnæus Syst. Nat.* p. 1226. *Gmelin*, p. 3563.

Murex annularis. *Gmelin*, p. 3561.

Murex Larva Erucæ. *Chemnitz*, x. p. 280. t. 164. f. 1575, and 1576. *Gmelin*, p. 3559. *Schreibers Conch.* i. p. 230.

Murex fuscus. *Gmelin*, p. 3561.

Murex, No. 41. *Schroeter Einl.* i. p. 558.

Turbo annulatus, &c. *Martini*, iv. p. 325. t. 157. f. 1486.

Cerithium torulosum. *Bruguiere Enc. Meth.* p. 482.

Lister Conch. t. 121. f. 16. *Klein Ost.* t. 2. f. 38. *Martini Berlin. Mag.* iv. t. 10. f. 54.

Inhabits the East Indies? *Bruguiere.*

Shell about an inch and a quarter long; and four lines and a half broad; the lower whirls are of a whitish colour, and the upper ones brown, or tinged with violet, except the longitudinal plaits, which are whitish; the three lower whirls have a thick rounded belt at the sutures, and the whole surface is finely striated transversely; the aperture has a small canal at the upper as well as the lower extremity.

RADULA. 52. Shell turreted, with four or five transverse tuberculated ribs, of which the uppermost but one has the tubercles larger than the others.

Murex fluviatilis, *Gmelin*, p. 3562.
Murex, No. 52. *Schroeter Einl.* i. p. 562.
Nerita aculeata. *Muller Verm.* ii. p. 193.
Cerithium Radula. *Bruguiere Enc. Meth.* p. 491.
Le Popel. *Adanson Senegal*, p. 152. t. 10. f. 1.
Lister Conch. t. 122. f. 20. *Martini Berl. Mag.* iv. t. 11. f. 58.
Junior. With the aperture imperfect.
Murex Radula. *Linnæus Syst. Nat.* p. 1226. *Born Mus.* p. 324. t. 11. f. 16. *Schroeter Einl.* i. p. 539. t. 3. f. 6. ? *Gmelin*, p. 3563.
Murex Terebella. *Gmelin*, p. 3562. *Schreibers Conch.* i. p. 206.
Murex Sinensis. Var. *γ*. *Gmelin*, p. 3542.
Murex, No. 39, and No. 60. *Schroeter Einl.* i. p. 557, and p. 562.
Lister Conch, t. 122. f. 18. *Gualter*, t. 58. f. F. *Klein Ost.* t. 2. f. 40. *Martini Berlin. Mag.* iv. t. 10. f. 57, and *Conch.* iv. t. 155. f. 1459.
Inhabits the African Seas. *Linnæus.* Muddy rivers of Senegal, within reach of the tide. *Adanson.*
Shell about three inches long and one-third as broad; it differs from *M. fuscatus* in having the transverse tuberculated ribs more numerous, and the spines on the uppermost rib but one, though larger than the others, are comparatively small; it has sixteen whirls, of which it is difficult to distinguish the separating line, and the base is flattish, though not so much so as in *M. fuscatus;* it is white or pale flesh-colour, covered with a brown epidermis, and Bruguiere says that young shells sometimes have a white band at the sutures; Born also mentions a Variety with red and brown transverse stripes.

MARGINATUS. **153.** Shell turreted with two granulated ribs below and a large compressed tuberculated belt at the upper extremity of the whirls; pillar with one plait.

Murex, No. 132, and No. 133. *Schroeter Einl.* i. p. 587.
Murex fuscatus. *Montagu Test.* p. 269. ? *Maton and Racket in Lin. Trans.* viii. p. 149. t. 4. f. 6. ? *Dorset Cat.* p. 47. ?
Cerithium marginatum. *Bruguiere Enc. Meth.* p. 493.
Gualter, t. 56. f. H. *Seba*, iii. t. 50. f. 32 to 34.
Inhabits the East Indies. *Bruguiere.*
Bruguiere has described this shell to be two inches long and ten lines broad, of a blackish colour below, which changes gradually to yellowish brown at the summit; he says it is distinguishable by a large compressed tuberculated belt at the sutures, besides which each whirl of the spire has two transverse granulated ribs, and on the body-whirl these ribs are more numerous. I never saw a specimen of the shell, which is said to have been ' found by Mr. Bryer, near Weymouth, after a violent storm in 1795,' but from the description, I think it is more likely to belong to the present species than to the Linnæan *M. fuscatus.*

SERRATUS. **154.** Shell turreted, with longitudinal striæ, and two transverse serrated ribs, of which the upper is largest, serratures spinous and compressed.

Murex Sinensis, Var. .. *Gmelin*, p. 3542.
Cerithium serratum. *Bruguiere Enc. Meth.* p. 482.
Clava Rubus. *Martyn Univ. Conch.* ii. t. 58.
Inhabits the coasts of the Friendly Islands, and is found fossil in France. *Bruguiere.*
Shell about two inches and a quarter long, and hardly one-third as broad, and is white without any coloured markings; the spire consists of about fifteen whirls, with fine longitudinal striæ, and two transverse ribs armed with pointed teeth, of which the lower rib is much smaller than the other; the body-whirl has four of these ribs, and they form corresponding grooves within the aperture.

ASPER. **155.** Shell turreted, with longitudinal plaited muricated ribs, and transverse striæ; pillar with one plait, and the beak ascending.

Murex asper.   *Linnæus Syst. Nat.* p. 1226.   *Gmelin,*
    p. 3563.
Murex granulatus.   *Martini,* iv. p. 322. t. 157. f. 1483.
    Schroeter Einl. i. p. 541.   *Gmelin,* p. 3563.   *Schrei-*
    bers Conch. i. p. 236.
Cerithium asperum.   *Bruguiere Enc. Meth.* p. 475.
Lister Conch. t. 1020. f. 84.   *Argenville,* t. 11. f. K. ?
    Seba, iii. t. 50. f. 20, and t. 51. f. 35.   *Favanne,* t. 39.
    f. C 18.
Inhabits the coasts of Guinea. *Linnæus.*   Isle of France,
    Lister.   Martinique and St. Domingo. *Bruguiere.*
Shell about two inches long, and not much more than one-
    fourth as broad; it has twelve whirls, with about the same
    number of longitudinal rib-like plaits, on each of which are
    three small pointed equidistant tubercles; the colour is
    white, and is sometimes marked with three transverse nar-
    row brown stripes on each whirl; the aperture is nearly
    similar to that of *M. Vertagus,* but the beak is considerably
    shorter.

GRANULATUS.   156. Shell turreted, with five trans-
    verse striæ, and three intermediate rows of
    tubercles, of which the upper row is largest;
    pillar one-plaited.

Murex granulatus.   *Linnæus Syst. Nat.* p. 1226.
Murex cingulatus.   *Gmelin,* p. 3561.
Murex Sinensis, Var. δ.   *Gmelin,* p. 3542.
Murex, No. 46, and No. 197.   *Schroeter Einl,* i. p. 561,
    and p. 610.
Strombus circulis, &c.   *Schroeter Fluss.* p. 380. t. 9. f. 9.
Turbo granulatus minor.   *Martini,* iv. p. 328. t. 157.
    f. 1492.
Cerithium granulatum.   *Bruguiere Enc. Meth.* p. 476.
Rumphius, t. 30. f. L.   *Petiver Amb.* t. 7. f. 12.   *Klein*
    Ost. t. 7. f. 119.   *Seba,* iii. t. 50. f. 45.
Inhabits the Asiatic Ocean.   *Linnæus.*
Shell about an inch and a half, or sometimes two inches long,
    and one-fourth as broad, of a yellowish brown colour, with
    the points of its tubercles whitish; M. Bruguiere says it has
    sixteen whirls, with five transverse striæ on each, and be-
    tween these are three rows of tubercles, of which the upper
    row is rather the largest; besides the plait on the pillar, he
    mentions four deep grooves on the lip near the umbilicus,
    which are peculiar to this species.

SULCATUS. 157. Shell ovate, turreted, with longitudinal plaits and transverse ribs; beak straight, very short, and tubular.

Murex sulcatus. *Born Mus.* p. 320.
Murex Moluccanus. *Gmelin,* p. 3563. *Schreibers Conch.* i. p. 236.
Murex, No. 40. *Schroeter Einl.* i. p. 558.
Strombus fuscus. *Gmelin,* p. 3523.
Strombus, No. 59. *Schroeter Einl.* i. p. 470.
Strombus palustris, Var. *β*. *Schreibers Conch,* i. p. 190.
Turbo Mangiorum. *Martini,* iv. p. 323. t. 157. f. 1484 and 1485.
Cerithium sulcatum. *Bruguiere Enc. Meth.* p. 486.
Bonanni Rec. and *Kirch.* 3. f. 68. *Lister Conch.* t. 1021. f. 85. *Rumphius,* t. 30. f. T. *Petiver Amb.* t. 13. f. 22. *Gualter,* t. 57. f. E. *Knorr,* v. t. 13. f. 8. *Favanne,* t. 40. f. A 3.
Inhabits marshes in the Molucca islands. *Rumphius.*
Shell about an inch and a half long, and half as broad, of an olive-brown colour, generally marked with two narrow chocolate bands on the body-whirl; the transverse ribs are broken into nodules on the body-whirl, and the spire is strongly plaited longitudinally; the aperture is dilated, and the outer lip united to the pillar so as to render the beak tubular.

LITERATUS. 158. Shell turreted, ventricose, with numerous dotted muricated striæ on each whirl, and the upper stria tuberculated.

Murex literatus. *Born Mus.* p. 323. t. 11. f. 14 and 15. *Gmelin,* p. 3548. *Schreibers Conch.* i. p. 238.
Murex, No. 175. *Schroeter Einl.* i. p. 600.
Cerithium litteratum. *Bruguiere Enc. Meth.* p. 499.
Gualter, t. 56. f. N.
Inhabits the coasts of Guadaloupe. *Bruguiere.*
Shell about an inch and a quarter long, and half an inch broad; it has nine or ten whirls, with numerous transverse muricated striæ, and a row of pointed tubercles at the upper extremity of each; the colour is white, and the transverse striæ prettily dotted with black.

HEXAGONUS. 159. Shell hexagonal, with three transverse granulated ribs, and the upper rib of the body-whirl nodulous; outer lip thickened.

Murex hexagonus.   *Chemnitz*, x. p. 261. t. 162. f. 1554
and 1555. *Gmelin*, p. 3548.
Cerithium hexagonum.   *Bruguiere Enc. Meth.* p. 494.
Inhabits the South Sea, and is frequently found fossil in Europe.   *Chemnitz.*
Shell about an inch and three quarters long, and three quarters
of an inch broad, of a yellowish colour, with the lips and
inside white; besides the transverse ribs, it has six strong
longitudinal ones, which give it an hexagonal appearance;
the aperture is nearly round, and the outer lip thickened,
expanded, and angulated at the margin.

RETICULATUS.   160. Shell turreted, with longitudinal ribs, and four transverse grooves on each
whirl, forming uniform tubercles all over.

Murex reticulatus.   *Montagu Test.* p. 272.   *Maton and
Racket, in Lin. Trans.* viii. p. 150.   *Dorset Cat.* p. 47.
t. 14. f. 13.
Strombiformis reticulatus.   *Da Costa Brit. Conch.* p. 117.
t. 8. f. 13.
Gualter, t. 58. f. G.
Inhabits the coasts of Great Britain.   *Borlase, &c.*
Shell half or sometimes three quarters of an inch long, and
about one-fifth as broad, of a brownish or pale chestnut-
colour; it has eleven or twelve whirls, and each whirl is in-
tersected by four transverse grooves, so as to form an equally
tuberculated surface; the aperture has only a very small
rather obsolete beak at the base.

TUBERCULARIS.   161. Shell turreted, with longitudinal ribs, and three transverse grooves on
each whirl, forming uniform tubercles all
over.

Murex tubercularis.   *Montagu Test.* p. 270, and *Supp.*
p. 116.   *Maton and Racket, in Lin. Trans.* viii. p. 150.
Inhabits the coasts of England.   *Montagu.*
Shell a quarter of an inch long, and not more than one-fourth
as broad, and has three rows of equal tubercles on each
whirl; the aperture is oval, and ends in a small beak,
which is nearly enclosed by the pillar-lip turning inwards;
the colour is chestnut-brown.

ADVERSUS.   162. Shell turreted, with the whirls reversed, and on each three transverse rows of
tubercles, of which the middle one is smallest.

Murex adversus. *Montagu Test.* p. 271, and *Supp.* p. 115. *Maton and Racket, in Lin. Trans.* viii. p. 151.
Turbo punctatus. *Walker's Minute Shells,* f. 48. *Adams's Microscope,* p. 638. t. 14. f. 21.
Turbo reticulatus. *Donovan,* v. t. 159.
Inhabits the coasts of England and the West Indies. *Montagu.*
Shell about three-eighths of an inch long, and one-fourth as broad, of an opake pale brown colour, or sometimes white; its reversed whirls and the middle row of its granules being smaller than the others, will readily distinguish this species.

SUBULATUS. **163.** Shell turreted, with two rows of tubercles divided by a depressed line in each whirl.

Murex subulatus. *Montagu Supp.* p. 115. t. 30. f. 6.
Found among sea-sand at Scalasdale, in the Sound of Mull; very rare. *Montagu.*
Mr. Montagu has described this shell to be about three-eighths of an inch long, slender, white, with about fifteen scarcely raised whirls, defined by a purplish brown spiral line; each whirl has two rows of beads, divided by a depressed line, which is marked with minute elevated longitudinal striæ.

DECOLLATUS. **164.** Shell turreted, and the summit truncated, with longitudinal plaited ribs, and transverse striæ.

Murex decollatus. *Linnæus Syst. Nat.* p. 1226. *Schroeter Einl.* i. p. 542. *Gmelin,* p. 3563.
Cerithium decollatum. *Bruguiere Enc. Meth.* p. 501.
Inhabits——
M. Bruguiere says, this singular shell is generally about an inch long, and half as broad; and the summit appears as if it had been broken off, in the same manner as in *Helix decollatus;* the whirls have numerous longitudinal plaits, and their interstices are very finely striated transversely; it is of a yellowish brown colour, and sometimes has two darker transverse lines round each whirl.

Genus XXVII.

TROCHUS:

SHELL UNIVALVE, SPIRAL, AND NEARLY CONICAL;
APERTURE SOMEWHAT QUADRANGULAR, OR ROUND-
ED WITH THE UPPER SIDE TRANSVERSE AND CON-
TRACTED; PILLAR OBLIQUE.

Subdivisions.†

* Umbilicated and erect.
** Imperforate and erect.
*** Tapering, with the pillar exserted, and the
    shell falling to one side when placed upon its
    base.

---

* Umbilicated and erect.

NILOTICUS. 1. Shell conical, nearly smooth, and
    the whirls flattish; inner lip nearly entire.

Trochus Niloticus. *Linnæus Syst. Nat.* p. 1227. *Born
    Mus.* p. 327. *Chemnitz,* v. p. 76. t. 167. f. 1605, and t.
    168. f. 1614. *Schroeter Einl.* i. p. 647. *Gmelin,* p.
    3565. *Schreibers Conch.* i. p. 239. *Lamarck Syst. des
    Anim.* p. 86. *Brookes's Introd.* p. 123. t. 7. f. 93.
Le Grand Sabot. *Favanne,* ii. p. 299, and p. 355. t. 12.
    f. B 1.
Bonanni Rec. and *Kirch.* 2. f. 102. *Lister Conch.* t. 617.

---

† *Trochus novus,* p. 3577, and *T. Pulligo,* p. 3585, of Gmelin, are too doubtful to be retained; and *T. Schroeteri,* p. 3575, and *T. ferrugineus,* p. 3577, are
fossil species.

f. 3. *Rumphius*, t. 21. f. **A.** *Petiver Amb.* t. 3. f. 12.
Gualter, t. 59. f. B and C. *Knorr*, ii. t. 5. f. 1, and t.
6. f. 1. *Geve.* t. 5. f. 34.
Variety **B.** With the whirls slightly granulated.
Le Bouton de la Chine. *Favanne*, ii. p. 300, and p. 370.
t. 12. f. B 2.
Lister Conch. t. 620. f. 6. *Rumphius*, t. 21. f. No. 4.
Argenville, t. 8. f. C. *Geve.* t. 6. f. 45. *Chemnitz*, v.
t. 167. f. 1606 and 1607.
Variety C. With the whirls slightly granulated and undulated
at the sutures.
Trochus tigrinus. *Hebenstreit Mus. Richt.* p. 324.
Trochus pyramidalis, &c. *Chemnitz*, v. p. 80. t. 167. f.
1608 and 1609.
Lister Conch. t. 619. f. 5. *Rumphius*, t. 21. f. 3. *Geve.*
t. 6. f. 44. *Regenfuss*, i. t. 4. f. 42.
Inhabits the Indian Ocean. *Linnæus.* Coasts of Amboyna.
Rumphius. Isles of France, New Zealand, and Otaheite.
Favanne. China, and Pulo Condore. *Humphreys.*
Shell three, four, or sometimes five inches long, and about
equally broad at the base, ponderous, with the whirls flattish
above, and slightly convex below; the colour is white, with
longitudinal red stripes extending over the base, and the shell
appears silvery when the outer coat is taken off. Favanne
erroneously considered *Le Bouton de la Chine* to be the
Linnæan *T. maculatus*, and Chemnitz has arranged both this
and the *T. tigrinus* of Hebenstreit as distinct species.

CONUS. 2. Shell conical, with transverse granulated striæ, and the whirls flattish; inner lip
entire.

Trochus Conus. *Gmelin*, p. 3569. *Schreibers Conch.* i.
p. 250.
Trochus acutangulus. *Chemnitz*, v. p. 81. t. 167. f. 1610.
Trochus Niloticus, Var. *Kæmmerer Cab. Rudolst.* p. 160.
Trochus, No. 41. *Schroeter Einl.* i. p. 696.
La Poire. *Favanne*, p. 360. t. 13. f. I.
Lister Conch. t. 631. f. 18.
Inhabits the East Indian Seas. *Chemnitz.* Coasts of New
Guinea, and is very rare. *Favanne.*
Shell about two inches long, and an inch and a half in diame-
ter at the base; the whirls are flattish, or only very slightly con-
vex, and marked with beaded transverse striæ; the inner lip
is entire, and the pillar sinuous, with the umbilicus extend-
ing only through the first whirl; the whole surface is white,

variegated with rose-colour, and transverse rows of darker red spots.

SPINOSUS. 3. Shell conical, with three transverse rows of granules on each whirl, and spinous at the sutures.

Trochus spinosus. *Chemnitz*, v. p. 82. t. 167. f. 1611.
Gmelin, p. 3570. *Schreibers Conch.* i. p. 250.
Trochus, No. 42. *Schroeter Einl.* i. p. 696.
Le Petit Cul de Lampe. *Favanne*, ii. p. 299, and p. 359.
t. 12. f. B 3.
Rumphius, t. 21. f. No. 10.
Inhabits the coasts of New Zealand. *Favanne.*
Shell about an inch long, and the diameter at the base rather exceeds the length; it has nine whirls, which are armed with spines on their lower margin, and have three rows of granules in the middle; the colour is reddish white irregularly variegated with dark olive or violet, and sometimes marked with a black and white tessellated band.

JUJUBINUS. 4. Shell conical, with the whirls rather concave, and granulated on both margins; base flat, and granulated in concentric rows.

Trochus Jujubinus. *Gmelin*, p. 3570.
Trochus zezyphinus umbilicatus. *Chemnitz*, v. p. 82. t. 167.
f. 1612 and 1613.
Trochus, No. 43. *Schroeter Einl.* i. p. 697.
Le Pavot. *Favanne*, ii. p. 297, and p. 349. t. 12. f. L.
Inhabits the coasts of the Isle of France. *Favanne.* West India Islands. *Chemnitz.*
Shell about an inch long, and the base is rather more than three quarters of an inch in diameter; it has eight or nine whirls of a reddish colour, and the lower part of each is tessellated with chestnut-colour.

MACULATUS. 5. Shell conical, tuberculated, and the umbilicus oblique; inner lip toothed, and the throat striated.

Trochus maculatus. *Linnæus Syst. Nat.* p. 1227. *Born Mus.* p. 329. *Chemnitz*, v. p. 83. t. 168. f. 1615 to 1618, and t. 169. f. 1623 and 1624. *Schroeter Einl.* i. p. 648. t. 3. f. 9. *Gmelin*, p. 3566. *Schreibers Conch.* i. p. 240.

Le Cardinal. *Favanne*, ii. p. 368. f. 13. f. C.
Bonanni Rec. and *Kirch.* 3. f. 96. *Lister Conch.* t. 632.
f. 20. *Gualter*, t. 61. f. D. *Geve*. t. 8. f. 58. *Regen-fuss*, ii. t. 4. f. 30.
Inhabits the coasts of America and Asia. *Linnæus.* Molucca
Islands. *Favanne.* Madagascar. *Humphreys.*
Shell about an inch and a half, or sometimes two inches long,
and an inch and a quarter, or an inch and a half broad at the
base; it is rather ponderous, and has nine whirls covered
with tubercles of different sizes; the colour is sometimes
greenish white mottled with red, or green with chestnut
spots, and the summit is generally more or less spotted with
crimson.

ALVEARE. 6. Shell conical, longitudinally plaited,
and irregularly granulated transversely; aperture rhomboidal, and the pillar toothed.

Trochus Alveare. *Gmelin*, p. 3570. *Schreibers Conch.* i.
p. 251.
Trochus fenestratus. *Chemnitz*, v. p. 85 and 87. t. 168. f.
1619 and 1622.
Trochus vellicatus. *Meuschen Mus. Gevers.* p. 286. No.
435.
Trochus, No. 44, and No. 47. *Schroeter Einl.* i. p. 697,
and p. 699.
Le Sabot boutonné. - *Favanne*, ii. p. 369.
Gualter, t. 60. f. P. *Regenfuss*, ii. t. 5. f. 45.
Inhabits the coasts of the Moluccas, and the Isle of France.
Favanne.
Shell about an inch and a half long, and the diameter at the
base rather exceeds the length; the colour is whitish mottled
with pale green, or sometimes with reddish brown, and the
summit is often dotted with crimson; it is slightly ribbed or
plaited longitudinally, and has eight or nine whirls, with four
transverse rows of granules on each, of which the lower ones
are largest.

CONCAVUS. 7. Shell conical, with longitudinal
somewhat undulated ribs; base concave,
with a large umbilicus in the centre, and
toothed at the margin.

Trochus concavus. *Gmelin*, p. 3570.
Trochus, No. 45. *Schroeter Einl.* i. p. 698.
Le Bonnet vert. *Favanne*, ii. p. 354.

Gualter, t. 63. f. A.? *Chemnitz*, v. p. 86. t. 168. f. 1620 and 1621.

Inhabits the coasts of Coromandel. *Chemnitz.* Isle of France, Amboyna, and New Zealand. *Favanne.*

Shell an inch and a half, or two inches long, and about equally broad at the base, of a greenish colour, more or less tinged with brown, and marked with longitudinal slightly undulated ribs; the base is remarkably hollow and concave, and has a large cavity at its centre.

VERNALIS. 8. Shell conical, with transverse nodulous belts, which are larger at the sutures, and the upper whirls white.

Trochus vernalis. *Chemnitz*, v. p. 89. t. 169. f. 1625 and 1626.

Trochus vernus. *Gmelin*, p. 3571. *Schreibers Conch.* i. p. 251.

Trochus Diadema. *Ulysses's Travels*, p. 468.

Trochus, No. 49. *Schroeter Einl.* i. p. 700.

Inhabits the East Indies. *Chemnitz.* Bay of Naples. *Ulysses.*

Shell about an inch and a quarter long, and the breadth, by Chemnitz's figure, appears rather to exceed the length; the transverse row of nodules next the sutures is larger than the others, and Ulysses says the sutures are marked by an elevation like a pearl necklace; the colour is greenish, except the upper whirls, which are white and speckled with brown.

CONSPERSUS. 9. Shell conical, with transverse rows of nodules, and the lower margin of the whirls glabrous; base flat, and white spotted with red.

Trochus conspersus. *Gmelin*, p. 3571.

Trochus, No. 50. *Schroeter Einl.* i. p. 700.

Le Pavot boutonné. *Favanne*, ii. p. 349.?

Gualter, t. 60, f. B. *Chemnitz*, v. p. 90. t. 169. f. 1627.

Inhabits the East Indian Seas. *Chemnitz.*

This shell appears, by Chemnitz's figure, to be fourteen lines long, and about equally broad, and the lower whirls are said to be covered with white, greenish, and buff-coloured spots, and the summit with black and red ones; the inside is pearly.

TENTORIUM. 10. Shell conical, green, somewhat obliquely ribbed, and the margin of the upper whirls nodulous; base flat, and the pillar slightly plaited.

Trochus Tentorium. *Chemnitz*, v. p. 90. t. 169. f. 1628. *Gmelin*, p. 3571. *Schreibers Conch.* i. p. 252. Trochus, No. 51. *Schroeter Einl.* i. p. 700.

Inhabits the East Indian Seas. *Chemnitz.*

Shell about an inch long, and equally broad, and is said to be covered with a hyaline epidermis, beneath which the colour is sea-green; the whirls have concatenated dots in the middle, and the two lower ones are much contracted at the sutures; base flat, white, with concentric granules, and a funnel-shaped umbilicus in the middle.

OCHROLEUCUS. 11. **Shell conical, with transverse obliquely crenated ribs; base flat, and white with red spots.**

Trochus ochroleucus. *Gmelin*, p. 3571. *Schreibers Conch.* i. p. 252.

Trochus, No. 52. *Schroeter Einl.* i. p. 701.

Apiarium Chinense. *Chemnitz*, v. p. 91. t. 169. f. 1629.

Variety. Variegated with white, dark red, blue, and green. Trochus, No. 55. *Schroeter Einl.* i. p. 702.

Trochus, No. 57. *Schreibers Conch.* i. p. 252.

Chemnitz, v. t. 169. f. 1632.

Inhabits the East Indian Seas. *Chemnitz.*

This shell appears, by Chemnitz's figure, to be an inch and a quarter long, and about equally broad, of a pale whitish brown colour, with oblique greenish stripes, and some crimson spots at the summit; the variety is more variegated, and has a row of dark red oblique broad longitudinal stripes on the body-whirl.

STELLATUS. 12. **Shell conical, greenish, with wrinkled plaits and concatenated dots, and the sutures of the upper whirls spinous and radiated.**

Trochus stellatus. *Chemnitz*, v. p. 91. t. 169. f. 1630. *Gmelin*, p. 3571. *Schreibers Conch.* i. p. 252.

Trochus, No. 53. *Schroeter Einl.* i. p. 701.

Inhabits the Indian Seas. *Chemnitz.*

Shell about thirteen lines long, and fourteen broad, and Chemnitz's figure is painted green, except the body-whirl, which is pale greenish white with crimson oblique broad longitudinal stripes.

SPENGLERI.   13. Shell conical, with transverse
    rows of equal ochraceous nodules, undulated
    with red, and one part of the base smooth.

Trochus Spengleri.   *Chemnitz,* v. p. 92. t. 169. f. 1631.
    Gmelin, p. 3571.   *Schreibers Conch.* i. p. 252.
Trochus, No. 54.   *Schroeter Einl.* i. p. 701.
Inhabits ———
This shell, in Chemnitz's figure, appears to be about eleven lines
    long, and near thirteen broad, of a whitish colour tinged with
    pale red in longitudinal waves, although the nodules are said
    to be ochraceous.

COSTATUS.   14. Shell conical, ribbed and grooved,
    with alternate red and white longitudinal
    stripes ; base flattish, white, and striped with
    rose-colour.

Trochus costatus.   *Gmelin,* p. 3571.   *Schreibers Conch.* i.
    p. 253.
Trochus, No. 56.   *Schroeter Einl.* i. p. 702.
Chemnitz, v. p. 93. t. 169. f. 1633 and 1634.
Inhabits ———
This appears to me to be a badly defined species, and the
    above description is taken from Chemnitz, whose figure re-
    presents a shell about fifteen lines long, and fourteen broad,
    with alternate red and whitish longitudinal stripes.   Gme-
    lin's description is quite different, and is as follows : " Shell
    with elevated concatenated dots ; whirls with oblong white
    nodules at the base, and intermediate purple grooves." Both
    these authors have quoted Argenville, t. 8. f. T. and this is
    said to be " remarquable par la quantité de ses boutons
    blancs, très saillans, et de différentes grosseurs ; on compte
    sur un fond minime, trois rangs de petits boutons entre les
    grands."   Favanne has referred to the same figure of Ar-
    genville's for his *Sabot cerclé,* and considered it to be the
    Trochus scaber of Linnæus.

INÆQUALIS.   15. Shell conical, with several trans-
    verse rows of nodules on each whirl, varying
    in size, and largest above the sutures.

Trochus inæqualis.   *Chemnitz,* v. p. 93. t. 170. f. 1635 and
    1636.   *Gmelin,* p. 3572.   *Schreibers Conch.* i. p. 253.
Trochus, No. 57.   *Schroeter Einl.* i. p. 702.
Le Sabot bourgeonné.   *Favanne,* ii. p. 364. t. 12. f. D.
Gualter, t. 60. f. O.

Inhabits the coasts of Mozambique and Zanguebar. *Favanne.*
Shell about an inch and a quarter, or an inch and a half long,
with five rows of irregular tubercles on each whirl, of which
the largest are those at the base immediately above the sutures; the colour is generally greenish white mottled with
red, and sometimes with olive; the base is granulated, and
white with reddish rays.

REGIUS. 16. Shell conical, and the whirls concave,
with transverse rows of granules, which are
larger both above and below the sutures;
umbilicus funnel-shaped.

Trochus regius. *Chemnitz,* v. p. 94. t. 170. f. 1637. *Gmelin,* p. 3572. *Schreibers Conch.* i. p. 253.
Trochus, No. 58. *Schroeter Einl.* i. p. 703.
Inhabits ——
Shell about an inch and a half long, and the breadth in Chemnitz's figure rather exceeds the length; the colour is white
clouded with rose-colour, and the base flat, with concentrical
rows of granulated striæ.

VERRUCOSUS. 17. Shell conical, with the whirls
nodulous at the sutures; umbilicus funnel-
shaped, and immaculate.

Trochus verrucosus. *Gmelin,* p. 3572. *Schreibers Conch.*
i. p. 253.
Trochus, No. 59. *Schroeter Einl.* i. p. 703.
Conus acutangulus. *Chemnitz,* v. p. 95. t. 170. f. 1638.
Inhabits the East Indian Seas. *Chemnitz.*
Chemnitz has described this species to be about sixteen lines
long, and twelve broad, with purplish spots and stripes on a
white ground: his having called it a *Conus* appears to have
been wholly accidental.

CYLINDRACEUS. 18. Shell sub-conical, with the
whirls convex and transversely striated, and
the summit obtuse; aperture round, and the
umbilicus crenulated.

Trochus cylindraceus. *Chemnitz,* v. p. 95. t. 170. f. 1639.
Trochus cylindricus. *Gmelin,* p. 3572. *Schreibers Conch.*
i. p. 254.
Trochus, No. 60. *Schroeter Einl.* i. p. 703.
Inhabits ——

Shell about eight lines long, and almost equally broad, of a pale brown colour, and has much the appearance of a Turbo; it is said to be a scarce shell.

RADIATUS. 19. Shell conical, with transverse rows of granulated striæ, and the umbilicus funnel-shaped.

Trochus radiatus. *Gmelin,* p. 3572. *Schreibers Conch.* i. p. 253.

Variety A. Shell elevated, with alternate greenish white and crimson longitudinal narrow stripes.
Trochus, No. 61. *Schroeter Einl.* i. p. 704.
Chemnitz, v. p. 96. t. 170. f. 1640.

Variety B. Shell elevated, with broad alternate white and crimson longitudinal stripes.
Trochus, No. 62. *Schroeter Einl.* i. p. 704.
La Renoncule. *Favanne,* ii. p. 361. t. 13. f. G 1.?
Lister Conch. t. 632. f. 19. *Geve.* t. 6. f. 41 and 42. *Regenfuss,* ii. t. 11. f. 51. *Chemnitz,* v. t. 170. f. 1641.

Variety C. Shell depressed, with broad alternate white and crimson longitudinal stripes.
Trochus, No. 63. *Schroeter Einl.* i. p. 704.
Trochus, No. 64. *Schreibers Conch.* i. p. 254.
Chemnitz, v. t. 170. f. 1642.

Inhabits the West Indian Seas. *Chemnitz.* Isles of France, and Madagascar. *Favanne.*

The two first Varieties are about an inch in length as well as in breadth, but C is figured only three-quarters of an inch long, and one inch broad; the longitudinal stripes are dark red, crimson, or purplish, and are somewhat interrupted at the sutures.

VIRIDIS. 20. Shell conical, with five transverse rows of nodules on the first, and four on the second whirl, and the upper ones glabrous.

Trochus viridis. *Gmelin,* 3572. *Schreibers Conch.* i. p. 254.
Trochus, No. 65. *Schroeter Einl.* i. p. 704.
Chemnitz, v. p. 97. t. 170. f. 1643 and 1644.
Inhabits the coasts of New Zealand. *Chemnitz.*
Chemnitz has figured this shell about ten lines long and nine broad, and says he received it from Mr. Humphreys, under the name of 'the Green and brown beaded Trochus from New Zealand;' the colour is described to be dusky green,

but the figure is pale green, and the description and figure do not in other respects well accord; the base is marked with concentrical striæ, and the umbilicus, which is funnel-shaped, is said not to follow the direction of the whirls.

ÆGRESTIS. 21. Shell sub-pyramidal, obtuse, smooth, of a blackish brown colour, and the inside pearly; pillar with one obsolete tooth.

Trochus agrestis. *Chemnitz*, v. p. 97. t. 170. f. 1645 and 1646.
Trochus rusticus. *Gmelin*, p. 3572. *Schreibers Conch.* i. p. 254.
Trochus, No. 66. *Schroeter Einl.* i. p. 705.
Inhabits the coasts of China. *Chemnitz.*
Shell about an inch and a quarter long, and an inch broad, coarse, unornamented, and of a blackish brown colour, except the base which is pale brown; the whirls are slightly ventricose, and obliquely wrinkled.

NIGER. 22. Shell turban-shaped, plaited, black, and the inside pearly; pillar with an obsolete tooth.

Trochus niger. *Chemnitz*, v. p. 98. t. 170. f. 1647.
Trochus nigerrimus. *Gmelin*, p. 3573.
Trochus, No. 67. *Schroeter Einl.* i. p. 705.
Inhabits the coasts of China. *Chemnitz.*
This shell, by Chemnitz's figure, appears to be three-quarters of an inch long, and about equally broad, consisting of five uniformly black whirls, and longitudinally ribbed or plaited; it somewhat resembles *T. argyrostomus,* but that species is not umbilicated.

FANULUM. 23. Shell pyramidal, with oblique plaits interrupted by a granulated transverse band at the sutures.

Trochus Fanulum. *Gmelin*, p. 3573. *Schreibers Conch.* i. p. 255.
Trochus, No. 68. *Schroeter Einl.* i. p. 706.
Le Boutonnier. *Favanne*, ii. p. 348. t. 13. f. O.
Sacellum Chineuse. *Chemnitz*, v. p. 98. t. 170. f. 1648 and 1649.
Bonanni Rec. 3. f. 396, and *Kirch.* f. 372. *Petiver Gaz.* t. 156. f. 15. *Geve.* t. 15. f. 138 and 139.

Inhabits the coasts of Fernambuca. *Favanne.*

This shell is said to be about the size of a filbert, with eight whirls rising above each other in the manner of a Chinese Pagoda, and white tinged with flesh-colour. Bonanni's, Petiver's, and Favanne's, are magnified figures. Both Chemnitz and Gmelin have quoted Knorr, iv. t. 25. f. 5, but it appears different, and is quoted by Favanne for his *Sabot tourné.*

STRIGOSUS. 24. Shell pyramidal, transversely striated, with the whirls flattish, and slightly tumid at the margin.

Trochus strigosus. *Gmelin,* p. 3573. *Schreibers Conch.* i. p. 255.

Trochus, No. 69, and No. 70. *Schroeter Einl.* i. p. 706. *Chemnitz,* v. t. 170. f. 1650? and 1651.

Inhabits the coasts of Morocco. *Chemnitz.*

This shell appears by Chemnitz's figure to be about five lines long, and nearly equally broad, and of a yellowish white colour variegated with red, and the summit black; the tumid margin is spotted with red in fig. 1651, but it is by no means clear that fig. 1650 belongs to the same species, and Schroeter has arranged them separately.

DUBIUS. 25. Shell pyramidal, with transverse granulated striæ, and the margin of the whirls nodulous and vaulted.

Trochus Pyramis. *Gmelin,* p. 3573. *Schreibers Conch.* i. p. 255.

Trochus, No. 71, and No. 72. *Schroeter Einl.* i. p. 707. *Chemnitz,* v. t. 170. f. 1652 and 1653.

Inhabits ——

The latter of the above-mentioned figures is about ten lines long and nine broad, but the former is much smaller, and narrower in proportion, and Chemnitz's description is so indefinite that it is impossible to ascertain whether Schroeter is right in placing them as separate species; the colour is whitish variegated with chestnut. *T. Pyramis* of Born is different, and is the *T. Obeliscus* of Gmelin.

DEPRESSUS. 26. Shell depressed, whitish, radiated with red, and the apex red; whirls girt with a belt of moniliform dots.

Trochus depressus. *Gmelin,* p. 3573.

Trochus, No. 76, and No. 77. *Schroeter Einl.* i. p. 709.
Trochus, No. 18, and No. 19. *Schreibers Conch.* i. p.
244, and p. 245.
Chemnitz, v. p. 107. t. 171. f. 1668 and 1669.
Inhabits ——
Chemnitz has here again described together two shells which
have a different appearance, and which have been arranged
separately both by Schroeter and Schreibers; the above
specific character is taken from Gmelin, and I regret my
inability to mend it; the figures are about eight lines long
and ten broad, and the former has been quoted by Ulysses
for his *T. ardens* (p. 469. t. 8. f. 9.), but he notwithstanding allows that there is an " essential difference."

LÆVIS. 27. Shell depressed, brown, with the whirls
smooth, and obsoletely striated transversely;
base rather convex, and the umbilicus funnel-
shaped and white.

Trochus lævis. *Chemnitz,* v. p. 108. t. 171. f. 1670.
Trochus lævigatus. *Gmelin,* p. 3573. *Schreibers Conch.*
i. p. 245.
Trochus, No. 78. *Schroeter Einl.* i. p. 709.
Inhabits ——
Shell about six lines long, and eight broad, of an uniform pale
brown colour, and Chemnitz says the umbilicus is sinuated.

GROENLANDICUS. 28. Shell somewhat turban-
shaped, flesh-coloured, with the whirls convex, and very finely striated transversely;
base convex, and the aperture sub-orbicular.

Trochus Groënlandicus. *Chemnitz,* v. p. 108. t. 171. f.
1671. *Gmelin,* p. 3574.
Trochus, No. 79. *Schroeter Einl.* i. p. 710.
Inhabits the coasts of Greenland. *Chemnitz.*
Shell about seven lines long and eight broad, of a pale dull
flesh-colour, and the inside pearly; the base is convex, and
the umbilicus large and deep.

CRUCIATUS. 29. Shell convex, with callo-punc-
tured transverse striæ, and the pillar one
toothed.

Trochus cruciatus. *Linnæus Syst. Nat.* p. 1228. *Chemnitz,* v. p. 113. t. 171. f. 1674. *Schroeter Einl.* i. p.

653. t. 3. f. 10.   *Gmelin*, p. 3567.   *Schreibers Conch.* i.
    p. 241.
Inhabits the Mediterranean. *Linnæus.*
Linnæus has described this shell to be smaller than a hazle-
    nut, and Chemnitz's figure is about eight lines long, and
    somewhat broader, but Schroeter's is almost twice as large;
    it consists of four or five whirls of a ferruginous colour, with
    four whitish or yellowish longitudinal rays, which form an
    irregular cross.

PHARAONIUS.   30. Shell sub-ovate, with crowded
    transverse rows of rounded beads ; aperture
    and pillar toothed, and the umbilicus cre-
    nated.

Trochus Pharaonius. *Linnæus Syst. Nat.* p. 1128.   *Born*
    Mus. p. 329.   *Chemnitz*, v. p. 109. t. 171. f. 1672 and
    1673.   *Schroeter Einl.* i. p. 653.   *Gmelin*, p. 3567.
    Schreibers Conch. i. p. 241.
Variety A.   With alternate rows of crimson and chequered
    black and white beads.
Le Bouton de Camisole. *Favanne,* ii. p. 415. t. 13. f. V 1,
    and magnified, V 2.
Bonanni Rec. and *Kirch.* 3. f. 222 and 223, magnified.
    Lister Conch. t. 638. f. 26.   *Petiver Gaz.* t. 14. f. 10.
    Gualter, t. 63. f. B.   *Argenville,* t. 8. f. Q.   *Knorr,* i.
    t. 30. f. 6, and iv. t. 26. f. 3 and 4.   *Geve.* t. 13. f. 102
    and 103.
Variety B.   With the beads smaller and 'mostly crimson, with
    three or four rows chequered, and of these only every third
    bead is black.
Le Vasset.   *Adanson Senegal,* t. 12. f. 3.
Le Bouton de Camisole de Goree.   *Favanne,* ii. p. 417.
    Lister Conch. t. 637. f. 25.   *Geve.* t. 13. f. 101.
Inhabits the Mediterranean. *Linnæus.*   Red Sea. *Petiver.*
    Coast of Brazil. *Bonanni.*   Goree. *Adanson.*   China,
    Philippine Islands, Bengal, Malabar, and in the Archipe-
    lago. *Favanne.*   Madagascar. *Humphreys.*
This beautiful species is generally about six or eight lines long,
    with the breadth sometimes exceeding the length, and it
    varies considerably in the relative number of its variegated
    and crimson rows of beads : Adanson says, ' Quand elle sort
    de la mer elle est ordinairement d'un cendré noir ;' and
    Chemnitz's fig. 1678, which Gmelin has quoted for a Va-
    riety, looks as this shell probably would when coated with a
    thin blackish epidermis.

CORALLINUS. 31. Shell turban-shaped, and the whirls ventricose, with crowded transverse rows of beads; aperture and pillar toothed, and the umbilicus crenated.

Variety A. Red, mottled with black or white.
Trochus corallinus. *Gmelin*, p. 3576.
Trochus, No. 178. *Schroeter Einl.* i. p. 747.
Le Fuget. *Adanson Senegal*, p. 183. t. 12. f. 4.
Variety B. Black, mottled with white, or some of the beads alternately white.
Trochus lugubris. *Gmelin*, p. 3583. *Schreibers Conch.* i. p. 265.
Trochus, No. 30. *Schroeter Einl.* i. p. 692.
Nodulus lugubris. *Chemnitz*, v. p. 54. t. 165. f. 1571.
Inhabits the coasts of the Magdalen Islands. *Adanson.*
Shell three or four lines long, and the length rather exceeds the breadth; it is nearly allied to *T. Pharaonius*, but differs in being constantly much smaller, and in having the whirls ventricose, and the base more convex. Chemnitz says that his *Nodulus lugubris* is a South Sea shell, but I have found it in a piece of coarse sponge, with the other Variety. It is a beautiful little species.

GUINEENSIS. 32. Shell convex, with numerous transverse rows of granules and nodules; aperture crenulated, and the umbilicus and pillar toothed.

Trochus Guineensis. *Gmelin*, p. 3574. *Schreibers Conch.* i. p. 246.
Trochus, No. 85. *Schroeter Einl.* i. p. 712.
Globulus asper Guinaicus. *Chemnitz*, v. p. 115. t. 171. f. 1680.
Inhabits the coasts of Guinea. *Chemnitz.*
Shell about six lines long, and eight broad, and is of a greyish ash-colour clouded with brown; in the aperture and umbilicus it somewhat resembles *T. Pharaonius*. Chemnitz's fig. 1681, which is *T. nodulus* of Gmelin, looks like the same shell coated with a thin greyish epidermis.

CARNEUS. 33. Shell depressed, somewhat flesh-coloured, with crowded rows of transverse beads; umbilicus large and one-toothed.

Trochus carneus. *Gmelin*, p. 3574. *Schreibers Conch. i.* p. 246.

Trochus, No. 87. *Schroeter Einl.* i. p. 712.

Globulus Indusii. *Chemnitz*, v. p. 116. t. 171. f. 1682.

Inhabits ———

Shell five or six lines long, and about eight broad, of an uniform dull flesh-colour; it is said to differ from Chemnitz's fig. 1681, in colour, and in being more depressed.

MAGUS. 34. Shell somewhat turban-shaped, striated transversely, and the whirls plaited on their upper margins; umbilicus oblique.

Trochus Magus. *Linnæus Syst. Nat.* p. 1228. *Pennant Zool.* iv. p. 127. t. 80. f. 107. *Born Mus.* p. 330. *Chemnitz*, v. p. 101. t. 171. f. 1656 to 1659, and xi. p. 163. t. 196. f. 1886 and 1887. *Schroeter Einl.* i. p. 655. *Gmelin*, p. 3567. *Schreibers Conch.* i. p. 242. *Donovan*, i. t. 8. f. 1. *Montagu Test.* p. 288. *Maton and Racket, in Lin. Trans.* viii. p. 151. *Dorset Cat.* p. 48. t. 16. f. 1.

Trochus tuberculatus. *Da Costa Brit. Conch.* p. 44. t. 3. f. 1.

La Sorcière ardente. *Favanne*, ii. p. 124. t. 8. f. I 4.

Bonanni Rec. 3. f. 170. *Lister Conch.* t. 641. f. 32. *Gualter*, t. 62. f. L. *Argenville*, t. 8. f. S. *Geve*, t. 12. f. 88 to 92. *Knorr*, vi. t. 27. f. 4.

Inhabits the Mediterranean. *Linnæus.* Coasts of England. *Lister, &c.* Red Sea. *Forskael.* Brittany, &c. *Favanne.* Naples. *Ulysses.*

Shell usually about three-quarters of an inch long, and near an inch broad, but is sometimes almost twice as large; it consists of five or six whirls, which are not convex, but marked by a deep suture; the colour is generally whitish, or pale ash, or flesh-colour, elegantly marked with broad undulated stripes of crimson, purple, or chestnut-brown, and sometimes prettily spotted all over with pink. Knorr, vi. t. 27. f. 4, is quoted by Born for this shell, and also for his *Turbo galeatus*, which is rather an uncertain species.

VARIEGATUS. 35. Shell depressed, white, variegated with chestnut; base convex, red, with darker rays and a white margin.

Trochus variegatus. *Chemnitz*, v. p. 104. t. 171. f. 1661 and 1662.

Trochus Capensis. *Gmelin*, p. 3573. *Schreibers Conch.* i. p. 244.
Trochus, No. 74. *Schroeter Einl.* i. p. 708.
Inhabits the sea at the Cape of Good Hope. *Chemnitz.*
The description given by Chemnitz is very imperfect, but the shell appears by the figures to be about five lines long, and seven broad; the colour is white mottled with red, and the lower margin of the whirls spotted with dark chestnut.

AFER. 36. Shell convex, depressed, and the whirls transversely grooved.

Trochus afer. *Gmelin*, p. 3577.
Trochus griseus. *Gmelin*, p. 3576.
Trochus, No. 180. *Schroeter Einl.* i. p. 748.
Le Lonier. *Adanson Senegal*, p. 184. t. 12. f. 6.
Inhabits the coasts about Cape Dakar in Senegal. *Adanson.*
Adanson describes this species to be about half an inch long, and says that the spire is much more elevated in some shells than in others, and that in the more depressed the breadth is almost double the length; it has twelve transverse grooves on the first whirl, six on the second, and four on the third; the colour is grey or brown, marbled with white spots.

MODULUS. 37. Shell depressed, transversely striated, and plaited longitudinally; body-whirl transversely carinated; base convex, and the pillar one-toothed.

Trochus Modulus. *Linnæus Syst. Nat.* p. 1228. *Chemnitz*, v. p. 105. t. 171. f. 1665. *Schroeter Einl.* i. p. 656. t. 3. f. 11. *Gmelin*, p. 3568. *Schreibers Conch.* i. p. 242.
Le Grenat. *Favanne*, ii. p. 150. t. 8. f. D.
Lister Conch. t. 653. f. 52.
Inhabits the coasts of Barbadoes. *Lister.* Red Sea. *Schroeter.* Isle of France, and the West Indies. *Favanne.*
Shell about five lines long and six broad, with dotted chestnut striæ; the throat as well as the base is grooved, and the aperture ovate.

DECLIVIS. 38. Shell sub-ovate, and transversely striated; whirls shelving, longitudinally plaited, and tuberculated on the margin; base convex, and the pillar one-toothed.

o 2

Trochus Ægyptius.   *Gmelin*, p. 3573. *Schreibers Conch.*
i. p. 244.
Trochus, No. 75.   *Schroeter Einl.* i. p. 708.
Turbo declivis.   *Forskael Desc. Anim.* p. 126.
Tectum declive.   *Chemnitz,* v. p. 104. t. 171. f. 1663 and
1664.
Inhabits the Red Sea about Suez. *Forskael.*
Shell three-quarters of an inch, or an inch long, and about
equally broad; a narrow space under the sutures is perpendicular, and marked with elevated nodulous striæ, below
which the whirls shelve abruptly off, and widen considerably;
the aperture is nearly round; the colour is white variegated
with red.

MURICATUS. 39. Shell ovate, sub-umbilicated, and
armed with pointed tubercles.

Trochus muricatus.   *Linnæus Syst. Nat.* p. 1229. *Schroe-*
ter Einl. i. p. 657.   *Gmelin,* p. 3568.
Gualter, t. 64. f. H.
Inhabits the Mediterranean. *Linnæus.*
Linnæus describes this species to be about as large as a hazel-
nut, white, with an ovate smooth aperture, and the umbilicus minute and oblong.   Gualter, t. 64. f. H, which he has
alone quoted, has much the appearance of a Turbo, and has
been referred to by Meuschen for a variety of *Turbo Co-*
chlus.

ROSEUS. 40. Shell convex, of an uniform rose-colour, and transversely grooved; umbilicus
very small.

Trochus roseus.   *Gmelin,* p. 3574.   *Ulysses's Travels,* p.
470.
Trochus, No. 80.   *Schroeter Einl.* i. p. 710.
Globulus roseus.   *Chemnitz,* v. p. 113. t. 171. t. 1675.
Inhabits the shores of the Cape of Good Hope. *Chemnitz.*
Bay of Naples. *Ulysses.*
This shell appears by the figure, for the size is not noticed in
the description, to be about a quarter of an inch long, with
the breadth rather exceeding the length, and of an uniform
rose-colour; Ulysses says it is slightly tuberculated, and has
the aperture toothed, and the lip serrated within.

PATHOLATUS. 41. Shell sub-conical, transversely
striated, and the whirls produced and flattish;
umbilicus small.

Trochus patholatus. *Gmelin,* p. 3574. *Schreibers Conch.* i. p. 245.

Trochus Nassaviensis. *Chemnitz,* v. p. 113. t. 171 f. 1676.

Trochus tumidus. *Montagu Test.* p. 280. t. 10. f. 4. *Maton and Racket, in Lin. Trans.* viii. p. 153. *Dorset Cat.* p. 48. t. 16. f. 9 and 10.

Trochus, No. 81. *Schroeter Einl.* i. p. 710.

Inhabits the coasts of the West of England. *Montagu, &c.*

Shell rather more than a quarter of an inch long, and the breadth rather exceeds the length; it has five transversely striated whirls separated by a deep suture, and the body-whirl has a somewhat carinated edge; the colour is generally cinereous-brown, sometimes tinged with yellow, and more or less streaked with fine obscure undulated longitudinal lines; Mr. Montagu says the colour is sometimes purplish, and is sometimes spotted with white at the top of each volution. *Trochus fuscus,* of Adams and other British Authors, probably belongs to this species, or is otherwise undeserving of notice.

VIRIDULUS. 42. Shell sub-conical, and the whirls convex, with a row of moniliform granulations; pillar toothed.

Trochus viridulus. *Gmelin,* p. 3574. *Schreibers Conch.* i. p. 245.

Trochus, No. 82. *Schroeter Einl.* i. p. 711.

Globulus ex viridi et rubro variegatus. *Chemnitz,* v. p. 114. t. 171. f. 1677.

Inhabits ——

Gmelin has described this species to be greenish, obliquely radiated with white, but Chemnitz's figure is painted greenish, with broad, somewhat oblique, blackish longitudinal stripes, and measures about ten lines, both in length and breadth.

URBANUS. 43. Shell sub-convex, with numerous transverse granulated striæ; aperture crenulated, and the umbilicus toothed.

Trochus urbanus. *Gmelin,* p. 3574. *Schreibers Conch.* i. p. 245.

Trochus, No. 84. *Schroeter Einl.* i. p. 711.

Globulus asper civicus. *Chemnitz,* v. p. 114. t. 171. f. 1679.

Inhabits ——

This shell appears by the figure to be about seven lines long, and ten broad, of a dirty white colour, with broad longitudinal dull purplish-brown stripes; some of the granules are said to be white, and others flesh-coloured, and the base as well as the sides is granulated.

SCABER. 44. Shell sub-ovate, with transverse beaded grooves, which are alternately larger.

Trochus scaber. *Linnæus Syst. Nat.* p. 1229.? *Chemnitz,* v. p. 107. t. 171. f. 1667. *Schroeter Einl.* i. p. 658. *Gmelin,* p. 3568.
Inhabits ——
Linnæus has described this shell to be the size of a pea, with four or five rounded whirls separated by a deep groove, and the aperture rounded and striated within: he has quoted Argenville, t. 11. f. T (i. e. t. 8. f. T of the second edition), but it differs materially from his description, and is probably either *T. costatus,* or *T. inæqualis;* the shell which Chemnitz has figured is half an inch long, and about equally broad.

QUADRATUS. 45. Shell sub-ovate, tessellated, and transversely striated, with the whirls shelving at the upper margin, and more perpendicular below.

Trochus tessellatus. *Chemnitz,* v. p. 116. t. 171. f. 1683. *Gmelin,* p. 3574. *Schreibers Conch.* i. p. 246.
Trochus, No. 88. *Schroeter Einl.* i. p. 712.
Inhabits the Mediterranean, frequent. *Chemnitz.*
Shell about eight lines long and nine broad, of a brownish ash-colour, with numerous transverse rows of dark square spots. Gmelin has another species with the name of *tessellatus.*

CROCEUS. 46. Shell convex, with convex whirls, of which the lower are pale chestnut and the upper of a saffron-colour.

Trochus croceus. *Gmelin,* p. 3584. *Schreibers Conch.* i. p. 247.
Trochus crocatus. *Chemnitz,* v. p. 116. t. 171. f. 1684.
Trochus, No. 89. *Schroeter Einl.* i. p. 713.
Inhabits the coasts of Morocco. *Chemnitz.*
This shell, as it is figured by Chemnitz, appears to be about eight lines long and nine broad, with the two lower whirls of a pale chestnut, and the upper ones of a saffron-colour. *T. crocatus* of Born is described to be imperforate.

VARIUS. 47. Shell convex, obliquely umbilicated, and the whirls slightly margined.

Trochus varius. *Linnæus Syst. Nat.* p. 1229. *Schroeter Einl.* i. p. 659. *Gmelin,* p. 3568.
Inhabits the Mediterranean. *Linnæus.*
Linnæus has not given any reference, and in addition to the above short character, only says that the shell is convex, surrounded with an obsolete margin, and of a pale colour with greyish bands.

OBLIQUATUS. 48. Shell depressed, rounded at the summit, and the whirls contiguous; umbilicus large.

Trochus obliquatus. *Gmelin,* p. 3575. *Schreibers Conch.* i. p. 247.
Trochus umbilicaris. *Pennant Zool.* iv. p. 126. t. 80. f. 106.
Trochus umbilicalis. *Da Costa Brit. Conch.* p. 46. t. 3. f. 7 and 8.
Trochus cinerarius. *Pulteney Dorset Cat.* p. 44. *Donovan,* iii. t. 74, three middle figures.
Trochus umbilicatus. *Montagu Test.* p. 286. *Maton and Racket, in Lin. Trans.* viii. p. 153. *Dorset Cat.* p. 48. t. 16. f. 7 and 8.
Trochus oblique radiatus. *Chemnitz,* v. p. 117. t. 171. f. 1685.
Trochus, No. 91. *Schroeter Einl.* i. p. 714.
Inhabits the Mediterranean. *Chemnitz.* Coasts of Great Britain. *Pennant, &c.*
Shell about five-eighths of an inch long, and three-quarters of an inch broad, of a whitish ash-colour, marked with rather broad and somewhat oblique violet undulated rays; it is nearly allied to *T. cinerarius,* but is much flatter, is of a different colour, and has the rays invariably much broader.

CINERARIUS. 49. Shell sub-conical, with the summit produced, and the whirls slightly convex; umbilicus small.

Trochus cinerarius. *Linnæus Syst. Nat.* p. 1229. *Pennant Zool.* iv. p. 127. *Chemnitz,* v. p. 117. t. 171. f. 1686. *Schroeter Einl.* i. p. 659. *Muller Zool. Dan.* iii. p. 35, t. 102. f. 1 to 4. *Gmelin,* p. 3568. *Donovan,* iii. t. 74, upper and lower figures. *Montagu Test.* p. 284.

Trochus lineatus. *Da Costa Brit. Conch.* p. 43. t. 3. f. 11 and 12. *Maton and Racket, in Lin. Trans.* viii. p. 152. *Dorset Cat.* p. 48. t. 16. f. 11 and 12.

La Livrée. *Favanne,* ii. p. 121. t. 8. f. I 2.

Lister Anim. Ang. t. 3. f. 15, and *Conch.* t. 641. f. 31.

Inhabits the Mediterranean, and coasts of Norway. *Linnæus.* Britain. *Lister, &c.* France, England, Provence, St. Domingo, and Martinique. *Favanne.*

Shell about five-eighths of an inch long, and equally broad, of a pale ash-colour, marked with fine approximated purplish brown rays; the umbilicus is deep and narrow. *T. cinerarius* of Born is different, and is the *T. albidus* of Gmelin.

NERITOIDEUS. 50. Shell sub-ovate, convex, depressed, smooth, glabrous, and of a reddish colour.

Trochus neritoideus. *Olaffsen Isl.* No. 1015. *Gmelin,* p. 3577.

Trochus helicinus. *Fabricius Fauna Groenl.* p. 393.

Inhabits the coasts of Greenland, frequent. *Fabricius.*

This shell is said to be only two lines long, and allied to *T. cinerarius,* but of a reddish colour; the animal which inhabits it is bluish black.

ALBIDUS. 51. Shell ovate-conical, with the whirls transversely striated, and the sutures channelled; aperture roundish, and the umbilicus nearly closed.

Trochus albidus. *Gmelin,* p. 3576. *Schreibers Conch.* i. p. 250.

Trochus cinerarius. *Born Mus.* p. 330. t. 11. f. 19 and 20.

Trochus, No. 175. *Schroeter Einl.* i. p. 746.

Inhabits ——

Born describes this species to be five lines long, and six broad, with seven whirls, and white, marked with undulated oblique black stripes.

VITTATUS. 52. Shell convex, sub-umbilicated, of a pale chestnut colour, with a red and white spotted transverse band below the sutures.

Trochus vittatus. *Gmelin,* p. 3375. *Schreibers Conch.* i. p. 247.

Trochus subumbilicatus. *Chemnitz,* v. p. 118. t. 171. f. 1687.

Trochus, No. 92. *Schroeter Einl.* i. p. 714.

Le Limaçon rubané. *Favanne,* ii. p. 108. t. 9. f. E 3. ?

Inhabits ——

Chemnitz has figured this shell about seven lines long, and six broad, of a pale chestnut-colour with remote narrow darker transverse stripes, and a white band spotted with chestnut at the upper margin of the whirls. Chemnitz says it resembles *Le Limaçon rubané,* which is described to be imperforated, but Favanne adds, 'Quoique sans ombilic, cette coquille en montre néanmoins quelquefois de légers vestiges.'

DIVARICATUS. 53. Shell ovate, sub-umbilicated, with the body-whirl more remote, and the umbilicus nearly consolidated.

Trochus divaricatus. *Linnæus Syst. Nat.* p. 1229. *Fabricius Fauna Groenl.* p. 392. *Schroeter Einl.* i. p. 660. *Gmelin,* p. 3568.

Inhabits the Mediterranean, and coasts of Norway. *Linnæus.* Linnæus has not given any reference, and to the above short character only adds, that the shell is green with rows of red spots, and that the lower whirls are more remote than the upper.

FUSCATUS. 54. Shell depressed-conical, with the whirls contiguous, and finely striated transversely; aperture roundish, and the umbilicus sub-cylindrical.

Trochus fuscatus. *Gmelin,* p. 3576. *Schreibers Conch.* i. p. 251.

Trochus umbilicaris. *Born Mus.* p. 331. t. 12. f. 1 and 2.

Trochus, No. 176. *Schroeter Einl.* i. p. 746.

Inhabits ——

Born describes this shell to be eighteen lines long, and only half as broad, which is obviously erroneous; and he says the shell is of an olive-colour, with transverse brown and white spotted bands, but there is not any white in his figure.

UMBILICARIS. 55. Shell conical-convex, with the whirls slightly margined, and the umbilicus cylindrical.

Trochus umbilicaris. *Linnæus Syst. Nat.* p. 1229. *Chemnitz,* v. p. 106. t. 171. f. 1666. *Schroeter Einl.* i. p. 660. *Gmelin,* p. 3568. *Schreibers Conch.* i. p. 243. Inhabits the Mediterranean. *Linnæus.*

Linnæus has not given any reference, and only says, in addition to his specific character, that the shell is often clouded with ferruginous, that the summit and inside are white, and that it has an exactly cylindrical pervious umbilicus. The shell figured by Chemnitz is about seven lines long and eight broad, of a blackish grey colour, with darker oblique longitudinal markings. Born, though he has quoted the *T. umbilicaris* of Linnæus and Pennant, has figured under this name a very different shell, which is the *T. fuscatus* of Gmelin.

AREOLA. 56. Shell ovate, transversely striated, white tessellated with red, and the spire somewhat produced; aperture round, and the umbilicus crenated.

Trochus Areola. *Chemnitz,* v. p. 134. t. 173. f. 1710 and 1711. *Gmelin,* p. 3575. *Schreibers Conch.* i. p. 248. Trochus, No. 99. *Schroeter Einl.* i. p. 718. Inhabits ——

Shell about six lines long and seven broad, white with reddish square spots in transverse rows, and the whirls separated by a white band.

CINEREUS. 57. Shell pyramidal, with flattish whirls separated by a very depressed suture, and the umbilicus cylindrical; base concave.

Trochus cinereus. *Da Costa Brit. Conch.* p. 42. t. 3. f. 9 and 10. *Montagu Test.* p. 289, and *Supp.* p. 119. *Donovan,* v. t. 155. f. 3. *Maton and Racket, in Lin. Trans.* viii. p. 152. Trochus, No. 116. *Schroeter Einl.* i. p. 725. Le Petit Entonnoir. *Favanne,* ii. p. 352. t. 13. f. M. *Lister Conch.* t. 633. f. 21. Inhabits the Mediterranean. *Montagu.*

Da Costa, who erroneously states that it is common on the British coasts, has given the following description of this species: "The shell is thick and strong, of the size of a cherry, shape obtusely pyramidal, or not quite tapering to a point. The base is very concave, with some circular furrows; the mouth roundish and capacious, within fine mother-of-pearl; the outer lip smooth and even; the inner

TROCHUS.   783

or pillar-lip has two jags or slight teeth, and two furrows
crossing it transversely : from hence it widens, runs oblique,
and forms a spacious cavity, at the bottom of which lies the
umbilicus, deep, cylindric, and so hollow as to admit the
head of a large pin.   All this part is of a dark ash, greatly
variegated with blackish lines or streaks, which run length-
ways and across; but the beginning of the umbilicus is generally pearly, and of a fine light greenish colour."

FASCIATUS.   58. Shell conical, smooth, with the
      whirls rounded, and flattened at their upper
      margins; umbilicus deep, and the outer lip
      crenulated.

Trochus fasciatus.   *Born Mus.* p. 331. t. 12. f. 2 and 3.
    Gmelin, p. 3576.   *Schreibers Conch.* i. p. 251.
Trochus, No. 177.   *Schroeter Einl.* i. p. 747.
Inhabits ——
Born has described this shell to be five lines long and six
    broad, with five rounded whirls, which become flattish to-
    wards their upper margins; umbilicus vertical and deep;
    colour white spotted with brown, and marked with a trans-
    verse white band.   The figure has a rounded aperture, and
    looks a good deal like a Turbo.

PERSPECTIVIUNCULUS.   59. Shell depressed, trans-
      versely striated and slightly granulated; um-
      bilicus pervious and crenated; aperture
      roundish.

Trochus perspectiviunculus.   *Meuschen in Mus. Gronov.*
    p. 125.? *Chemnitz,* v. p. 134. t. 173. f. 1708 and 1709.
Trochus variegatus.   *Gmelin,* p. 3575.   *Schreibers Conch.*
    i. p. 247.
Trochus, No. 98.   *Schroeter Einl.* i. p. 718.
Geve. t. 25. f. 275.
Inhabits ——
Shell about seven lines broad, and much depressed, with the
    umbilicus large, and crenated like that of *T. perspectivus;*
    it is white, with brown longitudinal broadish rays.

INFUNDIBULIFORMIS.   60. Shell depressed; whirls
      rounded, transversely striated and crenulated
      with granules; umbilicus very wide, sinu-
      ated, and minutely crenulated.

Trochus infundibuliformis. *Gmelin,* p. 3575. *Schreibers Conch.* i. p. 247.

Trochus planior infundibuliformis. *Chemnitz,* v. p. 133. t. 173. f. 1706 and 1707.

Trochus, No. 97. *Schroeter Einl.* i. p. 718.

Inhabits ——

Shell about seven lines broad, and appears by the figure to be almost as flat as some of the depressed Helices ; the colour is pale chestnut, without any variegations, and the aperture is round. Chemnitz says it is a very scarce species.

HYBRIDUS. 61. Shell convex, with the pillar two-toothed, and the umbilicus crenulated.

Trochus hybridus. *Linnæus Syst. Nat.* p. 1228. *Chemnitz,* v. p. 132. t. 173. f. 1702 to 1705. *Schroeter Einl.* i. p. 652. *Gmelin,* p. 3567. *Schreibers Conch.* i. p. 241.

Le Cadran flambé. *Favanne,* ii. p. 428.

Inhabits the Mediterranean. *Linnæus.*

Linnæus says that this shell resembles *T. perspectivus,* but is only one fourth as large, and has the superficies smooth but not shining, and is not marginated at the base.

PERSPECTIVUS. 62. Shell convex, and obtusely marginated at the base; umbilicus large, pervious, and crenulated.

Trochus perspectivus. *Linnæus Syst. Nat.* p. 1227. *Born Mus.* p. 329, and Vign. at p. 326. f. b. *Chemnitz,* v. p. 121. t. 172. f. 1691 to 1696, and xi. p. 162. t. 196. f. 1884 and 1885. *Schroeter Einl.* i. p. 650. *Gmelin,* p. 3566. *Schreibers Conch.* i. p. 241. *Brookes's Int.* p. 123. t. 7. f. 94.

Solarium perspectivus. *Lamarck Syst. des Anim.* p. 86.

Le Cadran. *Favanne,* ii. p. 422. t. 12. f. K.

Bonanni Rec. 3. f. 27 and 28, and *Kirch,* f. 26 and 27. *Lister Conch.* t. 636. f. 24. *Rumphius,* t. 27. f. L. *Petiver Amb.* t. 2. f. 14. *Gualter,* t. 65. f. O. *Argenville,* t. 8. f. M. *Seba,* iii. t. 40. f. 1, 2, 13, 14, 41, and 42. *Knorr,* i. t. 11. f. 1 and 2. *Geve.* t. 25. f. 266 and 267. *Regenfuss,* i. t. 6. f. 61.

Inhabits the shores of Asia, and is frequent about Alexandria. *Linnæus.* Amboyna. *Rumphius.* Tranquebar. *Regenfuss.* Bantam, the Moluccas, Borneo, and other islands in the Indian Ocean. *Favanne.* China. *Humphreys.*

Shell one, or sometimes near three inches broad, and less than

half as long, and beautifully variegated and spotted in transverse rows; the base is flat, and the umbilicus large, and
elegantly crenulated throughout; it is generally known by
the name of the *Staircase*, but Petiver has called it the
Spinnet Shell.

**STRAMINEUS. 63. Shell depressed; whirls round-
ed, with a distinct groove, and decussated
striæ; umbilicus pervious, and slightly crenulated.**

Trochus stramineus. *Gmelin*, p. 3575.
Trochus perspectivus stramineus. *Chemnitz*, v. p. 128.
t. 172. f. 1699.
Trochus perspectivus, Var. *Schreibers Conch.* i. p. 241.
Trochus, No. 96. *Schroeter Einl.* i. p. 717.
Le Cadran Américain. *Favanne*, ii. p. 426.?
Lister Conch. t. 635. f. 23.
Inhabits the coasts of Tranquebar. *Chemnitz.*
Shell about ten or eleven lines broad, and but little elevated,
with five or six whirls of a straw colour.

**INDICUS. 64. Shell depressed, slightly convex,
with minute oblique·striæ, and the base concave; umbilicus deep, sinuated, and slightly
striated.**

Trochus Indicus. *Gmelin*, p. 3575.
Trochus solaris Indiæ Orientalis. *Chemnitz*, v. p. 127.
t. 172. f. 1697 and 1698.
Trochus, No. 95. *Schroeter Einl.* i. p. 717.
Inhabits the East Indian Seas. *Chemnitz.*
Shell two inches and a quarter broad, and hardly an inch high,
consisting of fine whirls of a snow-white colour, and somewhat hyaline; the base is concave, and marked with lines
radiating from the umbilicus.

**PLANUS. 65. Shell depressed, with the whirls lon-
gitudinally plaited, and the umbilicus pervious.**

Trochus planus. *Gmelin*, p. 3576. *Schreibers Conch* i.
p. 248.
Trochus solaris complanatus. *Chemnitz*, v. p. 143. t. 174.
f. 1721 and 1722.
Trochus, No. 102. *Schroeter Einl.* i. p. 721.

Inhabits ——
Shell about an inch broad, and much depressed, of a straw-co-
lour, with the interstices of the longitudinal plaits brown ;
the longitudinal plaits, as in *T. solaris*, do not extend to the
lower margin of the whirls.

SOLARIS.　66. Shell convex-conical, with marginated
　　　spinous whirls, and the aperture semi-heart-
　　　shaped.

Trochus solaris. *Linnæus Syst. Nat.* p. 1229. *Schröeter
　Einl.* i. p. 661. *Gmelin*, p. 3569. *Schreibers Conch.* i.
　p. 243.
Variety A.　With long narrow spines.
　Trochus solaris Indiæ Orientalis. *Chemnitz*, v. p. 129. t.
　173. f. 1700 and 1701.
L'Eperon Soleil. *Favanne*, ii. p. 410. t. 13. f. C 1.
Rumphius, t. 20. f. K.
Variety B.　With short broader spines.
　Trochus solaris Indiæ Occidentalis. *Chemnitz*, v. p. 139.
　t. 174. f. 1716 and 1717.
Le Grand Eperon. *Favanne*, ii. p. 402. t. 13. f. C 2.
Bonanni Rec. 3. f. 366 and 367, and *Kirch.* f. 359 and 360.
　Lister Conch. t. 622. f. 9. *Klein Ost.* t. 1. f. 19. *Knorr*,
　vi. t. 26. f. 4. *Seba*, iii. t. 59. f. 1 to 4. *Geve.* t. 20.
　f. 192 and 193. *Regenfuss*, ii. t. 8. f. 13 and 14.
Inhabits the East Indies. *Linnæus.*　Coasts of Amboyna.
Rumphius.　Coromandel. *Chemnitz.*　Philippines, Moluccas, and Isle of France. *Favanne.*　New Zealand, China,
and Jamaica. *Humphreys.*
Shell one, or sometimes two inches broad, and hardly half so
long, and of a gold-colour, sometimes mottled with white ;
the lower part of the whirls is rather concave and marginated, and the margin spinous.　Mr. Humphreys, in the
Portland Catalogue, lot 3616, says it has the same faculty
of affixing extraneous bodies to itself as *T. conchyliophorus*,
and at lots 943 and 1846, two Varieties are mentioned, of
which the first has a double row of spines, and the other a
scarlet aperture.　Schreibers has quoted *Trochus helicinus*,
of Gmelin, as a Variety of *T. solaris*; and Knorr, iv. t. 6. f.
2, which has been quoted both by Chemnitz and Gmelin, is
also quoted for other shells as follows :——by Born for his
Turbo Tectum Persicum, which is *Trochus imbricatus*, and
by Gmelin for *Turbo Calcar*; neither of the figures or the
descriptions of *T. helicinus* appear to be sufficiently accu-

rate, and the species is every way involved in so much obscurity, that I have thought it undeserving of notice.

INERMIS. 67. **Shell convex-conical, with longitudinal plaits ending in short spines ; aperture compressed, and the umbilicus sub-consolidated and wrinkled.**

Trochus inermis. *Gmelin,* p. 3576.

Trochus solaris Indiæ Occidentalis. *Chemnitz,* v. p. 135. t. 173. f. 1712 and 1713.

Trochus, No. 100. *Schroeter Einl.* i. p. 719.

Inhabits the West Indian Seas. *Chemnitz.*

This shell, as it is figured by Chemnitz, appears to be about eleven lines long, and fifteen broad, of a greenish yellow colour, and the longitudinal plaits nearly white ; the longitudinal plaits terminate at the margin in short processes, which, however, have not much the appearance of spines.

IMPERIALIS. 68. **Shell convex-conical, sub-ventricose, with transverse somewhat scaly striæ, and the whirls spinous at their margins.**

Trochus imperialis. *Gmelin,* p. 3576. *Schreibers Conch.* i. p. 248.

Trochus solaris imperialis. *Chemnitz,* v. p. 135. t. 173. f. 1714 and t. 174. f. 1715.

Trochus Heliotropium. *Martyn Univ. Conch.* t. 30.

Trochus, No. 101. *Schroeter Einl.* i. p. 720.

Turbo echinatus, Var. *Gmelin,* p. 3591.

L'Eperon royal. *Favanne,* ii. p. 408.

Inhabits the coast of New Zealand. *Martyn, &c.*

Shell nearly two inches and a half long, and four inches broad, with seven whirls, of a dark olive-brown colour tinged with violet ; the base is whitish, with a large umbilicus, and the inside pearly ; the spines at the base and margins of the whirls are broad, and shaped somewhat like the serratures of a saw : Favanne says that the operculum is rather pearly on its outer surface.

CONCHYLIOPHORUS. 69. **Shell conical, sub-umbilicated, coarse, obtusely plaited, pellucid, and the whirls tiled ; base concave, and the pillarlip sickle-shaped.**

Trochus conchyliophorus. *Born Mus.* p. 333. t. 12. f. 21

and 22. *Chemnitz*, **v.** p. 118. t. 172. f. 1688 to 1690.
Gmelin, p. 3584. *Schreibers Conch.* i. p. 259.
Trochus onustus. *Humphreys Port. Cat.* p. 4, lot 31.
Trochus, No. 93. *Schroeter Einl.* i. p. 714.
Turbo trochiformis. *Born Index*, p. 355.
La Fripiere. *Favanne*, ii. p. 411. t. 12. f. C 1 and C 2.
D'Avila, t. 6. f. M.
Inhabits the coasts of St. Domingo. *Favanne.* China and
Guadaloupe. *Humphreys.*
Shell commonly about an inch and a half long, and an inch
and three-quarters broad, with six remarkably rugged whirls
which over-lie each other, and of a brownish white colour;
it has the singular faculty of affixing to its surface any ex-
traneous substance, and is often covered with stones, coral,
or the fragments of other shells, from whence it has derived
its common name of the *Carrier.*

TECTUM. 70. Shell striated transversely, and longitudinally plaited, with the body-whirl ventricose, and the spire depressed; aperture
sub-orbicular.

Trochus Tectum. *Gmelin*, p. 3569.
Trochus, No. 28. *Schroeter Einl.* i. p. 691.
Tectum declive. *Chemnitz*, v. p. 53. t. 165. f. 1567 and
1568.
Le Bossu. *Favanne*, ii. p. 153. t. 9. f. M 3.
Lister Conch. t. 653. f. 51. *Argenville*, t. 6. f. Q.
Variety. Of a chestnut-colour without spots.
Chemnitz, v. t. 165. f. 1569 and 1570.
Inhabits the West Indies. *Chemnitz.* Coasts of the Isle of
France. *Favanne.*
Shell about eight lines long and nine broad, and the body-whirl
is twice as large as the spire; the colour is white variegated
with transverse rows of purplish brown spots; a singular
arched extension of the pillar projects from the inner lip of
the aperture.

PERLATUS. 71. Shell depressed, with transverse
unequally granulated striæ, and the base
convex; aperture roundish, with the inner
lip toothed, and the throat striated.

Trochus perlatus. *Gmelin*, p. 3577.
Trochus Tectum, Var. *Gmelin*, p. 3569.
Trochus Modulus, Var. *Gmelin*, p. 3568.

Trochus, No. 128. *Schroeter Einl.* i. p. 728.
Trochus, No. 20. *Kæmmerer Cat. Rudolst.* p. 163. t. 12. f. 1.
Trochilus unidens. *Chemnitz*, x. p. 290. t. 165. f. 1583 and 1584.
Le Grenat. *Favanne*, ii. p. 150. t. 8. f. D.
Lister Conch. t. 654. f. 54.
Inhabits the coasts of the Isle of France, St. Domingo, Martinique, and Barbadoes. *Favanne.*
Shell a quarter, or sometimes half an inch long, and the breadth a little exceeds the length ; it has six whirls with strong elevated rather distant granulated striæ, of which one in the middle of the body-whirl is larger than the others ; the colour is reddish white, or sometimes white tinged and spotted with reddish brown.

PUMILIO. 72. Shell conical, with the whirls marginated at their bases, and the marginal ribs muricated ; base scabrous and slightly convex.

Trochus Pumilio. *Chemnitz*, xi. p. 164. t. 196. f. 1888 and 1889.
Inhabits Africa. *Chemnitz.*
This shell appears to be allied to *T. terrestris,* and nearly of the same size, but the margin at the base of the whirls is muricated, and the colour is reddish.

TERRESTRIS. 73. Shell conical, longitudinally striated, and the whirls marginated at their bases ; base flat and striated from the center.

Trochus terrestris. *Pennant Zool.* iv. p. 127. t. 80. f. 108. *Chemnitz*, ix. part 2. p. 47. t. 122. f. 1045. *Montagu Test.* p. 287. *Donovan,* iv. t. 111. *Maton and Racket, in Lin. Trans.* viii. p. 157.
Trochus terrestris, Var. C. *Da Costa Brit. Conch.* p. 36.
Trochus, No. 103. *Schroeter Einl.* i. p. 721.
Helix elegans. *Gmelin,* p. 3642.
Lister Anim. Ang. t. 3. f. 12, and *Conch.* t. 61. f. 58. *Petiver Gaz.* t. 22. f. 10. *Favanne,* t. 64. f. O.
Inhabits Great Britain in moss about the roots of trees. *Lister.* Plentiful about Montpellier. *Petiver.* Italy, Tunis, Tripoli, and Algiers. *Chemnitz.*
Shell about a quarter of an inch long, and the breadth rather exceeds the length ; it has five or six flattish sub-continuous

whirls, with a prominent transverse ridge at the base of each; the colour is livid white.

BIDENS. 74. Shell sub-conical, somewhat keeled, with eight finely striated whirls, and the base convex; aperture narrow, and the outer lip two-toothed and reflected.

Trochus bidens. *Chemnitz,* ix. part 2. p. 50. t. 122. f. 1052.
Helix bidentata. *Gmelin,* p. 3642.
Inhabits the Botanic Garden at Strasburgh. *Chemnitz.*
This shell appears, by Chemnitz's figures, to be rather smaller, but nearly of the same form as *T. terrestris,* and is said to be variegated with alternately white and pale yellowish transverse bands.

FRAGILIS. 75. Shell ovate, thin, brittle, and wax-coloured; body-whirl large, with a brown band in the middle, and the aperture sublunar.

Trochus fragilis. *Gmelin,* p. 3577.
Trochus globosus, anfractibus tribus acuminatus. *Schroeter Fluss.* p. 281.
Trochus, No. 190. *Schroeter Einl.* i. p. 754. t. 3. f. 16.
Inhabits ——
Schroeter's figure measures about a quarter of an inch both in length and breadth, and is shaped somewhat like *Turbo littoreus,* but has the body-whirl clasping the second at the aperture, in the manner of a Helix, to which Genus it probably ought to be removed.

CARINATUS. 76. Shell obtusely pyramidal, with four whirls, and a transverse rib round the body-whirl; aperture roundish, and the umbilicus pervious.

Trochus carinatus. *Chemnitz,* ix. part 2. p. 50. t. 122. f. 1501.
Trochus callosus. *Gmelin,* p. 3577.
Trochus elongatus carinatus et umbilicatus. *Schröeter Fluss.* p. 282. t. 6. f. 10.
Trochus, No. 191. *Schroeter Einl.* i. p. 754.
Helix pervia. *Gmelin,* p. 3640.
Helix, No. 241. *Schroeter Einl.* ii. p. 245.
Nerita Trochus. *Muller Verm.* ii. p. 176.

Inhabits Saxe-Weimar. *Muller.* Neighbourhood of Bayreuth, and is a fresh-water shell. *Schroeter.*
Shell three lines long, and two lines and a quarter broad, thick, of a chalky white colour, with five whirls, of which the body-whirl has a strong transverse rib or keel in the middle, and the others are tumid at their margins; the aperture is roundish and somewhat compressed.

FLUMINEUS. 77. Shell umbilicated, sub-pyramidal, smooth, with the whirls separated by a deep suture, and the aperture roundish.

Trochus flumineus. *Gmelin,* p. 3587.
Trochus callosus, Var. β. *Gmelin,* p. 3577.
Trochus elongatus et umbilicatus, absque carina. *Schroeter Fluss.* p. 282. t. 6. f. 12.
Trochus, No. 192, and No. 193.? *Schroeter Einl.* i. p. 755.
Helix lævissima. *Gmelin,* p. 3641.
Helix, No. 242. *Schroeter Einl.* ii. p. 245.
Argenville, t. 27. f. 4.?
Inhabits the river Huines. *Argenville.*
This shell is of about the same size, but rather longer in proportion to the breadth, as *T. carinatus,* of which it is probably only a Variety, and differs principally in wanting a transverse rib on the body-whirl; it is white, and the summit is said to be tinged with red.

** *Imperforate and erect.*

VESTIARIUS. 78. Shell conical, convex, smooth, with a gibbous callosity on the base, and the aperture somewhat heart-shaped.

Trochus vestiarius. *Linnæus Syst. Nat.* p. 1230. *Born Mus.* p. 334. *Chemnitz,* v. p. 70. t. 166. f. 1601, *a* to *h.* *Schroeter Einl.* i. p. 665. t. 3. f. 12. *Gmelin,* p. 3578. *Schreibers Conch.* i. p. 256.
L'Œil flambé. *Favanne,* ii. p. 429. t. 12. f. G.
Bonanni Rec. and *Kirch.* 3. f. 208 to 210, magnified. *Lister Conch.* t. 650. f. 45 and 46, t. 651. f. 48, and t. 652. f. 49 and 50. *Gualter,* t. 65. f. F to H. *Klein Ost.* t. 1. f. 13. *Knorr,* iv. t. 21. f. 4, and vi. t. 22. f. 7. *Geve.* t. 19. f. 184 to 191.
Variety. With the whirls somewhat nodulous.

Trochus vestiarius coronatus.   *Chemnitz*, xi. p. 168. t. 196. f. 1898 and 1899.

L'Œil goutteux.   *Favanne*, ii. p. 430.

Bonanni Rec. and *Kirch.* 3. f. 213, magnified.   *Lister Conch.* t. 649. f. 44, and t. 651. f. 47.   *Gualter*, t. 65. f. E.   *Klein Ost.* t. 1. f. 14 and 15.   *Schroeter Einl.* i. t. 3. f. 13.

Inhabits the Mediterranean and Asiatic Seas, and the coasts of China. *Linnæus.*   Molucca and Philippine Islands, Jamaica, Brazil, Cape of Good Hope, &c. *Favanne.*

Shell from two to four lines long, and almost twice as broad, and varies very much in its colour and markings ; some have a white transverse spotted or sub-reticulated belt, and others are ornamented with oblique undulated rays, but under all these different appearances it is distinguishable by its flattened form, and by a somewhat vitreous excrescence which covers the center of the base.   Gmelin has mistaken the Linnæan *Buccinum neriteum,* which it somewhat resembles, for a Variety of this species.

GRANDINATUS.   **79.** Shell conical, and granulated and tuberculated in transverse rows ; base convex, and the outer lip double and grooved.

Trochus grandinatus.   *Chemnitz*, x. p. 291. t. 169. f. 1639. *Gmelin*, p. 3585.

Trochus bullatus.   *Martyn Univ. Conch.* i. t. 38.

Inhabits the coasts of Palmerston's Islands in the South Seas. *Chemnitz.*

Shell about an inch and three-quarters long, and an inch and a quarter broad, and whitish with transverse pale reddish and yellowish brown transverse stripes ; the base is granulated in concentrical rows, and the outer lip is thick, and has an internal plaited border.

LABEO.   **80.** Shell ovate, with transverse granulated ribs, and a strong tooth on the pillar ; aperture silvery, and the outside double and grooved.

Trochus Labio.   *Linnæus Syst. Nat.* p. 1230.   *Born Mus.* p. 335. t. 12. f. 7 and 8.   *Chemnitz*, v. p. 60. t. 166. f. 1579 to 1581.   *Schroeter Einl.* i. p. 667.   *Gmelin*, p. 3578.   *Schreibers Conch.* i. p. 256.   *Brookes's Int.* p. 123. t. 7. f. 95.

Monodonta Labio.   *Lamarck Syst. des Anim.* p. 87.

La Bouche double granuleuse. *Favanne*, ii. p. 49. t. 8. f. A 2.

Lister Conch. t. 584. f. 42, and t. 645. f. 37. *Rumphius*, t. 21. f. E. *Petiver Amb.* t. 11. f. 2. *Argenville*, t. 6. f. N. *Adanson Senegal*, t. 12. f. 2. *Geve.* t. 18. f. 165, 167, and 168.

Variety. With the transverse ribs entire.

Trochus Labeo australis. *Chemnitz*, xi. p. 165. t. 196. f. 1890 and 1891.

Le Limaçon Ratelier. *Favanne*, ii. p. 48. t. 8. f. A 1. ?

Inhabits the coasts of Africa and Asia. *Linnæus.* Amboyna. *Rumphius.* Senegal. *Adanson.* Isle of France and Straights of Manilla. *Favanne.* South Sea. *Chemnitz.*

Shell generally about an inch long, and ten lines broad, of a whitish or reddish colour, tessellated with darker spots on the transverse bands, which are more or less broken into crowded nodules; the inside is grooved and beautifully pearly, and within the outer lip is a thick plaited border of fine ivory-white.

ASPER. 81. Shell ovate, with transverse nodulous ribs and intermediate striæ; aperture somewhat pearly, with the outer lip grooved and double, and the pillar toothed.

Trochus asper. *Chemnitz*, v. p. 63. t. 166. f. 1582. *Gmelin*, p. 3583. *Schreibers Conch.* i. p. 258.

Trochus, No. 31. *Schroeter Einl.* i. p. 692.

La Framboise. *Favanne*, ii. p. 50. t. 8. f. A 3.

Inhabits the coasts of New Zealand. *Favanne.*

This shell is of the same size, and is nearly allied to *T. Labeo*, of which it may be doubted whether it is more than a Variety; the transverse ribs are more strongly and more distinctly tuberculated, and more distant from each other, with narrow transverse striæ in their interstices; it is of an earthy or dull grey colour, with the tubercles somewhat alternately black.

QUADRICARINATUS. 82. Shell ovate, with transverse nodulous ribs, and four muricated transverse keels; aperture silvery, with the outer lip double and grooved, and the pillar toothed.

Trochus quadricarinatus. *Chemnitz*, xi. p. 167. t. 196. f. 1892 and 1893.

Inhabits the Mediterranean. *Chemnitz.*

This shell is of about the same size, and appears to be nearly allied to *T. Labeo;* it may, however, be readily distinguished by its having on each whirl four transverse muricated keels, which are much broader, and more elevated than the other ribs.

TESSELLATUS.   83. Shell ovate, with transverse distant ribs; outer lip somewhat double, and the pillar obsoletely toothed.

Trochus tessellatus.   *Chemnitz,* v. p. 63. t. 166. f. 1583. *Gmelin,* p. 3583.   *Ulysses's Travels,* p. 467.
Trochus tessulatus.   *Born Mus.* p. 332. t. 12. f. 5 and 6.
Trochus, No. 32.   *Schroeter Einl.* i. p. 693.
Inhabits the coasts of St. Croix. *Chemnitz.*   Bay of Naples. *Ulysses.*

Shell about nine lines long and eleven broad, and has the spire less elevated than in either of the three preceding species; the transverse ribs are channelled on the upper, and convex on the lower edge, with their interstices smooth; the colour is yellowish white, tessellated with blackish purple; the white inner margin of the lip is not grooved, and there is some faint appearance of an umbilicus.

TURBINATUS.   84. Shell ovate, smooth, and the whirls convex; outer lip somewhat double, and the pillar obsoletely toothed.

Trochus turbinatus.   *Born Mus.* p. 335.
Variety A.   More or less checquered with transverse rows of dark spots.
Trochus tessellatus, Var.   *Chemnitz,* v. p. 64. t. 166. f. 1584.   *Gmelin,* p. 3584.
Trochus, No. 36.   *Schroeter Einl.* i. p. 693.
Trochus, No. 78.   *Schreibers Conch.* i. p. 258.
Le Damier.   *Favanne,* ii. p. 107. t. 9. f. E 1.
Bonanni Rec. and *Kirch.* 3. f. 201.   *Lister Conch.* t. 642. f. 33 and 34.   *Gualter,* t. 63. f. D, E, and G.   *Klein Ost.* t. 2. f. 53 and 54.   *Knorr,* i. t. 10. f. 6.   *Adanson Senegal,* t. 12. f. 1.   *Geve.* t. 20. f. 198 and 200.
Variety B.   With dark longitudinal undulated stripes.
Trochus citrinus.   *Gmelin,* p. 3584.
Trochus Labio, Var. β.   *Gmelin,* p. 3578.
Trochus, No. 34.   *Schroeter Einl.* i. p. 694.
La Fraise sauvage.   *Favanne,* ii. p. 110. t. 9. f. E 2.
Knorr, i. t. 10. f. 7.

Inhabits the Adriatic. *Bonanni.* Mediterranean and coasts of the Canary Islands. *Lister.* Isle of France. *Favanne.*

Shell about eleven lines long and nine broad, and may be distinguished from *T. tessellatus* by its being destitute of either transverse ribs or nodules; the colour is reddish or brownish white, tessellated with darker transverse rows of spots, or longitudinal undulated irregular stripes.

ARGYROSTOMUS. 85. Shell ovate, black, with oblique somewhat undulated plaits, and the inside silvery ; pillar one-toothed.

Trochus argyrostomus. *Gmelin,* p. 3583. *Schreibers Conch.* i. p. 265.

Trochus Atramentarium. *Callone's Cat.* p. 26, No. 461.

Trochus Diadema. *Meuschen Mus. Gevers.* p. 282, No. 395.

Trochus, No. 25. *Schroeter Einl.* i. p. 689.

Trochus imperforatus niger argyrostomus. *Chemnitz,* v. p. 51. t. 165. f. 1562 and 1563.

L'Ecritoire. *Favanne,* ii. p. 304.

D'Avila, t. 5. f. K.

Inhabits the coasts of New Zealand. *Favanne.* Arabia Felix. *Humphreys.*

This shell is sometimes two inches long, and about equally broad, but my specimen is not much more than half so large. May not *T. agrestris* or *T. niger* be only this species in an earlier stage of growth, for it has been ascertained that the young shells of *T. crassus* are slightly umbilicated?

MERULA. 86. Shell ovate, obtuse, turban-shaped, black, with the summit and throat silvery, and the base reddish.

Trochus Merula. *Chemnitz,* v. p. 52. t. 165. f. 1564 and 1565.

Trochus Sinensis. *Gmelin,* p. 3583. *Schreibers Conch.* i. p. 265.

Trochus, No. 26. *Schroeter Einl.* i. p. 690.

Le Merle. *Favanne,* ii. p. 112. t. 9. f. B 1.

Knorr, v. t. 3. f. 1.

Variety. With distant scattered white dots.

Trochus tigrinus. *Chemnitz,* v. p. 53. t. 165. f. 1566.

Trochus, No. 27. *Schroeter Einl.* i. p. 690

Inhabits the coasts of China. *Chemnitz.* Otaheite. *Favanne.*

Shell about an inch long, and rather more than an inch and a quarter broad, and differs from the foregoing species in having the whirls more rounded and less produced, and in the surface being smooth; the lower whirls are said to be sometimes coated with a black crust, and in Knorr's figure the upper whirls are painted dark orange.

CRASSUS. 87. Shell sub-ovate, thick, coarse, with one tooth on the pillar, and the base by the inner lip white and flattened.

Trochus crassus. *Montagu Test.* p. 281. *Maton and Racket, in Lin. Trans.* viii. p. 154. *Dorset Cat.* p. 48. t. 17. f. 3 and 7.
Turbo lineatus. *Da Costa Brit. Conch.* p. 100. t. 6. f. 7. *Donovan,* ii. t. 71.
Inhabits the coasts of Great Britain. *Da Costa,* &c.
Shell about three-quarters of an inch, or an inch long, and nearly equally broad, of a greyish brown or blackish colour, obsoletely marked with crowded zic-zac transverse rather darker lines, and the inside pearly.

TUBER. 88. Shell somewhat depressed, with the whirls strongly plaited above, and nodulous at their lower margins.

Trochus Tuber. *Linnæus Syst. Nat.* p. 1230. *Chemnitz,* v. p. 55. t. 165. f. 1572 to 1576. *Schroeter Einl.* i. p. 668. *Gmelin,* p. 3578. *Schreibers Conch.* i. p. 256.
La Perruche verde. *Favanne,* ii. p. 98. t. 9. f. C.
Lister Conch. t. 646. f. 38. *Argenville,* t. 8. f. I. *Seba,* iii. t. 74. f. 12. *Knorr,* i. t. 3. f. 2. *Geve.* t. 15. f. 135 and 136. *Regenfuss,* i. t. 12. f. 76.
Variety. Smaller, and the whirls much depressed.
Trochus pantherinus. *Gmelin,* p. 3584.
Trochus, No. 182. *Schroeter Einl.* i. p. 749.
La Perruche aplatie. *Favanne,* ii. p. 14.
Le Kachin. *Adanson Senegal,* p. 187. t. 12. f. 9.
Geve. t. 15. f. 141 and 142. *Regenfuss,* i. t. 3. f. 27.
Inhabits the Mediterranean. *Linnæus.* Coasts of Barbadoes. *Lister.* Martinique and St. Domingo. *Favanne.*
Shell an inch and a quarter, or sometimes nearly two inches long, and about equally broad, and is remarkable for its strong longitudinal plaits, which occupy the upper part, and disappear on the middle of the whirls; the colour is green

variously shaded with brown, and the body-whirl is obsoletely studded with small white tubercles.

MELANOSTOMUS. 89. Shell pyramidal, obtuse, and
greenish with transverse rows of dark dots;
throat black.

Trochus melanostomus. *Gmelin.* p. 3581.
Trochus in fauce nigerrimus. *Chemnitz,* v. p. 29. t. 161. f.
1526.
Trochus, No. 12. *Schroeter Einl.* i. p. 683.
Inhabits the Southern Ocean. *Chemnitz.*
This shell, as figured by Chemnitz, appears to be eight lines
long, and equally broad, of a green colour with transverse
rows of blackish dots, and the throat black.

STRIATUS. 90. Shell conical, with strong transverse, and very minute oblique striæ; sutures
obsolete, and the inside pearly.

Trochus striatus. *Linnæus Syst. Nat.* p. 1230. *Chemnitz,* v. p. 29. t. 162. f. 1527 and 1528. *Schroeter Einl.*
i. p. 670. *Gmelin,* p. 3579. *Schreibers Conch.* i. p.
257. *Pulteney's Dorset Cat.* p. 44. *Montagu Test.* p.
278, and *Supp.* p. 119.
Trochus exasperatus. *Pennant Zool.* iv. p. 126.
Trochus parvus. *Da Costa Brit. Conch.* p. 41.
Trochus conicus. *Donovan Brit. Shells,* v. t. 155. f. 1.
Trochus erythroleucos. *Maton and Racket, in Lin. Trans.*
viii. p. 156. *Dorset Cat.* p. 48. t. 18. f. 2.
Lister Conch. t. 616. f. 2. *Gualter,* t. 61. f. N.
Inhabits the Mediterranean. *Linnæus.* Coasts of Britain.
Da Costa, &c.
Shell about three-eighths of an inch long, and a quarter of an
inch broad, generally of a pale greyish white, but sometimes
of a reddish colour, with interrupted longitudinal dark purplish brown lines; it has six flat whirls, and it is difficult to
discover the separating line. I think there can be no doubt
that this is the Linnæan *T. striatus,* and *T. exiguus* of
Montagu appears to me to be the *T. erythroleucos* of Gmelin,

MINUTUS. 91. Shell conical, with obsolete crenulated transverse striæ, and a broader rib at
the margin of the whirls; summit crimson,
and the inside white.

Trochus minutus.   *Chemnitz*, v. p. 30. t. 162. f. 1529.
Trochus erythroleucos.   *Gmelin*, p. 3581.
Trochus Conulus.   *Da Costa Brit. Conch.* p. 40. t. 2. f. 4.
   Donovan, i. t. 8. f. 2.
Trochus exiguus.   *Montagu Test.* p. 277.   *Maton and*
   Racket, in Lin. Trans. viii. p. 156.   *Dorset Cat.* p. 48.
   t. 21. f. 4.
Trochus, No. 13.   *Schroeter Einl.* i. p. 683.
Le Rubis.   *Favanne*, ii. p. 376. t. 12. f. N 2.
Lister Conch. t. 621. f. 8.
Junior.   Trochus parvus. *Adams, in Lin. Trans.* iii. p. 65.?
Inhabits the coasts of Great Britain. *Da Costa, &c.* Moroc-
co. *Chemnitz.* Brittany, Provence, Martinique, and St.
Domingo. *Favanne.*
Shell sometimes three-eighths of an inch long, and a quarter of
an inch broad, but is generally rather smaller ; the rib which
attends the margin of the whirls is much broader and more
elevated than the transverse striæ, and is obliquely crenulated ; the lower whirls are greyish brown, sometimes spotted
with white, and the summit is almost constantly of a fine
deep crimson colour.

PUNCTULATUS. 92. Shell conical, with crenulated
   transverse striæ, and a strong somewhat beaded rib at the margin of the whirls ; inside
   white.

Trochus punctulatus.   *Gmelin*, p. 3581.
Trochus minimus.   *Chemnitz*, v. p. 30. t. 162. f. 1530.
Trochus, No. 14.   *Schroeter Einl.* i. p. 683.
Le Fruit d'If.   *Favanne*, ii. p. 375. t. 12. f. N 1.
Inhabits the coasts of Morocco. *Chemnitz.* Brittany, Provence, St. Domingo, and Martinique. *Favanne.*
This shell much resembles *T. minutus*, and is of the same size,
but rather narrower in proportion to its length, and differs
in having the transverse striæ much stronger, and the rib at
the margins of the whirls more elevated, and so distinctly
striated crossways as to have a beaded appearance ; it is of a
rose-colour, more or less marked with white longitudinal
stripes, and the summit is not of a brighter red than the
other whirls.

CONULUS. 93. Shell conical, smooth, with a strong
   rib on the margin of the whirls.

Trochus Conulus. *Linnæus Syst. Nat.* p. 1230. *Born Mus.*
p. 336. *Chemnitz,* v. p. 65. t. 166. f. 1588 and 1589.
Schroeter Einl. i. p. 670. *Gmelin,* p. 3579. *Schreibers*
Conch. i. p. 257.
Le Sabot veiné. *Favanne,* ii. p. 383. t. 13. lower fig. A.
Bonanni Rec. and *Kirch.* 3. f. 93 and 99.
Inhabits the Mediterranean. *Linnæus.*
Linnæus considered it to be doubtful whether this is more than
a variety of *T. Zizyphinus,* and most of the figures to which
he has referred, probably belong either to that species, or to
T. papillosus; it differs in being rather smaller, less elevat-
ed, and less distinctly striated, and in having a stronger rib
on the margin of the whirls.

ZIZYPHINUS. 94. Shell conical, and the whirls
flat, with transverse striæ, of which those on
the margins are largest; base flattish.

Trochus Zizyphinus. *Linnæus Syst. Nat.* p. 1231. *Pen-*
nant Zool. iv. p. 126. t. 80. f. 103. *Da Costa Brit.*
Conch. p. 37. t. 3. f. 3 and 4. *Born Mus.* p. 337. *Chem-*
nitz, v. 66. t. 166. f. 1592 to 1594. *Schroeter Einl.* i. p.
672. *Gmelin,* p. 3579. *Donovan,* ii. t. 52. *Montagu*
Test. p. 274. *Maton and Racket, in Lin. Trans.* viii.
p. 155. *Dorset Cat.* p. 48. t. 16. f. 3 and 4.
Le Sabot panaché. *Favanne,* ii. p. 378. t. 70. f. H 2.
Lister Anim. Ang. t. 3. f. 14, and *Conch.* t. 616. f. 1.
Gualter, t. 61. f. C. *Argenville,* t. 8. f. N. *Klein Ost.*
t. 2. f. 36. *Knorr,* iii. t. 14. f. 2, and vi. t. 27. f. 5.
Geve. t. 9. f. 71 and 72.
Variety. Bluish grey, with longitudinal chestnut stripes.
Trochus Conulus Tranquebaricus. *Chemnitz,* v. p. 68. t.
166. f. 1595 and 1596.
Trochus, No. 37. *Schroeter Einl.* i. p. 695.
Le Sabot canellé. *Favanne,* ii. p. 379.
Inhabits the Mediterranean and European Seas. *Linnæus.*
Coasts of Great Britain. *Lister, &c.* France, England, and
the Red Sea. *Favanne.*
Shell about an inch long, and nearly equally broad, marked
with several strong transverse striæ, of which those at the
base of the whirls are broader and more prominent than the
others; the colour is livid or reddish, streaked longitudinally
with darker broad undulated irregular stripes, and spotted on
the ridge at the base of the whirls; the inside is pearly.

PAPILLOSUS. 95. Shell conical, thin, and the whirls rather convex, with transverse granulated striæ; base slightly convex.

Trochus papillosus. *Da Costa Brit. Conch.* p. 38. t. 3. f. 5 and 6. *Donovan,* iv. t. 127. *Maton and Racket, in Lin. Trans.* viii. p. 155. *Dorset Cat.* p. 48. t. 16. f. 5 and 6.

Trochus Zizyphinus, Var. *Linnæus Mus. Reg. Ulr.* p. 650.

Trochus granulatus. *Born Mus.* p. 337. t. 12. f. 9 and 10. *Chemnitz,* v. p. 68. t. 166. f. 1597 and 1598.

Trochus fragilis. *Pulteney's Dorset Cat.* p. 44.

Trochus tenuis. *Montagu Test.* p. 275. t. 10. f. 3.

Le Sabot grenu. *Favanne,* ii. p. 380.

Gualter, t. 61. f. G and M.

Inhabits the coasts of Britain. *Da Costa, &c.* Mediterranean. *Favanne.* East Indies. *Chemnitz.*

Shell about an inch and a quarter long, and the length rather exceeds the breadth; in shape and size it bears a considerable resemblance to *T. Zizyphinus,* but may be readily distinguished by its fine granulated striæ, by both the whirls and the base being more rounded, and by the striæ at the margin of the whirls being of the same size, or but very little thicker than the others.

GRANATUM. 96. Shell sub-conical, with transverse granulated striæ, and the whirls slightly ventricose; summit acuminated, and the base convex.

Trochus Granatum. *Chemnitz,* v. p. 100. t. 170. f. 1654 and 1655. *Gmelin,* p. 3584. *Schreibers Conch.* i. p. 259.

Trochus Tigris. *Martyn Univ. Conch.* ii. t. 75. *Gmelin,* p. 3585.

Trochus, No. 73. *Schroeter Einl.* i. p. 707.

La Pomme de Grenade. *Favanne,* ii. p. 343.

Inhabits the coasts of New Zealand. *Favanne, &c.*

Shell about two inches long, and equally broad, and is a beautiful and rare species; the colour is whitish, variegated with fine scarlet longitudinal irregular stripes, and the transverse rows of granules are whitish; the base is convex and oblique.

VIRGINEUS. 97. Shell sub-conical, slightly ventricose, with transverse rows of granules and red dots, and a band round the base of the whirls; base convex.

Trochus virgineus.　*Chemnitz*, x. p. 289. t. 165. f. 1581 and 1582.
Trochus annulatus.　*Martyn Univ. Conch.* i. t. 33.
Trochus cælatus, Var. β.　*Gmelin*, p. 3582.
Le Sabot Magellanique.　*Favanne*, ii. p. 342. t. 79. f. I.
Inhabits the Straights of Magellan, and coasts of the Falkland Islands. *Favanne.* New Zealand. *Martyn.* North West coast of America. *Humphreys.*
Shell about an inch long, and equally broad, of a yellowish colour, with a red dot between each of its minute transverse granules, and generally a purplish band at the base of the whirls; the granules extend in concentric rows over the base, and a purplish circle surrounds the pillar; the base is obliquely convex, and the inside pearly and iridescent. *Trochus annulatus* of Gmelin is a very minute species.

DIAPHANUS. 98. Shell sub-conical, thin, with crowded transverse rows of alternately white and chestnut granules; whirls and base convex, and the aperture large.

Trochus diaphanus.　*Gmelin*, p. 3580.
Trochus punctulatus.　*Martyn Univ. Conch.* t. 36.
Trochus asper, &c. *Chemnitz*, v. p. 26. t. 161. f. 1520 and 1521.
Trochus, No. 7.　*Schroeter Einl.* i. p. 681.
Turbo punctulatus.　*Gmelin*, p. 3589.
Spengler Naturf. ix. p. 152. t. 5. f. 2.
Inhabits the coasts of New Zealand. *Chemnitz.*
Shell an inch and a quarter long, and about equally broad, and beautifully granulated all over in crowded transverse rows; the colour is a pale dull chestnut, and in the specimen now before me the granules on the body-whirl are nearly all white, and those on the second whirl are only alternately so; the inside is striated, and very pearly and iridescent.

SELECTUS. 99. Shell sub-conical, with granulated transverse striæ, and the body-whirl broad and marginated at the base; spire acuminated, and the base convex.

Trochus selectus.　*Chemnitz*, xi. p. 168. t. 196. f. 1896 and 1897.
Inhabits the coasts of New Zealand. *Chemnitz.*
Shell about an inch and a quarter long, and near an inch and a

half broad, with transverse rows of concatenated red and white granules ; the body-whirl is much broader than the other whirls, and the summit of the spire is produced to a fine point ; the throat is silvery and grooved.

PURPURASCENS. 100. Shell conical, with plaited somewhat tuberculated whirls, and transverse rows of minute granules; aperture roundish.

Trochus purpurascens. *Chemnitz*, v. p. 35. t. 162. f. 1538 and 1539.
Trochus purpureus. *Gmelin*, p. 3582. *Schreibers Conch.* i. p. 264.
Trochus, No. 19. *Schroeter Einl.* i. p. 685.
Inhabits ———
This shell, as figured by Chemnitz, appears to be about thirteen lines long and twelve broad, of rather a bright purplish colour, and spotted with white round the margin of the whirls.

AMERICANUS. 101. Shell conical, with longitudinal grooves and transverse striæ ; inner lip toothed.

Trochus Americanus. *Gmelin*, p. 3581.
Trochus plicato-nodosus. *Chemnitz*, v. p. 33. t. 162. f. 1534 and 1535.
Trochus, No. 17. *Schroeter Einl.* i. p. 684.
Inhabits the coasts of the West India Islands. *Chemnitz.*
Shell about sixteen lines long, and fourteen lines broad, and is described by Chemnitz to be yellowish white, but his figures are painted yellow with a tinge of brown.

IMBRICATUS. 102. Shell conical, with strong oblique plaits forming vaulted tubercles on the base of the whirls, and the inner lip undulated.

Trochus imbricatus. *Gmelin*, p. 3581.
Trochus profundè sulcatus, &c. *Chemnitz*, v. p. 30. t. 160. f. 1531.
Trochus ramosus. *Meuschen Mus. Gevers.* p. 284. No. 418.
Trochus, No. 15. *Schroeter Einl.* i. p. 683.
Turbo Tectum Persicum. *Born Mus.* p. 344. t. 12. f. 19 and 20.
Le Concombre. *Favanne*, ii. p. 396. t. 13. f. D.

Lister Conch. t. 628. f. 14.   *Gualter,* t. 60. f. Q.

Inhabits the coasts of Jamaica and Barbadoes. *Lister.* St. Domingo and Martinique. *Favanne.*

Shell about an inch and three quarters long, and an inch and a half broad, with thick somewhat curved oblique plaits, which become thicker at the margin of the whirls, and at the base form obtuse short spines or vaulted tubercles; the colour is whitish, yellowish, or reddish brown, and generally marked with a paler band at the upper margin of the whirls.

CÆLATUS.   103.   Shell sub-conical, with oblique somewhat scaly plaits on the upper part, and transverse spinous ribs at the base of the whirls.

Trochus cælatus. *Chemnitz,* v. p. 33. t. 162. f. 1536 and 1537. *Gmelin,* p. 3581. *Schreibers Conch.* i. p. 263.

Trochus, No. 18. *Schroeter Einl.* i. p. 685.

La Raboteuse. *Favanne,* ii. p. 89. t. 8. f. M.

Bonanni Rec. and *Kirch.* iii. f. 167. *Lister Conch.* t. 647. f 40. *Seba,* iii. t. 60. f. 1 and 2. *Knorr,* v. t. 12. f. 3. *Geve.* t. 15. f. 133 and 134. *Regenfuss,* ii. t. 1. f. 9.

Inhabits the coasts of Jamaica. *Lister.* Isle of France, Madagascar, and St. Domingo. *Favanne.* St. Croix. *Chemnitz.*

The length, which rather exceeds the breadth, varies from one to three inches, and the colour is white, most commonly tinged with sea-green, and sometimes reddish; the upper part of the whirls is covered with narrow rather crowded oblique plaits, which become more or less broken below into granules or small vaulted scales, and near the margin there are two or three transverse rows of strong obliquely vaulted spines; the base is covered with concentric rows of small imbricated scales, and the pillar and inside are pearly: *Trochus virgineus,* which Gmelin has arranged as a variety of this species, is a totally different shell.

GIBBEROSUS.   104.   Shell conical, with oblique plaits on the upper part, a transverse row of tubercles below, and the margin of the whirls spinous.

Trochus gibberosus. *Chemnitz,* x. p. 287. Vign. 23. f. A and B.

Trochus inæqualis. *Martyn Univ. Conch.* i. t. 31. *Gmelin*, p. 3585.

La Raboteuse de la Nouvelle Zélande. *Favanne,* ii. p. 13, and p. 92.

Inhabits the coasts of New Zealand. *Favanne.* Friendly Islands. *Martyn.*

Shell about an inch and three-quarters long, and two inches broad, of an olive or greenish or reddish brown colour ; below the oblique plaits on the upper half of the whirls is a double row of granules, and the margin has only a single row of spines ; it is nearly allied to, and probably not specifically distinct from, *T. cælatus.*

MAURITIANUS. 105. Shell conical, with short, imbricated, transversely granulated whirls, obtusely plaited on their lower margins; base flat, and the pillar grooved and twisted.

Trochus Mauritianus. *Gmelin*, p. 3582.
Trochus muricatus. *Chemnitz*, v. p. 42. t. 163. f. 1547 and 1548.
Trochus, No. 22. *Schroeter Einl.* i. p. 687.
Le Clocher gothique. *Favanne,* ii. p. 390. t. 13. f. S.
Bonanni Rec. and *Kirch.* 3. f. 90. *Lister Conch.* t. 625. f. 11. *Gualter,* t. 61. f. D and F.

Inhabits the coasts of the Isle of France and New Guinea. *Favanne.*

Shell about an inch and three-quarters long, and an inch and a half broad, and differs from *T. fenestratus* principally in being larger and in having the whirls imbricated. Favanne doubts whether it is more than a variety of that species, and Kæmmerer considered both to be only varieties of *T. dentatus.* See Cab. Rudolst. p. 161.

FENESTRATUS. 106. Shell conical, with short transversely granulated whirls, obtusely plaited on their lower halves; base flat, and the pillar grooved and twisted.

Trochus fenestratus. *Gmelin,* p. 3582.
Trochus pyramidalis asper. *Chemnitz,* v. p. 44. t. 163. f. 1549 and 1550.
Trochus, No. 23. *Schroeter Einl.* i. p. 688.
Le Sabot ciselé. *Favanne,* ii. p. 389. t. 12. f. I.

Rumphius, t. 21. f. 7. *Gualter*, t. 60. f. N. *Geve*, t. 7.
f. 55 and 56. *Regenfuss*, i. t. 2. f. 13.
Inhabits the coasts of Amboyna. *Rumphius*. Frederick's Islands. *Regenfuss*.
Shell about an inch and a quarter long, and an inch broad, white, with the spaces between the obtuse plaits on the lower half of the whirls, and the row of granules above them, generally of a dull olive-green, but Favanne says it sometimes varies in colour; the base is white and spotless, with the inner lip incurved, and the pillar longitudinally grooved and twisted.

PYRAMIS. 107. Shell conical, with imbricated whirls, granulated in transverse rows, and crenated on their lower margins; base flat, and the pillar grooved and twisted.

Trochus Pyramis. *Born Mus.* p. 333. *Chemnitz*, v. p. 19. t. 160. f. 1510 to 1512.
Trochus Obeliscus. *Gmelin*, p. 3579. *Schreibers Conch.* i. p. 261.
Trochus, No. 1. *Schroeter Einl.* i. p. 678.
Le Sabot echancré. *Favanne*, ii. p. 384. t. 13. f. &.
Knorr, i. t. 12. f. 4. *Geve*, t. 9. f. 68.
Inhabits the coasts of Otaheite. *Favanne*. East Indian Seas. *Chemnitz*.
Shell an inch and a half, or sometimes two inches and a half long, and the length rather exceeds the breadth; it has twelve or thirteen short whirls, with four or five transverse rows of granules on the first, and their number decreases upwards; the colour is greenish clouded with brown, and the base is white; the inside is pearly.

DENTATUS. 108. Shell conical, with the whirls obliquely wrinkled, and remotely coronated at their lower margins; pillar grooved and twisted.

Trochus dentatus. *Forskael Descript. Anim.* p. 125. *Chemnitz*, v. p. 23. t. 161. f. 1516 and 1517.
Trochus foveolatus. *Gmelin*, p. 3580. *Schroeter Conch.* i. p. 262.
Trochus nodosus. *Meuschen Mus. Gevers.* p. 284. No. 421.
Trochus, No. 5. *Schroeter Einl.* i. p. 680.

Le Pain de Sucre tuberculé. *Favanne,* ii. p. 387. t. 13. f. A.

Lister Conch. t. 626. f. 11 a. *Gualter,* t. 59. f. A.

Variety. With the nodules on the base of the whirls more crowded.

Trochus dentatus duplex. *Chemnitz,* v. p. 24. t. 161. f. 1518 and 1519.

Inhabits the Red Sea. *Forskael.* Coasts of New Guinea. *Favanne.*

Shell about two inches and a half long, and rather more than two inches broad, and has from eight to sixteen obtuse tubercles on the lower margin of the whirls; the colour is whitish tinged with red, and a circle of green on the base surrounding the pillar; the inner lip is incurved, and the pillar longitudinally grooved and twisted.

VIRGATUS. 109. Shell conical, with transverse granulated striæ, and the whirls slightly undulated at the sutures; base concave.

Trochus virgatus. *Gmelin,* p. 3580. *Schreibers Conch.* i. p. 262.

Trochus pyramidalis granulatus, &c. *Chemnitz,* v. p. 22. t. 160. f. 1514 and 1515.

Trochus, No. 3. *Schroeter Einl.* i. p. 679.

Le Sabot flambé. *Favanne,* ii. p. 371.

Lister Conch. t. 631. f. 17. *Gualter,* t. 61. f. E.

Inhabits the East Indian Seas. *Chemnitz.* Coasts of Amboyna and Mindanao. *Favanne.*

Shell about an inch and a half long, and rather more than an inch and a quarter broad, white with irregular longitudinal rose-coloured stripes, and the base white with concentric rows of red spots.

COOKII. 110. Shell sub-conical, with the whirls ventricose, and numerous oblique crowded transversely wrinkled plaits armed with vaulted scales; outer lip crenated.

Trochus Cookii. *Gmelin,* p. 3582. *Schreibers Conch.* i. p. 264.

Trochus Cooksianus. *Chemnitz,* v. p. 36. t. 163. f. 1540, and 164. f. 1551.

Trochus sulcatus. *Martyn Univ. Conch.* t. 35. *Portland Cat.* p. 113. lot 2481.

Trochus, No. 20. *Schroeter Einl.* i. p. 686.

Turbo sulcatus. *Gmelin*, p. 3592.

Spengler's Naturf. ix. p. 155. t. 3. f. 5 and 6.

Inhabits Cook's Straights in New Zealand. *Chemnitz.* Friendly Islands. *Martyn.*

Shell three or four inches long, and about equally broad, of a greyish brown colour, and the inside pearly; it has a singular horny operculum, which Chemnitz has figured at t. 163. f. *a* and *b*.

IRIS. 111. Shell sub-conical, slightly ventricose and acuminated, smooth, and beautifully iridescent under the epidermis; aperture oblique.

Trochus Iris. *Gmelin*, p. 3580. *Schreibers Conch.* i. p. 262.

Trochus Iridis. *Chemnitz*, v. p. 27. t. 161. f. 1522 and 1523.

Trochus Opalus. *Martyn Univ. Conch.* i. t. 24.

Trochus, No. 8. *Schroeter Einl.* i. p. 681.

Elenchus Iris. *Humphreys Call. Cat.* p. 25. No. 434.

Le Grand Point d'Hongrie. *Favanne*, ii. p. 132. t. 79. f. G.

Walch Naturf. iv. p. 42. t. 1. f. 5 and 6. *Zorn Naturf.* vii. p. 161. t. 2. f. C 1 and C 2.

Variety. Shell smaller, and whitish, with reddish stripes.

Trochus rostratus. *Gmelin*, p. 3580.? *Schreibers Conch.* i. p. 263.

Trochus pyramidalis, &c. *Chemnitz*, v. p. 28. t. 161. f. 1524 and 1525.?

Trochus, No. 9. *Schroeter Einl.* i. p. 682.?

Inhabits the coasts of New Zealand. *Chemnitz, &c.*

Shell about an inch and a half long, and not much more than an inch broad, of a pale purplish or chocolate-colour, with darker longitudinal zic-zac stripes; when deprived of the outer coat the whole surface, as well as the inside, is beautifully pearly and iridescent, and a few rather remote transverse grooves then become more apparent. *T. rostratus* of Gmelin is most probably only a variety, but I am unacquainted with the shell, and the figures do not answer well to the description.

ELEGANS. 112. Shell sub-conical, elongated, acuminated, smooth, with a few slight remote transverse grooves, and iridescent under the epidermis; aperture oblique.

Trochus elegans. *Gmelin*, p. 3581.
Trochus, No. 11. *Schroeter Einl.* i. p. 682.
Zorn Naturf. vii. p. 167. t. 2. f. D 1, and D 2.
Inhabits the coasts of New Zealand. *Chemnitz*.
Shell about an inch, or an inch and a quarter long, and hardly
half as broad, of a reddish or yellowish brown, with the transverse grooves of a paler colour; there are about eight
grooves disposed somewhat in pairs on the body-whirl, and
four on the whirls of the spire; the whole surface when deprived of its epidermis, as well as the inside, is iridescent.

NOTATUS.   113.   Shell sub-conical, transversely
   grooved, and obliquely striated; aperture
   oblique, and the outer lip crenated.

Trochus notatus. *Gmelin*, p. 3581.   *Schreibers Conch.* i.
   p. 263.
Trochus, No. 10.   *Schroeter Einl.* i. p. 682.
L'Aigrette. *Favanne*, ii. p. 134.
Schroeter Journal, v. p. 438. f. 10 and 11.
Inhabits the South Sea.
Shell about three-quarters of an inch long, and half an inch
broad; it differs from *T. elegans* in being proportionably
broader, as well as in having oblique longitudinal striæ, and
in the whirls being more ventricose; the specimen now before me is reddish chestnut, marbled with irregular, pale
ash-coloured, somewhat zic-zac, longitudinal stripes; the inside is iridescent, and transversely grooved.

ZIC-ZAC.   114.   Shell sub-conical, transversely striated, and the body-whirl transversely flattened in the middle; aperture roundish.

Trochus zic-zac.   *Chemnitz*, v. p. 69.   *Gmelin*, p. 3587.
Turbo zic-zac.   *Maton and Racket, in Lin. Trans.* viii. p.
   160.   *Montagu Supp.* p. 135.
Variety A.   Whitish, with longitudinal zic-zac brown stripes.
Trochus, No. 38.   *Schroeter Einl.* i. p. 695.
Trochus, No. 80.   *Schreibers Conch.* i. p. 258.
L'Emouchette. *Favanne*, ii. p. 20.
Lister Conch. t. 583. f. 38.   *Chemnitz*, v. t. 166. f. 1599.
Variety B.   With the zic-zac lines interrupted by a transverse
bluish band in the middle of the whirls, and the shell smaller.
Trochus, No. 39.   *Schroeter Einl.* i. p. 695.
Trochus, No. 81.   *Schreibers Conch.* i. p. 259.
L'Epervier. *Favanne*, ii. p. 20.

Chemnitz, v. t. 166. f. 1600.
Variety C.   Greyish white, and the markings obsolete.
Lin. Trans. viii. t. 4. f. 14.
Inhabits the coasts of Jamaica and Barbadoes. *Lister.*   Virginia. *Favanne.*   The Var. C has been found on the coasts of Britain by Lady Wilson. *Maton and Racket.*
The Variety A is about three-quarters of an inch, the Variety B half an inch, and the Variety C only a quarter of an inch long, and the breadth is about half the length; in the Variety A, the body-whirl is transversely carinated towards the base, and also somewhat so towards the upper margin, but the latter is less obvious in the other Varieties.   Mr. Montagu considered a minute shell which I found in Bantry Bay to be another Variety, but it appears to me to be distinct and nondescript.   The Varieties A and B have been arranged as separate species both by Schroeter and Schreibers.

OBTUSUS.   115. Shell slightly conical, with the spire flattened, and the whirls marginated; base convex, with the aperture roundish, and the throat lead-coloured.

Trochus obtusus.   *Chemnitz,* xi. p. 167. t. 196. f. 1894 and 1895.
Inhabits the East Indian Seas. *Chemnitz.*
Shell about nine lines long, and ten broad, with the spire remarkably flattened, and something like that of a Nerite; it is prettily coloured with alternate longitudinal undulated purple and white stripes.

CROCATUS.   116. Shell sub-conical, smooth, with convex whirls which are spinous above, and the sutures grooved; base convex.

Trochus crocatus.   *Born Mus.* p. 338. t. 12. f. 11 and 12.
   Gmelin, p. 3584.
Trochus, No. 90.   *Schroeter Einl.* i. p. 713.
Inhabits ——
Born has described this shell to be seven lines long, and seven broad, with the base and epidermis white and shining, but he says that the spire, when stripped of the epidermis, is saffron-coloured; his figure, however, does not answer the description, but is ten or eleven lines long, with the base and body-whirl white, and the spire crimson; nor is there any appearance of spines, and it has much the appearance of a Helix.

HORTENSIS. 117. Shell sub-conical, with the whirls rounded, and the summit obtuse; base slightly convex.

Trochus hortensis. *Chemnitz*, ix. part 2. p. 52. t. 122. f. 1055 and 1056. *Gmelin*, p. 3587.
Helix Trochus. *Muller Verm.* ii. p. 79.
Inhabits gardens in southern climates. *Chemnitz.*
Shell an inch long, and ten lines broad, white, with a broad reddish brown band in the middle of the whirls, and has much more the appearance of a Garden Snail than of a Trochus; I am not acquainted with either, but from the figure it appears to me that both this and the preceding species, should be placed among the Helices.

*** *Tapering, with the Pillar exserted, and the shell falling to one side when placed upon its base.*

TELESCOPIUM. 118. Shell conical, turreted, with the whirls continuous and transversely grooved; pillar twisted.

Trochus Telescopium. *Linnæus Syst. Nat.* p, 1231. *Born Mus.* p. 338, and Vign. at p. 326. f. d. *Chemnitz*, v. p. 14. t. 160. f. 1507 to 1509, and Vign. at p. 3. f. *A* and *B. Schroeter Einl.* i. p. 672, and *Inn. Bau Conch.* p. 100. t. 5. f. 8. *Gmelin*, p. 3585. *Schreibers Conch.* i. p. 266.
Cerithium Telescopium. *Bruguiere Enc. Meth.* p. 483,
Le Télescope.. *Favanne*, ii. p. 317. t. 39. f. B 2.
Bonanni Rec. and *Kirch.* 3. f. 92. *Lister Conch.* t. 624. f. 10. *Rumphius*, t. 21. f. 12. *Petiver Amb.* t. 4. f. 10. *Gualter*, t. 60. f. D and E. *Argenville*, t. 11. f. B. *Seba*, iii. t. 50. f. 1 to 12. *Knorr*, iii. t. 22. f. 2.
Inhabits the East Indian Seas. *Seba.* Coasts of Tranquebar. *Humphreys.*
Shell three or four inches long, and two fifths as broad, of a brown or blackish colour, generally marked with a paler band round the lower margin of the body-whirl; the base is nearly flat, and the pillar protuberant and twisted.

TEREBELLUS. 119. Shell sub-conical, turreted, and glabrous; pillar-lip with three plaits, and the inside of the outer lip smooth.

Trochus dolabratus, Var. *λ. Gmelin,* p. 3586.
Helix terebella. *Muller Vermium,* ii. p. 123.
Helix, No. 141. *Schroeter Einl.* ii. p. 215.
Voluta dolabrata senior. *Solander's MSS.*
Bulimus terebellum. *Bruguiere Enc. Meth.* p. 355.
Bonanni Rec. 3. f. 379, and *Kirch.* f. 366. *Lister Conch.*
t. 844. f. 72. *Petiver Gaz.* t. 118. f. 15. *Gualter,* t. 4.
f. M.
Inhabits Barbadoes. *Lister.* And is a land shell. *Gualter.*
Shell about fourteen lines long and six broad, and differs from
T. dolabratus in being rather broader, in having the whirls
flatter, and the inside of the outer lip smooth.

DOLABRATUS. 120. Shell sub-conical, turreted,
glabrous, and umbilicated; pillar twisted,
with three plaits, and the outer lip ribbed.

Trochus dolabratus. *Linnæus Syst. Nat.* p. 1231. *Born
Mus.* p. 339. *Chemnitz,* v. p. 73. t. 167. f. 1603 and
1604. *Schroeter Einl.* i. p. 675. *Gmelin,* p. 3585.
Schreibers Conch. i. p. 266.
Helix dolabrata. *Muller Vermium,* ii. p. 121.
Voluta dolabrata. *Solander's MSS.*
Pyramidella dolabrata. *Lamarck Syst. des Anim.* p. 92.
Argenville, t. 11. f. L. *Knorr,* vi. t. 29. f. 2. *Favanne,*
t. 65. f. L.
Variety. With two or three transverse rows of brown spots.
Voluta notata. *Solander's MSS.*
Lister Conch. t. 844. f. 72 b.
Inhabits Africa, and is a land shell. *Linnæus.*
Shell three-quarters of an inch, or an inch long, and only one
third as broad, with the whirls slightly ventricose; it is
white, and marked with from one to five transverse yellowish
brown lines; the outer lip on its inside has about six ribs,
which are slightly toothed at a short distance from the margin, and by these ribs this species may be distinguished from
T. terebellus, which has the outer lip smooth.

PERVERSUS. 121. Shell turreted, minute, imperforate, glabrous, with the whirls reversed,
and a double row of excavated dots.

Trochus perversus. *Linnæus Syst. Nat.* p. 1231. *Gmelin,*
p. 3586.
Cerithium perversus. *Bruguiere Enc. Meth.* p. 496.?
Inhabits the Mediterranean. *Linnæus.*

Linnæus has described this shell to be small, of a horn-colour, and the whirls cylindrical, with a double row of excavated dots, besides which the margins also have crenated punctures; aperture four-sided, and the pillar prominent at the base, but not so as to form a canal. Bruguiere has conjectured that *T. ventricosus* may be this species.

PUSILLUS. 122. Shell turreted, minute, and the whirls reversed, with transverse obliquely decussated striæ; base flat, and the aperture compressed.

Trochus pusillus. *Gmelin*, p. 3586.
Trochus perversus, Var.? *Chemnitz*, ix. part 1. p. 126. t. 113. f. 966.
Trochus, No. 185. *Schroeter Einl.* i. p. 751.
Spengler n. Schrift. der Daen. i. p. 375. t. 1. f. 1.
Inhabits the sand on the East Indian Shores. *Spengler.*
Shell hardly a quarter of an inch long, and about one third as broad, of a reddish brown colour with darker spots.

UNDULATUS. 123. Shell turreted, minute, and the whirls reversed, with longitudinal ribs, and transverse undulated striæ; base flat, and the aperture semi-lunar.

Trochus undulatus. *Gmelin*, p. 3586.
Trochus, No. 186. *Schroeter Einl.* i. p. 752.
Spengler n. Schrift. der Daen. i. p. 376. t. 1. f. 2. *Chemnitz*, ix. part 1. p. 126. t. 113. f. 967.
Inhabits the sands on the East Indian Shores. *Spengler.*
This shell is said to be rather smaller than *T. pusillus*, and the figure is painted nearly of the same colour.

VENTRICOSUS. 124. Shell tub-turreted, minute, cancellated, with the whirls reversed, and the lower ones ventricose; base glabrous, and the aperture sub-ovate.

Trochus ventricosus. *Gmelin*, p. 3586.
Trochus, No. 187. *Schroeter Einl.* i. p. 752.
Cerithium perversus.? *Bruguiere Enc. Meth.* p. 496.
Spengler n. Schrift. der Daen. i. p. 376. t. 1. f. 3. *Chemnitz*, ix. part 1. p. 127. t. 113. f. 968.
Inhabits the sands on the East Indian shores. *Spengler.*
Shell about two lines long, and two fifths as broad, with the

lower half somewhat ventricose, and the summit acuminated; the figure is painted brown with darker spots.

ANNULATUS. 125. Shell turreted, minute, with the whirls reversed, and a transverse rib on each of the sutures; aperture somewhat four-sided.

Trochus annulatus. *Gmelin*, p. 3587.
Trochus, No. 188. *Schroeter Einl.* i. p. 753.
Spengler n. Schrift. der Daen. i. p. 376. t. 1. f. 4. *Chemnitz*, ix. part 1. p. 127. t. 113. f. 969.
Inhabits the sands on the East Indian shores. *Spengler.*
This shell is of about the same size as *T. ventricosus*, and has from twelve to fifteen whirls; the figure is painted of a pale brownish colour, with darker short longitudinal stripes between the transverse ribs which attend the sutures.

PUNCTATUS. 126. Shell turreted, imperforate, with a triple row of raised dots on each whirl.

Trochus punctatus. *Linnæus Syst. Nat.* p. 1231. *Gmelin*, p. 3587.
Ceritheum ferrugineum. *Bruguiere Enc. Meth.* p. 496.?
Inhabits Southern Europe. *Linnæus.* Mediterranean. *Bruguiere.*
Linnæus has described this shell to be ferruginous, and of the size of a barley-corn, with three rows of obtuse elevated dots on each whirl, of which the middle row is smaller than the others; aperture four-sided, with the pillar a little prominent and slightly channelled.

STRIATELLUS. 127. Shell turreted, imperforate, with longitudinal oblique parallel striæ.

Trochus striatellus. *Linnæus Syst. Nat.* p. 1232. *Gmelin*, p. 3587.
Cerithium zonale. *Bruguiere Enc. Meth.* p. 497.?
Murex minimus. *Gmelin*, p. 3564.?
Murex, No. 71. *Schroeter Einl.* i. p. 567.?
Lister Conch. t. 1018. f. 81.?
Inhabits the Mediterranean. *Linnæus.*
Linnæus, to the above specific character, has only added that the shell is small, subulate, and white, and has the apex of a violet colour. Bruguiere suspects that it is his *Cerithium zonale*, which he describes to be eight lines long, and three

broad, and ornamented alternately with black and white
bands.

LUNARIS.   128. Shell sub-ovate, minute, slightly
umbilicated, smooth, and the whirls reversed; summit very obtuse, and the aperture roundish.

Trochus lunaris. *Gmelin*, p. 3589.
Chemnitz, ix. part 1. p. 128. t. 113. f. 971.

Inhabits ———

This shell appears by the figures to be only a line long, and
about half as broad, of a pale horn-colour; it has five whirls,
and the summit is remarkably obtuse.

Genus XXVIII.

TURBO:

SHELL UNIVALVE, SPIRAL AND SOLID; APERTURE
COARCTATE, ORBICULAR AND ENTIRE.

Subdivisions.†

* Imperforate, and the Pillar-lip flat.
** Imperforate and solid.
*** Umbilicated and solid.
**** Cancellated.
***** Turreted.
****** Depressed.

---

* *Imperforate, and the Pillar-lip flat.*

OBTUSATUS.   1. Shell roundish, smooth, very obtuse and more ventricose above; margin of the pillar flat.

Turbo obtusatus. *Linnæus Syst. Nat.* p. 1232. *Chemnitz,*
v. p. 234. t. 185. f. 1854, *a* to *f. Schroeter Einl.* ii. p. 3.
Gmelin, p. 3588. *Schreibers Conch.* i. p. 268.

Turbo littoreus, Var. ? *Montagu Test.* p. 302.

Inhabits the Northern Ocean. *Linnæus.*

Linnæus has not given any reference, or made any addition to the above short description. The shells figured by Chemnitz are very like *T. littoreus,* which in an early stage of growth is very variable both in shape and markings.

---

† The following of Gmelin's species are omitted. *T. Afer,* p. 3602. *T. crystallinus,* p. 3609. *T. Ludus,* p. 3601. *T. marginellus,* p. 3602. *T. obtusus,* p. 3611. *T. planorbis,* p. 3602.

NERITOIDES.   2. Shell ovate, glabrous, rather obtuse, and the margin of the pillar flat.

Turbo neritoides.   *Linnæus Syst. Nat.* p. 1232.
    Gualter, t. 45. f. F.
Inhabits the Mediterranean. *Linnæus.*
Linnæus, to the above character and reference to Gualter, has
   only added that this shell is minute, and very much like a
   Nerite.   *T. neritoides* of Chemnitz and Gmelin is certainly
   the Linnæan *Nerita littoralis,* and Montagu says, " there is
   little doubt that the *Nerita littoralis* and *Turbo neritoides*
   are only Varieties of the same shell," but Gualter's figure is
   much more like a young shell of *Turbo littoreus.*

NICOBARICUS.   3. Shell thick, nearly smooth, and variegated with somewhat reticulated lines ; pillar-lip callous, and the throat yellow.

Turbo Nicobaricus.   *Gmelin,* p. 3596.   *Schreibers Conch.*
   i. p. 286.
Helix ambigua.   *Born Index,* p. 409.
Helix paradoxa.   *Born Mus.* p. 394. t. 13. f. 16 and 17.
Chrysostomus Nicobaricus.   *Chemnitz,* v. p. 216. t. 182. f.
   1822 to 1825.
Inhabits the coasts of the Nicobar Islands. *Chemnitz.*
Shell about eight or ten lines long, and equally broad, and is
   shaped somewhat like a Nerite with five convex whirls ; two
   of Chemnitz's figures are whitish, marked with minute dark
   brown closely reticulated lines ; and the others, as well as
   Born's, are clouded with short red and yellow streaks.

NIGERRIMUS.   4. Shell smooth, very black, with the whirls excavated at the sutures, and the pillar-lip flattish.

Turbo nigerrimus.   *Gmelin,* p. 3597.   *Schreibers Conch.*
   i. p. 275.
Turbo, No. 54.   *Schroeter Einl.* ii. p. 83.
Cochlea lunaris pernigra.   *Chemnitz,* v. p. 228. t. 185. f.
   1848.
Nerita glaucina, Var.   *Born Mus.* p. 397. t. 13. f. 20 and
   21.? *Gmelin,* p. 3671.?
Nerita, No 93.   *Schroeter Einl.* ii. p. 331.?
Variety.   With white dots.
Turbo, No. 55.   *Schroeter Einl.* ii. p. 83.
   Chemnitz, v. p. 229. t. 185. f. 1849.

Inhabits the South Sea on the coasts of New Zealand. *Chemnitz.*

This shell appears, by Chemnitz's figures, to be about eleven lines long and ten broad, with the outer surface very black, and the pillar-lip and inside white. The shell which Born has figured for a Variety of *Nerita glaucina*, is much more like the present species.

LITTOREUS. 5. Shell sub-ovate, acute, striated, and the margin of the pillar flat.

Turbo littoreus. *Linnæus Syst. Nat.* p. 1232. *Pennant Zool.* iv. p. 128. t. 81. f. 109. *Da Costa Brit. Conch.* p. 96. t. 6. f. 1. *Born Mus.* p. 341. t. 12. f. 13 and 14. *Chemnitz,* v. p. 230. t. 185. f. 1852. 1 to 8. *Schroeter Einl.* ii. p. 5. *Gmelin,* p. 3588. *Donovan,* i. t. 33. f. 1 and 2. *Montagu Test.* p. 301. *Maton and Racket in Lin. Trans.* viii. p. 158. t. 4. f. 8 to 11. *Dorset Cat.* p. 49. t. 17. f. 1, and t. 19. f. 2 and 3.

Nerita littorea. *Muller Zool. Dan. Prodr.* p. 244. *Fabricius Fauna Groenl.* p. 403.

Le Marron roti. *Favanne,* ii. p. 143. t. 9. f. K 1, and K 2. *Lister Anim. Ang.* t. 3. f. 9, and *Conch.* t. 585. f. 43. *Gualter,* t. 45. f. G. *Argenville,* t. 6. f. L. *Schroeter Fluss.* t. 8. f. 5, and t. 11 c. f. 5.

Junior. Shell smaller, and the colours brighter.

Turbo, No. 59. *Schroeter Einl.* ii. p. 85.

Cochlea lunaris Groenlandica. *Chemnitz,* v. p. 235. t. 185. f. 1855. *a* to *g.*

Schroeter Fluss. p. 344. t. 9. f. 16, 18, and 19.

Inhabits the European seas, and coasts of Norway and Sweden. *Linnæus.* Great Britain. *Lister, &c.* Denmark. *Muller.* Greenland. *Schroeter.* France. *Favanne.*

Shell three quarters of an inch or an inch long, and nearly equally broad, thick, with five or six whirls, of which the body-whirl is larger than all the others together; when full grown it is usually nearly smooth, and of an uniform brownish colour; but younger shells are more distinctly striated transversely, and variously marked with broad or narrow purplish brown, white, yellow, or reddish bands, and have the summit more acuminated. This species is extremely abundant on the coasts of Great Britain, and is commonly sold in our markets under the name of the *Periwinkle.*

TENEBROSUS. 6. Shell sub-conical, rather obtuse, with the body-whirl ventricose.

Turbo tenebrosus. *Montagu Test.* p. 303. t. 20. f. 4.
Maton and Racket, in Lin. Trans. viii. p. 160. *Dorset
Cat.* p. 49. t. 18. f. 15.

Inhabits the mud and rocks near high-water mark, and in
ditches subject to the daily influx of the tide, on the coasts
of England. *Montagu.*

Shell about a quarter of an inch long, and nearly equally broad,
of a dark chocolate colour, with five rather ventricose whirls,
of which the body-whirl occupies half the length. Mr.
Montagu, who entertained considerable doubts of its being
distinct, says, " it is faintly wrinkled across the spire, but
rarely spirally striated as in the young *littoreus*, and that in
an obsolete manner; add to this, the strength of the shell
seems to indicate its being formed and full grown."

RUDIS.   7. Shell sub-ovate, rather obtuse, and the
        whirls ventricose.

Turbo rudis. *Maton's Western Counties,* i. p. 277. *Donovan,* i. t. 33. f. 3.   *Montagu Test.* p. 304.   *Maton
and Racket, in Lin. Trans.* viii. p. 159. t. 4. f. 12 and 13.
Dorset Cat. p. 49. t. 18. f. 6.

Turbo littoreus, Var. β. *Gmelin,* p. 3588.

Turbo, No. 58. *Schroeter Einl.* ii. p. 84.

Cochlea lunaris, &c. *Chemnitz,* v. p. 233. t. 185. f. 1853.

Inhabits the coasts of Norway. *Chemnitz.* Great Britain,
Maton, &c.

Shell about three quarters of an inch long, and nearly equally
broad, of a pale brown or dirty yellowish white colour; it
has five whirls well defined by the suture, and they are much
more ventricose than those of *T. littoreus,* by which the
species may be at once discriminated. Mr. Montagu says,
that the animal is yellowish without any stripes, whereas the
animal of *T. littoreus* is striped with black.

MURICATUS.   8. Shell umbilicated, sub-ovate, acute,
        with transverse beaded striæ, and the margin
        of the pillar rather obtuse.

Turbo muricatus. *Linnæus Syst. Nat.* p. 1232. *Born
Mus.* p. 341. t. 12. f. 15 and 16. *Chemnitz,* v. p. 170.
t. 177. f. 1752 and 1753. *Schroeter Einl.* ii. p. 7. and
Inn. Bau Conch. p. 60. t. 2. f. 2. *Gmelin,* p. 3589.
Schreibers Conch. i. p. 268. *Burrow's Elements,* p. 167.
t. 19. f. 4.

Le Boson. *Adanson Senegal,* p. 171. t. 12. f. 2.

Le Limaçon à grains de petite Verole. *Favanne*, ii. p. 140. t. 9. f. I.

Lister Conch. t. 30. f. 28. *Petiver Gaz*. t. 70. f. 11. *Gualter*, t. 45. f. E. *Argenville*, t. 6. f. M. *Seba*, iii. t. 39. f. 28 and 29.

Inhabits Southern Europe. *Linnæus*. Jamaica. *Lister*. Coasts of Goree. *Adanson*. Normandy, Brittany, England, Spain, the Antilles, Provence, &c. *Favanne*.

Shell about an inch long, and nearly two-thirds as broad, of a pale lead-colour, with whitish beads in crowded transverse rows; it has seven whirls, which are much contracted at the sutures, and the throat is brown.

ÆTHIOPS. 9. **Shell obtuse, transversely grooved, and alternately crenulated with black and white; aperture effuse, white, and the pillar-lip bordered with brown.**

Turbo Æthiops. *Gmelin*, p. 3596.

Turbo, No. 43. *Schroeter Einl*. ii. p. 78.

Æthiopissa, cum dentibus suis candidissimis. *Chemnitz*, v. p. 215. t. 182. f. 1820 and 1821.

Born Mus. Vign. at p. 340. f. b.

Inhabits the coasts of New Zealand. *Chemnitz*.

This shell appears, by Chemnitz's figures, to be about an inch long, and ten lines broad, black, with transverse ribs, which are divided by white longitudinal narrow stripes; the summit is obtuse and white, with the appearance of having been worn. Born, without any description, has figured this shell as a specimen of a Turbo of the present division.

PUNCTATUS. 10. **Shell ovate, thick, with six depressed whirls, of which the two lower are very large, and the spire mucronated; inside brown.**

Turbo punctatus. *Gmelin*, p. 3597.

Turbo, No. 120. *Schroeter Einl*. ii. p. 104.

Le Marnat. *Adanson Senegal*, p. 168. t. 12. f. 1.

La Guignette Africaine. *Favanne*, ii. p. 146. t. 71. f. A 1, and A 2.

Inhabits the coasts of Goree. *Adanson*.

Shell seven or eight lines long and about five broad, very thick, smooth, shining, and composed of six slightly ventricose whirls, which are slightly marked by the suture; the outer surface is of a greyish lead-colour, or sometimes reddish, marked with oblique rows of white dots, and the inside is said to be of a coffee-brown colour.

JUGOSUS. 11. Shell sub-ovate, rather ventricose, transversely ribbed, and the margin of the pillar flattish.

Turbo jugosus. *Montagu Test.* p. 586. t. 20. f. 2. *Maton and Racket in Lin. Trans.* viii. p. 158. t. 4. f. 7. *Dorset Cat.* p. 49. t. 19. f. 1.

Turbo obtusatus. *Pulteney Dorset Cat.* p. 45.

Inhabits the western coasts of England. *Montagu, &c.* Not uncommon on the coasts of South Wales. *L. W. D.*

Shell a quarter or three-eighths of an inch long, and the length and breadth are about equal; it varies greatly in its colour from a dull orange yellow to dark purple, and differs from the infant state of *T. littoreus,* which it much resembles, in being ribbed transversely; the body-whirl has about eight or ten elevated sharp ribs, reflecting a little upwards, and there are three or four on the second whirl; it has four whirls, of which the body-whirl is thrice as large as all the others together.

CRASSIOR. 12. Shell conical, coarse, with five rounded obsoletely striated whirls, depressed at the suture, and the body-whirl slightly keeled.

Turbo crassior. *Mont. Test.* p. 309, and *Supp.* p. 127. t. 20. f. 1. *Maton and Racket, in Lin. Trans.* viii. p. 159.

Turbo pallidus. *Donovan,* v. t. 178. f. 4.

Inhabits the coasts of England. *Boys, &c.* Common in Langland Bay, near Swansea. *L. W. D.*

Shell about three-eighths or half an inch long, and rather more than half as broad, with five produced somewhat ventricose whirls, which are depressed at the suture, and the body-whirl has two more or less distinct transverse keels; the colour is very pale yellowish brown or brownish white.

PETRÆUS. 13. Shell conical, acute, and the aperture pear-shaped.

Turbo petræus. *Maton and Racket in Lin. Trans.* viii. p. 160. *Dorset Cat.* p. 49. t. 18. f. 13.

Helix petræa. *Montagu Test.* p. 403.

Inhabits the coasts of Dorset and Devon, on rocks a little below high-water mark. *Montagu.*

Mr. Montagu says, that upon a cursory view this shell might be mistaken for *T. tenebrosus*, being nearly of the same size and colour, but the shape is more conical and the aperture quite different.

FULGIDUS. 14. Shell sub-conical, minute, smooth, with three whirls, of which the body-whirl is very large, and the apex small and obtuse.

Turbo fulgidus. *Montagu Test*. p. 332. *Maton and Racket in Lin. Trans*. viii. p. 161.
Helix fulgidus. *Adams in Lin. Trans*. iii. p. 254.
Inhabits the coasts of Pembrokeshire. *Adams*. Cornwall. *Montagu*.
Shell half a line long, pellucid, smooth, glossy, variegated with white and bronze usually in bands ; aperture sub-orbicular, and the margin attenuated.

** *Imperforate and solid.*

CIMEX. 15. Shell oblong-ovate, with cancellated striæ, forming punctures by their interstices, and four whirls.

Turbo Cimex. *Linnæus Syst. Nat*. p. 1233. *Schroeter Einl*. ii. p. 8. *Gmelin*, p. 3589. *Donovan*, i. t. 2. f. 1. *Montagu Test*. p. 315. *Maton and Racket, in Lin. Trans*. viii. p. 161. *Dorset Cat*. p. 49. t. 14. f. 6 and 9.
Turbo cancellatus. *Da Costa Brit. Conch*. p. 104. t. 8. f. 6 and 9.
Buccinum Soni. *Bruguiere Enc. Meth*. p. 283. ?
Le Soni. *Adanson Senegal*. p. 151. t. 10. f. 6. ?
Gualter, t. 44. f. X.
Inhabits the Mediterranean. *Linnæus*. Coasts of Britain. *Da Costa*, &c.
Shell about one-sixth of an inch long, and rather more than half as broad, white, solid, with four whirls well defined by the suture, and strongly marked with decussated striæ, which give it an appearance of being punctured all over ; it has four whirls, and *Le Soni*, which has been quoted by Linnæus, is said to have eight, so that it may more probably belong to *T. Calathiscus*.

CALATHISCUS. 16. Shell sub-conical, with crowded cancellated striæ, forming punctures by their interstices, and six whirls.

Turbo Calathiscus. *Montagu Test. Supp.* p. 132. t. 30. f. 5.

Inhabits the coasts of Britain. *Montagu.*

Shell a quarter of an inch long, with elegantly cancellated whirls, and of a brown colour. Mr. Montagu says, " In its worn state it is so like *T. Cimex,* that several, which had been picked up on the western shores of England, had been placed in our cabinet with that shell; the striæ, however, that form the cancelli are much closer, leaving the depressions much smaller; the shape of the shell is also more slender, and may readily be known even in this state from *T. Cimex,* by having four series of cancelli on the second spire instead of two, which that shell invariably has.

PULLUS. 17. Shell ovate, glossy, with four convex variegated whirls; aperture sub-orbicular and somewhat produced at the base.

Turbo Pullus. *Linnæus Syst. Nat.* p. 1233. *Born Mus.* p. 342. t. 12. f. 17 and 18. *Gmelin,* p. 3589. *Donovan,* i. t. 2. f. 2 to 6. *Montagu Test.* p. 319. *Maton and Racket in Lin. Trans.* viii. p. 162. *Dorset Cat.* p. 49. t. 14. f. 1 to 3.

Turbo pictus. *Da Costa Brit. Conch.* p. 103. t. 8. f. 1 to 3.

Inhabits the Mediterranean. *Linnæus.* Coasts of Britain. *Da Costa, &c.*

Shell a quarter, or sometimes three-eighths, of an inch long, and half as broad, with four rounded glossy whirls, of which the body-whirl is larger than all the others; it is always beautiful, and in Langland Bay, near Swansea; it occurs in innumerable varieties, as Da Costa has described : ' The ground colour is white, prettily variegated with red ; sometimes only mottled or shaded with pale red ; sometimes with longitudinal broad waved stripes of a fine deep purple red, and sometimes the pale red runs in circular girdles, very regular, and adorned with short transverse streaks of dark brown.'

PERSONATUS. 18. Shell imperforate, convex, smooth, and the aperture produced.

Turbo personatus. *Linnæus Syst. Nat.* p. 1233. *Gmelin,* p. 3589. *Rumphius,* t. 19. f. No. 1.

Inhabits ——

The above figure of Rumphius, to which Linnæus has referred, is clearly only a Variety of *T. Petholatus,* and *T. personatus* is probably nothing more. Chemnitz, on the other hand, doubted whether it might not be *Helix aperta;* and Schreibers has quoted Knorr, i. t. 10. f. 3, which has more the look of a Nerite.

PETHOLATUS. 19. Shell ovate, smooth, and glossy, with the whirls transversely angulated, and somewhat concave towards their upper margins.

Turbo Petholatus. *Linnæus Syst. Nat.* p. 1233. *Born Mus.* p. 342. *Chemnitz,* v. p. 221. t. 183. f. 1830 to 1835, and t. 184. f. 1836 to 1839. *Schroeter Einl.* ii. p. 10. *Gmelin,* p. 3590. *Schreibers Conch.* i. p. 271. *Shaw Nat. Misc.* x. t. 359.

La Peau de Serpent. *Favanne,* ii. p. 69. t. 9. f. D 1.

Lister Conch. t. 584. f. 39. *Rumphius,* t. 19. f. D, and No. 5 to 7. *Petiver Amb.* t. 7. f. 15. *Gualter,* t. 64. f. F. *Klein Ost.* t. 2. f. 51. *Argenville,* t. 6. f. K. *Seba,* iii. t. 74. f. 25 to 29. *Knorr,* i. t. 3. f. 4; ii. t. 22. f. 1 and 2; iii. t. 3. f. 3, t. 23, f. 4, and t. 28. f. 2 to 5. *Geve,* t. 20. f. 202 to 204, and t. 21. f. 205 to 212. *Regenfuss,* i. t. 8. f. 18, t. 9. f. 27; and ii. t. 6. f. 54, and t. 9. f. 27.

Inhabits Barbadoes. *Linnæus.* Amboyna. *Rumphius.* Coasts of the Isle of France, New Zealand, New Guinea, the Philippines, Sumatra, &c. *Favanne.* China and the East Indies. *Humphreys.*

Shell about an inch and a half or two inches long, and the length rather exceeds the breadth; it has five or six rather ventricose whirls, of which the body-whirl is much larger than all the others, and has its upper third part next the suture slightly concave; it is commonly of a pale or dark chestnut colour, variously mottled and marked with particoloured bands and longitudinal stripes; the pillar is yellowish.

CIDARIS. 20. Shell smooth, with compressed roundish whirls, of which the body-whirl is ventricose; aperture compressed, silvery, and the pillar somewhat produced.

Turbo Cidaris. *Gmelin,* p. 3596. *Schreibers Conch.* i. p. 269.

Turbo, No. 47 to 53. *Schroeter Einl.* ii. p. 79 to 82.

Cidaris Persica. *Chemnitz,* v. p. 225. t. 184. f. 1840 to 1847.

Le Turban.   *Favanne*, ii. p. 52. t. 8. f. C 1, and C 2.
Argenville, t. 6. f. B and O.   *Seba*, iii. t. 74. f. 15.
Inhabits the coasts of China and India. *Chemnitz.* Isle of
France and the Moluccas. *Favanne.*
This species is about an inch and a half long, and an inch and
three quarters broad, and varies much in its colour and
markings ; some shells are greenish, clouded or spotted with
red, or striped with chestnut ; and others are yellow, orange,
or brown, variously marked with one or other of these
colours ; it has five or six convex more or less depressed
whirls, of which the body-whirl is comparatively very large,
and more ventricose ; the inside is silvery and iridescent.

HELICINUS.   21. **Shell sub-imperforate, glabrous,**
roundish, with convex whirls, and the pillar
thickened.

Turbo helicinus.   *Born Mus.* p. 348. t. 12. f. 23 and 24.
Gmelin, p. 3597.
Turbo, No. 116.   *Schroeter Einl.* ii. p. 102.
Inhabits ——
Born has described this shell to be eleven lines long and fifteen
broad, rounded, solid, spotted with green and purple, and
the throat silvery ; in form the figure a good deal resembles
Trochus Merula, and Born has quoted Knorr, v. t. 3. f. 1.
which belongs to that species.

COCHLUS.   22. Shell ovate, with very fine longitudinal striæ, and a transverse rib on the shoulder of the whirls.

Turbo Cochlus.   *Linnæus Syst. Nat.* p. 1233.   *Born
Mus.* p. 343.?   *Chemnitz*, v. p. 209. t. 182. f. 1805 and
1806.   *Schroeter Einl.* ii. p. 12. t. 3. f. 17.   *Gmelin*,
p. 3590.   *Schreibers Conch.* i. p. 271.   *Brookes's Int.*
p. 126. t. 8. f. 99.
Le Cameleopard.   *Favanne*, ii. p. 82.
Lister Conch. t. 584. f. 40.   *Klein Ost.* t. ii. f. 55.   *Knorr,*
i. t. 3. f. 5.   *Seba*, iii. t. 74. f. 30.   *Regenfuss*, i. t. 1.
f. 12.   *Geve*, t. 20. f. 194.
Inhabits the Asiatic Ocean, and the coasts of Alexandria and
Iceland. *Linnæus.*   East Indies, Guinea, and Frederick's
Island. *Regenfuss.*   Isle of France, Molucca Islands, Batavia, and China. *Favanne.*
Shell about an inch and three quarters long, and an inch and a
half broad, and is slightly angulated rather than ribbed on
the shoulder of the whirls ; it is of a green or greenish

brown colour, with white transverse bands, which are spotted with brown, and the inside is silvery. *Born* appears to have mistaken the species, for he has described it to be obsoletely grooved transversely, whereas Linnæus says it is like *T. argyrostomus,* but is without any transverse striæ.

SMARAGDUS. 23. Shell somewhat depressed, green, obliquely wrinkled, with five whirls, of which the body-whirl is very broad and ventricose; pillar grooved.

Turbo Smaragdus. *Gmelin,* p. 3595, and p. 3602.
Turbo, No. 40. *Schroeter Einl.* ii. p. 77.
Cochlea lunaris smaragdina. *Chemnitz,* v. p. 213. t. 182. f. 1815 and 1816.
Limax Smaragdus. *Martyn Univ. Conch.* ii. t. 73.
Zorn Naturf. vii. p. 157. t. 2. f. A 1, and A 2.
Inhabits New Zealand. *Chemnitz.*
Shell an inch and a half or two inches long, and about equally broad, of an uniform green colour, and the inside pearly; it has five rather depressed ventricose, obliquely wrinkled whirls, of which the body-whirl is larger, and more than usually broader than the others; the pillar-lip is broad, and slightly channelled longitudinally.

CHRYSOSTOMUS. 24. Shell sub-ovate, with transverse ribs, and two rows of vaulted spines.

Turbo chrysostomus. *Linnæus Syst. Nat.* p. 1233. *Born Mus.* p. 344. *Chemnitz,* v. p. 178. t. 178. f. 1766. *Schroeter Einl.* ii. p. 14, and *Inn. Bau Conch.* p. 61. t. 5. f. 4. *Gmelin,* p. 3591.
La Bouche d'Or. *Favanne,* ii. p. 59. t. 9. f. A 2.
Rumphius, t. 19. f. E. *Petiver Amb.* t. 5. f. 3. *Gualter,* t. 62. f. H. *Argenville,* t. 6. f. D. *Klein Ost.* t. 7. f. 126. *Seba,* iii. t. 74. f. 9 to 11. *Knorr* ii. t. 14. f. 2, and v. t. 13. f. 3. *Geve,* t. 18. f. 171 to 175. *Regenfuss,* ii. t. 12. f. 59.
Junior. Shell smaller, and the aperture somewhat silvery.
Turbo spinosus. *Gmelin,* p. 3594. *Schreibers Conch.* i. p. 281.
Turbo, No. 29. *Schroeter Einl.* ii. p. 73.
Chemnitz, v. p. 204. t. 181. f. 1797.
Inhabits the Asiatic Ocean. *Linnæus.* Red Sea. *Forskael.* Coasts of Amboyna. *Rumphius.* Moluccas and Isle of France. *Favanne.* Friendly Islands. *Humphreys.*
Shell about two inches and a quarter long, and one inch and

three quarters broad, of a yellowish white colour, sometimes tinged with green, and marbled in irregular longitudinal streaks with chestnut-brown; it has five or six produced transversely angulated whirls, which are ribbed transversely and longitudinally wrinkled; on the angulated keel of each whirl there is a row of vaulted spines, and also another row towards the base of the body-whirl; the inside in full-grown shells is of a rich glittering gold colour.

TECTUM-PERSICUM.　25.　Shell conical-ovate, with two rows of obtuse depressed spines, and the base granulated.

Turbo Tectum-Persicum. *Linnæus Syst. Nat.* p. 1234. *Chemnitz,* v. p. 41. t. 168. f. 1543 and 1544. *Schroeter Einl.* ii. p. 15. *Gmelin,* p. 3591. *Schreibers Conch.* i. p. 271.

La petite Pagode. *Favanne,* ii. p. 341. t. 13. f. F. *Gualter,* t. 60. f. M. *Argenville,* t. 8. f. P. *Geve,* t. 9. f. 66.

Inhabits the Asiatic Ocean. *Linnæus.*

Shell about an inch long, and nearly equally broad, of a somewhat pyramidal form, with transverse wrinkles and two rows of spines on each whirl; the colour is white, prettily mottled and somewhat fasciated with brown; the throat is white, somewhat silvery, and grooved. Chemnitz, although he has described this shell for the Linnæan *Turbo Tectum-Persicum,* has followed Favanne and Geve in placing it among the Trochi; and the shell, which Born has figured for this species, is *Trochus imbricatus.* Mr. Burrows has justly observed, that ' it is scarcely possible ' to define the boundary at which the Trochi with rounded apertures are supposed to end, and the Turbines with imperfectly circular mouths to begin their jurisdiction.'

TROCHIFORMIS.　26.　Shell conical-ovate, with two rows of white granules on the body-whirl, and one on each whirl of the spire.

Trochus nodulosus. *Gmelin,* p. 3582.
Trochus, No. 21. *Schroeter Einl.* i. p. 687.
Trochus duplici serie granulorum, &c. *Chemnitz,* v. p. 42. t. 163. f. 1545 and 1546.

Inhabits the Southern Ocean. *Chemnitz.*

This shell appears by Chemnitz's figures to be nearly of the same size and shape as *T. Tectum-Persicum,* and is distinguished by its transverse rows of white granules.

PAGODUS. 27. Shell conical, with obtuse concatenated spines, and the base marked with granulated striæ.

Turbo Pagodus. *Linnæus Syst. Nat.* p. 1234. *Born Mus.* p. 345. *Chemnitz*, v. p. 38. t. 163. f. 1541 and 1542. *Schroeter Einl.* ii. p. 16. *Gmelin*, p. 3591. *Schreibers Conch.* i. p. 272.
Le Toit Chinois. *Favanne*, ii. p. 339. t. 12. f. A.
Lister Conch. t. 644. f. 36. *Rumphius*, t. 21. f. D. *Petiver Amb.* t. 10. f. 8. *Gualter*, t. 62. f. B, C. *Argenville*, t. 8. f. A. *Knorr*, i. t. 25. f. 3 and 4. *Seba*, iii. t. 60. f. 3. *Geve*, t. 8. f. 62 and 63, and t. 9. f. 64 and 65. *Spengler Conch.* t. 2. f. K.
Inhabits the Asiatic Ocean. *Linnæus.* Coasts of Amboyna. *Rumphius.* Philippine Islands. *Favanne.* Moluccas. *Chemnitz.*
Shell about two inches long, and nearly equally broad, of a pale brownish grey, or whitish with irregular brown stripes, and the inside is of a pale ochre-colour; it has seven rugged transversely wrinkled whirls, with two spinous keels on the body-whirl, and one on each of the others; the name is derived from its supposed resemblance to the roof of a Chinese pagoda; Chemnitz has placed both this and the following species among the Trochi.

CALCAR. 28. Shell sub-imperforate, depressed, with the whirls scabrous, and armed with a row of narrow compressed vaulted spines.

Turbo Calcar. *Linnæus Syst. Nat.* p. 1234. *Born Mus.* p. 346. *Chemnitz*, v. p. 46. t. 164. lower figures 1552 and 1553. *Schroeter Einl.* ii. p. 18. *Gmelin*, p. 3592.
Le petit Eperon blanchatre. *Favanne*, ii. p. 405. t. 13. f. C 3.
Rumphius, t. 20. f. I. *Petiver Amb.* t. 9. f. 13. *Gualter*, t. 65. f. N. *Argenville*, t. 6. f. R. *Seba*, iii. t. 59. f. 5. *Knorr*, iv. t. 4. f. 2. *Geve*, t. 4. f. 31, and t. 8. f. 60 and 61.
Inhabits the coasts of Amboyna. *Rumphius.* China. *Favanne.*
Shell about half an inch long, and twice as broad, with five whitish whirls, and a stellated row of compressed vaulted horizontal spines on the margin of each; the inside is silvery. Chemnitz's references to his own figures for this and the fol-

lowing species are by no means clear, and he has not mentioned more than one, though there are two different figures under each of the above numbers.

· STELLARIS.　29. Shell sub-imperforate, with broad radiated spines, of which there are twelve on the base of the body-whirl.

Turbo stellaris. *Gmelin*, p. 3600.
Turbo, No. 1. *Schroeter Einl.* ii. p. 62.
Trochus stellatus. *Chemnitz*, v. p. 47. t. 164. upper figures 1552 and 1553.
Inhabits the South Sea. *Chemnitz.*
Shell about an inch and a quarter long, and an inch and three quarters broad, and of a greenish colour; it differs from *T. Calcar* in being larger and narrower in proportion to the length, and in having the spines much larger and more numerous.

ACULEATUS.　30. Shell sub-imperforate, with laciniated spines, of which there are nine very large and compressed on the body-whirl.

Turbo aculeatus. *Gmelin*, p. 3600.
Turbo, No. 2. *Schroeter Einl.* ii. p. 62.
Turbo Calcar, Var. b. *Schreibers Conch.* i. p. 272.
Variety. With the lower whirls more ventricose and produced.
Turbo, No. 3. *Schroeter Einl.* ii. p. 63.
Turbo Calcar, Var. c. *Schreibers Conch.* i. p. 272.
Calcar Spenglerianum. *Chemnitz*, v. p. 49. t. 164. f. 1556 and 1557.
Inhabits the coasts of the Nicobar Islands. *Chemnitz.*
This shell is of about the same size as *T. stellaris*, but has the spines shorter, broader, and laciniated, and is of a sea-green colour; in the Variety the body-whirl and the first whirl of the spire appear to be ventricose, and much more produced, and the base is more rounded.

STELLATUS.　31. Shell sub-pyramidal, yellowish, with the base flattened, and the lower margins of the whirls spinous.

Turbo stellatus. *Gmelin*, p. 3600.
Turbo, No. 4. *Schroeter Einl.* ii. p. 63.
Trochus leviter radiatus, seu muricatus. *Chemnitz*, v. p. 49. t. 164. f. 1558 and 1559,
Knorr, iv. t. 4. f. 5.

Inhabits ——
This is an imperfectly defined species, and the shell, by Chemnitz's figure, appears to be ten lines long, and about equally
broad, with the pyramidal form and general habit of a
Trochus. Knorr, iv. t. 4. f. 5, which all the above cited
authors have referred to, has been quoted by Favanne (vol. ii.
p. 305), for a variety of *Le Concombre,* which is *Trochus
imbricatus.*

ARMATUS. 32. Shell sub-imperforate, scabrous,
with the summit of the spire depressed, and
a spinous rib round the middle of the lower
whirls; base with concentrical somewhat
granulated striæ.

Turbo Calcar, Var. ζ. *Gmelin,* p. 3592.
Turbo Calcar Mediterraneo. *Ulysses's Travels,* p. 470.
Turbo Calcar. *Montagu Supp.* p. 137. t. 29. f. 3.
Turbo rugosus. *Chemnitz,* x. p. 297. t. 165. f. 1585 and
1586. ?
Turbo, No. 22. *Schroeter Einl.* ii. p. 70.
Calcar Maris Mediterranei. *Chemnitz,* v. p. 198. t. 180.
f. 1786 and 1787.
La fausse Raboteuse eperonnée. *Favanne,* ii. p. 96.
Lister Conch. t. 608. f. 46 a. *Martini, n. Mannigf.* t. 2.
f. 16 and 17.
Inhabits the Mediterranean. *Chemnitz.* Bay of Naples. *Ulysses.*
Coasts of the Western Islands of Scotland. *Montagu.*
Shell about eight lines long and ten broad; and Mr. Montagu
says, that a shell, which exactly corresponds with Chemnitz's
figures 1786 and 1787, has been taken by Mr. Laskey, in
Iona, one of the Western Islands, but is less than half as
large; it has about five whirls, and in the middle of the two
lower ones, which are more ventricose, is a transverse rib,
armed with large radiated smooth somewhat upright spines;
it is of a greenish colour, more or less tinged with pale
pink, especially on the spines, and the pillar-lip purplish
red. It is more likely to be a Variety of *T. rugosus* than of
T. Calcar, and principally differs from the former in having
much longer spines.

RUGOSUS. 33. Shell sub-ovate, transversely striated
and armed with small vaulted scales, which
become larger on the margins of the whirls;
pillar-lip flat, and the inside silvery.

Turbo rugosus.   *Linnæus Syst. Nat.* p. 1234.   *Born Mus.*
    p. 346.   *Chemnitz,* v. p. 195. t. 180. f. 1782 to 1785.
    Schroeter Einl. ii. p. 19.   *Gmelin,* p. 3592.   *Schreibers*
    Conch. i. p. 273.
La fausse Raboteuse.   *Favanne,* ii. p. 92. t. 9. f. O.
Bonanni Rec. 3. f. 12 and 13, and *Kirch,* f. 12.   *Lister*
    Conch. t. 647. f. 41.   *Gualter,* t. 63. f. F. and H.   *Ar-*
    genville, t. 8. f. O.   *Klein Ost.* t. 2. f. 50.   *Knorr,* iii.
    t. 20. f. 1.   *Geve,* t. 15. f. 144.
Variety? Of a reddish colour, with transverse granulated striæ.
Turbo, No. 34.   *Schroeter Einl.* ii. p. 74.
Turbo, No. 54.   *Schreibers Conch.* i. p. 282.
Cochlea lunaris rubicunda.   *Chemnitz,* v. p. 207. t. 181.
    f. 1803 and 1804.
Le Turban granuleux.   *Favanne,* ii. p. 9, and p. 54.
Inhabits India. *Linnæus.*   Mediterranean. *D'Avila.*   Red
    Sea, coasts of Provence, Languedoc, Spain, Italy, Isle of
    France, St. Domingo, and Barbadoes. *Favanne.*   Every
    where in the Mediterranean. *Ulysses.*
Shell an inch, or sometimes two inches long, and the breadth
    rather exceeds the length; it has six whirls separated by a
    slight groove at the suture, with longitudinal plaits above,
    and armed below with transverse striæ and vaulted scales,
    which become somewhat spinous on the larger striæ; the
    colour is olive or green, clouded with darker shades, or
    sometimes with white, and the pillar red.   *Le Turban gra-*
    nuleux of Favanne, which Gmelin has placed as a Variety,
    has not been well defined, but is probably a distinct species
    more allied to *T. Spenglerianus,* and is said to come from
    New Zealand.

MARMORATUS.   34. Shell sub-ovate, ventricose,
    smooth, with three transverse nodulous belts,
    and the outer lip dilated.

Turbo marmoratus.   *Linnæus Syst. Nat.* p. 1234.   *Born*
    Mus. p. 347.   *Chemnitz,* v. p. 188. t. 179. f. 1775 and
    1776. *Schroeter Einl.* ii. p. 21. *Gmelin,* p. 3592. *Schrei-*
    bers Conch. ii. p. 273.   *Lamarck Syst. des Anim.* p. 86.
Le Burgau appelé Princesse.   *Favanne,* ii. p. 81.
Lister Conch. t. 587. f. 46.   *Gualter,* t. 64. f. A.   *Seba,* iii.
    t. 74. f. 1.   *Knorr,* iii. t. 26. f. 1, and t. 27. f. 1 and 2.
    Geve, t. 14. f. 128 and 129.
Inhabits the Asiatic Ocean on the coasts of Ceylon and Java.
    Linnæus.   Molucca Islands. *Favanne.*   China and Mada-
    gascar. *Humphreys.*

Shell about three, or sometimes four and a half inches long, and the length rather exceeds the breadth ; it has four or five well defined whirls, of which the body-whirl has three transverse tuberculated belts, and of these the uppermost is the strongest ; it is of a dull green colour with spotted brown and white bands, and the inside is pearly. Favanne says it only differs in age from *T. Olearius.*

SARMATICUS. 35. Shell convex, obtuse, with the whirls nodulous above, and separated by a canal.

Turbo Sarmaticus. *Linnæus Syst. Nat.* p. 1235. *Born Mus.* p. 347. ? *Chemnitz,* v. p. 190. t. 179. f. 1777 and 1778, and t. 180. f. 1781. *Schroeter Einl.* ii. p. 22. *Gmelin,* p. 3593. *Schreibers Conch.* i. p. 274.
La Veuve perlée, ou le Coco. *Favanne,* ii. p. 87. t. 8. f. L. *Argenville,* t. 8. f. B. *Regenfuss,* i. t. 1. f. 7.
Inhabits the coasts of the Moluccas. *Favanne.* False Bay, near the Cape of Good Hope. *Humphreys.*
Shell three or four inches long, and the length rather exceeds the breadth ; it has four or five ventricose whirls, and several rows of nodules, of which the uppermost on the body-whirl are largest ; it is said to be black coated with yellowish orange, and the inside is silvery ; but it is so often played tricks with by the dealers, that it is difficult to know which is its natural appearance. The operculum is granulated.

OLEARIUS. 36. Shell ponderous, ventricose, and angulated with three somewhat nodulous keels on the body-whirl.

Turbo Olearius. *Linnæus Syst. Nat.* p. 1235. *Chemnitz,* v. p. 183. t. 178. f. 1771 and 1772. *Schroeter Einl.* ii. p. 24. *Gmelin,* p. 3593. *Schreibers Conch.* i. p. 274.
Le Grand Olearia. *Favanne,* ii. p. 75. t. 8. f. K 1 ; and operculum, f. K 2.
Bonanni Rec. and *Kirch.* 3. f. 9. *Rumphius,* t. 19. f. A and B. *Gualter,* t. 68. f. A. *Knorr,* ii. t. 9. f. 1. *Geve,* t. 16. f. 149.
Variety. With only one nodulous keel, and the spire more produced.
Turbo, No. 19. *Schroeter Einl.* ii. p. 69.
Turbo, No. 45. *Schreibers Conch.* i. p. 280.
Le Pot Verd. *Favanne,* ii. p. 80.

Regenfuss, i. t. 5. f. 52.   *Geve*, t. 14. f. 131.   *Chemnitz*, v.
    t. 178. f. 1773 and 1774.
Inhabits India. *Linnæus.*   Coasts of Coromandel. *Rumphius.*
Shell five or six inches long, and the breadth rather exceeds
    the length ; it has five whirls, of which the body-whirl is
    comparatively very large, flattish above, and has three some-
    what oblique tuberculated transverse keels ; it is of a brownish
    or yellowish olive colour, with paler bands, or is sometimes
    tinged with red, particularly on the keels, and the whole in-
    side is silvery.

CORNUTUS.   37.   Shell ponderous, longitudinally
    wrinkled, and obsoletely ribbed transversely,
    with two rows of large vaulted spines on the
    body-whirl.

Turbo cornutus. *Gmelin*, p. 3593.   *Schreibers Conch.* i.
    p. 280.
Turbo, No. 21.   *Schroeter Einl.* ii. p. 69.
Os argenteum cornutum.   *Chemnitz*, v. p. 193. t. 179. f.
    1779 and 1780.
La Bouche d'Argent cornue.   *Favanne*, ii. p. 54. t. 8. f.
    G 1.
D'Avila, t. 5. f. I.
Inhabits the coasts of China. *D'Avila.*
Shell two and a half, or three inches long, and about equally
    broad, of a greenish or dull yellowish brown colour; it has
    five ventricose whirls, somewhat depressed towards their
    summits, and marked with longitudinal wrinkles and irregular
    unequal transverse ribs ; the body-whirl has two transverse
    rows of large vaulted spines, pointing in rather opposite di-
    rections, and these disappear on the first whirl of the spire ;
    the pillar-lip is slightly channelled longitudinally, and the
    whole aperture is silvery.

RADIATUS.   38.   Shell longitudinally wrinkled, with
    the whirls rounded, distant, transversely
    striated, and armed with small imbricated
    spines.

Turbo radiatus. *Gmelin*, p. 3594.   *Schreibers Conch.* i.
    p. 280.
Turbo, No. 23.   *Schroeter Einl.* ii. p. 70.
Argyrostomus Maris Rubri.   *Chemnitz*, v. p. 199. t. 180.
    f. 1788 and 1789.
Inhabits the Red Sea. *Chemnitz.*

This shell appears, by Chemnitz's figures, to be nearly two inches long, and about equally broad; it is commonly of a brownish white colour with brown longitudinal stripes, or sometimes brown with white rays, and the aperture is silvery. *T. argyrostomus* of Forskael, p. 23, which Chemnitz has quoted with a mark of doubt, is described ' umbilico brevi.'

IMPERIALIS. 39. Shell with the whirls very convex, glabrous, and green with darker transverse lines; pillar-lip callous above.

Turbo imperialis. *Gmelin*, p. 3594. *Schreibers Conch.* i. p. 274.
Turbo, No. 24. *Schroeter Einl.* ii. p. 71.
Cochlea lunaris imperialis. *Chemnitz*, v. p. 200. t. 180. f. 1790.
Inhabits the coasts of China. *Chemnitz.*
Shell about three and a half inches long, and almost equally broad, with six glabrous ventricose whirls; the colour is glossy pale green, with a few darker transverse lines, and numerous character-like spots and longitudinal streaks; the inside is silvery.

CORONATUS. 40. Shell roundish, thick, ventricose, with the body-whirl transversely nodulous, and coronated with spines; spire depressed, and the pillar produced.

Turbo coronatus. *Gmelin*, p. 3594. *Schreibers Conch.* i. p. 274.
Turbo, No. 25. *Schroeter Einl.* ii. p. 71.
Corona reclusa. *Chemnitz*, v. p. 201. t. 180. f. 1791 and 1792.
La Couronne fermée. *Favanne*, ii. p. 84. t. 8. f. O.
Geve, t. 19. f. 176.
Variety. Shell rather smaller, more depressed, and spotted with brown.
Turbo, No. 26. *Schroeter Einl.* ii. p. 71.
Chemnitz, v. t. 180. f. 1793.
Inhabits the coasts of the Moluccas. *Favanne.* Straights of Malacca, and Nicobar Islands. *Chemnitz.*
Shell about an inch and a half long, and equally broad, of a dirty white colour generally clouded with green, and the summit is orange; the body-whirl has three transverse rows of large nodules, of which one row at the upper margin is more pointed than the others.

CANALICULATUS. 41. Shell ponderous, with very convex whirls, strongly grooved and striated transversely; aperture silvery, and the outer lip crenated.

Turbo canaliculatus. *Gmelin,* p. 3594. *Schreibers Conch.* i. p. 281.
Turbo, No. 27. *Schroeter Einl.* ii. p. 72.
Argyrostomus canaliculatus. *Chemnitz,* v. p. 202. t. 181. f. 1794.
Le Bouche d'Argent à rigole. *Favanne,* ii. p. 67. t. 9. f. A 4.
Regenfuss, i. t. 10. f. 44.
Inhabits the coasts of the Molucca and Philippine Islands. *Favanne.*
Shell about three or four inches long, and not much more than two-thirds as broad, of a greenish colour mottled with white, and variegated with a few chestnut spots; the summit of the spire is reddish, and the aperture silvery; the whirls are strongly grooved transversely, and slightly wrinkled longitudinally. I am not certain that I know this shell, but suspect it is only a Variety of *T. setosus.*

SETOSUS. 42. Shell ventricose, transversely grooved and striated; aperture somewhat produced at the base, and the inside grooved and silvery.

Turbo setosus. *Gmelin,* p. 3594. *Schreibers Conch.* i. p. 281.
Turbo, No. 28. *Schroeter Einl.* ii. p. 72.
Cochlea lunaris albo-nigra. *Chemnitz,* v. p. 203. t. 181. f. 1795 and 1796.
Le Leopard, ou la Bouche d'Argent marquetée. *Favanne,* ii. p. 66. t. 9. f. A 1.
Rumphius, t. 19. f. C. *Gualter,* t. 64. f. B.
Inhabits the coasts of the Isle of France. *Favanne.*
Shell about three inches long, and nearly two and a half broad, mottled with green, white, and dark chestnut spots, forming obsolete longitudinal stripes; it has six whirls which are slightly wrinkled longitudinally, and strongly grooved transversely, with the ribs becoming somewhat alternately smaller towards the middle of the body-whirl; the aperture is silvery, transversely grooved, and slightly produced at the base. Kæmmerer considered it to be a Variety of the foregoing species.

SPARVERIUS. 43. Shell with transverse smooth ribs, which are thicker and broader on the shoulders of the whirls.

Turbo Sparverius. *Gmelin,* p. 3594. *Schreibers Conch.* i. p. 275.
Turbo, No. 30. *Schroeter Einl.* ii. p. 73.
Sparverius, sive Nisus. *Chemnitz,* v. p. 204. t. 181. f. 1798.
Regenfuss, ii. t. 6. f. 63.
Inhabits the East Indies. *Chemnitz.*
This shell appears, by Chemnitz's figure, to be about two and a quarter inches long, and one inch and three-quarters broad, with the spire rather more produced and acuminated than in *T. setosus;* it is said to be yellowish spotted with brown, and the inside pearly.

MOLTKIANUS. 44. Shell granulated in transverse rows, and the whirls somewhat spinous, and longitudinally plaited above; inside pearly.

Turbo Moltkianus. *Gmelin,* p. 3595. *Schreibers Conch.* i. p. 281.
Turbo, No. 31. *Schroeter Einl.* ii. p. 73.
Cochlea lunaris Moltkiana. *Chemnitz,* v. p. 205. t. 181. f. 1799 and 1800.
Regenfuss, ii. t. 2. f. 20.
Inhabits ——
This shell appears, by Chemnitz's figure, to be an inch and three-quarters long, and nearly an inch and a half broad; it is marked transversely with moniliform striæ and small vaulted scales, and the upper part of the whirls is plaited longitudinally; the colour is silvery-grey with two transverse orange bands, and it is said to be an extremely rare, as well as beautiful species.

SPENGLERIANUS. 45. Shell transversely ribbed, with the whirls convex, and separated by a broad canal at the suture.

Turbo Spenglerianus. *Gmelin,* p. 3595. *Schreibers Conch.* i. p. 282.
Turbo, No. 32. *Schroeter Einl.* ii. p. 74.
Cochlea lunaris Spengleriana. *Chemnitz,* v. p. 206. t. 181. f. 1801 and 1802.
D'Avila, t. 7. f. P.
Inhabits the East Indian Ocean. *Chemnitz.*

Shell about two inches and a half long, and two inches broad, of a yellowish white colour mottled with brown, or sometimes variegated with chestnut spots; it is strongly ribbed transversely, and slightly wrinkled obliquely, and has five rather remarkably convex whirls, which are separated by a broad canal at the suture; the canal is more or less radiated with darker spots; the inside is white, and somewhat silvery.

CASTANEUS. 46. Shell transversely striated with larger and smaller rows of beaded granules, and of a chestnut-brown colour marbled with white.

Turbo castaneus. *Gmelin*, p. 3595. *Schreibers Conch.* i. p. 282.

Castanea Indiæ Occidentalis. *Chemnitz*, v. p. 211.

Variety A. Transversely striated, and some of the striæ granulated.

Turbo, No. 35. *Schroeter Einl.* ii. p. 75.

Chemnitz, v. t. 182. f. 1807 and 1808.

Variety B. With the beaded striæ crowded, and alternately larger.

Turbo, No. 36. *Schroeter Einl.* ii. p. 75.

Chemnitz, v. t. 182. f. 1809 and 1810.

Geve, t. 16. f. 150.

Variety C. With the beaded striæ alternately larger, and rather remote.

Turbo mammillatus. *Donovan*, v. t. 173. *Maton and Racket, in Lin. Trans.* viii. p. 166. *Montagu Supp.* p. 126.

Turbo, No. 37. *Schroeter Einl.* ii. p. 76.

Geve, t. 16. f. 152. *Chemnitz*, v. t. 182. f. 1813 and 1814.

Inhabits the West Indies. *Chemnitz.* Has been picked up on the Scilly Rocks, at the western extremity of Cornwall. *Da Costa.*

Shell about an inch long, and ten lines broad, of a chestnut-brown colour, marbled with a few irregular large white spots, and the aperture white; it has five convex whirls, studded with transverse rows of beaded granules, which are more or less regularly alternately larger.

CRENULATUS. 47. Shell with transverse crenulated striæ and rows of granules, and of a silvery grey colour with greenish spots.

Turbo crenulatus. *Gmelin,* p. 3595. *Schreibers Conch.*
i. p. 282.
Turbo, No. 40. *Schroeter Einl.* ii. p. 77.
Cochlea lunaris crenulata. *Chemnitz,* v. p. 212. t. 181. f.
1811 and 1812.
Inhabits ———
This shell appears to be very nearly of the same size, and so
nearly allied to *T. castaneus,* that it may be doubted whether
it is more than a Variety.

PAPYRACEUS. 48. Shell thin and brittle, with five
transverse striæ, and rows of white dots near
the suture which are alternated with red
ones ; aperture effuse, and the throat silvery.

Turbo papyraceus. *Gmelin,* p. 3596. *Schreibers Conch.*
i. p. 269.
Turbo, No. 41. *Schroeter Einl.* ii. p. 77.
Cochlea lunaris papyracea. *Chemnitz,* v. p. 215. t. 182. f.
1817 and 1818.
Inhabits the East Indian Seas. *Chemnitz.*
By Chemnitz's figures this shell appears to be about ten or
eleven lines long, and the breadth rather exceeds the length ;
the body-whirl is very large, and the aperture oval and effuse ; it is said to be of an agate-colour, with rows of white
dots which are alternated with red ones near the suture, and
to be a very scarce species.

SEMICOSTATUS. 49. Shell conical, minute, with four
or five rounded whirls, of which the body-
whirl is minutely striated transversely, and
on the upper part longitudinally ribbed.

Turbo semicostatus. *Montagu Test.* p. 326, and *Supp.* p.
129. t. 21. f. 5. *Maton and Racket, in Lin. Trans.*
viii. p. 162.
Inhabits the coasts of Devonshire and Scotland. *Montagu.*
Shell about half a line long, and half as broad, white, obtusely
pointed, with four or five rounded whirls, which are well
defined by the suture ; the ribs on the body-whirl do not extend to the base, where the transverse striæ are most conspicuous. Mr. Montagu says it somewhat resembles *Voluta pellucida,* but the ribs are coarser, the whirls more
rounded, and it has not any plait on the pillar.

VOL. II.

RUBER.   50. Shell minute, with five pellucid smooth glossy rounded whirls; aperture sub-orbicular, and the lip a little reflected on the pillar.

Turbo ruber.   *Adams in Lin. Trans.* iii. p. 66. t. 13. f. 21 and 22.   *Montagu Test.* p. 320.   *Maton and Racket, in Lin. Trans.* viii. p. 162.

Inhabits the coasts of Pembrokeshire. *Adams.* Cornwall. *Montagu.*

Shell about an eighth of an inch long, and scarcely one third as broad, of an uniform reddish brown colour, or sometimes white; it has five whirls separated by a fine suture, and the summit is pointed.

VITREUS.   51. Shell minute, sub-cylindrical, with four rounded whirls; aperture sub-oval and a little contracted at the upper end.

Turbo vitreus.   *Montagu Test.* p. 321. t. 12. f. 3.

Helix vitrea.   *Maton and Racket, in Lin. Trans.* viii. p. 213.

Inhabits the coasts of Cornwall. *Montagu.*

Shell one eighth of an inch long, and one third as broad, white, smooth, and pellucid; Mr. Montagu says, " this species must not be confounded for the white Variety of the last described; it is more slender, the volutions are stronger and run singularly more oblique; and it is at once distinguished by its sub-cylindric shape, in which it resembles more the *T. striatus;* but differs from that shell in being perfectly smooth, and in not having a marginated aperture. The columella of this species is visible through the shell."

PUNCTURA.   52. Shell conical, minute, glossy, with six rounded and finely reticulated whirls.

Turbo Punctura.   *Montagu Test.* p. 320. t. 12. f. 5.

Helix Punctura.   *Maton and Racket, in Lin. Trans.* viii. p. 214.

Inhabits the coasts of the West of England, but is very rare. *Montagu.*

Shell about one tenth of an inch long, and one third as broad, of a transparent yellowish white colour; Mr. Montagu says, " its beauty is only to be seen under a strong magnifier, as the work is extremely fine; by the aid of a microscope it appears delicately punctured all over, and wrought with extremely fine decussated striæ."

ARENARIUS. **53. Shell conical, minute, with five rounded whirls, and decussated striæ.**

Turbo pellucidus. *Adams in Lin. Trans.* iii. p. 66. t. 13. f. 33 and 34. ?
Turbo decussatus. *Montagu Test.* p. 322. t. 12. f. 4.
Helïx arenaria. *Maton and Racket, in Lin. Trans.* viii. p. 214.
Walker's Minute Shells, f. 52. ?
Found among sand in Salcomb Bay on the coast of Devon. *Montagu.*
Shell scarce one-eighth of an inch long, and not half so broad ; Mr. Montagu says, " it differs from *T. Punctura* in being more inclined to a cylindric form, more strongly striated longitudinally, and in the aperture being more oval."

UNIFASCIATUS. **54. Shell conical, minute, smooth, with five flattish transversely fasciated whirls ; outer lip expanded.**

Turbo unifasciatus. *Montagu Test.* p. 327. t. 20. f. 6. *Maton and Racket, in Lin. Trans.* viii. p. 163.
Turbo trifasciatus. *Adams in Lin. Trans.* v. p. 2. t. 1. f. 12 and 13.
Inhabits the coasts of Great Britain. *Adams, &c.*
Shell about one-eighth of an inch long, and more than one third as broad, white, with one, or sometimes two, purplish brown bands on the body, and one on the second whirl ; Mr. Montagu says that when there are two bands they occupy no more space than when there is but one ; aperture sub-oval, with the outer lip thin and turning outwards.

NIVOSUS. **55. Shell minute, sub-turreted, obtuse, smooth, with five or six rounded whirls, and the pillar quite smooth and even.**

Turbo nivosus. *Montagu Test.* p. 326. *Maton and Racket, in Lin. Trans.* viii. p. 163.
Inhabits the coasts of Devonshire, very rare. *Montagu.*
Shell scarcely one line long, and hardly one third as broad ; it is white and glossy, and has the aperture sub-oval ; in shape it somewhat resembles *Voluta interstincta,* but has not any ribs or tooth on the pillar.

s 2

LABIOSUS.   56. Shell sub-conical, small, with about seven flattish whirls, and slightly ribbed longitudinally; aperture expanded, and the outer lip somewhat reflected.

Turbo labiosus.   *Maton and Racket, in Lin. Trans.* viii. p. 164. *Dorset Cat.* p. 49. t. 18. f. 16.
Turbo costatus.   *Pulteney's Dorset Cat.* p. 45.
Turbo membranaceus.   *Adams in Lin. Trans.* v. p. 2. t. 1. f. 14 and 15.
Helix labiosa.   *Montagu Test.* p. 400. t. 13. f. 7.
Inhabits the coasts of Great Britain. *Pulteney, &c.*
Shell rather more than a quarter of an inch long, and nearly half as broad, whitish or horn-coloured, sometimes streaked longitudinally with pale brown, and covered when alive with a yellowish epidermis; the three lower whirls have most commonly about fifteen or eighteen rather obsolete longitudinal ribs, and the upper whirls are smooth; the aperture is rather large, and the outer lip expanded.

ULVÆ.   57. Shell conical, small, acuminated, with about six smooth flattish whirls, and the aperture sub-ovate.

Turbo Ulvæ.   *Pennant Zool.* iv. p. 132. t. 86. f. 120. *Da Costa Brit. Conch.* p. 105. *Montagu Test.* p. 318. *Maton and Racket, in Lin. Trans.* viii. p. 164. *Dorset Cat.* p. 49. t. 18. f. 12.
Helix Ulvæ.   *Pulteney Dorset Cat.* p. 49.
Frequent on Ulvæ and other marine plants, and on mud about high water mark on the coasts of Great Britain. *Pennant, &c.*
Shell a quarter, or sometimes three-eighths of an inch long, and one-third as broad, of an uniform dull dark or reddish brown colour; the inner lip is reflected, beneath which there is a small depression, but no umbilicus; Mr. Montagu has justly remarked that where this species is found in large quantities scarce any other convoluted shell is seen, except perhaps *T. littoreus*, and one or two other common species.

VENTROSUS.   58.   Shell conical, minute, with six glossy ventricose whirls, and the aperture sub-ovate.

Turbo ventrosus. *Montagu Test.* p. 317. t. 12. f. 13. *Maton and Racket, in Lin. Trans.* viii. p. 164. *Dorset Cat.* p. 49. t. 18. f. 12 a.

Turbo eburneus. *Walker's Min. Shells,* f. 36. *Adams's Micros.* p. 637. t. 14. f. 15.
Inhabits the coasts of Great Britain. *Walker, &c.*
Shell about an eighth of an inch long, and one-third as broad, of a pale horn-colour, but appears black when the animal is in it; it is nearly allied to *T. Ulvæ,* but differs in having the whirls ventricose, and in being glossy, much thinner, and of a paler colour.

SUBUMBILICATUS. 59. Shell conical, minute, sub-umbilicated, with four or five tumid whirls, and the aperture completely ovate.

Turbo subumbilicatus. *Montagu Test.* p. 316. *Maton and Racket, in Lin. Trans.* viii. p. 165. *Dorset Cat.* p. 50. t. 18. f. 12 b.
Inhabits the shore at Weymouth. *Montagu.*
Shell one-eighth of an inch long, and half as broad, smooth, somewhat glossy, and of a yellowish white colour; the greater proportional breadth of the body-whirl readily distinguishes this species from both *T. Ulvæ* and *T. ventrosus,* and the aperture is not contracted into an angle at the upper end.

CINGILLUS. 60. Shell conical, minute, with six flattish slightly striated contiguous whirls, and alternate bands of chestnut-brown and horn-colour.

Turbo cingillus. *Montagu Test.* p. 328. t. 12. f. 7. *Maton and Racket, in Lin. Trans.* viii. p. 165.
Turbo vittatus. *Donovan,* v. t. 178. f. 1.
Inhabits the coasts of Great Britain. *Montagu, &c.*
Shell rather more than an eighth of an inch long, and one-third as broad; it somewhat resembles *T. unifasciatus,* but Mr. Montagu says it may be distinguished "by having the outer lip in a line with the shell, not turning outward; by the brown bands being darker, and the intermediate space not so white; and in live shells by the spiral striæ, which are strong at the base; it is also a more slender shell."

INTERRUPTUS. 61. Shell conical, minute, acumi-nated, with five flattish whirls, and interrupt-ed longitudinal yellowish brown streaks.

Turbo interruptus. *Adams in Lin. Trans.* v. p. 3. t. 1. f.

16 and 17.   *Montagu Test*. p. 329. t. 20. f. 8.   *Donovan*, v. t. 178. f. 2.   *Maton and Racket, in Lin. Trans.* viii. p. 166.

Inhabits the coasts of England and Wales, not uncommon.

Shell one-eighth of an inch long, and one-third as broad, pellucid, glossy, and white, marked with interrupted longitudinal yellowish brown streaks, which on the body-whirl frequently form two rows of oblong spots; aperture sub-orbicular.

SEMISTRIATUS. 62. Shell conical, minute, with five or six slightly rounded whirls, which are transversely striated at both ends, and plain in the middle.

Turbo semistriatus.   *Montagu Supp*. p. 136.

Inhabits the South coasts of Devonshire, not common. *Montagu*.

Shell about one-eighth of an inch long, and half as broad, white, and is sometimes marked like *T. interruptus* with faint interrupted brown stripes; it is however readily distinguishable by its transverse striæ, and the aperture is more ovate.

ALBULUS. 63. Shell imperforate, minute, glabrous, with the whirls rounded and striated.

Turbo albulus.   *Fabricius Fauna Groenlandica*, p. 394. *Gmelin*, p. 3609.

Inhabits the deeps of the Greenland Seas, among branches of Sertulariæ. *Fabricius*.

Shell but little more than a line long, pellucid, whitish, brittle, with five ventricose longitudinally grooved whirls, which gradually taper to a point.

*** *Umbilicated and solid.*

PICA. 64. Shell conical, smooth, with the whirls rounded, and the umbilicus toothed on its margin.

Turbo Pica.   *Linnæus Syst. Nat.* p. 1235.   *Born Mus.* p. 349.   *Chemnitz,* v. p. 168. t. 176. f. 1750 and 1751, and Operculum, iv. t. 151. f. 1420 and 1421.   *Schroeter*

Einl. ii. p. 25. *Gmelin*, p. 3598. *Schreibers Conch.* i. p. 276. *Shaw Nat. Misc.* xiv. t. 562.

La Veuve, ou la Pie. *Favanne*, ii. p. 102. t. 9. f. F 1, and F 2.

Bonanni Rec. and *Kirch.* 3. f. 29 and 30. *Lister Conch.* t. 640. f. 30. *Petiver Gaz.* t. 70. f. 9. *Gualter*, t. 68. . f. B. *Argenville*, t. 8. f. G. *Klein Ost.* t. 2. f. 52. *Adanson Senegal*, t. 12. f. 7. *Knorr*, i. t. 10. f. 1, and ii. t. 21. f. 3. *Regenfuss*, i. t. 6. f. 66, and t. 11. f. 57. *Geve*, t. 10. f. 74 to 77, and t. 11. f. 78 to 82.

Inhabits the coasts of Sardinia. *Linnæus.* Malabar. *Bonanni.* Barbadoes. *Lister.* Jamaica. *Petiver.* Magdalen Islands. *Adanson.* Martinique and St. Domingo. *Favanne.*

Shell commonly about two inches long, and near two and a half broad; but it is sometimes larger, and Favanne has mentioned a specimen which measured four inches and four lines in length; it has five rounded whirls, with a broad obsolete groove below the suture; the colour is whitish, marbled with black spots and longitudinal zic-zac stripes, and the inside is silvery. Chemnitz says that his figures 1850 and 1851 were taken from a shell which Mr. Humphreys had named 'the little Magpie from New Zealand;' and these figures have been quoted by Gmelin for a variety of the present species, but they are said to be imperforate, and appear perfectly distinct.

NODULOSUS. 65. Shell conical-rounded, with the whirls nodulous and striated; umbilicus toothed.

Turbo nodulosus. *Born Mus.* p. 349. t. 13. f. 1 and 2. *Chemnitz*, v. p. 181. t. 178. f. 1769 and 1770.

Turbo, No. 18. *Schroeter Einl.* ii. p. 68.

Turbo Pica, Var. δ. *Gmelin*, p. 3598.

Inhabits the coasts of the West India Islands. *Chemnitz.*

Born has described this species to be ten lines long, and eleven broad, and says it differs from *T. Pica* in having the whirls transversely nodulous and striated.

DENTATUS. 66. Shell depressed, sub-orbicular, with the pillar-lip denticulated on its lower margin, and the outer lip double, and striated within.

Turbo dentatus. *Gmelin*, p. 3601. *Schreibers Conch.* i. p. 280.

Turbo, No. 17.   *Schroeter Einl.* ii. p. 67.
Cochlea lunaris trochiformis lævis.   *Chemnitz,* v. p. 181. t.
    178. f. 1767 and 1768.
Inhabits ———
This shell appears, by Chemnitz's figures, to be eight lines long,
    and about equally broad ; it is said to have five smooth whirls,
    and the umbilicus pervious; colour white, marbled with
    brown.

AURICULARIS.   67.   Shell conical, small, smooth,
    with the whirls much rounded, and the aper-
    ture oval-ear-shaped.

Turbo auricularis.   *Montagu Test.* p. 308.   *Maton and
    Racket, in Lin. Trans.* viii. p. 166.
Inhabits the Sea near Southampton. *Montagu.*
Shell three-eighths of an inch long, and nearly two-eighths
    broad, sub-pellucid, and of a pale horn colour ; it has five
    much rounded whirls, and the summit is moderately pointed,
    and usually darker ; aperture sub-oval, or rather ear-shaped,
    with the inner lip much reflected, and forming an angle about
    the middle, behind which is a narrow umbilicus ; Mr. Mon-
    tagu adds that it bears some resemblance to *Helix fossaria,*
    but besides being a smaller shell, it is essentially different in
    the aperture.

VINCTUS.   68.   Shell conical, small, smooth, with
    six rounded whirls, and the summit rather
    obtuse; aperture sub-orbicular.

Turbo vinctus. · *Montagu Test.* p. 307. t. 20. f. 3.   *Ma-
    ton and Racket, in Lin. Trans.* viii. p. 167.
Inhabits the coast of Devonshire, adhering to Algæ. *Mon-
    tagu.*
Mr. Montagu says this shell is rather more than three-eighths
    of an inch long, and the following is his description, " with
    a smooth, conic shell, with six rounded volutions of a sub-
    pellucid, rufous horn-colour ; the lower spire marked with
    four, and sometimes five, purplish brown, or chestnut co-
    loured bands, with a broad space between the three lower
    and the upper ones; in the second and third spires are only
    two bands: the apex is small, but not very pointed : aperture
    sub-orbicular ; outer lip very thin ; inner lip thick, white,
    furnished with a narrow channel, which terminates in a small
    umbilicus." He adds that it is sometimes of a pale horn-co-
    lour, and the bands faint, and sometimes quite plain, so that

it might be mistaken for *Helix Canalis*, were it not for the aperture of that shell being sub-angulated, and the pillar-lip with a much larger canal and umbilicus. It also bears some resemblance to *T. quadrifasciatus*, but is not near so thick and strong, and is destitute of the subcarinated edge at the base.

QUADRIFASCIATUS. 69. Shell sub-conical, small, smooth, with the body-whirl somewhat keeled at the base; pillar-lip broad, with a small groove ending in an umbilicus.

Turbo quadrifasciatus. *Montagu Test.* p. 328. t. 20. f. 7. *Maton and Racket, in Lin. Trans.* viii. p. 167.
Inhabits the coasts of Cornwall and Devon, adhering to Algæ. *Montagu.* Langland Bay near Swansea. *L. W. D.*
Shell about three-eighths, or a quarter of an inch long, and half as broad, with four whirls, of which the body-whirl is larger than the whole spire; colour white, usually marked with four pale brown narrow, or sometimes only two broader bands on the body-whirl, and half that number on the next whirl.

SANGUINEUS. 70. Shell conical, convex, sub-umbilicated, striated, smooth, and the whirls slightly grooved.

Turbo sanguineus. *Linnæus Syst. Nat.* p. 1235. *Chemnitz,* v. p. 172. t. 177. f. 1756 and 1757. *Schroeter Einl.* ii. p. 27. *Gmelin,* p. 3598. *Schreibers Conch.* i. p. 276.
Inhabits the Mediterranean on the shores of Algiers. *Linnæus.* Naples. *Ulysses.*
Linnæus has described this species to be as large as a pea, of a blood red colour, convex, with the whirls obtusely grooved, and sometimes umbilicated and sometimes not; the shell which Chemnitz has figured appears to be half an inch long, and equally broad, of a whitish colour, and somewhat tessellated with bright red spots.

ATRATUS. 71. Shell with crowded transverse rows of alternately cinereous and black granules; aperture silvery, striated within, and the pillar one-toothed.

Turbo atratus. *Gmelin,* p. 3601. *Schreibers Conch.* i. p. 279.

Turbo, No. 16. *Schroeter Einl.* ii. p. 67.
Cochlea lunaris trochiformis deuigrata. *Chemnitz,* v. p. 172.
t. 177. f. 1754 and 1755.
Inhabits the coasts of the Nicobar Islands. *Chemnitz.*
This shell appears by the figures to be eight lines long, and
about equally broad, and to possess some affinity with *Trochus Labeo;* the pillar is said to have one obsolete tooth.

ANGUIS. 72. Shell transversely grooved, green with
blackish longitudinal zic-zac stripes, and the
inside pearly.

Turbo Anguis. *Gmelin,* p. 3602. *Burrow's Elem.* p. 168.
t. 19. f. 6.
Limax Anguis. *Martyn's Univ. Conch.* ii. t. 70.
Inhabits the South Sea. *Martyn.*
Shell about an inch long, and an inch and a quarter broad, of
a pale green colour with dark broad zic-zac longitudinal
stripes; it has four whirls, which are marked with transverse
rather distant dotted grooves; the aperture is pearly, and
there is a canal along the pillar-lip which terminates in an
umbilicus.

DIADEMA. 73. Shell of a dull green colour variegated with brown, and marked with crowded
transverse striæ; whirls four, and the body
large.

Turbo Diadema. *Gmelin,* p. 3601. *Schreibers Conch.* i.
p. 280.
Turbo, No. 20. *Schroeter Einl.* ii. p. 69.
Chemnitz, v. p. 187. Vign. 43. at p. 145. f. *A* and *B.*
Inhabits the coasts of New Zealand. *Chemnitz.*
Chemnitz says this shell is called in England 'The knobbed
umbilicated Emerald Snail from New Zealand;' and from his
figure it appears to differ from *T. Anguis* principally in being more than twice as large, and in not having the longitudinal stripes undulated, or so strongly marked.

UNDULATUS. 74. Shell ovate-convex, obtuse, glabrous, green with undulated black stripes,
and the throat silvery.

Turbo undulatus. *Humphreys Portland Cat.* p. 18. lot
408, and p. 178. lot 3828. *Chemnitz,* x. p. 296. t. 169.
f. 1640 and 1641. *Gmelin,* p. 3597.

Limax undulatus. *Martyn Univ. Conch.* i. t. 29.
Inhabits the coasts of Van Diemen's Land, and New Holland. *Humphreys.*
Shell two inches and a quarter long, and two inches and a half broad, of a fine glossy green colour, with black undulated longitudinal stripes; it is nearly allied to the two foregoing species, but differs in not having any transverse grooves or striae.

ARGYROSTOMUS. 75. Shell sub-ovate, with transverse ribs, which are somewhat alternately larger, and longitudinally wrinkled; aperture silvery.

Turbo argyrostomus. *Linnæus Syst. Nat.* p. 1236. *Born Mus.* p. 350. *Chemnitz,* v. p. 174. t. 177. f. 1760, 1761, and 1763 to 1765. *Schroeter Einl.* ii. p. 28. *Gmelin,* p. 3599. *Schreibers Conch.* i. p. 276.
La Bouche d'Argent chagrinée. *Favanne,* ii. t. 9. f. A 3.
Rumphius, t. 19. f. No. 2. *Argenville,* t. 6. f. F. *Seba,* iii. t. 74. f. 6. *Knorr,* i. t. 3. f. 3. *Geve,* t. 17. f. 157 to 162. *Regenfuss,* i. t. 11. f. 50. *Da Costa Elements,* t. 3. f. 12.
Variety. With vaulted scales on the ribs.
Turbo echinatus. *Gmelin,* p. 3591.
Limax echinatus. *Martyn Univ. Conch.* i. t. 26.
Argyrostomus spinosus. *Chemnitz,* v. p. 173. t. 177. f. 1758 and 1759.
La vraie Bouche d'Argent epineuse. *Favanne,* ii. p. 58.
Inhabits the Indian Seas. *Linnæus.* Red Sea. *Forskael.* Coasts of the Isle of France, and Cape of Good Hope. *Favanne.* China. *Humphreys.*
Shell varying from an inch and a half to two inches and a half long, and the length rather exceeds the breadth; it has six whirls, of which the body-whirl is larger than the whole spire, and unequally ribbed transversely; the colour is whitish, variously stained or marbled with reddish or purplish brown and green, and the inside is silvery; the aperture is more entire, and not produced at the base as in *T. margaritaceus.*

MARGARITACEUS. 76. Shell sub-ovate, with transverse ribs, of which that on the shoulder is largest, and the aperture silvery.

Turbo margaritaceus. *Linnæus Syst. Nat.* p. 1236. *Born Mus.* p. 351.? *Chemnitz,* v. p. 176. t. 177. f. 1762. *Schroeter Einl.* ii. p. 29. t. 3. f. 18. *Gmelin,* p. 3599. La Bouche d'Argent des Iles Fredericiennes. *Favanne,* ii. p. 10.

Rumphius, t. 19. f. 3 and 4. *Argenville,* t. 6. f. A. *Seba,* iii. t. 74. f. 4. *Regenfuss,* i. t. 10. f. 43.

Inhabits the coasts of Frederick's Islands. *Regenfuss.*

Linnæus says that this shell differs from *T. argyrostomus* in being of a greenish white colour, with brown instead of ferruginous or purplish spots, and he has quoted Rumphius, t. 19. f. 3. for both these species; the shell which Born has described appears to me to be certainly nothing more than a variety of *T. argyrostomus,* and they approach each other in all respects so closely that I am unable to find any separating line.

PORPHYRITES. 77. Shell slightly granulated, with the spire flattish, and the base produced; aperture silvery.

Turbo Porphyrites. *Gmelin,* p. 3602.
Turbo versicolor. *Gmelin,* p. 3599.
Turbo argyrostomus, Var. *Schreibers Conch.* i. p. 276!
Turbo, No. 8. *Schroeter Einl.* ii. p. 64.
Cochlea lunaris complanata. *Chemnitz,* v. p. 163. t. 176. f. 1740 and 1741.
Limax Porphyrites. *Martyn Univ. Conch.* ii. t. 70.
Lister Conch. t. 576. f. 29.?

Inhabits the coasts of New Caledonia. *Martyn.*

Shell about an inch and a quarter long, and an inch and a half broad, mottled with green, brown, and white, forming irregular somewhat tessellated transverse stripes; the lip is much thickened and produced under the umbilicus, and has a groove near the middle.

MESPILUS. 78. Shell with four rounded whirls, and a tessellated brown and white band at the suture; aperture silvery, and the pillar white.

Turbo Mespilus. *Gmelin,* p. 3601.
Turbo, No. 9. *Schroeter Einl.* ii. p. 65.
Mespilum. *Chemnitz,* v. p. 164. t. 176. f. 1742 and 1743.

Inhabits the South Sea. *Chemnitz.*

This shell appears, by Chemnitz's figures, to be nearly an inch

and a half long, and equally broad, and of a dull brown colour like that of a medlar, with a tessellated band at the suture.

GRANULATUS. 79. Shell with granulated transverse
ribs, and the body-whirl coronated; aperture
silvery, and the pillar-lip spotted with orange.

Turbo granulatus. *Gmelin*, p. 3601. *Schreibers Conch*. i.
p. 279.
Turbo, No. 10, and No. 11. *Schroeter Einl*. ii. p. 65.
Cochlea lunaris, &c. *Chemnitz*, v. p. 164. t. 176. f. 1744
to 1746.
Inhabits the coasts of the Nicobar Islands, and the South Sea.
Chemnitz.
This shell appears to be nearly of the same form and size as *T.
Mespilus*, but is transversely ribbed and granulated; the bo-
dy-whirl is dull brown, and the spire issues from the row of
tubercles with which it is crowned, of a pale reddish colour;
I have a specimen which in other respects answers, but it is
imperforate, and has the spire flatter than in Chemnitz's figures.

CINEREUS. 80. Shell roundish, smooth, with the
whirls convex, and the pillar produced at the
base; throat grooved, and the umbilicus
broad.

Turbo cinereus. *Born Mus*. p. 349. t. 12. f. 25 and 26.
Gmelin, p. 3601. *Schreibers Conch*. i. p. 283.
Turbo, No. 117. *Schroeter Einl*. ii. p. 103.
Inhabits ——
Born has described this shell to be eight lines long, and eleven
broad, of a grey colour with black and white spots and lines;
whirls six, flattened at the suture, and obsoletely striated
transversely; aperture semicircular.

TORQUATUS. 81. Shell with a keel, and row of
nodules, transversely ribbed, and strongly
wrinkled obliquely; pillar-lip broad and
white, with a large umbilicus, and the throat
silvery.

Turbo torquatus. *Gmelin*, p. 3597. *Schreibers Conch*. i.
p. 278.

Turbo singularis.   *Portland Cat.* p. 164. lot 3559.
Turbo collari præditus.   *Chemnitz,* x. p. 295. Vign. 24 at
    p. 293. f. *A* and *B.*
Limax staminea.   *Martyn Univ. Conch.* ii. t. 71.
Inhabits the coasts of New Zealand. *Humphreys.*
This singular shell is about two inches long, and two and a half
    broad, with the summit of the spire and the upper part of
    the lower whirls flattish ; the whole of the outer surface is
    covered with strong oblique crowded wrinkled striæ, of a
    dull brown colour sometimes variegated with olive.

DELPHINUS. 82. Shell depressed, and foliated ;
    umbilicus large, and armed with small vault-
    ed scales in spiral rows.

Turbo Delphinus.   *Linnæus Syst. Nat.* p. 1236.   *Born*
    Mus. p. 351.   *Chemnitz,* v. p. 153. t. 175. f. 1727 to
    1735.   *Schroeter Einl.* ii. p. 30.   *Gmelin,* p. 3599.
    Schreibers Conch. i. p. 277.   *Brookes's Introd.* p. 126.
    t. 8. f. 98.
Turbo distortus.   *Born Mus.* p. 352.
Cyclostoma Delphina.   *Lamarck Syst. des Anim.* p. 87.
Le Dauphin.   *Favanne,* ii. p. 155. t. 9. f. G 1, and G 2.
Bonanni Rec. and *Kirch.* 3. f. 31.   *Lister Conch.* t. 608.
    f. 45.   *Rumphius,* t. 20. f. H.   *Petiver Amb.* t. 3. f. 1.
    Gualter, t. 68. f. C and D.   *Argenville,* t. 6. f. H. *Se-*
    ba, iii. t. 59. f. 12 to 27. *Knorr,* i. t. 22. f. 4 and 5, iv.
    t. 7. f. 2 and 3, and t. 8. f. 1.   *Geve,* t. 4. f. 24 to 30.
    Regenfuss, i. t. 8. f. 14.
Variety B. With perpendicular spines on the keel of the body-
    whirl.
Turbo, No. 7.   *Schroeter Einl.* ii. p. 64.
    Chemnitz, v. t. 175. f. 1736.
Variety C. With the whirls detached.
Turbo Delphinus monstrosus. *Chemnitz,* xi. p. 292. t. 211.
    f. 2090 and 2091.
Inhabits the Asiatic Ocean. *Linnæus.*   Amboyna. *Rumphius.*
    Coasts of the Philippines, Isle of France, and New Guinea.
    Favanne.   China and Tranquebar.   *Humphreys.*
Shell about an inch and a half long, and two inches broad, of
    a blackish brown, red, or yellowish colour, variegated with
    white ; the whirls are flattened above, and transversely keel-
    ed, and marked with granulated striæ; there are two or three
    keels on the body-whirl, and one on the spire, and the for-
    mer are armed with more or less crowded, or sometimes
    distant, foliated spines ; the umbilicus is very large and deep,

and is strongly marked transversely with somewhat spinous striæ.

EXASPERATUS. 83. Shell depressed, nodulous, with transverse granulated striæ, and an unequally tuberculated keel on the body-whirl.

Turbo nodulosus. *Gmelin*, p. 3600.
Turbo Delphinus, Var. *Schreibers Conch.* i. p. 277.
Turbo, No. 5. *Schroeter Einl.* ii. p. 69.
Delphinus supra et infra tuberculato-nodosus. *Chemnitz*, v. p. 160. t. 174. f. 1723 and 1724.
Inhabits the East Indian Seas. *Chemnitz.*
It appears, by Chemnitz's figures, that this shell is about an inch in diameter, nearly equally convex on both sides, longitudinally plaited, and marked all over with transverse granulated striæ; the colour is red variegated with white, and the inside pearly; Gmelin's name of *nodulosus* has been preoccupied by Born for another species.

DISTORTUS. 84. Shell slightly mucronated, and covered all over with smooth spines.

Turbo distortus. *Linnæus Syst. Nat.* p. 1296. *Chemnitz*, v. p. 161. t. 175. f. 1737 to 1739. *Schroeter Einl.* ii. p. 32. *Gmelin*, p. 3600. *Schreibers Conch.* i. p. 278.
Inhabits ———
Both Born and Favanne have considered a variety of *T. Delphinus* to be the *T. distortus*, and Linnæus has strongly expressed his doubts of keeping them separate. In the shell figured by Chemnitz the length rather exceeds the breadth, and the two lower whirls of the spire are much produced; each whirl has a transverse nodulous keel, and is separated by a channelled suture; it is of a rose-colour, and the base nearly white, with crowded rows of concentric red granules.

**** *Cancellated.*

CRENELLUS. 85. Shell flattish, with a spreading umbilicus, and rounded striated whirls; striæ crenated.

Turbo crenellus. *Linnæus Syst. Nat.* p. 1236. *Gmelin*, p. 3602.
Inhabits ———

Linnæus, without any reference, has described this shell to be of the size of a lupine seed, red, solid, plano-convex above, and concave beneath, with a wide-spreading umbilicus; whirls rounded, striated longitudinally, and the striæ crenated.

THERMALIS.   86.   Shell umbilicated, rather oblong, and obtuse; whirls rounded and smooth.

Turbo thermalis.   *Linnæus Syst. Nat.* p. 1237.   *Schroeter Einl.* ii. p. 34.   *Gmelin,* p. 3603.

Turbo fontinalis.   *Donovan,* iii. t. 102.   *Montagu Test.* p. 348. t. 22. f. 4.   *Maton and Racket, in Lin. Trans.* viii. p. 168.   *Dorset Cat.* p. 50. t. 18. f. 3 and 4.

Nerita piscinalis.   *Muller Verm.* ii. p. 172.

Trochus cristatus.   *Schroeter Fluss.* p. 280. t. 6. f. 11.

Helix piscinalis.   *Gmelin,* p. 3627.

Helix fascicularis.   *Gmelin,* p. 3641.

Helix, No. 246.   *Schroeter Einl.* ii. p. 246.

Petiver Gaz. t. 18. f. 2.   *Walker's Minute Shells,* f. 56.

Junior.   Shell smaller.

Nerita pusilla.   *Muller Verm.* ii. p. 171.

Helix, No. 319.   *Schroeter Einl.* ii. p. 268.

Helix pusilla.   *Gmelin,* p. 3627.

Martini Berl. Mag. iv. t. 8. f. 26.

Inhabits rivers and other fresh waters.   Great Britain.   *Petiver, &c.*   Near the Baths of Pisa.   *Linnæus.*   Lakes of Germany, and fish-ponds in Denmark.   *Muller.*

Shell about a quarter of an inch long, and equally broad, thin, sub-pellucid, horn-coloured, and composed of four or five much rounded prominent whirls, which are slightly striated longitudinally.   There does not appear to be much doubt that this is the *T. thermalis* of the Systema Naturæ, and though so dissimilar in appearance, Linnæus may probably have placed it next to *T. scalaris* on account of its deep umbilicus, and of its having the whirls only slightly attached to each other.   The animal belongs to Muller's genus Valvata.

SCALARIS.   87.   Shell conical, with eight rounded detached whirls connected together by longitudinal ribs.

Turbo scalaris.   *Linnæus Syst. Nat.* p. 1237.   *Martini,* iv. p. 263. t. 152. f. 1426, 1427, 1430, and 1431; and t. 153. f. 1432 and 1433.   *Born Mus.* p. 354. *Schroeter*

Einl. ii. p. 36. t. 3. f. 20. *Gmelin,* p. 3603. *Schreibers Conch.* i. p. 284. *Burrow's Elements,* p. 168. t. 19. f. 7. Scalaria conica. *Lamarck Syst. des Anim.* p. 88. Aciona scalaris. *Leach Zool. Misc.* ii. p. 80. t. 87. Le Tuyau scalata. *Favanne,* i. p. 655. t. 5. f. A. *Rumphius,* t. 49. f. A. *Petiver Amb.* t. 2. f. 9. *Gualter,* t. 10. f. ZZ. *Argenville,* t. 11. f. V. *Knorr,* iv. t. 20. f. 2 and 3, and v. t. 23. f. 1, and t. 24. f. 6. *Regenfuss,* ii. t. 5. f. 44.

Inhabits the coasts of Tranquebar. *Born.* Batavia. *D'Herbigny.* Ceylon. *Martini.* Amboyna and Philippine Islands. *Favanne.* Japan. *Humphreys.*

This valuable and elegant shell is about an inch and a half, or sometimes two inches and a half long, and the breadth is about three-fifths of the length; it has eight sub-cylindrical whirls, which, without being attached to each other, are connected only by much elevated somewhat membranaceous longitudinal ribs, and of these there are about eight on the body-whirl; the aperture is marginated; it is generally of a snowy white or sometimes pale flesh-colour. Gualter and Favanne have placed it among the Serpulæ, and Dr. Leach has separated it from Lamarck's genus Scalaria, on account of its want of a pillar. Large and perfect specimens formerly sold at very high prices, and one which now belongs to Mr. Bullock has been valued at two hundred guineas. Da Costa, in his Elements of Conchology, relates that in 1753, at the sale of Commodore Lisle's collection, four of these *Wentletraps* were sold as follows:

First day, Lot 96, one not quite perfect - - £16 16   0
Third day, Lot 98, a very fine and perfect one 18 18   0
Fourth day, Lot 101, one for - - - - - 16 16   0
Sixth day, Lot 83, one for - - - - - - 23   2   0

PRINCIPALIS. 88. Shell imperforate, turreted, with ten contiguous cancellated whirls, and longitudinal ribs.

Turbo principalis. *Pallas Spicel. Zool.* Fasc. x. p. 33. t. 3. f. 5 and 6. *Martini,* iv. p. 273. t. 152. f. 1428 and 1429. *Chemnitz,* xi. p. 153. t. 195 A. f. 1876 and 1877. Turbo, No. 123. *Schroeter Einl.* ii. p. 105. Turbo scalaris, Var. *β. Gmelin,* p. 3603. Inhabits the coasts of Coromandel. *Gmelin.*

Shell rather more than two inches long, and about three quarters of an inch broad, and appears from the descriptions of Pallas and Chemnitz to be entirely distinct from *T. scalaris;*

it is much narrower in proportion to the length, is imperforate, and has the whirls contiguous and marked with cancellated striæ; the colour is said to be white, but Chemnitz's figures are slightly tinged with brown.

CLATHRUS. 89. Shell imperforate, turreted, with rounded sub-contiguous whirls, and thick longitudinal distant ribs.

Turbo Clathrus. *Linnæus Syst. Nat.* p. 1237. *Pennant Zool.* iv. p. 129. t. 81. f. 111. *Born Mus.* p. 354. *Martini,* iv. p. 275. t. 153. f. 1434. *Schroeter Einl.* ii. p. 36. *Gmelin,* p. 3603. *Schreibers Conch.* i. p. 284. *Chemnitz,* xi. p. 155. t. 195 A. f. 1878 and 1879. *Donovan,* i. t. 28. *Montagu Test.* p. 296. *Maton and Racket, in Lin. Trans.* viii. p. 170. *Dorset Cat.* p. 50. t. 15. f. 11.
Turbo scalaris. *Brookes's Introd.* p. 126. t. 8. f. 100.
Strombiformis clathratus. *Da Costa Brit. Conch.* p. 115. t. 7. f. 11.
Bonanni Rec. and *Kirch.* 3. f. 111. *Lister Conch.* t. 588. f. 51. *Rumphius,* t. 29. f. W. *Petiver Amb.* t. 13. f. 10. *Plancus,* t. 5. f. 7 and 8. *Gualter,* t. 58 H. *Klein Ost.* t. 3. f. 66. *Knorr,* i. t. 11. f. 5, iv. t. 20. f. 5, and vi. t. 39. f. 3. *Favanne,* t. 39. f. M 1, and M 2.
Inhabits the coasts of Europe and America; the Mediterranean, Iceland, and Sweden. *Linnæus.* Adriatic. *Bonanni.* England. *Pennant.* Italy. *Plancus.* France, Spain, Holland, Norway, and Greenland. *Chemnitz.* Naples. *Ulysses.*
Shell about an inch and a quarter long, and one third as broad; white, with nine or ten much elevated longitudinal ribs, and the interstices generally more or less distinctly striped transversely with brown; it has nine or ten extremely rounded whirls, which taper gradually to a fine point; the ribs are extended over the base, and united to the thickened lip of the aperture.

CLATHRATULUS. 90. Shell imperforate, turreted, with rounded sub-contiguous whirls, and thin longitudinal crowded ribs.

Turbo Clathratulus. *Montagu Test.* p. 297, and *Supp.* p. 124.
Turbo Clathrus, Var. *Pennant Zool.* iv. p. 129. t. 81. f. 111 A. *Martini,* iv. p. 275. t. 153. f. 1437. *Maton and Racket, in Lin. Trans.* viii. p. 171. t. 5. f. 1.

Junior. **Shell minute.**
Turbo Clathratulus. *Adams's Micros.* p. 627. t. 14. f. 19.
Walker's Minute Shells, f. 45.
Inhabits the coasts of Great Britain. *Pennant, &c.*
Shell half, or sometimes three quarters of an inch long, and is
rather broader in proportion to the length, and much thinner
and more pellucid than *T. Clathrus;* it also differs in having
the longitudinal ribs very thin and slender, and almost twice
as numerous; a specimen now before me has eight much
rounded whirls, and eighteen very thin longitudinal membranaceous ribs on the body-whirl, and is of a snowy white,
without any coloured markings.

PULCHER. 91. Shell imperforate, turreted, with contiguous whirls, and strong distant longitudinal ribs, ending in a transverse keel on the
body-whirl.

Turbo Clathrus, Var. β. *Gmelin,* p. 3603.
Turbo, No. 133. *Schroeter Einl.* ii. p. 108.
Lister Conch. t. 588. f. 50.
Inhabits the coasts of the West India Islands.
This shell is nearly allied to *T. Clathrus,* and of about the
same size, but differs in having the whirls less rounded, and
the longitudinal ribs, instead of being extended over the
base, terminating in a strong transverse rib on the body-
whirl, and the base is flattened; the colour is often wholly
white, or sometimes brownish or of a lead-colour, with the
ribs white.

AMBIGUUS. 92. Shell umbilicated, turreted, cancellated, with the whirls smooth and contiguous.

Turbo ambiguus. *Linnæus Syst. Nat.* p. 1237. *Gmelin,*
p. 3604.
Inhabits the Mediterranean. *Linnæus.*
Linnæus says that this shell very much resembles *T. Clathrus,*
except in its being perforated, and in having twice as many
longitudinal ribs; he has described it to be of a pale colour,
with two or three transverse ferruginous lines on each whirl.

CRENATUS. 93. Shell turreted, with the whirls crenated at their upper margins; aperture marginated, and the base keeled.

Turbo crenatus. *Linnæus Syst. Nat.* p. 1238. *Gmelin,*

p. 3604. *Chemnitz*, xi. p. 156. t. 195 A. f. 1880 and 1881.
Lister Conch. t. 588. f. 52.
Inhabits the coasts of the West India Islands. *Chemnitz.*
Shell about ten lines long, and one third as broad; white, with eight somewhat rounded contiguous whirls, which are crenated, or rather nodulous on their upper margins; the keel near the base of the body-whirl is also nodulous.

LACTEUS. 94. Shell turreted, with longitudinal elevated crowded striæ.

Turbo lacteus. *Linnæus Syst. Nat.* p. 1238. *Gmelin,* p. 3604.
Ginanni Adr. t. 6. f. 55.
Inhabits the Mediterranean. *Linnæus.*
Linnæus, with the above reference to Ginanni, has described this shell as being of the size of a grain of barley, white, and resembling *T. Clathrus*, from which it differs in being much smaller, and in the striæ not being membranaceous.

ELEGANTISSIMUS. 95. Shell turreted, with the whirls obliquely ribbed, and the aperture somewhat angulated at both ends.

Turbo elegantissimus. *Montagu Test.* p. 298. t. 10. f. 2.
Turbo acutus. *Donovan,* v. t. 179. f. 1.
Helix elegantissima. *Maton and Racket, in Lin. Trans.* viii. p. 209.
Walker's Minute Shells, f. 39.
Inhabits the coasts of England. *Walker, &c.*
Shell about a quarter or three-eighths of an inch long, and one-fourth as broad, white, glossy, and semi-pellucid. Mr. Montagu says it has from nine to thirteen whirls, which are defined by the suture, and cut longitudinally into regular equidistant furrows; aperture sub-orbicular, but is a little angulated at the upper and lower ends.

SIMILLIMUS. 96. Shell turreted, with longitudinal straight ribs, and the aperture sub-ovate.

Turbo simillimus. *Montagu Supp.* p. 136.
Inhabits the shores of the Island of Jura. *Mr. Laskey.*
Mr. Montagu says, " This shell has much the habit of *Turbo elegantissimus*, but is not so slender, the ribs are less numerous, and consequently more distant, the sulci, or depressions, being larger than the elevations. Those who have an op-

portunity of comparing these two shells, will also observe that the ribs in *elegantissimus* do not run straight, but oblique to the right, are not so much arched, and are larger than the interstices." The length is about three-eighths of an inch.

PARVUS. 97. Shell turreted, minute, with five or six whirls, and distant elevated ribs.

Turbo parvus. *Da Costa Brit. Conch.* p. 104. *Montagu Test.* p. 310. *Maton and Racket, in Lin. Trans.* viii. p. 171. *Dorset Cat.* p. 50. t. 19. f. 4.
Turbo æreus. *Adams in Lin. Trans.* iii. p. 66. t. 13. f. 29 and 30.
Turbo lacteus. *Donovan,* iii. t. 90.
Variety. Ribbed only on the lower whirls.
Turbo subluteus. *Adams, in Lin. Trans.* iii. p. 66.
Inhabits the shores of Great Britain and Guernsey. *Da Costa.* Shell about one-eighth of an inch long, with the colour very various, and is sometimes white or chestnut, or pale rufous, with the ribs white; the aperture is sub-orbicular and thickened. Da Costa, as well as Mr. Donovan, considered this to be the Linnæan *Turbo lacteus;* but there are only from nine to eleven rather distant ribs, whereas that species is described to have the ribs crowded.

STRIATULUS. 98. Shell turreted, minute, sub-cancellated, with the whirls contiguous, and interrupted by varicose belts.

Turbo striatulus. *Linnæus Syst. Nat.* p. 1238. *Schroeter Einl.* ii. p. 40. *Gmelin,* p. 3604. *Montagu Test.* t. 10. f. 5. *Maton and Racket, in Lin. Trans.* viii. p. 172. *Dorset Cat.* p. 50. t. 14. f. 10.
Turbo carinatus. *Da Costa Brit. Conch.* p. 102. t. 8. f. 10.
Inhabits the Mediterranean. *Linnæus.* Shores of England. *Da Costa.*
Linnæus has described this species to be " of the size of a barley-corn, white, and the whirls surrounded with small membranaceous striæ and convex callous wrinkles; aperture somewhat ob-ovate, and rather angular beneath." The shell, supposed by all British authors to be the same, has four or five whirls, each terminating in a flat top, which marks their divisions, and besides the elevated spiral ridges, is striated longitudinally, which gives an elegant cancellated appearance; it is about two-tenths of an inch long, and nearly equally broad, and is rather a strong shell for its size.

RETICULATUS. 99. Shell conical, minute, with rounded whirls, and strongly reticulated.

Turbo reticulatus. *Adams in Lin. Trans.* iii. p. 66. t. 13. f. 19 and 20. *Montagu Test.* p. 322. t. 21. f. 1. *Maton and Racket, in Lin. Trans.* viii. p. 172.

Inhabits the coasts of Pembrokeshire. *Adams.* Kent. *Mr. Boys.*

Shell scarcely one-tenth of an inch long, and about half as broad, white, or whitish brown, with six distended very convex whirls, and strongly reticulated; aperture sub-orbicular, with a thick margin, and the inner lip spreading on the pillar.

BRYEREUS. 100. Shell turreted, small, and glossy, with contiguous ribs, and the aperture ovate.

Turbo Bryereus. *Montagu Test.* p. 313. t. 15. f. 8. *Maton and Racket, in Lin. Trans.* viii. p. 172. *Dorset Cat.* p. 50. t. 19. f. 7.

Turbo costatus. *Donovan,* v. t. 178. f. 3.

Inhabits the coasts of Britain and the West Indies. *Montagu.*

Shell near a quarter of an inch long, and about one-third as broad, strong, conical, and composed of seven whirls, with seventeen or eighteen continuous longitudinal ribs; aperture oval, and the outer lip strong; it is twice as large as *T. costatus,* and is destitute of the transverse striæ and marginated lip which are conspicuous in that species. Mr. Montagu mentions a variety with stronger and fewer ribs, not exceeding ten or twelve in number.

CONIFERUS. 101. Shell turreted, small, with continuous undulated ribs, and the whirls somewhat crenulated at the suture.

Turbo coniferus. *Montagu Test.* p. 314. t. 15. f. 2. *Maton and Racket, in Lin. Trans.* viii. p. 173. *Dorset Cat.* p. 50. t. 19. f. 6.

Inhabits the sea at Weymouth. *Mr. Bryer.*

Shell a quarter of an inch long, and one-third as broad, strong, white, and formed of six whirls; it has about twelve undulated continuous ribs, interrupted only by a fine suture, and the interstices at the top of each whirl are formed into small cavities, giving that part a scolloped or denticulated appearance; the whirls are also faintly striated transversely, and the apex is rather obtuse; aperture oval, oblique, strongly marginated, and the pillar-lip not reflected.

DENTICULATUS. **102.** Shell conical, small, with oblique ribs, and the whirls denticulated at the suture.

Turbo denticulatus. *Montagu Test.* p. 315. *Maton and Racket, in Lin. Trans.* viii. p. 173.

Inhabits the sea at Weymouth. *Montagu.*

Shell nearly a quarter of an inch long, and about half as broad. Mr. Montagu says it very much resembles *T. coniferus,* from which it differs in having the aperture more round and not properly marginated, but only thickened by a rib; it is much more conical, the ribs are stronger, less numerous, and form deeper denticulations at the suture; the interstices of the ribs are also destitute of striæ, and sub-pellucid, and what seems the strongest specific distinction is, that the ribs do not undulate, but run oblique to the left, from the aperture to the apex.

ARCUATUS. **103.** Shell turreted, small, with continuous ribs, which are arched over at the suture.

Turbo marginatus. *Montagu Supp.* p. 128.

Inhabits the coasts of Guernsey. *Montagu.*

Shell about three-eighths of an inch long, and one-fourth as broad. Mr. Montagu says it somewhat resembles *T. coniferus,* but is more slender, and the ribs are regularly arched over each volution, and not abruptly finished at the top, as in that shell. The name of *marginatus* was before occupied by Chemnitz for another species.

STRIATUS. **104.** Shell sub-turreted, minute, with the whirls faintly ribbed at the upper end, and thickly and regularly striated transversely.

Turbo striatus. *Adams in Lin. Trans.* iii. p. 66. t. 13. f. 25 and 26. *Montagu Test.* p. 312. *Maton and Racket, in Lin. Trans.* viii. p. 173.

Turbo Shepeianus. *Adams's Micros.* p. 638. t. 14. f. 22. *Walker's Minute Shells,* f. 49.

Inhabits the roots of marine Algæ on the shores of Cornwall and Devon. *Montagu.* Pembroke. *Adams.* Neighbourhood of Swansea, and South of Ireland; not uncommon. *L. W. D.*

Shell about one-eighth of an inch long, and the breadth is ra-

ther more than one-third of the length; live shells are covered with a brown epidermis, but when stripped of this
they are white, glossy, and pellucid; the upper part of each
whirl is faintly ribbed longitudinally, and the whole shell is
regularly and beautifully covered with crowded transverse
striæ.

COSTATUS.   105. Shell sub-turreted, minute, with
    the whirls obliquely ribbed and transversely
    striated; aperture with a thick sulcated
    rim.

Turbo costatus. *Adams in Lin. Trans.* iii. p. 65. t. 13.
    f. 13 and 14. *Montagu Test.* p. 311. t. 10. f. 6. *Maton
    and Racket, in Lin. Trans.* viii. p. 174. *Dorset Cat.*
    p. 51. t. 19. f. 5.
Turbo crassus. *Adams's Micros.* p. 638. t. 14. f. 20.
Walker's Minute Shells, f. 47.
Inhabits the coasts of England and Wales. *Montagu, &c.*
Shell scarcely one-eighth of an inch long, and one-third as
    broad, solid, glossy, sub-pellucid, white, and rather obtuse
    at the apex; it is composed of four or five rounded whirls,
    with strong oblique undulated ribs, and finely striated trans-
    versely; aperture nearly orbicular, oblique, and bordered
    with a thick sulcated rim.

UNICUS.   106. Shell acuminated, minute, with nine
    very convex whirls, and cancellated striæ.

Turbo unicus. *Montagu Test.* p. 299. t. 12. f. 2. *Maton
    and Racket, in Lin. Trans.* viii. p. 174.
Turbo albidus. *Adams's Micros.* p. 637. t. 14. f. 17.
Walker's Minute Shells, f. 40.
Inhabits the sea at Sandwich. *Mr. Boys.*
Shell one-fifth of an inch long, very narrow in proportion to
    the length, and tapering to a fine point; it has nine glossy
    pellucid white very convex whirls, with slender longitudinal
    ridges, and crossed by extremely fine striæ; aperture sub-
    orbicular.

INDISTINCTUS.   107. Shell sub-cylindrical, minute,
    with longitudinal ribs and punctured trans-
    verse striæ in their interstices.

Turbo indistinctus. *Montagu Supp.* p. 129.
Inhabits ——
Shell one-tenth of an inch long, and one-third as broad.   Mr.

Montagu describes this species to be sub-cylindrical, glossy, white, with five or six nearly flat but well-defined whirls, which are finely striated longitudinally, and slightly and indistinctly punctured in the furrows, which gives the shell somewhat of a cancellated appearance when examined with a glass, but the transverse striæ are confined to the hollows between the ribs, and do not cross them; apex obtuse; aperture sub-ovate, with the inner lip smooth and a little spread on the pillar; it has the habit of *Voluta interstincta*, but differs in having a cancellated appearance, and in being destitute of a tooth on the pillar.

UVA. **108.** Shell cylindrical, obtuse, with straight longitudinal ribs, and about nine whirls; aperture semi-ovate and one-toothed.

Turbo Uva. *Linnæus Syst. Nat.* p. 1238. *Born Mus.* p. 354. and Vign. at p. 340. f. e. *Schroeter Inn. Bau Conch.* p. 54. t. 2. f. 7. *Gmelin*, p. 3604.
Turbo Fusus. *Gmelin*, p. 3610.
Helix Fusus. *Muller Verm.* ii. p. 108.
Bulimus Uva. *Bruguiere Enc. Meth.* p. 349.
Pupa Uva. *Lamarck Syst. des Anim.* p. 88.
Bonanni Rec. and *Kirch.* 3. f. 140. *Petiver Gaz.* t. 27. f. 2. *Gualter*, t. 58. f. D. *Seba*, iii. t. 55. f. 21, the seventh in the upper right-hand corner. *Favanne*, t. 65. f. B 11.
Inhabits the sea on the coasts of Bretagny. *Bruguiere.*
Shell hardly an inch long, and one-third as broad in the middle, and becoming somewhat narrower towards both extremities; it has about nine or ten whirls, which, excepting the two at the summit, are longitudinally ribbed, and the ribs rather thicker and more upright than in *T. Mumia*; the colour is greyish white, or sometimes tinged with red, and the inside white.

MUMIA. **109.** Shell sub-cylindrical, and obliquely ribbed; aperture semi-ovate, two-toothed, and the inside tawny.

Turbo Uva. *Martini*, iv. p. 281. t. 153. f. 1439. *Brookes's Introd.* p. 126. t. 8. f. 101.
Bulimus Mumia. *Bruguiere Enc. Meth.* p. 348.
Lister Conch. t. 588. f. 48.
Junior? Smaller, and the aperture toothless.
Turbo Fusulus. *Gmelin*, p. 3610.

Helix Fusulus. *Muller Verm.* ii. p. 109.

Inhabits the coasts of America. *Bruguiere.*

Shell nearly an inch and a half long, and one-third as broad, with twelve obliquely ribbed whirls; it has been generally confounded with *T. Uva*, from which it differs in being larger, and more acuminated towards the summit, 'in having the whirls more numerous, the ribs more oblique, and in the aperture, which has a second tooth placed nearly at the bottom of the pillar-lip.

ALVEARIA. 110. **Shell cylindrical, obtuse at both ends, and obliquely striated; aperture semi-ovate, one-toothed, and the inside white.**

Bulimus Fusus. *Bruguiere Enc. Meth.* p. 348.

Lister Conch. t. 588. f. 49. *Seba,* iii. t. 55. f. 21, the furthest shell on the right hand.

Inhabits St. Domingo and Guadaloupe, and is a land shell. *Bruguiere.*

Shell about an inch or an inch and a quarter long, and not much more than one-fourth as broad, with seven or eight obliquely striated whirls; the colour is white, both inside and out.

CYLINDRUS. 111. **Shell sub-umbilicated, cylindrical, obtuse, and longitudinally striated, with contiguous equal whirls.**

Turbo Cylindrus. *Chemnitz,* xi. p. 279. t. 209. f. 2061 and 2062.

Lister Conch. t. 21. f. 17. *Browne's Jamaica,* t. 40. f. 8.

Inhabits Jamaica, and is a land shell. *Lister.*

This shell appears by the figures to be about an inch long, and rather more than one-fourth as broad, and seems to be very nearly allied to *T. Alvearia,* but Chemnitz has not noticed any tooth in the aperture; it is moreover said to have eleven whirls, and to be whitish with a tinge of purple.

CROCEUS. 112. **Shell cylindrical, glabrous, obtuse, of an orange colour, and the lip marginated.**

Helix crocea. *Gmelin,* p. 3655.

Helix cylindracea glabra. *Chemnitz,* ix. part 2, p. 166. t. 135. f. 1233.

Inhabits ———

Chemnitz has given only a very short description of this species, which appears by his figure to be about sixteen times long, and one-third as broad, of an orange colour, with a white border at the suture and round the aperture.

SULCATUS. 113. Shell ovate, obtuse, umbilicated, obliquely striated, and white; aperture semi-ovate, and the outer lip reflected.

Turbo sulcatus. *Gmelin,* p. 3610.
Helix sulcata. *Muller Verm.* ii. p. 108. *Chemnitz,* ix. part 2. p. 165. t. 135. f. 1231 and 1232.
Bulimus sulcatus. *Bruguiere Enc. Meth.* p. 300.
Lister Conch. t. 588. f. 47.
Inhabits Ceylon, and is a land shell. *Bruguiere.*
Shell about an inch long, and rather more than half as broad, with eight whirls, of which the three uppermost form an obtuse summit; the shell which Lister has figured is described with two teeth at the aperture, but in other respects it exactly resembles the present species.

CORNEUS. 114. Shell umbilicated, rounded, and rather acute; whirls convex, with decussated striæ, and the aperture reflected.

Turbo corneus. *Linnæus Syst. Nat.* p. 1238. *Gmelin,* p. 3605.
Inhabits ———
Linnæus has only given a reference to his description of the Museum of the Queen of Sweden, and the size is no where mentioned; it is said to be of a brown or horn-colour, with the inside white, and the umbilicus pervious.

REFLEXUS. Shell umbilicated, convex, and rather prominent; whirls rounded, and slightly striated, and the aperture reflected.

Turbo reflexus. *Linnæus Syst. Nat.* p. 1238. *Gmelin,* p. 3605.
Inhabits ———
No reference, nor any addition to the above short character, has been made in the Systema Naturæ; and unless some specimen which Linnæus has named should happen to exist, it must remain very doubtful whether or not *T. elegans* belongs to this species.

ELEGANS. 116. Shell ovate, with five ventricose whirls, and strong transverse crossed by fine longitudinal striæ.

Turbo elegans. *Gmelin*, p. 3606. *Montagu Test.* p. 342. t. 22, f. 7. *Maton and Racket, in Lin. Trans.* viii. p. 167. *Dorset Cat.* p. 50. t. 21. f. 9.

Turbo tumidus. *Pennant Zool.* iv. p. 128. t. 82. f. 110.

Turbo striatus. *Da Costa Brit. Conch.* p. 86. t. 5. f. 9. *Donovan*, ii. t. 59.

Turbo Lincina, Var. *Chemnitz*, ix. part 2. p. 55. t. 123. f. 1060, *d* and *e*.

Turbo, No. 61. *Schroeter Einl.* ii. p. 85.

Nerita elegans. *Muller Verm.* ii. p. 177. *Schroeter Fluss.* p. 366. t. 9. f. 15.

L'Elegante striée. *Geoffroy*, p. 108. No. 1. t. 3.

Lister Anim. Ang. t. 2. f. 5, and *Conch.* t. 27. f. 25. *Gualter*, t. 4. f. A. *Ginanni Op. Post.* t. 3. f. 25. *Martini Berl. Mag.* ii. t. 1. f. 4 and 6.

Inhabits Italy, England, and France; and is a land shell. *Muller*.

Shell about five-eighths of an inch long, and three-eighths broad, of a pale purplish grey colour, and the summit is generally purple; it differs from *T. Lincina* principally in the want of a marginated aperture. It is one of the few land shells which have a testaceous operculum, and this probably misled Schroeter to consider it a fresh-water species.

LINCINA. 117. Shell ovate, obtuse, with five convex whirls, and decussated striæ; aperture marginated with a broad crenated border.

Turbo Lincina. *Linnæus Syst. Nat.* p. 1239. *Chemnitz*, ix. part ii. p. 54. t. 123. f. 1060, *b* and *c*. *Schroeter Einl.* ii. p. 43. *Gmelin*, p. 3605. *Schreibers Conch.* ii. p. 286.

Nerita Licinia. *Muller Verm.* ii. p. 178. *Schroeter Fluss.* p. 365.

Lister Conch. t. 26. f. 24. *Petiver Gaz.* t. 118. f. 11. *Klein Ost.* t. 3. f. 71. *Sloane's Jamaica*, t. 240. f. 12 and 13. *Martini Berl. Mag.* iii. t. 5. f. 54.

Variety. Smoother, and transversely striped.

Chemnitz, ix. part 2. t. 123. f. 1060, *a*.

Inhabits Jamaica, and is a land shell. *Lister, &c.*

Shell about eight or nine lines long, and two-thirds as broad, of a pale flesh-colour; the whirls in some shells are rough with decussated striæ, and in others nearly smooth; the aperture has a broad reflected border with a crenated edge, and by this border only it is attached to the whirl; the umbilicus is rather large, deep, and striated.

LABEO. 118. Shell oblong, umbilicated, brown, with elevated dotted decussated striæ, and a white dilated outer lip.

Turbo Labeo. *Gmelin*, p. 3605.
Turbo Lincina. *Born Mus.* p. 355. t. 13. f. 5 and 6.
Turbo Lincina magna. *Chemnitz*, ix. part 2. p. 56. t. 123. f. 1061 and 1062.
Turbo dubius. *Gmelin*, p. 3606.
Turbo, No. 60. *Schroeter Einl.* ii. p. 85.
Nerita Labeo. *Muller Verm.* ii. p. 180. *Schroeter Fluss.* p. 364.
Lister Conch. t. 25. f. 23. *Browne's Jamaica*, t. 40. f. 5 to 7.
Inhabits Jamaica, and is a land shell. *Browne.*
Shell about sixteen lines long and twelve broad, of a brownish colour, or sometimes white spotted with brown, and the marginated outer lip white; it appears by his references that Linnæus considered this to be a Variety of *T. Lincina,* from which it principally differs in being much larger, and in having the aperture more dilated.

LUNULATUS. 119. Shell cylindrical, with reticulated striæ, and the aperture detached.

Turbo lunulatus. *Gmelin*, p. 3605.
Nerita lunulata. *Muller Verm.* ii. p. 180. *Schroeter Fluss.* p. 363.
Inhabits ——
Shell seven lines long and four and a half broad, white, with four or five transverse rows of reddish moon-shaped spots; from Muller's description it appears to be most allied to *T. elegans,* but the body-whirl is extended so that the aperture is detached from the spire, and the reticulated striæ are less prominent.

LIGATUS. 120. Shell sub-umbilicated, nearly globular, acuminated, and the aperture thickish in the margin.

Turba ligatus. *Chemnitz*, ix. part 2. p. 60. t. 123. f. 1071 to 1074.
Nerita ligata. *Muller Verm.* ii. p. 181.
Inhabits ——
Shell varying from six to eleven lines in length, and the breadth

is very nearly equal; it has five or six convex minutely striated whirls, and is of a whitish or yellowish colour, with
reddish transverse bands. Muller has mentioned the following Varieties: A, with two nearly equal bands. B, with
four bands, of which the lowest is remote. C, with five
bands, of which the lowest is remote and broader. Gmelin
appears to have accidentally omitted this species.

FOLIACEUS. 121. Shell trochiform, with foliaceous
wrinkles, variegated with white and rose-
colour, and the umbilicus large.

Turbo foliaceus. *Chemnitz*, ix. part 2. p. 59. t. 123. f. 1069
and 1070. *Gmelin*, p. 3602.
Inhabits ———
Chemnitz has not mentioned the size, but by his figure this
shell appears to be about sixteen lines long and fourteen
broad, and is strongly wrinkled longitudinally, especially on
the body whirl; it is said to be white variegated with rose-
colour, and has much the habit of a Helix, but the aperture
is round.

MARGINATUS. 122. Shell umbilicated, sub-ovate,
wrinkled, and the whirls transversely marginated and striated.

Turbo marginatus. *Chemnitz*, ix. part 2. p. 60. t. 123.
f. 1075.
Turbo limbatus. *Gmelin*, p. 3606.
Inhabits Coromandel. *Chemnitz*.
This shell appears by the figures to be about a quarter of an
inch long, and equally broad, and is of a whitish colour;
the body-whirl has rather a broad rib round the shoulder,
and the aperture is somewhat marginated.

CARINATUS. 123. Shell rounded, sub-pyramidal,
with three keels on each whirl, and a large
spiral umbilicus.

Turbo carinatus. *Born Mus.* p. 353. t. 13. f. 3 and 4.
Gmelin, p. 3601.
Turbo, No. 118. *Schroeter Einl.* ii. p. 103.
Helix tricarinata. *Chemnitz*, ix. part 2. p. 85. t. 126.
f. 1103 and 1104.? *Gmelin*, p. 3621.?
Le Cabestan. *Favanne Cat. Rais.* p. 7.
Lister Conch. t. 28. f. 26.
Inhabits Jamaica, and is a land shell. *Chemnitz*.

Born has described this shell to be fourteen lines long, and fifteen broad, thin, sub-pellucid, and whitish, with five rounded whirls, and three or six lamellated transverse keels, of which the upper are thinner; umbilicus broad and spiral. Chemnitz has quoted Born's species for his *Helix tricarinata*, but his figure is more produced, and without any appearance of an umbilicus; in which respect it differs from his own description. Born's figure has more of the form of *Helix Gualteriana*.

SEPARATISTA. 124. Shell rather smooth, with three keels, and the lower end of the body-whirl detached; aperture triangular.

Turbo Separatista. *Chemnitz*, x. p. 298. t. 165. f. 1589 and 1590.
Turbo helicoides. *Gmelin*, p. 3598.
Inhabits the Indian Seas. *Chemnitz.*
This shell appears, by the figures, to be about seven lines long and nine broad, with the spire depressed, and the lower extremity of the body-whirl and the aperture quite detached; it is of a yellowish horn-colour. Gmelin has another Turbo with the name of *helicoides*.

NIVEUS. 125. Shell depressed, pellucid, white, with three transversely striated and slightly attached whirls; umbilicus large.

Turbo niveus. *Chemnitz*, x. p. 298. t. 165. f. 1587 and 1588. *Gmelin*, p. 3598.
Helix circinata. *Gmelin*, p. 3652.
Kæmmerer Cab. Rudolst. t. 8. f. 4 and 5.
Inhabits the Nicobar Islands. *Chemnitz.*
Shell about six lines long, and seven and a half lines broad, white, and has the whirls only slightly attached to each other, somewhat in the same manner as those of *Turbo scalaris*; the whirls are said to be sometimes distorted, and the margin of the aperture acute.

HELICOIDES. 126. Shell depressed, white with longitudinal brown zic-zac stripes, and the whirls rounded; umbilicus deep and wide.

Turbo helicoides. *Gmelin*, p. 3602.
Turbo helicinus. *Chemnitz*, ix. part 2. p. 59. t. 123. f. 1067 and 1068.
Inhabits ――――

By Chemnitz's figures this shell appears to be about eight lines in diameter, and to belong to the same family as *Helix Oculus Capri*; the upper surface is marked with longitudinal zic-zac stripes, and the base and umbilicus are entirely white.

***** *Turreted.*

IMBRICATUS. **127. Shell turreted, with the whirls imbricated downwards.**

Turbo imbricatus. *Linnæus Syst. Nat.* p. 1239. *Martini,* iv. p. 259. t. 152. f. 1422. *Born Mus.* p. 356. *Schroeter Einl.* ii. p. 45. t. 3. f. 21. *Gmelin,* p. 3606. *Schreibers Conch.* ii. p. 286.
  Gualter, t. 58. f. E. *Seba,* iii. t. 56. f. 26, 31, 33 and 34. *Knorr,* vi. t. 25. f. 2.
Inhabits Jamaica. *Linnæus.*
Shell three or four inches long, and not quite one-fourth as broad, of a pale brownish white colour, clouded and dotted with reddish brown; it has twelve or thirteen transversely striated flattish whirls, and some of the larger striæ are slightly granulated; the whirls are slightly gibbous at their lower, and contracted at their upper extremities, which gives them some appearance of being imbricated downwards.

REPLICATUS. **128. Shell turreted, smooth, with the whirls imbricated upwards.**

Turbo replicatus. *Linnæus Syst. Nat.* p. 1239. *Martini,* iv. p. 248. t. 151. f. 1412. *Schroeter Einl.* ii. p. 46. *Gmelin,* p. 3606. *Schreibers Conch.* ii. p. 287.
  Bonanni Rec. and *Kirch.* 3. f. 24. *Petiver Amb.* t. 127. f. 6. *Argenville,* t. 11. f. E.
Inhabits the coasts of Tranquebar. *Martini.*
Linnæus has so described this shell, but the whirls in Argenville's figure, which he has quoted, have not, any more than Martini's, the least appearance of being imbricated upwards; and it appears to me to be altogether rather a doubtful species: in the Mus. Lud. Ulrica it is described, "Testa cornea pallide umbrosa, lævis non glabra. Anfractus sursum imbricati, margine angusto, apertura ovata integra." Martini's figure is nearly two inches and three-quarters long, and one third as broad, of a pale brownish flesh-colour, with ten or twelve whirls, which have the appearance of being keeled in the middle, and slightly striated transversely.

ACUTANGULUS. 129. Shell turreted, with one acute transverse keel in the middle of each whirl.

Turbo acutangulus. *Linnæus Syst. Nat.* p. 1239. *Martini,* iv. p. 250. t. 151. f. 1413. *Born Mus.* p. 356. *Schroeter Einl.* ii. p. 47. *Gmelin,* p. 3607. *Schreibers Conch.* i. p. 287.

Lister Conch. t. 591. f. 59.? *Gualter,* t. 58. f. B. *Knorr,* iii. t. 19. f. 5.

Inhabits the coasts of Tranquebar. *Martini.*

Shell three or four inches long, and rather more than one fourth as broad, of a pale horn or greyish colour, and somewhat transparent; it has about fifteen transversely striated whirls, with a sharp keel in the middle, and a smaller one at the base of each. Lister's figure has been quoted by Martini, Born, Schroeter, and Gmelin, for this species, and is marked by Lister to have been found on the coasts of England.

DUPLICATUS. 130. Shell turreted, with two prominent acute ribs in the middle of each whirl.

Turbo duplicatus. *Linnæus Syst. Nat.* p. 1239. *Pennant Zool.* iv. p. 129. t. 81. f. 112. *Martini,* iv. p. 251. t. 151. f. 1414. *Born Mus.* p. 356. *Schroeter Einl.* ii. p. 48. *Gmelin,* p. 3607. *Schreibers Conch.* i. p. 287. *Donovan,* iv. t. 112. *Maton and Racket, in Lin. Trans.* viii. p. 175.

Strombiformis bicarinatus. *Da Costa Brit. Conch.* p. 110. t. 6. f. 3.

Bonanni Rec. and *Kirch.* 3. f. 114. *Lister Anim. Ang.* t. 3. f. 7, and *Conch.* t. 591. f. 58. *Petiver Gaz.* t. 102. f. 20. *Gualter,* t. 58. f. C. *Seba,* iii. t. 56. f. 7 and 8.

Inhabits the European Seas. *Linnæus.* Coasts of Persia. *Bonanni.* England. *Lister, &c.* Mediterranean. *Seba.*

Shell four or five inches long, and one-fourth as broad, with about fifteen whirls, and of a whitish or greyish horn-colour; each whirl has several transverse striæ, of which two are much larger and more elevated than the others. Lister says this shell has been found on the coasts of Scarborough, and Dr. Leach has lately taken it by dredging on those of the West of England.

TORCULARIS. 131. Shell turreted, with the whirls transversely striated, and an obtuse rib near both the extremities of each.

Turbo torcularis. *Born Mus.* p. 358. t. 13. f. 8.
Turbo duplicatus, Var. β.   *Gmelin*, p. 3607.
Turbo, No. 168.   *Schroeter Einl.* ii. p. 118.
Inhabits ———
Born has described this shell to be two inches and four lines
  long, and only four lines broad, but in his figure the breadth
  is about one-fourth of the length ; it is said to have twelve
  transversely striated whirls, with two distant obtuse ribs on
  each, and to be white with longitudinal undulated brown
  spots.

OBSOLETUS.   132. Shell turreted, with the whirls
    longitudinally wrinkled, and an obtuse rib
    on each side of the suture.

Turbo obsoletus. *Gmelin*, p. 3612.
Turbo, No. 167.   *Schroeter Einl.* ii. p. 118.
Turbo exoletus.   *Born Mus.* p. 357. t. 13. f. 7.
Bonanni Rec. 3. f. 113.   *Lister Conch.* t. 589. f. 53. *Ar-
    genville,* t. 11. f. C.
Inhabits ———
Born has described this shell to be two inches and two lines
  long, and seven lines broad, and the colour whitish ; the
  whirls are not rounded as in *T. exoletus,* but the prominent
  contiguous ribs at their extremities give them a concave ap-
  pearance.

EXOLETUS.   133. Shell turreted, with two obtuse
    ribs in the middle of each whirl, and the
    whirls variegated in longitudinal streaks.

Turbo exoletus.   *Linnæus Syst. Nat.* p. 1239. *Maton
    and Racket, in Lin. Trans.* viii. p. 176.
Turbo marmoreus.   *Martini,* iv. p. 260. t. 152. f. 1423.
Turbo variegatus.   *Schroeter Einl.* ii. p. 52. *Gmelin,* p.
    3607.
Turbo cinctus.   *Donovan,* i. t. 22. f. 1.   *Montagu Test.*
    p. 295.
Strombiformis cinctus.   *Da Costa Brit. Conch.* p. 114. t.
    7. f. 8.
Lister Conch. t. 592. f. 60.   *Seba,* iii. t. 56. f. 30, 37 and
    38.   *Knorr,* t. 16. f. 8.
Inhabits Southern Europe. *Linnæus.* Coasts of England.
  Da Costa, &c.
Shell about two inches and a half long, and one-fourth as
  broad, of a whitish or pale purplish colour, variegated with

chestnut in longitudinal somewhat undulated streaks; it has twelve or fourteen transversely striated whirls, with two obtuse slightly elevated ribs near the middle of each, and the intervening space is slightly channelled.

TEREBRA. 134. Shell turreted, with about fifteen whirls, and six elevated sharp transverse striæ on each.

Turbo Terebra. *Linnæus Syst. Nat.* p. 1239. *Pennant Zool.* iv. p. 130. t. 81. f. 113. *Martini,* iv. p. 254. t. 151. f. 1418 and 1419. *Schroeter Einl.* ii. p. 50. *Gmelin,* p. 3608. *Donovan,* i. t. 22. f. 2. *Montagu Test.* p. 293. *Maton and Racket, in Lin. Trans.* viii. p. 176. *Dorset Cat.* p. 51. t. 15. f. 5 and 6.
Turbo ungulinus. *Pulteney Dorset Cat.* p. 45.
Strombiformis Terebra. *Da Costa Brit. Conch.* p. 112. t. 7. f. 5 and 6.
Turritella Terebra. *Lamarck Syst. des Anim.* p. 89.
Le Ligar. *Adanson Senegal,* p. 158. t. 10. f. 6.?
Bonanni Rec. and *Kirch.* 3. f. 115. *Lister Anim. Ang.* t. 3. f. 8, and *Conch.* t. 591. f. 57. *Gualter,* t. 58. f. A. *Argenville,* t. 11. f. D. *Seba,* iii. t. 56. f. 32 and 40. *Knorr,* i. t. 8. f. 6. *Favanne,* t. 39. f. E.
Inhabits the European Seas, and coasts of Sweden. *Linnæus.* England. *Lister, &c.*
Shell an inch and a half, or two inches long, and about one-fourth as broad, of a pale or reddish brown colour; it has from twelve to fifteen slightly convex whirls, and each has about six transverse elevated striæ; the number of the striæ in the Systema Naturæ is stated to be six, but Linnæus has elsewhere described it differently; in the Fauna Suecica it is said to be only five, whereas in the History of the Museum of the Queen of Sweden, it is stated to be seven; the outer lip is thin and brittle. Martini's figures from 1415 to 1417 probably belong to *T. Archimedis.*

ARCHIMEDIS. 135. Shell turreted, with about twenty-four rounded whirls, and ten sharp ribs on each.

Turbo Terebra. *Born Mus.* p. 358. *Chemnitz,* x. p. 299. t. 165. f. 1591.
Turbo Terebra, Var. γ. *Gmelin,* p. 3608.
La Vis d'Archimede. *Favanne Cat. Rais.* p. 277.

Regenfuss, ii. t. 12. f. 57.   *Martyn Univ. Conch.* i. Title page.

Inhabits the coasts of China. *Chemnitz.*

Shell five or six inches long, and one-fourth as broad, of a greyish or brownish white or pale chestnut-colour; it has from twenty-four to thirty, or according to Born, thirty-six much rounded whirls, with about ten sharp ribs on each, and fine intermediate striæ; it was probably confounded by Linnæus with *T. Terebra,* from which it differs in having the whirls much more numerous and more rounded, and the ribs more elevated and crowded, as also in its larger size.

VARIEGATUS.   136.   Shell turreted, with flattish whirls, and seven obsolete striæ on each.

Turbo variegatus.   *Linnæus Syst. Nat.* p. 1240.   *Gmelin,* p. 3608.

Inhabits ———

Linnæus, without any description besides that which is contained in the above short character, has quoted Bonanni Rec. 3. f. 122, and Seba, iii. t. 56. f. 26, 34, 31 and 33, which have much the appearance of *T. imbricatus.* Martini doubted whether his *T. marmoreus* might not be this species, but that shell agrees better with the Linnæan definition of *T. exoletus.*

UNGULINUS.   137.   Shell turreted, with ten obsolete striæ on each whirl.

Turbo ungulinus.   *Linnæus Syst. Nat.* p. 1240.   *Muller Zool. Dan. Prod.* No. 2930. ? *Schroeter Einl.* ii. p. 53. *Gmelin,* p. 3608.

Inhabits the European Ocean. *Linnæus.*

Linnæus has not given any reference, or made any addition to the above short description, so that this must probably remain a doubtful species. Schroeter and Gmelin under this name have described a shell, which they say varies from two and a half to four inches in length, with about twenty-four whirls, white variegated with yellowish brown or chestnut, and intermediate between *T. Terebra* and *T. variegatus.* Dr. Pulteney considered the former of these shells to be the Linnæan *T. ungulinus,* and Montagu has conjectured that they may be only Varieties of the same species.

TEREBELLUM.   138.   Shell turreted, umbilicated, glabrous, and yellowish, with a keel on each whirl.

Turbo Terebellum. *Chemnitz,* x. p. 302. t. 165. f. 1592 and 1593.
Turbo Terebra, Var. *d.* *Gmelin,* p. 3608.
Inhabits the coasts of the Nicobar Islands. *Chemnitz.*
Shell about ten lines long and three broad, with fourteen glabrous whirls, and is of a yellowish colour, with a dark band at the suture.

ANNULATUS. 139. Shell turreted, with a prominent marginated suture. -

Turbo annulatus. *Linnæus Syst. Nat.* p. 1240. *Schroeter Einl.* ii. p. 55. *Gmelin,* p. 3609.
Gualter, t. 58. f. L.
Inhabits ——
Linnæus, without having made any addition to the above character, has referred to Gualter's figure, which appears to be rather more than a quarter of an inch long, and two-fifths as broad. Gualter has called it the entire small white Turbo, with an acuminated spire environed by a belt.

TURRIS THOMÆ. 140. Shell turreted, with about twelve reversed whirls, and two rows of nodules on each.

Turbo Turris Thomæ. *Chemnitz,* xi. p. 310. t. 213. f. 3022, *a* to *d.*
Inhabits the coasts of the Island of St. Thomas, in the West Indies. *Chemnitz.*
Shell about three lines and a half long, and one line broad, and has twelve white reversed whirls, with a double row of nodules and a transverse reddish stripe on each.

BIDENS. 141. Shell turreted, pellucid, with the whirls reversed, and the suture slightly crenated ; pillar-lip two-toothed.

Turbo bidens. *Linnæus Syst. Nat.* p. 1240. *Born Mus.* p. 359. *Schroeter Einl.* ii. p. 55. t. 3. f. 22. *Maton and Racket, in Lin. Trans.* viii. p. 178. t. 5. f. 3.
Turbo papillaris. *Chemnitz,* ix. part 1. p. 121. t. 112. f. 963 and 964.
Turbo bidens, Var. *β.* *Gmelin,* p. 3609.
Helix papillaris. *Muller Verm.* ii. p. 120.
Bulimus papillaris. *Bruguiere Enc. Meth.* p. 353.
Bonanni Rec. and *Kirch.* 3. f. 41, magnified. *Gualter,* t.

4. f. D and E.   *Ginanni Op. Posth.* ii. t. 3. f. 23.   *Fa-vanne,* t. 65. f. E 9.

Inhabits Southern Europe, and is a land shell. *Linnæus.* Italy, Dauphiny, and Languedoc. *Bruguiere.*

Shell about half, or sometimes nearly three-quarters of an inch long, and the breadth is nearly one-fifth of the length; it is smooth, of a brownish colour, with ten or eleven whirls, and may be readily distinguished from its congeners by the delicately crenulated suture.   Gmelin has confounded *T. laminatus* and *T. biplicatus* with this species, and several of Born's references are erroneous, though his description is good.   *T. bidens* of Montagu is *T. nigricans.*

LAMINATUS.   142. Shell turreted, pellucid, and the whirls reversed ; pillar-lip not detached, and furnished with two large white teeth.

Turbo laminatus. *Pulteney's Dorset Cat.* p. 46. *Montagu Test.* p. 359. t. 11. f. 4.   *Maton and Racket, in Lin. Trans.* viii. p. 179.   *Dorset Cat.* p. 51. t. 19. f. 9.

Turbo bidens.   *Chemnitz,* ix. part i. p. 119. t. 112. f. 960. No. 1.

Helix bidens.   *Muller Verm.* ii. p. 116.

Bulimus bidens.   *Bruguiere Enc. Meth.* p. 352.

Lister Conch. t. 41. f. 39 A. *Gualter,* t. 4. f. C. *Schroeter Erdconch.* t. 1. f. 4 a.   *Favanne,* t. 65. f. E 11.

Inhabits woods in England, and other parts of Europe.

Shell about three-quarters of an inch long, and one sixth as broad ; it is smooth, of a reddish horn-colour, and has ten obsoletely wrinkled whirls ; the aperture is roundish, but contracted at the upper outward margin, and sometimes has three or four prominent ridges in the throat; the pillar-lip is not detached from the body-whirl, and is white, with two large laminated teeth.

BIPLICATUS.   143. Shell turreted, opake, longitudinally striated, and the whirls reversed ; pillar-lip slightly detached, and furnished with two approximated teeth.

Turbo biplicatus.   *Montagu Test.* p. 361. t. 11. f. 5. *Maton and Racket, in Lin. Trans.* viii. p. 179.

Turbo bidens, Var.   *Chemnitz,* ix. part 1. p. 120. t. 112. f. 960, No. 2.

Schroeter Erdconch. t. 1. f. 4 b.

Inhabits woods in England and other parts of Europe.

This shell is of the same size with *T. laminatus,* which it resembles, but is regularly striated longitudinally, and has the pillar-lip prominent and detached.

CORRUGATUS. 144. Shell turreted, with the whirls reversed, and the base plaited and wrinkled; aperture with two teeth.

Turbo corrugatus. *Chemnitz,* ix. part 1. p. 120. t. 112. f. 961 and 962.
Turbo bidens, Var. γ. *Gmelin,* p. 3609.
Bulimus corrugatus. *Bruguiere Enc. Meth.* p. 354.
Inhabits Spain, Languedoc, and Provence; and is a land shell. *Bruguiere.*
Shell an inch or an inch and a quarter long, and one sixth as broad, of an ash-colour, except at the summit which is brownish, and the inside is brown; it is opake, shining, thicker than either of its congeners, and marked with a few longitudinal striæ; the body-whirl is wrinkled, and plaited to its base at the back of the aperture.

NIGRICANS. 145. Shell turreted, opake, minutely striated, and the whirls reversed; aperture with two rather distant teeth.

Turbo nigricans. *Pulteney Dorset Cat.* p. 46. *Maton and Racket, in Lin. Trans.* viii. p. 180. *Dorset Cat.* p. 51. t. 19. f. 10.
Turbo perversus. *Pennant Zool.* iv. p. 130. t. 82. f. 116. *Donovan,* ii. t. 72.
Turbo perversus, Var. *Chemnitz,* ix. part 1. p. 116. t. 112. f. 959 b. ?
Turbo bidens. *Montagu Test.* p. 357. t. 11. f. 7.
Helix perversa. *Muller Verm.* ii. p. 118.
La Nompareille. *Geoffroy,* p. 63, No. 23. t. 2.
Lister Anim. Ang. t. 2. f. 10, and *Conch.* t. 41. smaller fig. 39.
Inhabits trunks of trees and moss in England and France.
Shell about half an inch long, and resembles *T. laminatus,* but is smaller, of a darker colour, and the teeth at the aperture are not so close together. Schroeter, Chemnitz, Gmelin, and Bruguiere appear to have confounded it with *T. perversus.*

LABIATUS. 146. Shell turreted, opake, striated, and the whirls reversed; aperture with two teeth, and a thick dilated white margin.

Turbo labiatus. *Montagu Test.* p. 362. t. 11. f. 6. *Maton and Racket, in Lin. Trans.* viii. p. 180. *Dorset Cat.* p. 51. t. 21. f. 15.

Turbo tridens. *Chemnitz,* ix. part 1. p. 115. t. 112. f. 957. ?

Turbo perversus, Var. ε. *Gmelin,* p. 3610. ?

Strombiformis perversus. *Da Costa Brit. Conch.* p. 107. t. 5. f. 15.

Bulimus perversus. *Bruguiere Enc. Meth.* p. 351. ?

Inhabits the neighbourhood of London; and is very rare. *Montagu.*

Shell about five-eighths of an inch long, and one-eighth broad, of a pale brown colour, and is distinguished by a broad thick marginated white aperture.

PERVERSUS. 147. Shell turreted, pellucid, with the whirls reversed, and the aperture toothless.

Turbo perversus. *Linnæus Syst. Nat.* p. 1240. *Chemnitz,* ix. part 1. p. 116. t. 112. f. 959 a. *Montagu Test.* p. 355. t. 11. f. 12. *Maton and Racket, in Lin. Trans.* viii. p. 181. t. 5. f. 2. *Dorset Cat.* p. 51. t. 19. f. 11. *Lister Anim. Ang.* t. 2. f. 11. *Schroeter Erdconch.* t. 1. f. 5. *Favanne,* t. 65. f. E 4.

Inhabits moss on the trunks of trees, in Wiltshire, Dorsetshire, Devonshire, and Cornwall; but is not common. *Maton and Racket.*

Shell about a quarter of an inch long, and one-fourth as broad, pellucid, yellowish, and obsoletely striated, with the summit rather obtuse, and the whirls separated by a deep suture; in old shells a round obsolete tooth is sometimes observable, but the aperture is generally toothless. Gmelin's description answers best to *T. laminatus,* but in his references, as well as in Schroeter's, those belonging to *T. nigricans* and to the present species are mixed indiscriminately. Muller, in the Zoologia Danica, t. 102, has figured a shell which is more like *T. labiatus,* but has not any teeth. Bruguiere has probably confounded this species, with *T. nigricans* and *T. labiatus,* under the name of *Bulimus perversus.*

QUINQUEDENTATUS. 148. Shell turreted, slightly striated longitudinally, and of a greyish white colour; aperture ovate with five teeth.

Turbo quinquedentatus. *Born Mus.* p. 378. t. 13. f. 9. *Gmelin,* p. 3612. *Olivi Adriatica,* p. 171.

Turbo, No. 169. *Schroeter Einl.* ii. p. 119.

Bulimus similis. *Bruguiere Enc. Meth.* p. 355.

L'Anti-nompareille. *Geoffroy*, p. 55. t. 2.

Gualter, t. 4. f. G. *Argenville*, t. 32. f. 16. *Favanne*, t. 65. f. E 12.

Inhabits walls, and among moss in woods about Paris. *Geoffroy*. South of France. *Bruguiere*. Italy. *Olivi*.

Shell about half an inch long, and one-fourth as broad, with nine or ten whirls, which are nearly smooth, or only slightly striated longitudinally ; the aperture has a whitish border, and is armed with three teeth at its summit, and two at the base. Bruguiere has not quoted Born for this species, but I think there can be no doubt that his *Bulimus similis* is the same.

TRIDENS. 149. Shell sub-cylindrical, smooth, pellucid, and the aperture three-toothed.

Turbo tridens. *Gmelin*, p. 3611. *Montagu Test.* p. 338. t. 11. f. 2. *Maton and Racket, in Lin. Trans.* viii. p. 181. *Dorset Cat.* p. 51. t. 19. f. 12.

Helix tridens. *Muller Verm.* ii. p. 106.

Bulimus tridens. *Bruguiere Enc. Meth.* p. 350.

Gualter, t. 4. f. F.

Inhabits Italy. *Muller*. In the wood of Boulogne near Paris. *Bruguiere*. On water plants by the river Stour. *Pulteney*. In Carline park near Leith. *Mr. Laskey*.

Shell about four lines long, and one-third as broad, pellucid, smooth, and of a pale horn-colour ; it has six or seven whirls, which are nearly flat, and scarcely distinguishable but by the suture ; the aperture is small and curved, with one tooth near the base, and two opposite each other in the upper part. Mr. Montagu says there were also two other smaller intermediate teeth in the specimens which he examined, and which might have been easily overlooked.

JUNIPERI. 150. Shell conical-turreted, brown, with the aperture sub-ovate, and seven-toothed.

Turbo Juniperi. *Montagu Test.* p. 340. t. 12. f. 12. *Maton and Racket, in Lin. Trans.* viii. p. 182. *Dorset Cat.* p. 51. t. 19. f. 11.*

Turbo Muscorum, Var. β. *Schroeter Einl.* ii. p. 59. *Gmelin*, p. 3611.

Turbo multidentatus. *Olivi Adriatica*, p. 171. t. 5. f. 2.

Bulimus avenaceus. *Bruguiere Enc. Meth.* p. 355.

Helix granum avenaceum referens. *Chemnitz*, ix. part 2. p. 167. t. 135. f. 1236.

Le Grain d'Avoine. *Geoffroy*, p. 53. t. 2.

Schroeter Erdeonch. t. 1. f. 6.

Inhabits the neighbourhood of Paris. *Geoffroy.* Saxony. *Schroeter.* Great Britain, frequently about the roots of Juniper bushes. *Montagu.*

Shell about a quarter of an inch long, and nearly half as broad, of a dull brown colour, with eight flattish whirls, which are well defined by the suture ; the aperture has a marginated reflected white border, with three teeth on the outer, and four on the pillar-lip. *Le Grain d'Avoine* is described to be only two lines long, but in other respects it answers, and I have no doubt of its belonging to the present species ; Chemnitz, on the other hand, has made his figure rather too large.

MUSCORUM. 151. Shell ovate, obtuse, pellucid, with six whirls ; aperture with a white margin, and nearly toothless.

Turbo Muscorum. *Linnæus Syst. Nat.* p. 1240. *Chemnitz,* ix. part 2. p. 61. t. 123. f. 1076. *Schroeter Einl.* ii. p. 58. *Gmelin,* p. 3611. *Montagu Test.* p. 335. t. 22. f. 3. *Donovan,* iii. t. 80. *Maton and Racket, in Lin. Trans.* viii. p. 182. *Dorset Cat.* p. 51. t. 21. f. 16.

Turbo cylindraceus. *Da Costa Brit. Conch.* p. 89. t. 5. f. 16.

Helix Muscorum. *Muller Verm.* ii. p. 105.

Bulimus Muscorum. *Bruguiere Enc. Meth.* p. 334.

Le petit Barillet. *Geoffroy,* p. 58. t. 2.

Lister Anim. Ang. t. 2. f. 6. *Petiver Gaz.* t. 35. f. 6. *Schroeter Erdconch.* t. 1. f. 7.

Variety. With eight whirls.

Bulimus Doliolum. *Bruguiere Enc. Meth.* p. 351.

Le grand Barillet. *Geoffroy,* p. 57. t. 2.

Inhabits most parts of Europe, under moss at the roots of trees, or in the crevices of the bark, and on old walls.

Shell rather more than one-eighth of an inch long, and about one-third as broad, of a yellowish or whitish brown, or horn-colour ; it has commonly six whirls, of which the four lowermost are nearly of the same size ; the aperture is generally toothless, but the pillar-lip is sometimes furnished with a single tooth. *Bulimus Doliolum* appears, from the descriptions of Geoffroy and Bruguiere, to differ only in having eight whirls, and in being nearly twice as large.

QUADRIDENS. 152. Shell sub-cylindrical, and the whirls reversed; aperture with four white teeth.

Turbo quadridens. *Gmelin*, p. 3610.
Turbo Uva, Var. γ. *Gmelin*, p. 3604.
Turbo Uva terrestris. *Chemnitz*, ix. part 1. p. 123. t. 112. f. 965.
Helix quadridens. *Muller Verm.* ii. p. 107.
Bulimus quadridens. *Bruguiere Enc. Meth.* p. 351.
L'Anti-barillet. *Geoffroy*, p. 65. t. 2.
Lister Conch. t. 40. f. 38.

Inhabits walls, old trees, moss, and under stones, about Paris. *Geoffroy.* Narbonne. *Lister.* Italy. *Muller.* Among moss on the rocks of La Valette near Montpellier. *Bruguiere.*

Shell from three and a half to five lines long, glabrous, and of a yellowish colour, with seven minutely striated whirls; the aperture is oval, contracted, marginated, and white; it has two teeth within the margin of the outer lip, one at the base of the pillar, and a fourth in the middle of the pillar-lip. Gmelin has strangely cited the same figure of Lister's for *T. perversus*, for a variety of *T. Uva*, and for the present species.

SEXDENTATUS. 153. Shell sub-cylindrical, smooth, with five rounded whirls, and the aperture six-toothed.

Turbo sexdentatus. *Montagu Test.* p. 337. t. 12. f. 8. *Maton and Racket, in Lin. Trans.* viii. p. 183. *Dorset Cat.* p. 52. t. 19. f. 12.*
Helix minuta. *Muller Verm.* ii. p. 101.? *Gmelin*, p. 3660.?
Bulimus minutus. *Bruguiere Enc. Meth.* p. 310.?

Inhabits marshy places in the West of England, sometimes on Iris pseudacorus. *Montagu.*

Shell about a line long, and half as broad, of a brownish or reddish horn-colour, with five rounded convex whirls; the aperture is somewhat angulated, with four teeth on the outer, and two on the pillar-lip. It might at first sight be mistaken for the young of *T. Muscorum*, but the more distorted and toothed aperture affords specific marks of distinction. As Mr. Montagu has justly remarked, if Muller had not expressly said 'Apertura ovali edentula,' we should have had no doubt of this being his *Helix minuta*; the teeth, however,

are not always readily discoverable till the aperture has been
well cleaned, and even that accurate naturalist may possibly
have overlooked this circumstance. Bruguiere says his *Bulimus minimus* differs in being a water species, and in having
the aperture marginated.

**VERTIGO. 154. Shell oval, with five reversed and
faintly striated whirls; aperture sub-triangular, slightly marginated, and toothed.**

Turbo Vertigo. *Montagu Test.* p. 363. t. 12. f. 6. *Maton
and Racket, in Lin. Trans.* viii. p. 183.
Helix Vertigo. *Gmelin,* p. 3664.
Vertigo pusilla. *Muller Verm.* ii. p. 124. *Schroeter Fluss.*
p. 349.
Inhabits rotten wood in Denmark. *Muller.* Walls covered
with ivy at Sandwich. *Montagu.*
Shell not much more than half a line long, opake, and of a
brownish colour; the aperture is irregular, and somewhat
triangular with rounded angles. Sometimes there are three
teeth, one on the outer and two on the inner lip, and there
are sometimes six with the rudiment of a seventh. This species is nearly allied to *T. sexdentatus,* but is more cylindrical, and more obtuse at the summit, and the whirls are reversed.

**CARYCHIUM. 155. Shell conical, polished, and pel-
lucid; aperture marginated, with two teeth
on the inner, and a knob on the outer lip.**

Turbo Carychium. *Montagu Test.* p. 339. t. 22. f. 2. *Ma-
ton and Racket, in Lin. Trans.* viii. p. 184. *Dorset
Cat.* p. 52. t. 19. f. 13.
Carychium minimum. *Muller Verm.* ii. p. 125. *Schroeter Fluss.* p. 324. *Leach Zool. Misc.* i. p. 84.
Helix Carychium. *Gmelin,* p. 3665.
Inhabits woods and mossy banks, frequent in Great Britain.
Montagu.
Shell hardly a line long, and only one-third as broad, white,
glossy, and pellucid; it has five convex whirls, which are
finely striated longitudinally, and well defined by the suture;
the pillar-lip is furnished with two teeth, and there is some-
times the rudiment of a third above the other two.

**AURISCALPIUM. 156. Shell turreted, white, and
very smooth; aperture with a flattish concave obtuse reflected lip.**

Turbo Auriscalpium. *Linnæus Syst. Nat.* p. 1240. *Schroeter Einl.* ii. p. 59. *Gmelin*, p. 3611.
Inhabits the Mediterranean. *Linnæus.*
Linnæus, with a mark of doubt, has erroneously quoted Argenville, t. 32. f. 19, which is a land Turbo, and has given the following description : " Shell subulate, milk-white, extremely glabrous, with seven or eight whirls, and the aperture dilated so as to resemble an ear-picker ; outer lip produced, obtuse, concave, and marginated." My friend Mr. Brookes has sent me a shell four lines long, and rather more than one-fourth as broad, which in other respects answers the description, but is minutely cancellated; it has an ovate aperture like that of *T. politus.*

POLITUS. 157. Shell imperforate, extremely glabrous, and the aperture ovate.

Turbo politus. *Linnæus Syst. Nat.* p. 1241. *Schroeter Einl.* ii. p. 60. *Gmelin*, p. 3612.
Turbo lævis. *Pennant Zool.* iv. p. 130. t. 79. upper fig.
Turbo albus. *Donovan*, v. t. 177.
Helix polita. *Pulteney Dorset Cat.* p. 49. *Montagu Test.* p. 398. *Maton and Racket, in Lin. Trans.* viii. p. 210. *Dorset Cat.* p. 51. t. 19. f. 15.
Strombiformis albus. *Da Costa Brit. Conch.* p. 116.
Inhabits the Mediterranean. *Linnæus.* Western coasts of England. *Borlase, &c.*
Shell sometimes five-eighths of an inch long, and two-tenths of an inch broad, but is most commonly smaller; it has ten or eleven highly polished flat whirls, separated by an obsolete suture, and gradually tapering to a fine point ; it is generally white, but Mr. Montagu says that young shells, when the animal is alive, frequently appear mottled with pink and pale green.

SUBULATUS. 158. Shell subulate, extremely glabrous, white, with yellowish transverse lines, and the aperture ovate.

Turbo subulatus. *Donovan*, v. t. 172.
Helix subulata. *Maton and Racket, in Lin. Trans.* viii. p. 210. *Montagu Supp.* p. 142. *Dorset Cat.* p. 55. t. 19. f. 14.
Strombiformis glaber. *Da Costa Brit. Conch.* p. 117.
Inhabits the coasts of Great Britain. *Da Costa, &c.*
Shell about three-quarters of an inch long, and one-fourth as

broad, thin, of a white or pale flesh-colour, with two or more transverse yellowish lines becoming obsolete towards the summit; it has ten whirls, which are not well defined by the sutures, and the aperture is narrow, and contracted at the interior angle.

DECUSSATUS. 159. Shell subulate, decussated with longitudinal and very minute transverse striæ; aperture sub-oval, and contracted at both ends.

Helix decussatus. *Montagu Test.* p. 399. t. 15. f. 7. *Maton and Racket, in Lin. Trans.* viii. p. 209. *Dorset Cat.* p. 55. t. 19. f. 17.
Inhabits the coast of Dorsetshire. *Montagu.*
Shell three-tenths of an inch long, and one-third as broad, and is white and slender; Mr. Montagu says that in shape it is similar to *T. politus,* " except in the aperture which stands rather more oblique, the outer lip more expanded in the middle, and contracted at the lower angle."

****** *Depressed.*

NAUTILEUS. 160. Shell flattish, with the whirls annulated and crested on the back.

Turbo Nautileus. *Linnæus Syst. Nat.* p. 1241. *Schroeter Fluss.* p. 238, and *Einl.* ii. p. 60. *Chemnitz,* ix. part 2. p. 63. t. 123. f. 1077. *Gmelin,* p. 3612. *Shaw Nat. Misc.* xxii. t. 964. *Maton and Racket, in Lin. Trans.* viii. p. 169. t. 5. f. 4. *Dorset Cat.* p. 50. t. 19. f. 16.
Nautilus crista. *Linnæus Syst. Nat.* (edit. 10.) p. 709.
Planorbis imbricatus. *Muller Verm.* ii. p. 165.
Helix Nautileus. *Montagu Test.* p. 464. t. 25. f. 5.
Helix carinata. *Adams's Microscope,* p. 635. t. 14. f. 10.
Helix spinosa. *Adams's Microscope,* p. 636. t. 14. f. 11, and t. 22. f. 39.
Le Planorbe tuillé. *Geoffroy,* p. 97. t. 3.
Walker's Minute Shells, f. 20 and 21.
Inhabits fresh water, on aquatic plants. Germany and Switzerland. *Linnæus.* France. *Geoffroy.* Denmark. *Muller.* England. *Walker, &c.*
Shell about half an inch in diameter, of a pale transparent horn-colour, and covered with a brown epidermis; it is furnished with distant annulations, or elevated transverse ribs, which

are more or less produced on the marginal keel into short spines. It has the general habit of the depressed Helices, but the aperture is round, or sometimes rather oval, and the latter variety was considered by Walker and Adams to be a distinct species.

CRISTATUS. 161. Shell flattish above, and umbilicated beneath, with three or four rounded whirls.

Turbo cristatus. *Maton and Racket, in Lin. Trans.* viii. p. 169.
Valvata cristata. *Muller Verm.* ii. p. 198. *Schroeter Fluss.* p. 240. t. 5. f. 26. a and b.
Nerita valvata. *Gmelin,* p. 3675.
Helix cristata. *Montagu Test.* p. 460. Vign. 1. f. 7 and 8.
Le Porte Plumet. *Geoffroy,* p. 115. t. 3.
Helix. *Walker's Minute Shells,* f. 18.
Inhabits marshy places in Denmark. *Muller.* Pools of water, and small streams about Paris. *Geoffroy.* In the Avon, and ditches near Wedhampton, Wilts. *Montagu.*
Shell about one-tenth of an inch in diameter, sub-pellucid, of a pale horn-colour, and composed of three, or sometimes of four whirls, which are slightly wrinkled transversely, and flattened at the summit; the aperture is perfectly orbicular, and is attached to, but not interrupted by, the body-whirl. Muller, on account of the singular structure of the animal, (which is well described in the Testacea Britannica by Mr. Montagu) has placed this species in a separate genus, to which he has given the name of Valvata.

DEPRESSUS. 162. Shell minute, depressed, with four slightly wrinkled whirls, and umbilicated beneath.

Turbo depressus. *Maton and Racket, in Lin. Trans.* viii. p. 170.
Helix depressa. *Montagu Test.* p. 439. t. 13. f. 5.
Inhabits the sea on the coasts of Cornwall and Devon. *Montagu.*
Shell scarcely one line in diameter, and of a pale brown colour; Mr. Montagu says it resembles *T. cristatus,* but is much smaller, stronger, and more opake, and has the apex rather more prominent.

SERPULOIDES. 163. Shell minute, depressed, with three smooth whirls, and umbilicated beneath.

Helix serpuloides. *Montagu Supp.* p. 147. t. 21. f. 3.

Inhabits the Devonshire coast, and is extremely rare. *Montagu.*

Montagu says this shell is about the size of *T. depressus,* but is white, and differs in the slight connexion of the whirls, and their more cylindrical appearance underneath.

Genus XXIX.

HELIX:

SHELL UNIVALVE, SPIRAL, SUBDIAPHANOUS, AND BRITTLE; APERTURE CONTRACTED, AND SOMEWHAT SEMILUNAR.

Subdivisions.†

* Whirls longitudinally angulated on both sides.
** With a carinated margin on the body-whirl.
*** Umbilicated, and the whirls rounded.
**** Imperforate, and the whirls rounded.
***** Turreted.
****** Ovate, and imperforate.

---

* Whirls longitudinally angulated on both sides.

SCARABÆUS. 1. Shell ovate, longitudinally angulated on both sides, and the aperture seven-toothed.

---

† Gmelin's *H. atra*, p. 3655, *H. coriacea*, p. 3641, *H. Folliculus*, p. 3654, *H. muralis*, p. 3664, *H. neritina*, p. 3638, *H. rufescens*, p. 3640, *H. spadicea*, p. 3616, *H. splendidula*, p. 3655, *H. strigosula*, p. 3634, *H. trigonostoma*, p. 3667, *H. tumida*, p. 3668, and *H. turgida*, p. 3641, appear to me to be undeserving of notice. *H. coccinea*, p. 3651, *H. Nucleus*, p. 3651, and *H. purpurea*, p. 3656, are omitted, because Martyn's figures are unaccompanied with descriptions; and so many of his plates are without any engraved numbers, that it is almost impossible to distinguish or quote them with sufficient certainty. *H. cinerea*, *H. cærulescens*, and *H. turgida*, p. 3367, are minute shells, and most probably only the larvæ of some other species.

Helix Scarabæus. *Linnæus Syst. Nat.* p. 1241. *Bornæ Mus.* p. 365, and Vign. at p. 364. f. a. *Schroeter Einl.* ii. p. 122. *Chemnitz,* ix. part 2. p. 179. *Gmelin,* p. 3613. *Schreibers Conch.* i. p. 290. *Burrow's Elements,* p. 169. t. 20. f. 1.

Helix Pythia. *Muller Verm.* ii. p. 88.

Bulimus Scarabæus. *Bruguiere Enc. Meth.* p. 340.

Scarabæus Imbrium. *De Montfort Conch.* ii. p. 306. *Leach Zool. Misc.* p. 96. t. 42.

Variety A. Shell larger, of a yellowish white colour dotted with brown.

Bonanni Rec. 3. f. 385, and *Kirch.* f. 370. *Lister Conch.* t. 577. f. 31. *Gualter,* t. 4. f. S. *Argenville,* t. 9. f. T. *Klein Ost.* t. 1. f. 23. *Knorr,* vi. t. 19. f. 2 and 3. *Favanne,* t. 65. f. D 1. *Chemnitz,* ix. t. 136. f. 1249 and 1250.

Variety B. Shell smaller, of an uniform reddish or purplish brown colour.

Bonanni Rec. and *Kirch.* 3. f. 44. *Lister Conch.* t. 577. f. 32. *Klein Ost.* t. 1. f. 24. *Favanne,* t. 65. f. D 2, and D 4. *Chemnitz,* ix. t. 136. f. 1251 to 1253.

Junior. Without any teeth on the outer lip.

Rumphius, t. 27. f. I. *Petiver Amb.* t. 12. f. 8.

Inhabits the mountains of Asia. *Linnæus.* Among rotten leaves and wood, on the mountains and coasts of Amboyna. *Rumphius.* China. *Humphreys.*

Shell commonly about an inch and a quarter long, and three quarters of an inch broad, but the Variety B is three or four lines shorter, and is broader in proportion to the length; it has two obsolete opposite longitudinal keels, which on each whirl have generally a brown spot on one side, and a white one on the other; the aperture is white and narrow, and has four strong teeth on the outer, and three on the pillar-lip.

AFRA. 2. Shell ovate, thick, obliquely striated, and the aperture five-toothed.

Helix Afra. *Gmelin,* p. 3651.

Bulimus pedipes. *Bruguiere Enc. Meth.* p. 340.

Le Pietin. *Adanson Senegal,* p. 11. t. 1. f. 4.

Inhabits salt-water pools among the rocks on the coasts of Goree. *Adanson.*

Shell about a quarter of an inch long, and three-fourths as broad, with six whirls, of which the body-whirl is ventricose, and the spire very small; on the body-whirl there are said to be twenty-one oblique striæ, which are reduced to eight on the next, and the number becomes gradually smaller on the other whirls; the aperture has two irregular teeth on each lip, and a fifth larger one at the upper angle. I am far from certain that it is entitled to a place in the present division, but it appears to have more natural affinity with *H. Scarabæus* than with any other species.

** *With a carinated margin on the body-whirl.*

LAPICIDA. 3. Shell umbilicated, keeled, and convex on both sides, with a transverse ovate marginated aperture.

Helix lapicida. *Linnæus Syst. Nat.* p. 1241. *Muller Verm.* ii. p. 40. *Pennant Zool.* iv. p. 132. t. 83. f. 121. *Born Mus.* p. 366. t. 14. f. 1 and 2. *Schroeter Einl.* ii. p. 124. *Chemnitz,* ix. part 2. p. 88. t. 126. f. 1107. *Gmelin,* p. 3613. *Donovan,* ii. t. 39. f. 2. *Montagu Test.* p. 435. *Maton and Racket, in Lin. Trans.* viii. p. 187. *Dorset Cat.* p. 52. t. 20. f. 9.
Helix acuta. *Da Costa Brit. Conch.* p. 55. t. 4. f. 9.
Helix affinis. *Gmelin,* p. 3621.
Helix, No. 198. *Schroeter Einl.* ii. p. 231.
La Lampe. *Geoffroy,* p. 41. No. 10. t. 2.
Lister Anim. Ang. t. 2. f. 14, and *Conch.* t. 69. f. 68. *Petiver Gaz.* t. 92. f. 11. *Martini Berl. Mag.* iv. t. 3. f. 36. *Schroeter Erdconch.* t. 2. f. 23.
Inhabits most parts of Europe among rocks, in woods and hedges, and about the trunks of old trees.
Shell about three-eighths of an inch long, and three-quarters of an inch in diameter, and consists of five much depressed whirls, so as to be almost equally convex above and below; the whirls are all transversely striated, and the body-whirl sharply keeled; the aperture has a white margin, and the outer lip is reflected.

MARGINATA. 4. Shell umbilicated, keeled, obliquely striated, and the spire depressed; base convex, and the aperture ear-shaped with a white margin.

Helix marginata. *Born Mus.* p. 367. t. 14. f. 7 and 8. *Chemnitz,* ix. part 2. p. 80. t. 125. f. 1097.
Helix marginella. *Gmelin,* p. 3622.
Helix, No. 199. *Schroeter Einl.* i. p. 232.

Variety? Smaller, with three red bands on the body-whirl, and two on the other.

Helix marginata.   *Muller Verm.* ii. p. 41.   *Gmelin*, p. 3622.

Inhabits Jamaica.   *Chemnitz.*

Shell about eight lines long, and fifteen in diameter, white, with five depressed obliquely striated whirls, and a broad brownish red band in the middle of each; the umbilicus is large and deep, and the aperture has a white reflected margin. In the Hist. Vermium, *H. marginata* is described to be nine lines in diameter, with four whirls, of which the body-whirl has three, and the others two red bands; in other respects the present shell answers to Muller's description, but his meaning is rendered more obscure by his having compared it with *H. indistincta*, and I cannot find that any Helix has ever been described under this name.

CICATRICOSA.   5.   Shell umbilicated, keeled, depressed, with the whirls convex, reversed, and obliquely wrinkled; base convex, and the aperture effuse.

Helix cicatricosa.   *Muller Verm.* ii. p. 42.   *Chemnitz*, ix. part 1. p. 90. t. 109. f. 923, and Vign. No. 19. at p. 1. f. A, and xi. p. 305. t. 213. f. 3012 and 3013.   *Gmelin*, p. 3614.

Helix, No. 49.   *Kœmmerer Cab. Rudolst.* p. 173. t. 11. f. 6.

Argenville App. t. 1. f. C.   *Favanne*, t. 63. f. K.

Inhabits Jamaica, China, Botany Bay, and the South Sea Islands; and is a land shell. *Chemnitz.*

Shell about an inch long, and an inch and a half in diameter, of a pale brownish yellow colour with several narrow reddish transverse stripes; it has six obliquely wrinkled convex whirls, and a distinct transverse keel on the body-whirl; the aperture is somewhat four-sided and white, and has the outer lip reflected so as partly to cover the umbilicus.

CORNU.   6.   Shell umbilicated, convex, with the body-whirl very broad, and the summit obtuse; aperture oval-earshaped, and the outer lip marginated; base flattish.

Helix Cornu giganteum. *Chemnitz*, xi. p. 274. t. 208. f. 2051 and 2052.

Inhabits New Zealand; and is a land shell. *Chemnitz.*
Shell an inch and three-quarters long, and three inches broad, of a pale reddish yellow colour with a narrow white band round the keel, and covered with a brown epidermis; the outer lip is white, and has a reflected margin. It appears from the figures to possess some affinity with *H. Oculus-Capri,* but I am otherwise unacquainted with this shell.

OCULUS-CAPRI. 7. Shell umbilicated, slightly keeled, with the summit mucronated, and the whirls convex and variegated; aperture roundish and marginated.

Helix Oculus-Capri. *Linnæus Syst. Nat.* p. 1242. *Gmelin,* p. 3615.
Helix Volvulus. *Muller Verm.* ii. p. 82. *Gmelin,* p. 3638.
Helix Lituus. *Martyn Univ. Conch.* i. t. 27.
Helix, No. 44. *Schroeter Einl.* ii. p. 188.
Turbo Volvulus. *Chemnitz,* ix. part 2. p. 57. t. 123. f. 1064 to 1066.
Turbo Lituus. *Gmelin,* p. 3589.
Bonanni Rec. Supp. f. 31 and 32. *Lister Conch.* t. 75. f. 75. *Petiver Gaz.* t. 76. f. 6. *Seba,* iii. t. 40. f. 18 and 19.
Variety. Smaller, and more depressed at the summit.
Helix Volvulus. *Born Mus.* p. 379. t. 14. f. 23 and 24.
Helix Lituus brevis. *Martyn Univ. Conch.* i. t. 28.
Turbo Jamaicensis. *Chemnitz,* xi. p. 277. t. 209. f. 2057 and 2058. ?
Lister Conch. t. 55. f. 51. *Petiver Gaz.* t. 61. f. 7.
Inhabits Asia, on trees. *Linnæus.* Pulo Condore. *Petiver.* China, Sumatra, and Coromandel. *Humphreys.*
Shell about an inch and a quarter long, and an inch and a half broad; the spire is depressed, except at the summit, which is somewhat mucronated, and the whirls are convex; the aperture is oblique, nearly circular, and slightly pointed at its upper extremity; the colour is chestnut variegated with brownish yellow, and the base, as well as the aperture, is generally brownish white; the umbilicus is large and pervious. The Variety is only about half as large, and has the summit of its spire less produced. Muller has retained the name of *Oculus-Capri* for *L'Œil de Bouc* of Argenville, t. 6. f. E, which is different, and appears to be *H. Algira.* *H. Turbo* of Gmelin (which is *Trochus Turbo* of Chemnitz,) is not sufficiently well defined, but appears from the figure to belong

to this family, and to differ principally from the Variety B in having a much narrower umbilicus.

INVOLVULUS. 8. **Shell umbilicated, slightly keeled, white, with the summit mucronated, and the whirls convex and striated; aperture roundish, and the lip reflected.**

Helix Involvulus. *Muller Verm.* ii. p. 84. *Gmelin*, p. 3638.

Inhabits ——

Shell about eleven lines long, and thirteen broad; it is nearly allied to *H. Oculus-Capri,* from which it may be known by having the base as far as the keel marked with very fine longitudinal striæ, and the shell above this is also more strongly striated transversely; the outer-lip is reflected, but not so broadly marginated as in *H. Oculus-Capri;* all the specimens which I have seen are wholly white, but Muller mentions a Variety with a red band and spots.

ALBELLA. 9. **Shell umbilicated, flat above, and the base gibbous; aperture semi-heartshaped.**

Helix albella. *Linnæus Syst. Nat.* p. 1242.
Helix explanata. *Muller Verm.* ii. p. 26.
Helix Planorbis marginatus. *Chemnitz*, ix. part 2. p. 84. t. 126. f. 1102.
Helix Rhenana. *Gmelin*, p. 3622.
Helix, No. 325. *Schroeter Einl.* ii. p. 270.
Helix Planorbis, Var. β. *Gmelin*, p. 3617.
Helix umbilicaris. *Olivi Adr.* p. 177.
Planorbis umbilicatus margine interrupto. *Schroeter Fluss.* p. 244. t. 5. f. 31.
Lister Conch. t. 64. f. 62, and t. 140. f. 46. *Petiver Gaz.* t. 17. f. 1. *Gualter*, t. 3. f. N.
Variety? With a reddish band in the middle of the whirls, and the umbilicus small.
Helix albella. *Chemnitz*, ix. part 2. p. 87. t. 126. f. 1105 and 1106. *Gmelin*, p. 3615. *Schreibers Conch.* i. p. 291.
Junior. With three whirls, and a dotted band in the middle.
Helix maculata. *Muller Verm.* ii. p. 25. *Gmelin*, p. 3615.
Inhabits rocks in Europe. *Linnæus.* France, Virginia, and Jamaica. *Lister.* Neighbourhood of Rome. *Petiver.*
Shell half, or sometimes three-quarters of an inch in diameter,

of an uniform whitish colour, except the ultimate whirl, which forms a brown spot in the center; the upper surface is very flat, and bordered by a sharp keel, below which the base is remarkably gibbous and umbilicated; it has four or five whirls, which are prettily striated transversely. Petiver has not unaptly called it ' a small Trochus reverst, having its head flat and belly raised.' I have received it from the West Indies. The shell which Chemnitz has figured for *H. albella* does not accord well either with the description of the Linnæan *H. albella* or Muller's *H. explanata*, and is most probably a distinct species. Linnæus has erroneously quoted Gualter, t. 3. f. Q, which is *H. rotundata.*

ALBINA. 10. Shell umbilicated, keeled, with the spire flattened, and the base gibbous; aperture quadrangular.

Helix albina. *Muller Verm.* ii. p. 25. *Gmelin,* p. 3615. *Lister Conch.* t. 86. f. 86. ?
Inhabits ——
Shell about three lines in diameter, with three whirls, and white without any spots; the body-whirl has an acute keel, and Muller says it resembles *H. albella,* but may be distinguished by a tooth in the middle of the whirl opposite the aperture.

ROTUNDATA. 11. Shell umbilicated, slightly keeled, with six strongly striated whirls; spire depressed, and the base convex.

Helix rotundata. *Muller Verm.* ii. p. 29. *Gmelin,* p. 3633.
Helix radiata. *Da Costa Brit. Conch.* p. 57. t. 4. f. 15 and 16. *Montagu Test.* p. 432. t. 24. f. 3. *Maton and Racket, in Lin. Trans.* viii. p. 199. *Dorset Cat.* p. 54. t. 20. f. 15 and 16.
Helix, No. 275. *Schroeter Einl.* ii. p. 256.
Le Bouton. *Geoffroy,* p. 39, No. 9. t. 2.
Lister Conch. t. 1058. f. 11 A. *Petiver Gaz.* t. 31. f. 5. *Gualter,* t. 3. f. Q. *Argenville Zoom.* t. 9. f. 10. *Schroeter Erdconch.* t. 2. f. 25.
Inhabits Denmark and Norway; under the bark of decayed trees, and among moss in woods. *Muller.* France. *Geoffroy.* England. *Da Costa, &c.*
Shell about a quarter of an inch in diameter, of a pale dull brown colour, prettily rayed or rather tessellated with chest-

nut; it has six whirls, which are well defined by the suture, and marked transversely with close set regular striæ; the umbilicus is very large, and exhibits all the whirls of the spire. This shell is undoubtedly the *H. rotundata* of Muller, and the *H. radiata* of Muller and Gmelin is different.

STRIATULA.   12. Shell umbilicated, slightly keeled, convex, and striated; base rather gibbous, and the aperture roundish-lunated.

Helix striatula. *Linnæus Syst. Nat.* p. 1242. *Gmelin*, p. 3615.
Helix polita. *Muller Verm.* ii. p. 33.?
Inhabits Algiers. *Linnæus.*   Banks of rivers in Lombardy. *Muller.*

Linnæus, without any reference, had only added to the above specific character, that this shell is grey, with the whirls transversely striated, and the umbilicus large. With this description *H. rotundata* tolerably well agrees, but in describing *H. Algira*, Linnæus has mentioned its resemblance to *H. striatula*, which, he says, differs in having only four whirls. Muller suspects that *H. striatula* is the same as his *H. polita*, which he says resembles *H. hispida*, and it therefore cannot be at all allied to *H. Algira*, if Born and Chemnitz have figured the right shell for that species.

ALGIRA.   13. Shell umbilicated, slightly keeled, and convex, with six whirls; umbilicus pervious.

Helix Algira. *Linnæus Syst. Nat.* p. 1242. *Born Mus.* p. 366. t. 14. f. 3 and 4. *Chemnitz*, ix. part 2. p. 77. t. 125. f. 1093 and 1094. *Gmelin*, p. 3615.
Helix Oculus Capri. *Muller Verm.* ii. p. 39.
Helix Ægophthalmos. *Gmelin*, p. 3614.
Helix Olivetorum. *Gmelin*, p. 3639.?
Helix, No. 279. *Schroeter Einl.* ii. p. 258.
Lister Conch. t. 79. f. 80. *Rumphius*, t. 25. f. P. *Petiver Gaz.* t. 21. f. 6.? and *Amb.* t. 12. f. 12. *Gualter*, t. 3. f. G. *Argenville*, t. 6. f. E. *Browne's Jamaica*, t. 40. f. A and B. *Favanne*, t. 63. f. L 1.
Inhabits Africa. *Linnæus.* Amboyna. *Rumphius.* Jamaica. *Browne.*

Shell about three-quarters of an inch long, and an inch and three-quarters in diameter, whitish, without any variegations, and coated with a yellowish epidermis; it has six, or according to Muller, sometimes seven whirls, which are finely striated transversely, and Muller says that young shells have also concentrical lines with minute elevated dots, which are only discoverable with the help of a glass; young shells have a distinct keel on the body-whirl, which afterwards becomes obsolete; the umbilicus is very large and deep.

LEUCAS. 14. Shell umbilicated, slightly keeled, convex, and smooth; base gibbous, and the umbilicus very minute; aperture roundish-lunated.

Helix Leucas. *Linnæus Syst. Nat.* p. 1242. *Gmelin*, p. 3616.

Inhabits Africa. *Linnæus.*

Linnæus has not mentioned the size, or given any reference, but to the above character has only added that the shell is white, with a band above and purplish lines below. Chemnitz doubts whether it may not be a Variety of *H. vermiculata.*

LÆVIPES. 15. Shell umbilicated, slightly depressed, and somewhat keeled; whirls reversed, and minutely striated transversely.

Helix lævipes. *Muller Verm.* ii. p. 22. *Chemnitz*, ix. part 1. p. 83. t. 108. f. 915 and 916. *Gmelin*, p. 3616.

Helix candida. *Gmelin*, p. 3616.

Helix hyalina. *Gmelin*, p. 3640.

Helix, No. 45. *Kæmmerer Cab. Rudolst.* p. 172.

Martini n. Mannigf. iv. p. 423. t. 3. f. 22 and 23.

Inhabits Guinea. *Martini.* Tranquebar. *Chemnitz.*

Shell about an inch in diameter, thin, pellucid, and of a whitish colour, generally with a red band united to a white one at the margin of the whirls; it has five reversed whirls, minutely striated transversely, and the aperture has its margin acute. Muller says it somewhat resembles *H. exilis,* but has the spire less depressed, the whirls reversed, and the base more convex. Chemnitz has quoted Martini's figures for this species.

EXILIS. 16. Shell umbilicated, flattish, slightly keeled, and the whirls sharply striated transversely; aperture with an acute margin.

Helix exilis. *Muller Verm.* ii. p. 22. *Chemnitz*, ix. part 2. p. 121, t. 129. f. 1149. *Gmelin*, p. 3616.

894 HELIX.

Inhabits Tranquebar. *Chemnitz.*
Muller has described the diameter to be ten lines, and Chemnitz says it varies from an inch to an inch and a quarter; it has five thin pellucid whirls of a pale yellowish colour, with a reddish band joined to a white one at the margin, and the base is convex; Chemnitz says it is only slightly keeled, and that it has longitudinal as well as transverse striæ; the umbilicus is deep, and of a reddish brown colour. Gmelin has also given the name of *H. exilis* to Kæmmerer's t. 12. f. 3, which is too uncertain a species to be retained.

VERMICULATA. 17. Shell imperforate, nearly globular, rough with minute elevated dots, and very slightly keeled; outer lip white and reflected.

Helix vermiculata. *Muller Verm.* ii. p. 20. *Chemnitz,* ix. part 2. p. 120. t. 129. f. 1148. *a, b,* and *c.* *Gmelin,* p. 3616.
Petiver Gaz. t. 52. f. 11. *Gualter,* t. 1. f. G. *Favanne,* t. 64. f. K 2, and K 3.
Inhabits the sandy banks of rapid streams in Italy. *Muller.* Pisa. *Petiver.* Portugal. *Chemnitz.*
This shell is described to vary from ten to thirteen lines in diameter, is generally rough with minute white dots, and variegated with from one to four reddish bands; both sides are convex, but the apex is a little flattened, and there are five whirls. Chemnitz doubts whether it is more than a Variety of *H. Lucorum.*

INCARNATA. 18. Shell umbilicated, sub-globular, slightly keeled, with six horn-coloured whirls, and the outer lip flesh-coloured.

Helix incarnata. *Muller Vermium,* ii. p. 63. *Chemnitz,* ix. part 2. p. 151. t. 133. f. 1206. *Gmelin,* p. 3617.
Schroeter Erdconch. t. 2. f. 18.
Inhabits woods in Denmark; rare. *Muller.* Germany. *Schroeter.*
Shell half an inch in diameter, and very convex above, with six whirls; it is of a horn-colour, with a whitish keel on the body-whirl, and becomes of a yellowish red towards the aperture; umbilicus narrow.

CANTIANA. 19. Shell umbilicated, very slightly keeled, rather depressed, with six striated whirls, and the umbilicus small.

Helix Cantiana. *Montagu Test.* p. 422. t. 23. f. 1. *Ma-ton and Racket, in Lin. Trans.* viii. p. 197. *Dorset Cat.* p. 53. t. 19. f. 21.

Helix pallida. *Donovan,* v. t. 157. f. 2.

Inhabits hedges and old walls in several parts of Britain, and particularly Kent. *Montagu, &c.*

Shell about half an inch long, and three-quarters of an inch broad; it differs from *H. rufescens* in being larger, of a paler colour, and in having the body-whirl much less distinctly keeled, but in the place of the keel there is a faint white line. Lister has mentioned it in his Historia Anima-lium Angliæ, and considered it to be a variety of *H. ru-fescens.*

RUFESCENS. 20. Shell umbilicated, keeled, some-what depressed, with six transversely stri-ated whirls, and the umbilicus rather large.

Helix rufescens. *Pennant Brit. Zool.* iv. p. 134. *Da Costa Brit. Conch.* p. 80. t. 4. f. 6. *Montagu Test.* p. 420. t. 23. f. 2. *Donovan,* v. t. 157. f. 1. *Maton and Racket, in Lin. Trans.* viii. p. 196. *Dorset Cat.* p. 53. t. 20. f. 6.

Helix Turturum. *Gmelin,* p. 3639.

Helix, No. 100. *Schroeter Einl.* ii. p. 205.

Lister Anim. Ang. t. 2. f. 12, and *Conch.* t. 71. lower fig. A.

Junior. Covered with short hairs.

Helix sericea. *Muller Verm.* ii. p. 62. *Gmelin,* p. 3617.

Helix hispida. *Donovan,* v. t. 151. f. 1.

Schroeter Erdconch. t. 2. f. 24.?

Inhabits moist woods and shady places in England. *Pennant, &c.* Saxony. *Schroeter.*

Shell a quarter of an inch long, and about twice as broad, of a pale reddish brown colour, with six slightly raised trans-versely striated whirls, which are well defined by the suture; the body-whirl has a slightly carinated margin, which is ren-dered more conspicuous by being of a lighter colour, the aperture has the outer lip thin, and a little reflected over the umbilicus, which is large and deep.

CRENULATA. 21. Shell umbilicated, slightly de-pressed, somewhat keeled and striated, and the summit brown.

Helix crenulata. *Muller Verm.* ii. p. 68. *Gmelin,* p. 3617.

Helix caperata. *Montagu Test.* p. 430. t. 11. f. 11. *Maton and Racket, in Lin. Trans.* viii. p. 196. *Dorset Cat.* p. 53. t. 19. f. 20.

Helix variegata. *Chemnitz,* ix. part 2. p. 152. t. 133. f. 1207.

Helix naevia. *Gmelin,* p. 3623.
Lister Conch. t. 85. f. 85.

Inhabits France. *Lister.* Neighbourhood of Lyons. *Muller.* St. Croix. *Chemnitz.* Dry places in Great Britain. *Montagu, &c.*

The diameter is about three-eighths, or sometimes half an inch, and the length is rather more than half the diameter; it has six whirls, which are closely striated transversely, and the striæ, by crossing the sub-carinated margin of the body-whirl, give it a crenulated appearance when examined with a glass, as Muller describes; it varies much in its markings, some shells being fasciated with brown and white, others dark brown with a single white belt, and minutely spotted with white, and some have brown bands running into each other, and are elegantly spotted with white; there are generally several small interrupted bands on the base, and the umbilicus is rather large.

ANNULATA. 22. Shell umbilicated, slightly depressed, white, with four whirls; body-whirl gibbous, and doubly keeled.

Helix annulata. *Gmelin,* p. 3622:
Helix, No. 324. *Schroeter Einl.* ii. p. 270.
Planorbis umbilicatus costatus. *Schroeter Fluss.* p. 244. t. 5. f. 30.

Inhabits ――――

Shell two lines in diameter, of a dirty white colour, with five slightly depressed whirls, of which the body-whirl is said to have two transverse keels; aperture oval and large.

PLANORBIS. 23. Shell flat, concave above, and the whirls depressed on both sides, with a keel round the middle of the body-whirl.

Helix Planorbis. *Linnæus Syst. Nat.* p. 1242. *Schroeter Fluss.* p. 226. t. 5. f. 13, and *Einl.* ii. p. 128. *Born Mus.* p. 367. t. 14. f. 5 and 6. *Gmelin,* p. 3617.

Helix carinata. *Montagu Test.* p. 451. t. 25. f. 1.

Helix planata. *Maton and Racket, in Lin. Trans.* viii. p. 189. t. 5. f. 14. *Dorset Cat.* p. 52. t. 20. f. 18.

Planorbis carinatus. *Muller Verm.* ii. p. 157.

Le Planorbe à quatre spirales. *Geoffroy*, p. 90. No. 4. t. 3.

Inhabits stagnant water in Europe. *Linnæus.* France. *Geoffroy.* Denmark and Italy. *Muller.* Great Britain. *Montagu, &c.*

Shell about half an inch in diameter, of a pale horn-colour, covered with a dusky green epidermis; it has four transversely striated whirls, which are coiled horizontally, and the body-whirl is not so convex on the upper side as in *H. complanata*, and has a keel round the middle, so as to be equally conspicuous on both sides.

COMPLANATA. 24. Shell flat, concave above, and the whirls convex on the upper side, with a keel on the lower edge of the body-whirl.

Helix complanata. *Linnæus Syst. Nat.* p. 1242. *Schroeter Fluss.* p. 239. t. 5. f. 22 to 25, and t. 11 C. f. 4; and *Einl.* ii. p. 129. *Chemnitz,* ix. part 2. p. 96. t. 127. f. 1121 and 1122. *Gmelin,* p. 3617. *Montagu Test.* p. 450. t. 25. f. 4.

Helix limbata. *Da Costa Brit. Conch.* p. 63. t. 4. f. 10, and t. 8. f. 8.

Helix Planorbis. *Pennant Zool.* iv. p. 133. t. 83. f. 123. *Maton and Racket, in Lin. Trans.* viii. p. 188. t. 5. f. 13. *Dorset Cat.* p. 52. t. 14. f. 8, and t. 20. f. 10.

Planorbis umbilicatus. *Muller Verm.* ii. p. 160.

Lister Anim. Ang. t. 2. f. 27, and *Conch.* t. 138. f. 42. *Petiver Gaz.* t. 10. f. 11. *Gualter,* t. 4. f. EE. *Klein Ost.* t. 1. f. 8. *Favanne,* t. 61. f. B 5.

Inhabits fresh waters in Europe. *Linnæus.* Great Britain, *Lister, &c.*

Shell about five-eighths of an inch in diameter, of a pale reddish brown or horn-colour, not unfrequently covered with a dark brown epidermis; it has five transversely striated whirls, which are coiled horizontally, and considerably convex on the upper, but flat on the lower side; the body-whirl has its keel on the lower margin, and not in the middle, as in *H. Planorbis.* Although most, or perhaps all the figures which Linnæus has quoted for *H. Planorbis* belong to the shell which is here described, yet it appears to me that it must be his *H. complanata,* from the description of that species in the Fauna Suècica: "Margo qui anfractus cingit respicit latus planus seu inferius." Of *H. Planorbis* he only says, "Margo anfractuum prominet."

FONTANA. **25. Shell compressed, obtusely keeled, convex on both sides, with three whirls, and the base umbilicated.**

Helix fontana. *Lightfoot in Phil. Trans.* lxxvi. t. 2. f. 1 to 4. *Montagu Test.* p. 462. t. 6. f. 6. *Maton and Racket, in Lin. Trans.* viii. p. 193. *Dorset Cat.* p. 53. t. 19. f. 19.

Inhabits ditches in England. *Lightfoot, &c.*

Shell about two lines in diameter, of a glossy horn-colour, extremely flat, and nearly equally convex on both sides, but depressed in the center; the keel is formed by the gradual slope of the outer whirl on both sides; externally it much resembles *Nautilus lacustris*, but is flatter, and not so convex on the upper side, and may be at once distinguished by its want of any internal partitions.

RINGENS. **26. Shell imperforate, slightly keeled and convex; aperture turned upwards, and toothed on both sides.**

Helix ringens. *Linnæus Syst. Nat.* p. 1243. *Muller Verm.* ii. p. 17. *Schroeter Einl.* ii. p. 130. *Born Mus.* p. 369. t. 14. f. 11 and 12. *Chemnitz,* ix. part 1. p. 86. t. 109. f. 919 and 920. *Gmelin,* p. 3618. *Shaw Nat. Misc.* x. t. 374.

Tormigeres ringens. *De Montfort Conch.* ii. p. 359. *Leach Zool. Misc.* ii. p. 128. t. 107.

Grew Mus. t. 11. f. 8. *Bonanni Rec.* 3. f. 330 and 331, and *Kirch,* f. 331 and 332. *Lister Conch.* t. 99. f. 100. *Argenville,* t. 28. f. 13 and 14. *Martini Berl. Mag.* ii. t. 4. f. 42. *Favanne,* t. 63. f. F 10.

Inhabits Brazil; and is a land shell. *Humphreys.*

Shell about three-quarters of an inch long, and an inch and three quarters broad, of a whitish colour, mottled and dotted with yellowish brown; the aperture is turned up, and the number of its teeth is said to vary from four to eight. It is a very scarce shell, and was first described in an account of the curiosities belonging to the Royal Society, by Dr. Grew, who says, "both of the corners of the mouth are placed on the circumference of the utmost round, whereby, contrary to all other shells I ever yet saw, it hath the turban or whirle made before. The assertion of Aristotle, that the turban always stands behind, is here proved false." Its usual name is the *Grinner.*

PLICATA. 27. Shell umbilicated, keeled, and rather depressed; aperture ear-shaped, distorted, toothed, and sinuated, with a prominent marginated lip.

Helix plicata. *Born Mus.* p. 368.
Helix Otis. *Portland Cat.* p. 38. Lot 925, and p. 53. Lot 1260.
Helix Labyrinthus. *Chemnitz,* xi. p. 271. t. 208. f. 2048.
Seba, iii. t. 40. f. 24 and 25. *Knorr,* v. t. 26. f. 5. *Favanne,* t. 63. f. F 11.
Inhabits the East Indies. *Knorr.*
Shell about an inch and a half in diameter, of a bright fawn-colour, with six slightly convex whirls; the aperture appears in the figures to be very strangely formed, with a labyrinth of projecting teeth and sinuosities which commence at the umbilicus, and project considerably beyond the contour of the body-whirl.

PUNCTATA. 28. Shell sub-umbilicated, slightly keeled, roundish; inner lip thickened and sinuated, and the outer three-toothed.

Helix punctata. *Born Mus.* p. 372. t. 14. f. 17 and 18. *Gmelin,* p. 3622.
Helix isognomostomos. *Gmelin,* p. 3621.
Helix sinuata, Var. γ. *Gmelin,* p. 3618.
Helix, No. 62, and No. 202. *Schroeter Einl.* ii. p. 194, and p. 233.
Lister Conch. t. 93. f. 93. *Petiver Gaz.* t. 105. f. 5. *Klein Ost.* t. 1. f. 22. *Favanne,* t. 63. f. F 2.
Inhabits Virginia. *Lister.* Saxony and Alsace. *Schroeter.*
Shell about half an inch long, and ten lines broad, of a brown colour, with a row of distant yellowish spots in the middle of the whirls; it is roundish, but slightly depressed, and has five whirls; aperture ear-shaped, contracted, and has the pillar-lip projecting over the umbilicus. Lister's f. 94 is probably distinct, and is the same as Petiver, t. 105. f. 6, and Favanne, t. 63. f. F 3.

SINUATA. 29. Shell imperforate, sub-globular, and slightly keeled; lip reflected, with four teeth on the inside, and three-plaited behind.

Helix sinuata. *Muller Verm.* ii. p. 18. *Born Mus.* p. 370. t. 14. f. 13 and 14. *Chemnitz,* ix. part 2. p. 91. t. 126. f. 1110 to 1112. *Gmelin,* p. 3618.

Helix sinuosa. *Gmelin*, p. 3622.
Helix, No. 66, and No. 200. *Schroeter Einl.* ii. p. 195,
and p. 232.
Lister Conch. t. 97. f. 98. *Favanne*, t. 63. f. F 8.
Inhabits America. *Muller.* Barbadoes and Jamaica. *Lister.*
Shell about half an inch long, and nearly twice as broad, of a
pale yellowish brown colour, generally fasciated with white
on the keel of the body-whirl, and the base white; it has
five whirls, forming a rounded obtuse spire, and they are
every where covered with minute elevated dots; the aperture is ear-shaped, with four strong teeth on the lip, and corresponding indentures behind.

LUCERNA. 30. Shell imperforate, keeled, and rather
depressed; lip reflected, with two teeth on
the inside, and corresponding impressions behind.

Helix Lucerna. *Muller Verm.* ii. p. 13. *Chemnitz*, ix.
p. 90. t. 126. f. 1108 and 1109. *Gmelin*, p. 3619.
Helix sinuata, Var. β. *Gmelin*, p. 3618.
Lister Conch. t. 96. f. 97.
Inhabits the East Indies. *Lister.* Jamaica. *Chemnitz.*
Shell about half an inch long, and an inch and a quarter broad,
of a pale purplish brown colour, with a whitish band on the
keel; it has six whirls without any striæ, but studded all
over with minute elevated dots; the aperture is tinged with
violet, and in form resembles that of *H. sinuata.*

LYCHNUCUS. 31. Shell imperforate, keeled, trochiform, and white with yellowish brown bands;
aperture transverse, with two teeth.

Helix Lychnucus. *Muller Verm.* ii. p. 81. *Gmelin,*
p. 3619.
Lister Conch. t. 90. f. 90.
Inhabits Jamaica. *Lister.*
This shell appears, from Muller's description, to be very nearly
allied to *H. Lucerna,* and he says it is intermediate between *H. Lucerna* and *H. Lychnus*; but there is not any
species under the latter name, and in this, as well as many
other instances, he unfortunately appears to have changed
the name after the descriptions were written; he says it may
be known by its having the apex only, and not any of the
other whirls, depressed, and by having a band both on the
margin and the keel.

CEPA. **32. Shell imperforate, sub-globular, and slightly keeled ; aperture transverse, with one tooth on each lip, and a corresponding depression behind.**

Helix Cepa. *Muller Verm.* ii. p. 74. *Gmelin*, p. 3619. *Lister Conch.* t. 88. f. 89.

Inhabits Jamaica. *Lister.*

Shell about fifteen lines in diameter, of a pale reddish colour, with a white band, and convex on both sides ; it has five contiguous whirls. Muller says it is nearly allied to *H. sinuata*, but the difference in their teeth will serve readily to distinguish them.

LAMPAS. **33. Shell imperforate, keeled, flattish above, and gibbous below ; whirls scarred, and the outer lip orange-coloured.**

Helix Lampas. *Muller Verm.* ii. p. 12. *Gmelin*, p. 3619. Helix Carocolla. *Chemnitz*, ix. p. 267. t. 208. f. 2044 and 2045.

Inhabits ———

Shell about two inches and three-quarters in diameter, of a yellowish brown colour, and has the upper surface flatter and the lower more convex than *H. Carocolla*, to which it is most nearly allied ; the whirls are finely striated, and have a scarred appearance ; the throat is yellowish, and the outer lip orange-coloured.

CAROCOLLA. **34. Shell imperforate, keeled, and convex on both sides ; outer lip white.**

Helix Carocolla. *Linnæus Syst. Nat.* p. 1243. *Muller Verm.* ii. p. 77. *Born Mus.* p. 370. *Schroeter Einl.* ii. p. 132. *Chemnitz*, ix. part 2. p. 75. t. 125. f. 1090 to 1092. *Gmelin*, p. 3619. *Shaw Nat. Misc.* x. t. 374. larger figures.

Helix tornata. *Born Mus.* p. 369. t. 14. f. 9 and 10. *Lister Conch.* t. 63. f. 61, and t. 1055. f. 5. *Gualter,* t. 3. f. I. *Argenville*, t. 8. f. D. *Seba*, iii. t. 40. f. 16, 17, 22, and 23. *Knorr*, iv. t. 5. f. 2 and 3. *Favanne.* t. 63. f. F 12.

Inhabits America, and is a land shell. *D'Avila.*

Shell about an inch long, and an inch and three-quarters or two inches broad, of a dark blackish olive or chestnut-brown colour, with a paler obsolete band by the keel ; it has six slightly convex transversely striated whirls; the aperture is transverse and ear-shaped, and has a white reflected lip.

UNIDENTATA.   35. Shell sub-umbilicated, slightly keeled, with six whirls ending in an obtuse summit; base convex, with one tooth on the pillar-lip, and the outer lip marginated.

Helix unidentata.   *Chemnitz,* xi. p. 273. t. 208. f. 2049 and 2050.

Inhabits Ceylon.   *Chemnitz.*

Shell about an inch and three-quarters in diameter, of a dull brownish red colour, with a narrow white band on the keel of the body-whirl which is continued along the sutures of the spire; the summit of the spire is depressed and obtuse, and the base is convex.

CORNU-MILITARE.   36. Shell imperforate, solid, slightly keeled, ventricose, and the summit obtuse; aperture wide, and the outer lip thick and reflected.

Helix Cornu-militare. *Linnæus Syst. Nat.* p. 1243.? *Born Mus.* p. 371.   *Schroeter Einl.* ii. p. 133.   *Chemnitz,* ix. part ii. p. 116. t. 129. f. 1142 and 1143.   *Gmelin,* p. 3620.

Helix gigantea.   *Scopoli del Insul.* i. p. 66. t. 25. f. A. *Gmelin,* p. 3646.

Knorr, vi. t. 32. f. 2.   *Favanne,* t. 64. f. C 2.

Inhabits Germany. *Knorr.*   And is a land shell. *Linnæus.*

Linnæus has given only a short description, and Gualter, t. 3. f. I, which he has quoted, appears to me to be *H. Caro-colla,* so that it is almost impossible to ascertain his meaning with any certainty.   The present shell is about two and a half inches broad, of a whitish colour, except on the pillar-lip and margin of the pillar, which are generally dark brown, and live shells are entirely covered with a brown epidermis; it has five whirls, of which the upper ones are depressed, and form a flattish obtuse spire; the aperture is dilated, and the outer lip thick, white, and expanded.

MACULOSA.   37. Shell umbilicated, slightly keeled, depressed, obliquely striated, and the base convex.

Helix maculosa.   *Born Mus.* p. 371. t. 14. f. 15 and 16. *Gmelin,* p. 3622.

Helix, No. 201. *Schroeter Einl.* ii. p. 232.
Inhabits ——
Born has described this shell to be six lines long and eleven
broad, rounded, depressed, and the base convex, of a
whitish colour, with tawny spots varying in breadth ; it has
six rounded obliquely striated whirls, and an obsolete keel on
the body-whirl ; aperture lunated, with the outer lip marginated, and the pillar perforated.

CORRUGATA. 38. Shell umbilicated, keeled, with
longitudinal-oblique wrinkled striæ ; aperture lunated, with the outer lip reflected
within.

Helix corrugata. *Gmelin*, p. 3623.
Helix scabra. *Chemnitz*, ix. part 2. p. 152. t. 133. f. 1208.
Ulysses's Travels, p. 472.
Inhabits Otaheite. *Chemnitz.* Lucania. *Ulysses.*
Shell about seven lines in diameter, of a whitish ash-colour,
with six convex whirls ; the lip is rose-coloured, and the umbilicus is said to be large, round, and wide, though it is
made to appear rather small in the figure. I am unacquainted with this species, and a shell which I have procured from
Mrs. Mawe, with the name of *H. corrugata*, is quite different.

PELLIS SERPENTIS. 39. Shell umbilicated, keeled,
with six slightly convex whirls, and alternate
yellowish brown and white spotted bands ;
aperture ear-shaped, and the outer lip reflected.

Helix Pellis Serpentis. *Chemnitz*, ix. part 2. p. 79. t. 125.
f. 1095 and 1096, and xi. p. 268. t. 208. f. 2046 and
2047. *Gmelin*, p. 3620. *Shaw Nat. Misc.* xxii. t. 950.
Helix undata. *Portland Cat.* p. 177. lot 3802.
Bonanni Rec. Supp. f. 33 and 34. *Lister Conch.* t. 66.
f. 64, and t. 76. f. 76. *Favanne*, t. 63. f. G 3.
Inhabits South America. *Chemnitz.*
Shell about an inch and a half in diameter, of a yellowish or
reddish colour, with red and white spotted bands on the
upper surface, and concentrical rows of red dots round the
umbilicus ; it consists of six slightly convex whirls, which
form an obtuse depressed spire ; the aperture is ear-shaped,
and the lip has a reflected white margin. Petiver, Gaz.
t. 156. f. 1, which has been generally quoted for this species,

appears to be a magnified sketch of *Trochus vestiarius,* which has been copied from Bonanni.

VORTEX. 40. Shell keeled, with the upper side concave, and the base flat; aperture oval and compressed.

Helix Vortex. *Linnæus Syst. Nat.* p. 1243. *Pennant Zool.* iv. p. 133. t. 83. f. 124. *Schroeter Fluss.* p. 228. t. 5. f. 16 and 17, and *Einl.* ii. p. 134. *Chemnitz,* ix. part 2. p. 100. t. 127. f. 1127. *Gmelin,* p. 3620. *Donovan,* iii. t. 75. *Montagu Test.* p. 454. t. 25. f. 3. *Maton and Racket, in Lin. Trans.* viii. p. 189. *Dorset Cat.* p. 52. t. 20. f. 12.
Helix Planorbis. *Da Costa Brit. Conch.* p. 65. t. 4. f. 12.
Planorbis Vortex. *Muller Verm.* ii. p. 158.
Le Planorbe à six spirales a arrête. *Geoffroy,* p. 93. No. 5. t. 3.
Lister Anim. Ang. t. 2. f. 28, and *Conch.* t. 138. f. 43. *Petiver Gaz.* t. 92. f. 6. *Gualter,* t. 4. f. G G. *Klein Ost.* t. 1. f. 9. *Martini Berl. Mag.* iv. t. 8. f. 19.
Inhabits pools in Europe. *Linnæus.* England. *Lister.* France. *Geoffroy.* Denmark. *Muller.*
Shell about three-eighths of an inch in diameter, of a horn-colour, or sometimes reddish brown, and covered when alive with a blackish or greenish epidermis; it has six or seven narrow whirls, which are horizontally coiled, and the body-whirl is keeled at its lower edge.

SCABRA. 41. Shell ovate, imperforate, acuminated, keeled, and striated transversely; aperture roundish, and the outer lip acute.

Helix scabra. *Linnæus Syst. Nat.* p. 1243. *Gmelin,* p. 3620. *Chemnitz,* xi. p. 283. t. 210. t. 2074 and 2075. *Lister Conch.* t. 583. f. 37.
Inhabits Jamaica. *Lister.*
Shell about sixteen lines long, and half as broad, of a yellowish or brownish colour, with transverse broken bands or darker spots and crowded striæ; the shell which Chemnitz has figured for this species in form somewhat resembles *Trochus zic-zac.*

GOTHICA. 42. Shell keeled, convex on both sides, of a horn-colour, with sub-ferruginous bands.

Helix Gothica. *Linnæus Syst. Nat.* p. 1243. *Gmelin,* p. 3621.

Inhabits the woods of Sweden. *Linnæus.*

Linnæus has not made any addition to the above short character, and the species has not since been ascertained.

GUALTERIANA. 43. Shell imperforate, depressed, with decussated striæ, and the whirls keeled at their margins ; aperture ear-shaped.

Helix Gualteriana. *Linnæus Syst. Nat.* p. 1243. *Chemnitz,* v. p. 273. Vign. 44, at p. 237. f. *A, B,* and *C ;* and ix. part 2. p. 83. t. 126. 1100 and 1101. *Schroeter Einl.* ii. p. 136. t. 4. f. 2 and 3. *Gmelin,* p. 3621. Helix obversa. *Born Mus.* p. 368. t. 13. f. 12 and 13. *Gualter,* t. 68. f. E.

Inhabits Spain. *Favanne ?* And is a land shell. *Chemnitz.*

This scarce shell is about an inch and three-quarters in diameter, of a greyish colour, with oblique and transverse decussated striæ; it has five whirls, with a pale carinated margin at the suture, and flat on the upper side ; the base is very convex.

TURCICA. 44. Shell umbilicated, depressed, rough with elevated dots, and the keel crenated ; aperture somewhat four-sided.

Trochus Turcicus. *Chemnitz,* xi. p. 280. t. 209. f. 2065 and 2066.

Inhabits Mogadore and Morocco ; and is a land shell. *Chemnitz.*

This shell appears by the figures to be about ten lines in diameter, much depressed, and of a whitish colour. Chemnitz has mentioned its resemblance in some respects to *H. Gualteriana,* from which it may be at once distinguished by its crenated keel, and large marginated umbilicus, which exhibits the whole spire in the manner of *H. Oculus-Capri.*

AVELLANA. 45. Shell umbilicated, roundish, thick, with a broad rib-like keel on the body-whirl, and the spire depressed ; outer lip notched near the upper end.

Helix Avellana. *Gmelin,* p. 3640.

Helix crenata. *Martyn Univ. Conch.* ii. t. 69. *Gmelin,* p. 3623.

Nerita Nux Avellana. *Chemnitz,* v. p. 272. t. 188. f. 1919 and 1920.

Nerita, No. 18. *Schroeter Einl.* ii. p. 310.

Bulimus Avellana. *Bruguiere Enc. Meth.* p. 297.

Spengler Besch. Berl. Naturf. i. p. 395. t. 9. f. 4 and 5.

Inhabits New Zealand. *Martyn.*

Shell about an inch long, and equally broad, frequently of a dull greyish ash-colour, or sometimes whitish mottled with brown ; the body-whirl is ventricose, and the spire only a little produced at the summit ; the outer lip has a deep notch at the angle formed by the keel of the body-whirl, and the pillar-lip has a protuberance near its middle and opposite the umbilicus, which is placed remarkably high ; in colour and shape it bears some resemblance to a hazel-nut.

FABA. 46. Shell imperforate, smooth, of a saffron-colour, with the margin of the whirls and the base brown ; aperture blue.

Helix Faba, *Martyn Univ. Conch.* ii. t. 67. *Gmelin,* p. 3623.

Inhabits Otaheite. *Martyn.*

I am unacquainted with this shell, and unable to find the plate referred to in Martyn's work; the only perfect copy I ever saw is in All Souls' Library at Oxford, and the memoranda which I made on looking it over have been unfortunately mislaid.

*** *Umbilicated, and the Whirls rounded.*

CORNEA. 47. Shell with four rounded striated whirls, and the upper side umbilicated, and more concave than the base.

Helix cornea. *Linnæus Syst. Nat.* p. 1243. *Pennant Zool.* iv. p. 134. t. 83. f. 126. *Born Mus.* p. 372. *Schroeter Fluss.* p. 233. t. 5. f. 19 and 20, and t. 11 C. f. 7; and *Einl.* ii. p. 137. *Chemnitz,* ix. part 2. t. 127. f. 1113 to 1115. *Gmelin,* p. 3623. *Donovan,* ii. t. 39. f. 1. *Montagu Test.* p. 448. *Maton and Racket, in Lin. Trans.* viii. p. 190. *Dorset Cat.* p. 52. t. 20. f. 13. *Burrow's Elements,* p. 170. t. 20. f. 2.

Helix Cornu Arietis. *Da Costa Brit. Conch.* p. 60. t. 4.
f. 13.

Planorbis Purpura. *Muller Verm.* ii. p. 154.

Le Grand Planorbe. *Geoffroy,* p. 86. No. 1. t. 3.

Bonanni Rec. and *Kirch.* 3. f. 316, and *Kirch.* 312. *Lister Anim. Ang.* t. 2. f. 26, and *Conch.* t. 137. f. 41. *Petiver Gaz.* t. 92. f. 5. *Gualter,* t. 4. f. D D. *Argenville,* t. 27. f. 8, and *Zoom.* t. 8. f. 7. *Seba,* iii. t. 39. f. 17. *Knorr,* v. t. 22. f. 6. *Geve,* t. 3. f. 18 and 19. *Martini Berl. Mag.* iv. t. 8. f. 17.

Junior. Shell smaller.

Helix nana. *Pennant Zool.* iv. p. 134. t. 83. f. 125.

Inhabits slow rivers and stagnant waters; common throughout Europe.

Shell about an inch in diameter, and the aperture is about one-third as broad; it is of a reddish brown, or frequently of a horn-colour, with four rounded transversely wrinkled whirls, coiled horizontally, and separated by a deep suture; the upper side is strongly umbilicated, and more concave than the lower.

SIMILIS. 48. Shell umbilicated on the upper side, with four rounded striated whirls, and elevated dots on the striæ.

Helix similis. *Gmelin,* p. 3625.

Helix, No. 320. *Schroeter Einl.* ii. p. 269.

Planorbis similis. *Muller Verm.* ii. p. 166.

Martini Berl. Mag. iv. p. 265. t. 11. f. 64 B.

Inhabits ditches in Denmark, and about Berlin. *Muller.*

Shell three lines in diameter, and the diameter of the whirls is two lines; Muller says it resembles at first sight a young shell of *H. cornea,* but is of a yellower colour, and under a glass the striæ appear to be armed with minute elevated dots.

SPIRORBIS. 49. Shell concave on both sides, with six rounded whirls.

Helix spirorbis. *Linnæus Syst. Nat.* p. 1244. *Schroeter Fluss.* p. 236. t. 5. f. 18, and *Einl.* ii. p. 138. *Gmelin,* p. 3624. *Montagu Test.* p. 455. t. 25. f. 2. *Maton and Racket, in Lin. Trans.* viii. p. 191. *Dorset Cat.* p. 53. t. 20. f. 17.

Planorbis spirorbis. *Muller Verm.* ii. p. 161.

Le petit Planorbe. *Geoffroy,* p. 87. No. 2. t. 3.

Martini Berl. Mag. iv. p. 258. t. 8. f. 20.

Inhabits stagnant water in Sweden, and other parts of Europe. *Linnæus.* France. *Geoffroy.* Denmark. *Muller.* Germany. *Schroeter.* England. *Montagu.*

Shell about three-tenths of an inch in diameter, of a horn-colour, and differs from *H. Vortex*, to which it is most nearly allied, in being equally concave on both sides, and in the want of a keel on the outer whirl.

POLYGYRATA. 50. Shell concave above and flat below, with ten rounded whirls, and the aperture marginated.

Helix polygyrata.   *Born Mus.* p. 373. t. 14. f. 19 and 20. *Chemnitz,* ix. part 2. p. 98. t. 127. f. 1124 and 1125. Helix polygyra.   *Gmelin,* p. 3624. Helix, No. 308.   *Schroeter Einl.* iii. p. 266.

Shell about an inch and three-quarters in diameter, of a yellowish colour, with ten rounded and obliquely striated whirls ; aperture ovate, and marginated.

CONTORTA. 51. Shell flattish above, and deeply umbilicated below, with six whirls.

Helix contorta. *Linnæus Syst. Nat.* p. 1244.   *Chemnitz,* ix. part 2. p. 98. t. 127. f. 1126. *Gmelin,* p. 3624. *Donovan,* iii. t. 99. *Montagu Test.* p. 457. t. 25. f. 6. *Maton and Racket, in Lin. Trans.* viii. p. 191.   *Dorset Cat.* p. 53. t. 20. f. 11. Helix crassa.   *Da Costa Brit. Conch.* p. 66. t. 4. f. 11. Helix umbilicata.   *Pulteney Dorset Cat.* p. 47. Helix, No. 314.   *Schroeter Einl.* ii. p. 267. Planorbis contortus.   *Muller Verm.* ii. p. 162.   *Schroeter Fluss.* p. 243. t. 5. f. 29. Le Planorbe à six spirales rondes. *Geoffroy,* p. 89. No. 3. t. 3. *Petiver Gaz.* t. 92. f. 8.   *Martini Berl. Mag.* iv. t. 8. f. 21.

Inhabits stagnant waters in Sweden, and other parts of Europe. *Linnæus.* England. *Petiver.* France. *Geoffroy.* Denmark. *Muller.* Germany. *Schroeter.*

Shell about two tenths of an inch in diameter, and the whirls nearly one-tenth thick; the colour is brown, reddish chestnut, or horn-colour, and when alive it is covered with a dusky epidermis ; it has five or six whirls separated by a small deep suture, and is distinguished from the other depressed Helices by their greater thickness in proportion to the size, and by its singularly large spiral umbilicus.

ALBA. 52. Shell umbilicated on both sides, with four whirls, and slightly decussated striæ; aperture dilated.

Helix alba. *Gmelin,* p. 3625. *Montagu Test.* p. 459. t. 25. f. 7. *Maton and Racket, in Lin. Trans.* viii. p. 192. *Dorset Cat.* p. 53. t. 9. f. 18.
Helix, No. 316. *Schroeter Einl.* ii. p. 267.
Planorbis albus. *Muller Verm.* ii. p. 164. *Schroeter Fluss.* p. 225. t. 5. f. 12.
Petiver Gaz. t. 92. f. 7. *Martini Berl. Mag.* iv. t. 8. f. 23.
Inhabits rivers, ditches, and ponds in England. *Petiver, &c.* On aquatic plants in Denmark. *Muller.*
Shell about a quarter of an inch in diameter, of a pale horn-colour, covered when alive with a brown epidermis; the apex is depressed into a sub-umbilicus, and the base somewhat convex, and distinctly umbilicated, the whole shell, when slightly magnified, appears to be marked with very fine longitudinal and transverse decussated striæ.

CRYSTALLINA. 53. Shell with four rounded whirls, rather convex above, and the base largely umbilicated; aperture nearly orbicular, with a reflected margin.

Helix crystallina. *Muller Verm.* ii. p. 23. *Gmelin,* p. 3635.
Helix paludosa. *Da Costa Brit. Conch.* p. 59. *Montagu Test.* p. 440. *Maton and Racket, in Lin. Trans.* viii. p. 193. t. 5. f. 5. *Dorset Cat.* p. 53. t. 19. f. 25.
Variety. With the whirls regularly striated transversely, and the striæ membranaceous.
Helix crenella. *Montagu Test.* p. 441. t. 13. f. 3.
Turbo helicinus. *Lightfoot, in Lin. Trans.* lxxvi. t. 3. f. 1 to 4.
Inhabits Denmark among moss. *Muller.* Boggy places in England. *Da Costa, &c.*
Shell about one-tenth of an inch in diameter, white, with four well-defined rounded whirls; Mr. Montagu says that every gradation is to be seen, from the most strongly striated specimens of *H. crenella,* to those of *H. crystallina,* which are entirely destitute of any such mark.

CORNU-ARIETIS. 54. Shell smooth, concave on one side, and strongly umbilicated, with four striated whirls; aperture oval, dilated.

Helix Cornu Arietis. *Linnæus Syst. Nat.* p. 1244. *Born Mus.* p. 373. *Schroeter Fluss.* p. 230. t. 9. f. 13, and *Einl.* ii. p. 139. *Chemnitz,* ix. part 1. p. 109. t. 112. f. 952 and 953. *Gmelin,* p. 3625.

Planorbis contrarius. *Muller Verm.* ii. p. 152.

Planorbis Arietis. *Lamarck Syst. des Anim.* p. 94.

Lister Conch. t. 136. f. 40. *Petiver Gaz.* t. 92. f. 4. *Klein Ost.* t. 1. f. 7. *Knorr,* i. t. 2. f. 4 and 5. *Seba,* iii. t. 39. f. 14 and 15. *Geve,* t. 3. f. 9 to 13.

Inhabits rivers in Europe. *Linnæus.* Brazil. *Petiver.* Molucca Islands, Amboyna, and China. *Chemnitz.* Madrass and Jamaica. *Humphreys.*

Shell about three-quarters of an inch, or sometimes an inch and a quarter, in diameter, of a whitish colour, with from three to seven red bands of different breadths; it has four striated whirls, of which the outer increases rather rapidly in size towards the aperture. The shell which my friend Mr. Brookes, in his Introduction to Conchology, has figured under this name, appears to me to be more nearly allied to *H. Oculus-Capri.*

CORNU-VENATORIUM. 55. Shell with the spire much depressed, and the base strongly umbilicated; aperture roundish, and the lip notched near its junction with the whirl.

Helix Cornu venatorium. *Chemnitz,* ix. part 2. p. 104. t. 127. f. 1132 and 1133. *Gmelin,* p. 3641.

Le Veritable Cornet de Chasseur. *Favanne Cat. Rais.* p. 5. t. 1. f. 14.

Lister Conch. t. 1055. f. 4. ?

Inhabits ——

Shell about an inch in diameter, white, and the center whirl of the depressed spire is brownish; the base is concave, and forms a broad umbilicus, in which every whirl is visible; the aperture is round, and there is a notch on the margin of the outer lip near its lower junction with the body-whirl.

ERICETORUM. 56. Shell flattish above, with six wrinkled whirls, which are convex, and strongly umbilicated at the base.

Helix Ericetorum. *Muller Verm.* ii. p. 33. *Chemnitz,*

ix. part 2. p. 143. t. 132. f. 1193. *Gmelin*, p. 3632.
Montagu Test. p. 437. t. 24. f. 2. *Donovan*, v. t. 151.
f. 2. *Maton and Racket, in Lin. Trans.* viii. p. 194.
Dorset Cat. p. 53. t. 20. f. 8.
Helix albella. *Pennant Zool.* iv. p. 132. t. 85. f. 122.
Helix Erica. *Da Costa Brit. Conch.* p. 53. t. 4. f. 8.
Le Grand Ruban. *Geoffroy*, p. 47, No. 13. t. 2.
Lister Anim. Ang. t. 2. f. 13, and *Conch.* t. 78. f. 78.
Gualter, t. 3. f. P. *Martini Berl. Mag.* ii. t. 4. f. 46.
Inhabits heaths and other dry places, in England. *Lister, &c.*
France. *Geoffroy.* Italy. *Muller.*
Shell not much more than a quarter of an inch long, and thrice
as broad, of a whitish or pale yellowish brown colour, with
a purplish brown band on the upper part of the body-whirl,
which extends along the suture and more distinctly marks
the divisions of the other whirls; the body-whirl has also
most commonly from one to five other smaller bands at the
base; it is distinguished by its flattish upper surface, and
capacious umbilicus below.

STRIGATA. **57. Shell umbilicated, with five striated
whirls; spire rather convex, obtuse, and flattened at the summit; aperture roundish.**

Helix strigata. *Muller Verm.* ii. p. 61. *Gmelin*, p. 3632.
Helix cingenda. *Montagu Test.* p. 418. t. 24. f. 4. *Ma-
ton and Racket, in Lin. Trans.* viii. p. 195. t. 5. f. 6.
Dorset Cat. p. 53. t. 18. f. 5.
Helix zonaria. *Pennant Zool.* iv. p. 137. t. 85. f. 133.
Gualter, t. 2. f. H, and t. 3. f. C.
Inhabits Italy. *Muller.* Sandy shores of England, on maritime plants; but not common. *Maton and Racket.*
Shell about half an inch long, and three-quarters of an inch
broad, of a yellowish white colour, with several small purplish brown or chestnut bands, which are frequently interrupted, especially towards the summit of the spire; the base
has generally one broad band and several concentric lines;
with a glass the whole surface is seen to be minutely striated both longitudinally and transversely; the three upper
whirls of the spire are but little elevated, and the umbilicus
is nearly closed by the overlapping of the pillar-lip.

PISANA. **58. Shell umbilicated, with six whirls,
and the spire rather produced; aperture sub-
lunated, with a narrow rib round the inner
margin of the outer-lip.**

Helix Pisana. *Muller Verm.* ji. p. 60. *Chemnitz,* ix. part 2. p. 139. t. 132. f. 1186 and 1187.? *Gmelin,* p. 3631.

Helix media. *Gmelin,* p. 3640.

Helix virgata. *Montagu Test.* p. 415. t. 24. f. 1. *Maton and Racket, in Lin. Trans.* viii. p. 195. *Dorset Cat.* p. 53. t. 20. f. 7.

Helix lineata. *Olivi Adr.* p. 177.

Helix zonaria, Var. *Pennant Zool.* iv. p. 138. t. 85. f. 133 A.

Helix zonaria. *Donovan,* ii. t. 65.

Helix, No. 229. *Schroeter Einl.* ii. p. 241. t. 4. f. 8.?

Cochlea virgata. *Da Costa Brit. Conch.* p. 79. t. 4. f. 7. *Lister Conch.* t. 59. f. 56. *Petiver Gaz.* t. 17. f. 6, and t. 52. f. 12. *Gualter,* t. 2. f. L. *Schroeter Erdconch.* t. 2. f. 22, and f. 22 a.

Inhabits England, on dry banks and in sandy barren places. *Pennant, &c.* France and Spain. *Lister.* Italy. *Muller.*
Shell about half, or five-eighths of an inch long, and the length and breadth are equal; it is white with three or four reddish brown bands on the body-whirl, of which the broadest is extended to the summit of the spire; the uppermost whirls are produced, by which and its smaller size this species may be distinguished from *H. strigata;* the outer lip has more or less of a reddish tinge, and the rib which runs parallel to its inner margin is narrow, and considerably elevated.

NITIDA. 59. Shell umbilicated, pellucid, glossy, and convex on both sides, with five depressed horn-coloured whirls; aperture large.

Helix nitida. *Muller Verm.* ii. p. 32. *Chemnitz,* ix. part 2. p. 103. t. 127. f. 1130 and 1131.?

Helix lucida. *Pulteney's Dorset Cat.* p. 47. *Montagu Test.* p. 425. t. 23. f. 4.

Helix nitens. *Gmelin,* p. 3633. *Maton and Racket, in Lin. Trans.* viii. p. 198. t. 5. f. 7. *Dorset Cat.* p. 54. t. 19. f. 22.

La Luisante. *Geoffroy,* p. 36, No. 7. t. 2.

Petiver Gaz. t. 93. f. 14. *Gualter,* t. 2. f. G. *Argenville,* t. 28. f. 4. *Ström. in Act. Nidros.* iii. t. 6. f. 16.

Inhabits dry banks as well as moist woods, in England; common. *Petiver, &c.* France. *Geoffroy.* Denmark. *Muller.*
Shell sometimes two-tenths of an inch long, and half an inch in diameter, but is generally smaller; it is remarkably pel-

lucid, smooth and glossy, of a horn-colour, and has five whirls, with the spire nearly flat ; the base is more opake, and of a lighter colour, and has a large deep umbilicus. *H. ni-tida* of Gmelin is *Nautilus lacustris.*

TENUIS. **60.** Shell umbilicated, pellucid, and glossy, with six transversely striated whirls ; aperture semi-lunar.

Helix tenuis. *Gmelin*, p. 3641.
Helix lucida, Var. *Montagu Test.* p. 426.
Helix, No. 326. *Schroeter Einl.* ii. p. 270.
Cochlea terrestriformis, &c. *Schroeter Fluss.* p. 258. t. 5. f. 32.
Variety. Shell smaller, with only four whirls.
Helix pellucida. *Pennant Zool.* iv. p. 138.
Helix, No. 220. *Schroeter Einl.* ii. p. 239.
Cochlea umbilicata quatuor spirarum teretium. *Schroeter Fluss.* p. 259. t. 5. f. 33.
Martini Berl. Mag. iv. t. 7. f. 16.
Inhabits a water-course near Penzance ; and the Variety in wet drains and ditches about Newbury. *Montagu.*
Shell four or five lines in diameter, pellucid, and of a horn-colour : there can be no doubt that the Variety which Montagu has described of his *H. lucida* is the *H. tenuis* of Gmelin, and the following is Mr. Montagu's observation : " We have found a shell inhabiting wet places, and once alive under water, so much like this, that we dare not give it a distinct place, though we are much inclined to believe it a different species. It is more pellucid, the apex more produced, and never has that opake greenish-colour at the base : that found under water (which we take to be Schroeter's figure 32.) was of a light transparent horn-colour. Others taken in wet drains and ditches, about Newbury in Berkshire, are of a rufous horn-colour (which is probably Schroeter's fig. 33) : the animal is black, and when alive gives the shell a deep chocolate brown colour. This last we have also found on a swamp, in a wood belonging to Lord Clifford in Devonshire, rather lighter in colour."

CELLARIA. **61.** Shell umbilicated, pellucid, glossy, with five depressed yellowish whirls, and the base white ; aperture large.

Helix cellaria. *Muller Verm.* ii. p. 28. *Chemnitz*, ix.

part 2. p. 102. t. 127. f. 1129, No. 1, and No. 2. *Gmelin*, p. 3634.

Helix tenella. *Gmelin*, p. 3640.

Schroeter Erdconch. p. 200. t. 2. f. 26.

Inhabits cellars. *Muller.*

Muller says this shell is three lines and a half in diameter, and I cannot find from his description that it differs in any respect besides colour from *H. nitida*, and Schroeter's figure which he has quoted, as well as those of Chemnitz, are very like that species.

OBVOLUTA. **62.** Shell umbilicated, depressed on both sides with six reversed whirls ; aperture sub-triangular with a white reflected lip.

Helix obvoluta. *Muller Verm.* ii. p. 27. *Chemnitz*, ix. part 2. p. 101. t. 127. f. 1128. *Gmelin*, p. 3634.

Helix holosericea. *Gmelin*, p. 3641.

Gualter, t. 2. f. S. *Martini Berlin. Mag.* ii. t. 3. f. 37. *Schroeter Erdconch.* t. 2. f. 24 a.

Junior. With the whirls hairy.

La Veloutée à bouche triangulaire. *Geoffroy*, p. 46, No. 12. t. 2.

Inhabits wet places in the park of Meudon. *Geoffroy.* Saxony. *Schroeter.* Italy. *Muller.*

Shell four or five lines in diameter, white, glabrous, with the upper side flattish, particularly in the center, and the base convex ; it has six or seven reversed whirls, which when magnified appear to be minutely striated ; the umbilicus is wide and deep ; the aperture is triangular, with the outer lip reflected, white, polished, and slightly sinuated.

MINIMA. **63.** Shell sub-imperforate, ovate, conical, with the two last whirls placed in the center of the first ; aperture orbicular.

Helix minima. *Gmelin*, p. 3666.

Helix, No. 247. *Schroeter Einl.* ii. p. 247.

Buccinum trochiforme. *Schroeter Fluss.* p. 321. t. 7. f. 18.

Inhabits ——

Shell about a line and a half in diameter, with an obtuse spire consisting of two whirls placed in the center of the body-whirl ; it is not well defined, and is probably the Larva of some other species.

HISPIDA. 64. Shell umbilicated, convex, hispid, diaphanous, with five whirls; aperture roundish-lunated.

> Helix hispida. *Linnæus Syst. Nat.* p. 1244. *Muller Verm.* ii. p. 73. *Da Costa Brit. Conch.* p. 58. t. 5. f. 10. *Schroeter Einl.* ii. p. 141. *Chemnitz,* ix. part 2. p. 52. t. 122. f. 1057 and 1058. *Gmelin,* p. 3625. *Montagu Test.* p. 423. t. 23. f. 3. *Maton and Racket,* in *Lin. Trans.* viii. p. 198. *Dorset Cat.* p. 54. t. 21. f. 10.
> La Veloutée. *Geoffroy,* p. 44, No. 11. t. 2.
> *Petiver Gaz.* t. 93. f. 13.
> Inhabits Sweden. *Linnæus.* France. *Geoffroy.* Denmark and Switzerland. *Muller.* Wet and shady places in England. *Da Costa, &c.*
> Shell about a quarter of an inch long, and the breadth somewhat exceeds the length; it is sub-globular, thin, brittle, and diaphanous, with five rounded whirls covered with fine thick-set short downy hairs; umbilicus small. The young of *H. rufescens* are hairy, and have been mistaken for this species; but they differ in being flatter and more opaque, and in having the umbilicus much larger.

UMBILICATA. 65. Shell sub-conical, semi-pellucid, with five rounded striated whirls, and the umbilicus very large.

> Helix umbilicata. *Montagu Test.* p. 434. t. 13. f. 2. *Maton and Racket, in Lin. Trans.* viii. p. 200. *Dorset Cat.* p. 54. t. 19. f. 24.
> Inhabits walls near Tenby, and some parts of England, under tiles and loose stones. *Montagu.*
> Shell about one-tenth of an inch in diameter, and the length not quite so much; it differs from young shells of *H. rotundata* in having the spire more produced, in not having the least appearance of a carinated edge on the body-whirl, and in being much less strongly striated; it is of an uniform dark horn-colour.

COSTATA. 66. Shell slightly depressed, with four transversely plaited whirls; aperture round, and the outer lip white and reflected.

> Helix costata. *Muller Verm.* ii. p. 81. *Gmelin,* p. 3633.
> Inhabits high places about Frederickdal in Denmark, very rare. *Muller.*

Muller says this shell is one line in diameter, rather convex, brownish above, and of a paler colour below; it has four whirls with transverse ribs or plaits, of which thirty may be counted on the body-whirl.

PULCHELLA. 67. Shell umbilicated, rather depressed, with hardly four whirls; aperture roundish, and the outer lip reflected.

Helix pulchella. *Muller Verm.* ii. p. 30. *Gmelin*, p. 3633.
La petite Striée. *Geoffroy*, p. 35, No. 6. t. 2.
Inhabits woods, under stones in moist places and among moss, about Paris; common. *Geoffroy*. Frequent in Denmark. *Muller*.
Shell one line in diameter, white, smooth, rather pellucid, flattish above and somewhat convex beneath; it has three and a half or four whirls, which are minutely striated, but the striæ are not visible to the naked eye.

TROCHULUS. 68. Shell sub-umbilicated, somewhat conical, with six glabrous whirls; aperture sub-lunated, and transversely compressed.

Helix Trochulus. *Muller Verm.* ii. p. 79.
Helix trochiformis. *Montagu Test.* p. 427. t. 11. f. 9. *Maton and Racket, in Lin. Trans.* viii. p. 200.
Trochus terrestris, Var. A. *Da Costa Brit. Conch.* p. 35.
Trochilus sylvestris, &c. *Lister Anim. Ang.* p. 123.
Inhabits England, in decayed wood, and among the moss about the roots of trees. *Lister, &c.* Decayed wood in Denmark. *Muller*.
Shell about an eighth of an inch both in length and breadth, thin, pellucid, and of a horn-colour with sometimes a reddish tinge; it has six produced rounded whirls, strongly marked by the sutures, and Mr. Montagu says it has much the habit of a Trochus, but the lunated aperture forbids it being placed in that Genus.

ACULEATA. 69. Shell umbilicated, somewhat conical, with five transversely striated whirls, and the striæ membranaceous and bristly.

Helix aculeata. *Muller Verm.* ii. p. 81. *Chemnitz*, ix. part 2. p. 153. t. 133. f. 1209. *Gmelin*, p. 3638.
Helix spinulosa. *Lightfoot in Phil. Trans.* lxxvi. p. 166,

t. 2. f. 1 to 5. *Montagu Test.* p. 429. t. 11. f. 10. *Maton and Racket, in Lin. Trans.* viii. p. 201. *Dorset Cat.* p. 54. t. 19. f. 23.

Inhabits woods in Denmark. *Muller.* England. *Lightfoot, &c.*

Shell about one-tenth of an inch both in length and breadth, thin, semi-pellucid, and of a brownish horn-colour, with regular membranaceous bristly striæ; Mr. Montagu says " the striæ of this species seem to be formed by the epidermis, which rises in some parts into thin flat bristly processes that give it a spinous appearance; neither these nor the striæ are testaceous, but flexible and easily rubbed off."

LACUNA. 70. Shell sub-globular, with four ventricose smooth whirls; aperture large and nearly orbicular, and the pillar sinuated.

Helix Lacuna. *Montagu Test.* p. 428. t. 13. f. 6. *Maton and Racket, in Lin. Trans.* viii. p. 201.

Inhabits the sea-shore near Southampton, and other parts of the West of England, about high-water mark. *Montagu.*

Shell about a quarter of an inch both in length and breadth, of a pale horn-colour, sometimes marked with two reddish brown bands. Mr. Montagu says, " the first whirl is large, the two uppermost very small, and placed somewhat lateral; pillar-lip thick, white, and grooved with a long canal or gutter, which terminates in a small but deep umbilicus." Maton and Racket say it is nearly allied to *H. ampullacea,* and Montagu has compared it with *Nerita pallidula,* from which it strikingly differs in the much smaller size of the body-whirl and aperture.

AMPULLACEA. 71. Shell sub-umbilicated, ventricose, nearly globular, and the spire depressed; aperture ovate-oblong, and the umbilicus small.

Helix ampullacea. *Linnæus Syst. Nat.* p. 1244. *Born Mus.* p. 374. *Schroeter Fluss.* p. 249. t. 6. f. 2, and t. 9. f. 14; and *Einl.* ii. p. 142. *Chemnitz,* ix. part 2. p. 105. t. 128. f. 1133 to 1135. *Gmelin,* p. 3626. *Shaw Nat. Misc.* xiv. t. 576. *Brookes's Introd.* p. 129. t. 8. f. 112.

Nerita ampullacea. *Muller Verm.* ii. p. 172.

Bulimus ampullaceus. *Bruguiere Enc. Meth.* p. 297.

Lister Conch. t. 130. f. 30. *Rumphius,* t. 27. f. Q. *Pe-*

tiver Amb. t. 12. f. 14.   *Gualter,* t. 1. f. R.   *Seba,* iii.
t. 38. f. 1 to 7.   *Geve,* t. 27. f. 289 and 291.   *Knorr,*
v. t. 5. f. 2.   *Martini Berlin. Mag.* iii. t. 6. f. 68. *Favanne,* t. 61. f. D 8.

Inhabits Asia. *Linnæus.*   Amboyna. *Rumphius.*   Jamaica.
Lister.   Rivers in St. Domingo and Guadaloupe.   *Bruguiere.*

Shell about two inches long, and the breadth rather less, of a bluish or greyish white colour, with several darker transverse bands, and covered with rather a pale green epidermis; the body-whirl is ventricose, especially toward the upper end, at which it becomes more or less flattened; the spire is comparatively small, depressed, and consists of five whirls; the inside is striped with brown; the umbilicus is small, and nearly covered by the lip, which on that side is reflected.

URCEUS.   72.   Shell ventricose, nearly globular, wrinkled longitudinally, and the spire somewhat produced; aperture ovate-oblong, and the umbilicus large.

Helix plicata.   *Portland Cat.* p. 18, lot 400, and p. 91, lot 2027.
Helix maxima.   *Chemnitz,* ix. part 2. p. 108. t. 128. f. 1136.
Helix ampullacea, Var. β.   *Schroeter Einl.* ii. p. 143. *Gmelin,* p. 3626.
Nerita Urceus.   *Muller Verm.* ii. p. 174.   *Schroeter Fluss.* p. 253.
Bulimus Urceus.   *Bruguiere Enc. Meth.* p. 298.
Ampullaria rugosa.   *Lamarck Syst. des Anim.* p. 93.
Lister Conch. t. 125. f. 25.   *Favanne,* t. 61. f. D 10.

Inhabits the Indian Islands. *Muller.*   River Mississippi. *D'Avila.*   Barbadoes. *Humphreys.*

Shell four or five inches long, and almost equally broad, thick, and of a whitish colour covered with a dark brown epidermis; the body-whirl is strongly wrinkled or somewhat plaited longitudinally, and the spire is more produced, and the umbilicus much larger than in *H. ampullacea;* the inside is white. It is generally known by the name of the *Cocoa Nut Snail.*

GLAUCA.   73.   Shell ventricose, with the spire depressed, and the summit pointed; aperture effuse, with the inside orange, and the umbilicus very large.

Helix glauca. *Linnæus Syst. Nat.* p. 1245. *Born Mus.*
p. 377. *Kæmmerer Cab. Rud.* p. 185. *Schroeter Einl.*
ii. p. 145. *Gmelin*, p. 3628.
Helix effusa. *Chemnitz*, ix. part 2. p. 118. t. 129. f. 1144
and 1145.
Helix ampullacea, Var. γ. *Gmelin*, p. 3626.
Helix Oculus communis. *Gmelin*, p. 3621.
Helix, No. 182. *Schroeter Einl.* ii. p. 226 and No. 82.
p. 201.
Nerita effusa. *Muller Verm.* ii. p. 175. *Schroeter Fluss.*
p. 255.
Bulimus effusus. *Bruguiere Enc. Meth.* p. 296.
Limax vittatus. *Martyn Univ. Conch.* iii. t. 118.
Lister Conch. t. 129. f. 29. *Seba*, iii. t. 40. f. 3 to 5.
Knorr, v. t. 5. f. 3. *Geve*, iii. t. 3. f. 20. *Martini Na-*
turf. ii. t. 79. f. 1.
Inhabits rivers in Guadaloupe. *Bruguiere.* West Indies.
Martyn.
Shell commonly about an inch long, and an inch and a half
broad, but is sometimes, though rarely, almost twice as
large; the colour is whitish or pale greyish brown, with
from two to twelve transverse brown bands on the body-
whirl, and the inside is of a bright orange colour; the aperture is oval and effuse, and the umbilicus remarkably large.
Linnæus has described *H. glauca* 'labro marginato,' but
the present shell answers so well in other respects to his description, that I have no hesitation in considering it the same.
Meuschen, in the Museum Geversianum, p. 256, No. 169,
erroneously asserts that the Linnæan *H. ampullacea* and
H. glauca are the same species.

VITREA. 74. Shell sub-umbilicated, ovate, ventricose, smooth, with the whirls of the spire
keeled in the middle.

Helix vitrea. *Born Mus.* p. 383. t. 15. f. 15 and 16.
Gmelin, p. 3622. *Chemnitz*, xi. p. 282. t. 210. f. 2072
and 2073.
Helix, No. 204. *Schroeter Einl.* ii. p. 234.
Bulimus vitreus. *Bruguiere Enc. Meth.* p. 282.
Inhabits ——
Shell twenty-two lines long and eighteen broad, sub-pellucid,
brittle, of a yellowish brown colour, with longitudinal undulated sulphur-coloured bands; it has five whirls, of which
the body-whirl is ventricose, and those of the spire trans-
z 2

versely keeled in the middle; the aperture is oval-oblong,
and a good deal resembles that of *H. ampullacea.*

SULTANA.   75. Shell ventricose, very thin, much
     variegated, and minutely striated longitudi-
     nally; aperture patulous and ovate.

Helix Gallina Sultana.   *Chemnitz,* xi. p. 281. t. 210. f. 2070
     and 2071.
La Poule Sultane.   *Favanne Cat. Rais.* p. 13. t. 1. f. 47.
Inhabits New Zealand.   *Favanne.*
Shell two inches and a half long, and an inch and a half broad,
     prettily variegated with brown, white, and yellow spots and
     bands both inside and out; it has five whirls, and Favanne
     says they are marked with almost imperceptible reticulated
     striæ.

POMATIA.   76. Shell sub-umbilicated, ovate, ventri-
     cose, with five longitudinally wrinkled whirls;
     aperture roundish-lunated, and the outer lip
     slightly marginated.

Helix Pomatia.   *Linnæus Syst. Nat.* p. 1244.   *Muller
     Verm.* ii. p. 243.   *Pennant Zool.* iv. p. 134. t. 84. f. 128.
     Da Costa Brit. Conch. p. 67. t. 4. f. 14.   *Born Mus.*
     p. 375.   *Schroeter Einl.* ii. p. 143.   *Chemnitz,* ix. part 2.
     p. 111. t. 128. f. 1138.   *Gmelin,* p. 3627.   *Donovan,* iii.
     t. 84.   *Montagu Test.* p. 405.   *Maton and Racket, in
     Lin. Trans.* viii. p. 201.   *Lamarck Syst. des Anim.* p. 94.
     Dorset Cat. p. 54. t. 20. f. 14.   *Brookes's Introd.* p. 130.
     t. 8. f. 110.
Le Vigneron.   *Geoffroy,* p. 24. No. 1. t. 2.
Lister Conch. t. 48. f. 46.   *Gualter,* t. 1. f. A.   *Argenville,*
     t. 28. f. 1.   *Geve,* t. 29. f. 334 and 335.   *Martini Berl.
     Mag.* ii. t. 1. f. 1.   *Schroeter Erdconch.* t. 1. f. 10.
Variety.   With the whirls reversed.
Helix Pomaria.   *Muller Verm.* ii. p. 45.   *Born Mus.*
     p. 376. t. 14. f. 21 and 22.   *Schroeter Inn. Bau Conch.*
     p. 63. t. 5. f. 11.   *Chemnitz,* ix. part 1. p. 77. t. 108.
     f. 908 to 912.
Lister Anim. Ang. t. 2. f. 1, and *Conch.* t. 33. f. 32.   *Fa-
     vanne,* t. 63. f. E.
Junior?   Smaller, with four whirls, and five reddish bands.
Helix ligata.   *Muller Verm.* ii. p. 58.   *Gmelin,* p. 3631.
Helix Pomatia, Var.   *Chemnitz,* ix. part 2. p. 110. t. 128.
     f. 1137.

Inhabits woods in France, England, and Sweden. *Linnæus.* Italy. *Da Costa.* Germany. *Born.*

Shell two or three inches long, and nearly equally broad, of a pale dull yellowish brown colour, with three obsolete darker bands on the body-whirl, or sometimes nearly white without any bands; it has five rounded whirls, with a rather small but produced spire; the margin of the outer lip is thickened, and the inner lip much reflected over the umbilicus. The animals were formerly bred and fattened in stews for the table by the Romans; and Varro says they were made to attain such a size, that a shell has been known to hold ten quarts! *H. ligata* is said to be only fourteen lines long, and is probably a young shell of this species.

SCALARIS. 77. Shell conical, sub-turreted, with five ventricose almost unconnected whirls, and the aperture ovate.

Helix scalaris. *Muller Verm.* ii. p. 113. *Chemnitz,* ix. part 2. p. 114. t. 128. f. 1139. No. 1. *Gmelin,* p. 3652. *Argenville Zoom.* t. 9. f. 8. *Martini Berl. Mag.* ii. t. 1. f. 5. *Ginanni Op. Post.* ii. t. 1. f. 6. *Favanne,* t. 76. f. L.

Inhabits France. *Argenville.* Italy. *Ginanni.*

This shell is of the same size and colour as *H. Pomatia,* from which it differs in having the whirls remarkably produced and almost unconnected, and is probably nothing more than an accidental variety. Chemnitz has given three figures, of which the two smaller appear to me to belong to a perfectly distinct species.

CINCTA. 78. Shell sub-umbilicated, nearly globular, with five transversely fasciated whirls, and parallel striæ; outer lip red.

Helix cincta. *Muller Verm.* ii. p. 58. *Gmelin,* p. 3630.

Inhabits ——

This shell is said to be eighteen lines in diameter, and to differ from *H. ligata* of Muller in being larger, with five instead of four whirls; it has also the outer lip red instead of white, and the transverse reddish bands are accompanied by very fine parallel striæ, which are however not discernible without a glass.

ROSACEA. 79. Shell sub-umbilicated, nearly globular, of a flesh-colour, and transversely striated; outer lip reflected and brown.

Helix rosacea. *Muller Verm.* ii. p. 76. *Gmelin*, p. 3636.
Inhabits ——

Shell nineteen lines in diameter, ventricose, and flesh-coloured
both inside and out; it has five whirls which are finely
striated transversely, and is said to be nearly allied to *H.
Pomatia*, but has the umbilicus much larger, so as to expose the first whirl of the spire.

EXTENSA. 80. Shell sub-umbilicated, nearly globular, with four whirls, of a pale colour without spots, and the aperture large.

Helix extensa. *Muller Verm.* ii. p. 60. *Gmelin*, p. 3631.
Inhabits ——

Shell sixteen lines in diameter, and is nearly allied to *H. Pomatia*, from which it is said to differ in having the spire less
prominent, the aperture larger, and only four, instead of five
whirls.

CITRINA. 81. Shell smooth, convex, with five whirls,
and the spire rather depressed; umbilicus
nearly closed by the reflected lip.

Helix citrina. *Linnæus Syst. Nat.* p. 1245. *Muller Verm.*
ii. p. 63. *Born Mus.* p. 377. t. 13. f. 14 and 15, and
t. 15. f. 1 to 10. *Schroeter Einl.* ii. p. 146. *Chemnitz,*
ix. part 2. p. 130. t. 131. f. 1167 to 1175. *Gmelin,*
p. 3628. *Shaw Nat. Misc.* xvi. t. 633.

Lister Conch. t. 54. f. 50, and t. 60. f. 57. *Gualter,* t. 3.
f. D and E. *Argenville,* t. 28. f. 10. *Seba,* iii. t. 39.
f. 1 to 5. *Geve,* t. 26. f. 277 to 279, 281, 282, and
285. *Martini Berl. Mag.* ii. t. 4. f. 38. *Favanne,* t. 63.
f. I 1.

Inhabits Jamaica; and is a land shell. *Linnæus.* China.
D'Avila.

Shell about three-quarters of an inch long, and almost twice as
broad, convex on both sides, and minutely wrinkled longitudinally; it is a pretty species, and Muller has enumerated
the following varieties: A. Sulphur-coloured, with a single
white band; B. Sulphur-coloured, with two broad white
bands; C. Sulphur-coloured, with two white bands, of
which the upper is narrowest; D. Sulphur-coloured, with a
red and white band united; E. Sulphur-coloured, with a very
broad white band divided by a red line; F. Yellow, with a
white-and red band united, and the base white; G. Tawny,
with the band and base white; H. Tawny, with a red band

divided by a white one; I. Reddish brown, with a single white band; K. Reddish brown, with a red and white band united; L. Blackish brown, with the band and base white; M. Pale purple, with a sulphur-coloured band; N. Pale brown, with the band and base white; O. White, with a blackish red broad band; P. White, with two narrow red bands; Q. Yellowish green, with scattered black dots and a red band and lines, and the base white.

RAPA. 82. Shell sub-umbilicated, nearly globular, with minute excavated dots, and the upper part of the body-whirl deflected inwardly.

Helix Rapa. *Muller Verm.* ii. p. 67. *Chemnitz*, ix. part 2. p. 134. t. 131. f. 1176. *Gmelin*, p. 3629.
Inhabits ———
Shell about an inch and a half in diameter, white, becoming yellowish upwards, and marked transversely with a red band; it is punctured all over with excavated dots, which are so minute, that the naked eye can hardly discover them. Muller suspects that young shells have an open umbilicus, and that it becomes with age almost entirely closed over by the reflected pillar-lip.

CASTANEA. 83. Shell umbilicated, sub-globular, with seven striated whirls; aperture without any pillar-lip.

Helix castanea. *Muller Verm.* ii. p. 67. *Chemnitz*, ix. part 2. p. 135. t. 131. f. 1177 and 1178. *Gmelin*, p. 3629.
Inhabits ———
Shell about an inch and a half in diameter, thick, of a dull yellowish chestnut-colour, with a darker and white united band round the middle of the body-whirl, which is continued along the sutures of the spire; it is said to have seven whirls, with crowded striæ, and a moon-shaped aperture without any pillar-lip. According to Muller, it is an intermediate species between *H. citrina* and *H. Algira*.

GLOBULUS. 84. Shell sub-umbilicated, thickish, somewhat globular, white, with five whirls, and the spire produced; outer lip marginated.

Helix Globulus. *Muller Verm.* ii. p. 68. *Chemnitz*, ix.

part 2. p. 126. t. 130. f. 1159 and 1160.   *Gmelin,*
p. 3629.

Lister Conch. t. 44. f. 41, and t. 46. f. 44. *Gualter,* t. 2. f. C.
Inhabits Tranquebar. *Chemnitz.* Madagascar. *Humphreys.*
Shell about an inch long, and almost equally broad, generally
of a milk-white colour, but sometimes slightly tinged with
red; it has five whirls, which are irregularly wrinkled transversely, and the spire produced; the lip is strongly marginated, and reflected over the umbilicus.

INCISA.   85. Shell umbilicated, depressed, white,
    deeply cut at the margin, and the aperture
    roundish.

Helix incisa.   *Chemnitz,* ix. part 2. p. 129. t. 130. f. 1166.
    Gmelin, p. 3630.
Favanne, t. 64. f. S.
Inhabits the East Indies; and is a land shell.   *Chemnitz.*
This shell appears, by Chemnitz's figure, to be about half an
inch in diameter; but neither the figure or description is
sufficiently clear, and I am quite unacquainted with the
species.

LUCANA.   86. Shell umbilicated, nearly globular,
    with five whirls, and the base gibbous; outer
    lip white and reflected.

Helix Lucana.   *Muller Verm.* ii. p. 75.   *Chemnitz,* ix.
    part 2. p. 124. t. 130. f. 1155.
Helix Lucena.   *Gmelin,* p. 3636.
Helix, No. 269.   *Schroeter Einl.* ii. p. 253. t. 4. f. 9.
Argenville, t. 28. f. 7. ?   *Favanne,* t. 64. f. M 1.
Inhabits Tranquebar.   *Schroeter.*
Shell rather more than an inch in diameter, glabrous, pellucid, without striæ, band, or spot; the colour is either
wholly white, or yellowish above; it has five whirls, of
which the body-whirl is very convex, and the spire obtuse;
aperture lunated, with a thick reflected somewhat silvery
outer lip, and the umbilicus pervious.

ARBUSTORUM.   87. Shell sub-umbilicated, some-
    what globular, with five transversely wrinkled
    whirls, and the umbilicus nearly concealed
    by a white reflected lip.

Helix Arbustorum.   *Linnæus Syst. Nat.* p. 1245.   *Muller Verm.* ii. p. 55.   *Pennant Zool.* iv. p. 136. t. 85.

f. 130. *Schroeter Einl.* ii. p. 147. *Chemnitz* ix. part 2.
p. 148. t. 133. f. 1202. *Gmelin*, p. 3630. *Montagu
Test.* p. 413. *Donovan*, iv. t. 136. *Maton and Racket,
in Lin. Trans.* viii. p. 202. *Dorset Cat.* p. 54. t. 2. f. 6.
Helix unifasciata. *Da Costa Brit. Conch.* p. 75. t. 17. f. 6.
Lister Conch. t. 2. f. 4, and *Conch.* t. 56. f. 53. *Seba*, iii.
t. 38. f. 68. *Geve*, t. 30. f. 354 to 356. *Martini Berl.
Mag.* ii. t. 3. f. 23.

Inhabits shrubs in Europe. *Linnæus.* Wet shady places in
England. *Lister, &c.* Denmark. *Muller.*

Shell sometimes three-quarters of an inch long, and an inch
broad, but is generally smaller; the colour is usually pale
grey, mottled with brown in streaks and lines, or approach-
ing to chestnut, with most commonly a darker band in the
middle of the body-whirl, which extends along the suture of
the spire; in old full-grown shells the umbilicus is almost
entirely concealed by the lip, which has a white reflected
margin all round.

FRUTICUM. **88. Shell umbilicated, nearly globular,
with six whirls; umbilicus large, and the
inner margin of the outer lip white.**

Helix Fruticum. *Muller Verm.* ii. p. 71. *Chemnitz,* ix.
part 2. p. 150. t. 133. f. 1203. *Gmelin*, p. 3635.
Helix terrestris. *Gmelin*, p. 3639.
Schroeter Erdconch. p. 178. t. 2. f. 19.

Inhabits hedges in Denmark; generally on Urticæ, Rubi, or
Serratulæ. *Muller.*

Shell about seven lines and a half in diameter, and is said to
resemble *H. nemoralis*, from which it differs in being more
brittle and more pellucid, and in having six whirls. Muller
has described the following varieties: A. Pale yellow, with
scattered gold-coloured spots; B. Pale yellow, with black
spots and dots; C. Sulphur-coloured, with black spots and
dots; D. Sulphur-coloured, with a purple band and a few
black spots; E. Yellowish, with a purple band and crowded
black dots; F. Brown, with black spots and dots; G. Deep
red, with black spots and dots; H. Deep red, with a purple
band and dark spots and dots.

FULVA. **89. Shell sub-umbilicated, globular, pellucid, with seven transversely striated whirls of
a tawny colour, and the lip white.**

Helix fulva. *Muller Verm.* ii. p. 56. *Gmelin*, p. 3630.

Inhabits marshy woods about Fredericksdal, in Denmark, among decaying beech leaves; very rare. *Muller.*

Shell varying from half a line to three lines in diameter, of an amber colour, pellucid, glabrous, with seven or eight whirls, and very fine transverse striæ; aperture arcuated and narrow, with the margin of the lip acute; the lip, when full grown, is white, but in young shells is of the same colour as the whirls.

EPISTYLIUM. 90. Shell sub-umbilicated, rather pyramidal, with eight whirls, and the summit obtuse; umbilicus nearly closed by the pillar-lip.

Helix Epistylium. *Muller Verm.* ii. p. 57. *Gmelin*, p. 3630. *Brookes's Introd.* p. 163. t. 8. f. 114.?

Helix Cookiana. *Gmelin*, p. 3642.

Trochus australis. *Chemnitz*, ix. part 2. p. 49. t. 122. f. 1049 and 1050.

Trochus alveolatus. *Portland Cat.* p. 52. No. 1240.

Lister Conch. t. 62. f. 60. *Favanne*, t. 64. f. O 4.

Inhabits Jamaica. *Lister.* South Sea Islands. *Chemnitz.*

Shell nearly an inch long, and almost equally broad, of an uniform whitish colour, with eight obliquely striated whirls; the base is rather convex, less strongly striated than the spire, and the umbilicus is nearly closed by the pillar-lip; in the throat of my specimen there are three large teeth in a row. It is called the *Bee-hive Snail* in the Portland Catalogue.

PAPILLA. 91. Shell sub-umbilicated, obtusely pyramidal, wrinkled, and the summit flat and glabrous; aperture transverse and marginated.

Helix Papilla. *Muller Verm.* ii. p. 100. *Gmelin*, p. 3660.

Trochus Papilla. *Chemnitz*, ix. part 2. p. 51. t. 122. f. 1053 and 1054.

Inhabits ——

Shell about eleven lines long and ten broad, with oblique longitudinal striæ, and alternate reddish brown and white stripes; it has five or six whirls, which form an obtuse flattish summit, not wrinkled or striated like the lower whirls; the aperture is ovate, nearly horizontal, and has a thick marginated whitish reflected lip.

NEMORENSIS. 22. Shell umbilicated, nearly globular, thick, shining, with five whirls, and the summit depressed.

Helix nemorensis. *Muller Verm.* ii. p. 62. No. 257 (see Errata). *Gmelin,* p. 3632.
Helix cretacea. *Born Mus.* p. 376. t. 16. f. 1 and 2. *Chemnitz,* ix. part 2. p. 119. t. 129. f. 1146 and 1147. *Gmelin,* p. 3639.
Helix, No. 206. *Schroeter Einl.* ii. p. 234.
Inhabits the East Indies. *Chemnitz.*
Shell ten lines long and fourteen broad, white, or tinged with citron or flesh-colour, and marked with from one to three brown bands on the body-whirl; it differs from *H. nemoralis* in being larger, thicker, stronger, and more polished, and the spire is more produced than in *H. citrina;* it has five and a half or six whirls.

VITTATA. 93. Shell umbilicated, nearly globular, with five transversely striped whirls; outer lip reflected and white, and the throat blackish.

Helix vittata. *Muller Verm.* ii. p. 76. *Chemnitz,* ix. part 2. p. 142. t. 132. f. 1190 to 1192. *Gmelin,* p. 3636. *Lister Conch.* t. 67. f. 66.? *Knorr,* v. t. 21. f. 4.
Inhabits Coromandel. *Muller.* Tranquebar. *Chemnitz.*
Shell about nine lines in diameter, white, with several brown bands of different breadths and shades, which Muller has compared to the concentric veins of a piece of dried wood; it has five whirls, and the summit of the spire is blackish blue; the aperture is dark brown or blackish, and the umbilicus exposes the first whirl.

ZONARIA. 94. Shell umbilicated, with a convex rather depressed spire; aperture oblong, with a broad white reflected margin.

Helix zonaria. *Linnæus Syst. Nat.* p. 1245. *Muller Verm.* ii. p. 35. *Born Mus.* p. 378. *Schroeter Einl.* ii. p. 148. *Chemnitz,* ix. part 2. p. 140. t. 132. f. 1188 and 1189. *Gmelin,* p. 3632.
Gualter, t. 3. f. L. *Seba,* iii. t. 40. f. 52, 53, and 55. *Knorr,* v. t. 21. f. 3.
Inhabits the South of Europe; and is a land shell. *Linnæus.* Barbary. *Gmelin.*

Shell about an inch in diameter, and the breadth exceeds the
length, but not in half the proportion of eleven to four, as
Born has described; Linnæus says it has only four whirls,
but a shell, which in other respects answers the description,
has five, and the umbilicus is large.   Muller has enumerated the following varieties: A. White, with a dorsal red
band; B. White and spotless, with two dorsal red bands;
C. White with milky spots, and two dorsal red bands; D.
White, with three dorsal red bands; E. White, with five dorsal red bands; F. White, with two ventral and one dorsal
red band; G. White, with two ventral and two dorsal red
bands; H. White, with four red bands, of which the uppermost is very narrow; I. White, with five red bands, of which
the dorsal are narrowest; K. Uniformly yellow; L. Yellow,
with a red band; M. Pale flesh-colour, with marginal white
and red alternate bands; N. Pale flesh-colour above, and the
base white, with red and white alternate bands; O. White
with red spots, and six unequal red bands; P. Grey spotted
with yellow, and three red bands.

STRIATA.   95. Shell umbilicated, and striated, with
    six whirls, and the spire depressed; umbili-
    cus distinct and pervious.

Helix striata. *Muller Verm.* ii. p. 38.   *Gmelin,* p. 3632.
Helix nivea. *Gmelin,* p. 3639.
Helix, No. 60.   *Schroeter Erdconch,* p. 183. t. 2. f. 20.
Inhabits Saxony; and is a land shell.   *Schroeter.*
Muller says this shell is half an inch in diameter, altogether
    white, and sharply striated; it is convex below and rather so
    above, with nearly six rounded whirls without any keel, and
    the aperture lunated.

UNGULINA.   96. Shell convex, with the summit of
    the spire and the base umbilicated; outer
    lip white and reflected.

Helix ungulina. *Linnæus Syst. Nat.* p. 1245.   *Muller
    Verm.* ii. p. 69.   *Born Mus.* p. 379. t. 15. f. 11 and 12.
    Schroeter Einl. ii. p. 149.   *Chemnitz,* ix. part 2. p. 81.
    t. 125. f. 1098 and 1099. *Gmelin,* p. 3635.
Helix badia. *Gmelin,* p. 3639.
Helix, No. 203.   *Schroeter Einl.* ii. p. 233.
Rumphius, t. 27. f. R.   *Petiver Amb.* t. 12. f. 13.   *Klein
    Ost.* t. 1. f. 11.   *Seba,* iii. t. 40. f. 11.   *Geve,* t. 3. f. 15.
Inhabits India. *Linnæus.*   Ceylon. *Humphreys.*

Shell about an inch long, and an inch and a half broad, of a shining yellowish red or chestnut-colour, often marked with a white band round the middle of the body-whirl; the body-whirl is large, and the others small, forming a depressed sub-umbilicated spire; the base is convex, and has a large umbilicus; aperture lunated, with a white reflected lip, and the throat purplish.

ITALA. 97. Shell umbilicated, convex, obtuse, with five rounded whirls; umbilicus wide, and the aperture sub-orbicular.

Helix Itala. *Linnæus Syst. Nat.* p. 1245. *Gmelin,* p. 3636.
Inhabits the South of Europe; and is a land shell. *Linnæus.*
Linnæus has not given any reference, and to the above specific character has only added the following short description: " Shell as large as a hazel-nut, and white, with one brown band." Da Costa and Montagu have conjectured that *H. Ericetorum* may be this species.

LUSITANICA. 98. Shell convex, obtuse, with five rounded reversed whirls, and the body-whirl ventricose; umbilicus large.

Helix Lusitanica. *Linnæus Syst. Nat.* p. 1245. *Gmelin,* p. 3636.
Helix Guineensis. *Chemnitz,* ix. part 1. p. 80. t. 108. f. 913 and 914, and x. p. 367. t. 173. f. 1684 and 1685.
Helix Varica. *Muller Verm.* ii. p. 70. *Gmelin,* p. 3635.
Helix crepuscularis. *Gmelin,* p. 3640.
La Prune de Reine Claude. *Favanne Cat. Rais.* p. 26. t. 2. f. 107.
Gualter, t. 2. f. T. *Martini n. Mannig.* iv. p. 422. t. 5. f. 20.
Inhabits the South of Europe; and is a land shell. *Linnæus.* Marshes in Guinea. *Chemnitz.*
Shell about one and a half or two inches long, and nearly globular, of a yellowish brown or whitish colour, or sometimes brown marbled with white, and coated with a dark green epidermis; the aperture is roundish, with the lip acute, and the umbilicus large; it is furnished with a leathery diaphanous operculum. Linnæus has quoted Gualter, t. 2. f. T, and this shell answers his description in other respects, though he has not noticed the reversed direction of the whirls.

MAMMILLARIS. 99. Shell umbilicated, ovate, with three striated whirls, and the body-whirl very large; aperture ovate, and large.

Helix mammillaris. *Linnæus Syst. Nat.* p. 1246. *Born Mus.* p. 380. t. 15. f. 13 and 14. *Gmelin*, p. 3636.
Nerita adusta. *Chemnitz*, v. p. 278. t. 189. f. 1926 and 1927. *Museum Leskianum*, p. 290.
Nerita melanostoma. *Gmelin*, p. 3674.
Nerita, No. 21. *Schroeter Einl.* ii. p. 311.
Le Téton de Negresse. *Favanne*, ii. p. 290. t. 11. f. H 3.
Lister Conch. t. 566. f. 15. *Seba*, iii. t. 41. f. 20.
Inhabits rivers in Guinea. *Linnæus.* Philippines, Isle of France, St. Domingo, and Martinique. *Favanne.*
Shell about an inch and a half long, and thirteen lines broad, of a whitish colour, with three broad obsolete purplish brown bands, which are visible in the throat, and the inner lip is brown; the body-whirl comparatively with the spire is very large, and is marked with fine transverse crowded striæ, and more remote longitudinal ones; the aperture is large, and has the pillar-lip reflected over the umbilicus. The present shell would stand better with the Nerites, but I do not think there can be any doubt that it is the Linnæan *H. mammillaris.*

HISPANA. 100. Shell convex, with five rounded whirls, and the aperture sub-orbicular; umbilicus narrow, and perforated.

Helix Hispana. *Linnæus Syst. Nat.* p. 1246. *Gmelin*, p. 3637.
Inhabits the South of Europe. *Linnæus.*
To the above description Linnæus, without any reference, has only added that the shell is of a horn-colour; and this must therefore most probably always remain a very uncertain species.

LUTARIA. 101. Shell umbilicated, ovate-oblong, more strongly coloured internally, and the aperture sub-ovate.

Helix lutaria. *Linnæus Syst. Nat.* p. 1246. *Gmelin*, p. 3637.
Inhabits ——
This is another doubtful species, which no subsequent author has noticed; and Linnæus, in the Mus. Lud. Ulr., describes

it to be ovate-conical, minutely striated, with the umbilicus oblique, and the aperture white within.

OVATA. 102. Shell sub-umbilicated, ovate, ventricose, longitudinally wrinkled, with the summit and outer lip rose-coloured, and the pillar white.

Helix ovata. *Muller Verm.* ii. p. 85.
Helix ovalis. *Gmelin*, p. 3637.
Helix, No. 91. *Schroeter Einl.* ii. p. 203.
Bulla ovata. *Chemnitz,* ix. part 2. p. 28. t. 119. f. 1020 and 1021.
Bulimus ovatus. *Bruguiere Enc. Meth.* p. 318.
Lister Conch. t. 1055. f. 1. *Seba*, iii. t. 71. f. 18 and 19.
Inhabits the East Indies ; and is a land shell. *Muller.* Rio Janeiro. *D'Avila.* Tranquebar, Ceylon, and the Nicobar Islands. *Chemnitz.*
Shell about four inches and a half long, and two inches and a half broad, of a reddish white colour, covered with a yellow epidermis, and the summit and outer lip are rose-coloured ; it has six rounded whirls, of which the body-whirl is by far the largest, and the spire is produced ; the aperture is marginated, and in a full-grown shell measures two inches and a half long, and is half as broad in the middle ; the outer lip is thickest in the middle, and the umbilicus is nearly covered by a reflection of the pillar-lip.

OBLONGA. 103. Shell sub-umbilicated, ovate-oblong, longitudinally striated, with the pillar and outer lip rose-coloured.

Helix oblonga. *Muller Verm.* ii. p. 86. *Born Mus.* p. 381. t. 15. f. 21 and 22. *Gmelin,* p. 3637. *Shaw Nat. Hist.* viii. t. 294.
Helix, No. 8. *Schroeter Einl.* ii. p. 180.
Helix ovipara. *Portland Cat.* p. 87. lot 1930.
Bulla oblonga. *Chemnitz,* ix. part 2. p. 30. t. 119. f. 1022 and 1023.
Bulimus hæmastomus. *Scopoli Insub.* i. p. 67. t. 25. f. B. *Lamarck Syst. des Anim.* p. 90. *Leach Zool. Misc.* i. p. 68. t. 29.
Bulimus oblongus. *Bruguiere Enc. Meth.* p. 318.
Turbo hæmastomus. *Gmelin,* p. 3597.
Lister Conch. t. 23. f. 21. *Seba,* iii. t. 71. f. 17. *Favanne,* t. 65. f. I 1.

Inhabits Cayenne, Surinam, Guinea, St. Domingo, and Jamaica. *Chemnitz.* St. Vincent's. *Humphreys.* Forests of
Cayenne, and the Island of St. Thomas on the coast of Africa. *Bruguiere.*

Shell about three and a half inches long, and rather more than
half as broad, and of a pale reddish brown or tawny colour,
becoming paler at the sutures; it is nearly allied to *H. ovata,* but may be at once distinguished by the aperture, which
is smaller in proportion to the size of the shell, and has the
lip of an equal thickness throughout; *H. ovata* is also more
ventricose, and has the pillar white. The egg of this species
is remarkably large, and is about three-quarters of an inch
long, and two-thirds as broad; it has been figured by Lister,
and also by Mr. Mawe in his Account of the Brazils; and
Favanne, t. 80. f. P, exhibits the young shell bursting from
the egg.

FLAMMEA. 104. Shell umbilicated, conical-oblong,
   smooth, with brown longitudinal stripes, and
   the aperture dilated at the base.

Helix flammea. *Muller Verm.* ii. p. 87. *Gmelin,* p. 3637.
Helix, No. 3. *Schroeter Einl.* ii. p. 179.
Bulla flammea. *Chemnitz,* ix. part 2. p. 32. t. 119. f. 1024
   and 1025.
Bulimus flammeus. *Bruguiere Enc. Meth.* p. 322.
Lister Conch. t. 578. f. 33.   *Martini Berl. Mag.* iii. t. 5.
   f. 49.
In Guinea, about the Fort of Christiansburg; and is a land
shell. *Muller.* Country of the Hottentots. *Bruguiere.*
Shell about three inches long, and sixteen lines broad, glabrous,
white, and prettily marked with irregularly undulated longitudinal reddish brown stripes; it has nine whirls, and a produced spire; the aperture oval, somewhat dilated at the base,
and the throat bluish; the pillar is said to be slightly truncated, which indicates a close affinity with the *Bulla Zebra.*

KAMBEUL. 105. Shell umbilicated, oblong-ovate,
   white, with decussated striæ, and brown longitudinal stripes; aperture contracted at the
   base.

Bulimus Kambeul. *Bruguiere Enc. Meth.* p. 322.
Le Kambeul. *Adanson Senegal,* p. 14. t. 1. f. 1.
Inhabits Senegal; and is a land shell. *Adanson.*
Shell three inches and a half long, and an inch and a half broad,

and owing to its having been described to be perfectly smooth by Adanson, has been erroneously confounded with the preceding species; Bruguiere, on examining Adanson's specimens however, found them to differ in being marked with fine decussated striæ, and in having the aperture longer, and more contracted at the base.

PILEUS. 106. Shell sub-umbilicated, pyramidal, smooth, and white with transverse bands; aperture transverse, with the outer lip reflected.

Helix Pileus. *Muller Verm.* ii. p. 80. *Born Mus.* p. 380. t. 16. f. 11 and 12. *Gmelin*, p. 3637.
Helix pileata. *Gmelin*, p. 3639.
Helix, No. 208. *Schroeter Einl.* ii. p. 235.
Bulla bifasciata. *Gmelin*, p. 3431.
Bulla, No. 10. *Schroeter Einl.* i. p. 190.
Trochus Pileus. *Chemnitz*, ix. part 2. p. 48. t. 122. f. 1046 to 1048.
Lister Conch. t. 16. f. 11.
Inhabits ――――

Shell an inch and a quarter long, and one inch broad, with seven whirls, and transverse reddish and yellowish bands; it bears some resemblance to *Bulla fasciata*, but may be at once distinguished by its reflected lip and the want of a truncated pillar, and it is broader in proportion to the length.

TRIFASCIATA. 107. Shell sub-umbilicated, ovate-conical, with three transverse bands on the lower part of the body-whirl, and the outer lip white and reflected.

Helix trifasciata. *Chemnitz,* ix. part 2. p. 155. t. 134. f. 1215. *Gmelin*, p. 3642.
Bulimus trifasciatus. *Bruguiere Enc. Meth.* p. 317.
Inhabits Tranquebar ; and is a land shell. *Chemnitz.*

Shell an inch and a quarter long, and three quarters of an inch broad, white with three transverse brown bands on the lower half of the body-whirl, of which the middle one is broadest, and two on the whirls of the spire ; it has six smooth flattish whirls,. of which the body-whirl is almost as large as all the others, and the aperture is roundish-ovate. *Bulimus trifasciatus* of Dr. Leach's Zoological Miscellany is different, and probably only a variety of *H. Guadaloupensis.*

BONTIA. 108. Shell umbilicated, ovate-conical, ventricose and yellowish white with three brown bands; aperture ovate, broad, and somewhat truncated at the base.

Helix Bontia. *Chemnitz*, ix. part 2. p. 156. t. 134. f. 1216 and 1217. *Gmelin*, p. 3642.

Favanne, t. 65. upper f. L.

Inhabits Tranquebar. *Chemnitz*.

Shell about ten lines long and six broad, brittle, pellucid, of a yellowish white colour, with three transverse reddish brown bands on the body-whirl, and the summit blackish; the uppermost of the three bands is narrower than the others, and sometimes broken into dots.

LABIOSA. 109. Shell sub-cylindrical, polished, diaphanous, and white; aperture semi-ovate, marginated, and one-toothed.

Helix labiosa. *Muller Verm.* ii. p. 96. *Gmelin*, p. 3645.

Helix cylindracea acuta. *Chemnitz*, ix. part 2. p. 166. t. 135. f. 1234.

Bulimus labiosus. *Bruguiere Enc. Meth.* p. 347.

Turbo, No. 153. *Schroeter Einl.* ii. p. 114.

Gualter, t. 4. f. R.

Inhabits India. *Muller*.

Shell about an inch long, and rather more than one third as broad, and in the aperture possesses some affinity with *Turbo Uva*; it has nine smooth whirls, of which the body-whirl is almost as long as the others put together; the aperture is semi-ovate, and has one tooth on the pillar-lip; Muller probably described a young shell, and has not mentioned this tooth, which Bruguiere says is always to be found when the shell has attained to maturity; it is umbilicated, and Gmelin has erroneously placed it among the imperforate species.

TROCHOIDES. 110. Shell umbilicated, trochiform, shining, longitudinally wrinkled, with the whirls reversed, and the body-whirl keeled; aperture angulated, and the pillar brown.

Helix trochoides. *Chemnitz*, x. p. 369. t. 173. f. 1686. *Gmelin*, p. 3642.

Inhabits the East Indies. *Chemnitz*.

Chemnitz has not mentioned the size, but this shell appears by his figure to be about fourteen lines long, and nine broad, of a brownish white colour tinged with green, especially towards the summit.

OTAHEITANA. 111. Shell umbilicated, oblong-ovate, thick, with the whirls reversed, and the outer lip emarginated and white.

Helix perversa, in rivulis Insulæ Australis Otaheite reperta. *Chemnitz,* ix. part 1. p. 108. t. 112. f. 950 and 951.
Helix perversa, Var. ε. *Gmelin,* p. 3643.
Bulimus Otaheitanus. *Bruguiere Enc. Meth.* p. 347.
Inhabits rivers in Otaheite. *Chemnitz.*
Shell about eleven lines long, and six broad, with five reversed whirls, of an uniform brown or coffee-colour; the inner lip has a tooth-like callosity in the middle, and the shell may probably belong to the same natural family as *Voluta Auris Midæ,* or *Voluta fasciata,* which it most resembles.

LÆVA. 112. Shell umbilicated, oblong, diaphanous, with the whirls reversed, and transversely banded; aperture elongated at the base.

Helix læva. *Muller Verm.* ii. p. 95. *Chemnitz,* ix. part 1. p. 103. t. 111. f. 940 to 949. *Gmelin,* p. 3644.
Bulimus lævus. *Bruguiere Enc. Meth.* p. 317.
Bulla, No. 22. *Schroeter Einl.* i. p. 194.
Lister Conch. t. 33. f. 31. *Knorr,* vi. t. 29. f. 3. *Favanne,* t. 65. f. A 3.
Inhabits the East Indies. *Chemnitz.*
Shell about sixteen lines long and seven broad; it is more oblong and pellucid, and more produced, but less angulated at the base than *H. interrupta,* and differs from all the following species in having the umbilicus less covered by the reflected pillar-lip. Bruguiere has enumerated the following Varieties: A. White, with two bluish or blackish brown transverse bands; B. White, with three brown or bluish bands; C. White, with six brown bands, which appear of a darker colour within the aperture than on the outside; D. Yellowish white, with five bands, of which the three larger are bluish and the others reddish; E. Yellow, with five bluish broad bands and a narrower red one at the sutures; F. Yellow, with three broad purple bands; G. Variegated with greenish, yellow, and reddish spots, and green bands.

**** *Imperforate, and the Whirls rounded.*

AUREA.   113. Shell sub-umbilicated, ovate-oblong, obtuse, smooth, yellow, with the outer lip white and marginated.

Helix dextra. *Muller Vermium*, ii. p. 89. *Chemnitz*, ix. part 2. p. 153. t. 134. f. 1210 to 1212. *Gmelin*, p. 3643.
Bulimus citrinus, Var. D. *Bruguiere Enc. Meth.* p. 314.
Variety.   With the whirls reversed.
Helix perversa. *Linnæus Syst. Nat.* p. 1246. *Born Mus.* p. 381. *Schroeter Fluss.* p. 294. t. 10 A. f. 2 and 3, and *Einl.* ii. p. 153. t. 4. f. 4. *Chemnitz*, ix. part 1. p. 95. t. 110. f. 928 to 933, and t. 111. f. 934 to 937. *Gmelin*, p. 3642.
Helix sinistra. *Muller Verm.* ii. p. 90.
Bulimus citrinus. *Bruguiere Enc. Meth* p. 313.
Limax aureus. *Martyn Univ. Conch.* iii. t. 115.
Lister Conch. t. 34. f 33, and t. 35. f. 34. *Gualter*, t. 5. f. P. *Argenville*, t. 9. f. G. *Seba*, iii. t. 40. f. 37. *Knorr*, i. t. 16. f. 5, iv. t. 28. f. 4 and 5, and v. t. 23. f. 4. *Martini Berl. Mag.* iii. t. 5. f. 50. *Favanne*, t. 65. f. A 8.
Inhabits the West Indian Islands, and the forests of Cayenne and Guinea. *Bruguiere.*   Pulo Condore, Prince's Island, and China. *Humphreys.*
Shell about an inch and three-quarters long, and half as broad, and has six or seven whirls, of which the body-whirl is somewhat ventricose, and the spire conical, but ending in an obtuse summit; the whirls are most commonly reversed, but this cannot be considered the natural state, and between *H. dextra* and *H. sinistra* of Muller, there is no other difference : young shells are slightly umbilicated, but with age the umbilicus becomes closed by the pillar-lip; it varies much in the markings, but yellow is the prevailing colour, and the following are some of the leading Varieties : A. Pale or darker yellow, without spots ; B. Yellow, with one or two reddish or blackish longitudinal stripes ; C. Yellow, with grey clouds and bands ; D. Green, with two or three yellow transverse bands.   Muller also mentions two other Varieties, of which one has red, and the other party-coloured spots, but I suspect that these belong to the following species. Linnæus erroneously considered it to be a river shell.

RECTA. **114.** Shell oblong-ovate, acuminated, with longitudinal stripes interrupted by a white band on the body-whirl; outer lip reflected.

Helix recta. *Muller Verm.* ii. p. 93. *Gmelin,* p. 3643.
Variety. With the whirls reversed.
Helix inversa. *Muller Verm.* ii. p. 93. *Chemnitz,* ix. part 1. p. 93. t. 110. f. 925 and 926. *Gmelin,* p. 3644.
Helix perversa, Var. γ. *Born Mus.* p. 382.
Bulimus inversus. *Bruguiere Enc. Meth.* p. 315.
Grew Mus. t. 10. *Petiver Gaz.* t. 76. f. 5. *Gualter,* t. 5. f. O. *Knorr,* v. t. 23. f. 5.
Inhabits ———
Shell about two inches long, and nearly half as broad, with seven or eight whirls, of which the body-whirl is ventricose, and the spire conical; it differs from *H. aurea* in having the upper whirls less ventricose, the spire rather longer and more acuminated, and the base less rounded; it is of a pale fawn-colour tinged with blue, and variegated with reddish brown or blackish longitudinal stripes interrupted by a transverse white band.

INTERRUPTA. **115.** Shell oblong-ovate, smooth, whitish, with transverse rows of oblong spots on the body-whirl, and the summit black; outer lip white and reflected.

Helix interrupta. *Muller Verm.* ii. p. 94. *Chemnitz,* ix. part 2. p. 154. t. 134. f. 1213 and 1214. *Gmelin,* p. 3644.
Bulimus interruptus. *Bruguiere Enc. Meth.* p. 316.
Variety. With the whirls reversed.
Helix contraria. *Muller Verm.* ii. p. 95. *Gmelin,* p. 3644.
Helix interrupta sinistrorsa. *Chemnitz,* ix. part 1. p. 101. t. 111. f. 938 and 939.
Favanne, t. 65. f. A 6.
Inhabits ———
Shell about an inch and three-quarters long, and three-quarters of an inch broad, of a yellowish white colour, with pale chocolate broad longitudinal stripes, which are interrupted by several different coloured bands, so as to give the whole surface somewhat of a tessellated appearance; it has seven whirls, and is nearly allied to *H. aurea,* but is much more prettily marked, and all the specimens which I have seen have had a black spot on the summit.

ARENARIA. 116. Shell minute, glossy, with very thin longitudinal striæ ; spire sub-hemisphærical and the whirls reversed.

Helix arenaria. *Gmelin*, p. 3644.
Spengler Nov. Act. Dan. i. t. 1. f. 6. *Chemnitz*, ix. part 1. p. 129. t. 113. f. 972 and 973.
Found among the sea-sand at Rimini in Italy. *Spengler*.
This shell appears by the figures to be about a line long, and rather more than half as broad, and is white and glossy.

JANTHINA. 117. Shell nearly imperforate, roundish, obtuse, diaphanous, and extremely brittle ; aperture sub-triangular, with a notch in the margin of the outer lip.

Helix janthina. *Linnæus Syst. Nat*. p. 1246. *Born Mus*. p. 382. *Schroeter Einl*. ii. p. 155, and *Inn. Bau Conch*. p. 62. t. 5. f. 1. *Gmelin*, p. 3645. *Brookes's Introd*. p. 129. t. 8. f. 107.
Trochus janthinus. *Chemnitz*, v. p. 57. t. 166. f. 1577 and 1578.
Janthina fragilis. *Lamarck Syst. des Anim*. p. 89.
Lister Conch. t. 572. f. 24. *Rumphius*, t. 20. f. 2. *Gualter*, t. 64. f. O. *Argenville*, t. 6. f. S. *Knorr*, ii. t. 30. f. 2 and 3. *Favanne*, t. 64. f. K 1.
Inhabits the coasts of Europe, Asia, and Africa ; and is common in the Mediterranean. *Linnæus*. Red Sea. *Forskael*.
A few of these shells have been picked up on the sea-shore at Oxwich in Glamorganshire. *L. W. D.*
Shell about an inch long, and the breadth rather exceeds the length ; it is remarkably brittle, and has four obliquely striated whirls of a violet colour, which becomes paler towards the summit ; the animal is said to shine by night, and to stain the hand with a purple colour.

JAMAICENSIS. 118. Shell sub-globular, with the body-whirl ventricose, and the spire short and convex; aperture large, with the outer lip marginated, and notched at the pillar.

Helix Jamaicensis. *Gmelin*, p. 3644.
Helix pulla. *Gmelin*, p. 3650.
Helix terrestris Jamaicensis. *Chemnitz*, ix. part 2. p. 115. t. 129. f. 1140 and 1141.
Helix, No. 16. *Schroeter Einl*. ii. p. 183.

Lister Conch. t. 42. f. 40, and t. 43.   *Knorr*, i. t. 21. f. 3.
Favanne, t. 63. f. M.

Junior? Smaller, and of a yellowish colour, with a white band bordered with red.

Helix venusta.   *Gmelin*, p. 3650.

Kæmmerer Cab. Rudolst. p. 176. t. 11. f. 4 and 5.

Inhabits Jamaica. *Lister, &c.*

Shell about two inches long, and two and a quarter broad, of a brown colour with two or three white bands on the body-whirl; it has four whirls, of which the body-whirl is very large, and the others form a small convex obtuse spire; the aperture, when measured somewhat obliquely, is an inch and a half long, and an inch and a quarter broad, with a white reflected margin on the outer lip, and the inner lip minutely striated.

RHODIA.   119. Shell sub-globular, depressed, slightly keeled, and the base convex; aperture lunated.

Helix Rhodia.   *Chemnitz*, ix. part 2. p. 136. f. 1179 and 1180.   *Gmelin*, p. 3645.

Inhabits the Island of Rhodes; and is a land shell. *Chemnitz.*

Chemnitz has not mentioned the size, but this shell appears by his figures to be about thirteen lines in diameter, and is of a chalk-white colour, with a row of oblong spots on the base, and coated with a yellowish brown epidermis; the aperture has a white reflected margin.

VIVIPARA.   120. Shell ovate, ventricose, obtuse, and longitudinally wrinkled; aperture sub-orbicular, and the pillar-lip slightly reflected.

Helix vivipara.   *Linnæus Syst. Nat.* p. 1247.   *Pennant Zool.* iv. p. 137.   *Da Costa Brit. Conch.* p. 81. t. 6. f. 2.   *Born Mus.* p. 383.   *Schroeter Flussconch.* p. 330. t. 8. f. 1 and 2, and t. 11 C. f. 6; and *Einleit.* ii. p. 156. *Chemnitz*, ix. part 2. p. 136. t. 132. f. 1182 to 1185. *Gmelin*, p. 3646.   *Donovan*, iii. t. 87.   *Montagu Test.* p. 386.   *Maton and Racket, in Lin. Trans.* viii. p. 205. *Dorset Cat.* p. 54. t. 17. f. 2.

Nerita vivipara.   *Muller Verm.* ii. p. 182.

Le Vivipare.   *Geoffroy*, p. 110. No. 2. t. 3.

Lister Anim. Ang. t. 2. f. 18, and *Conch.* t. 1055. f. 6. *Gualter*, t. 5. f. A.   *Seba*, iii. t. 38. f. 12, and t. 39. f.

33 and 34. *Knorr*, v. t. 17. f. 4. *Berlin. Mag.* iv. t. 7. f. 4 to 9. *Favanne*, t. 61. f. D 9.

Variety. White, with three yellowish brown bands.

Helix vivipara. *Muller Zool. Dan.* iii. p. 33. t. 101. f. 5 and 6.

Junior. Less ventricose, and the whirls less produced.

Helix compactilis. *Pulteney's Dorset Cat.* p. 48.

Pennant Zool. iv. t. 85. upper fig.

Inhabits stagnant waters in Europe, especially in a clayey soil. *Linnæus.* Britain. *Lister.* France. *Geoffroy.* Denmark. *Muller.*

Shell about an inch and a quarter long, and an inch broad, of an olive-colour, with three brown bands on the body-whirl, which become gradually obliterated on the spire ; it has six ventricose whirls, and some of the wrinkles are stronger than others, which mark the growth of the shell ; behind the reflected pillar-lip is the slight appearance of an umbilicus ; it is unusually strong for a fresh-water shell, and the animal is viviparous.

FASCIATA. 121. Shell ovate, ventricose, mucronated, striated, shining, and white with three red bands ; aperture ovate.

Helix fasciata. *Gmelin,* p. 3646.

Helix ventricosa. *Olivi Adr.* p. 178.

Nerita fasciata. *Muller Verm.* ii. p. 182. *Schroeter Fluss.* p. 369.

Chiocciola maggiore. *Ginanni Op. Post.* ii. p. 49. t. 1. f. 6.

Gualter, t. 5. f. M.

Inhabits Italy and Saxony. *Muller.*

Muller says that this shell varies in size from nine and a half to fifteen lines long, and from seven and a half to ten broad, and that it is white, pellucid, and glabrous, and has three red bands on the body-whirl, and two on the first whirl of the spire ; he says it has the aperture less rounded than in *H. vivipara*, to which it is nearly allied, and the shell, which in the Zoologia Danica he has figured for a Variety of that species, appears to me to be a link which connects them. Mr. Adams, in the Fifth Volume of the Transactions of the Linnæan Society, has figured a perfectly distinct microscopic species under the name of *H. fasciata*, but his descriptions are rarely sufficiently clear, and I have not in general thought it necessary to notice them.

DISSIMILIS.　122.　Shell sub-ovate, acuminated, of a yellowish white colour, and the outer lip black.

Helix dissimilis.　*Gmelin*, p. 3647.
Helix, No. 266.　*Schroeter Einl.* ii. p. 253. t. 4. f. 10.
Nerita dissimilis.　*Muller Verm.* ii. p. 184.
Inhabits Tranquebar. *Schroeter.*
Muller says this shell is intermediate between *H. vivipara* and *H. tentaculata*, and has described it to be nine and a half lines long and seven and a half broad, pellucid, glabrous, and of a brownish or yellowish white colour, with the base whiter than the other parts, so as to give the body-whirl somewhat of a banded appearance; it has six whirls, and their convexity is between that of *H. vivipara* and *H. fasciata*; the aperture has a black shining margin, and the operculum is pellucid, yellowish brown, and shining.

ANGULARIS.　123.　Shell sub-conical, with three transverse angulated keels, and of a greenish colour ; aperture roundish and sub-angulated.

Helix angularis.　*Chemnitz*, ix. part 2. p. 160. t. 134. f. 1222 and 1223.　*Gmelin*, p. 3661.
Nerita angularis.　*Muller Verm.* ii. p. 187.
Inhabits the rivers about Canton. *Muller.*
Shell about an inch long, and half as broad, opake, of a greenish colour, and the inside white; it has five produced very slightly convex whirls, which are minutely striated transversely, and on the body-whirl are three transverse angulated keels, which decrease in number on the spire : Muller has quoted Lister, t. 127. f. 7, which is stated to be a native of Virginia, and appears to me to be a different species ; Petiver Gaz. t. 106. f. 13, is the same.

NEMORALIS.　124.　Shell sub-globular, smooth, diaphanous, with five transversely banded whirls; aperture roundish-lunated, and the outer lip slightly reflected.

Helix nemoralis.　*Linnæus Syst. Nat.* p. 1247.　*Pennant Zool.* iv. p. 137.　*Da Costa Brit. Conch.* p. 76. t. 5. f. 1 to 4, 8, 14, and 19.　*Schroeter Einl.* ii. p. 158. *Donovan*, i. t. 13.　*Maton and Racket, in Lin. Trans.* viii. p. 206.　*Dorset Cat.* p. 54. t. 21. f. 1, 6, 14, and 19.

Variety A. With the margin of the outer lip brown.
  Helix nemoralis. *Muller Verm.* ii. p. 46. *Born Mus.* p.
    384. t. 16. f. 3 to 8. *Chemnitz,* ix. part 2. p. 144. t.
    132. f. 1196 to 1198. *Gmelin,* p. 3647. *Montagu Test.*
    p. 411.
  Lister Anim. Ang. t. 2. f. 3, and *Conch.* t. 57. f. 54. *Pe-*
    tiver Gaz. t. 91. f. 9 to 12, and t. 92. f. 9 and 10. *Gual-*
    ter, t. 1. f. P, and left Q. *Seba,* iii. t. 39. f. 12, 13, 18,
    and 19. *Geve,* t. 32. f. 393 to 411, and t. 33. f. 412 to
    420. *Favanne,* t. 63. f. H.
Variety B. With the outer lip white.
  Helix hortensis. *Muller Verm.* ii. p. 52. *Born Mus.* p.
    385. t. 16. f. 18 and 19. *Chemnitz,* ix. part 2. p. 146. t.
    133. f. 1199 to 1201. *Gmelin,* p. 3649. *Montagu Test.*
    p. 412.
  Gualter, t. 1. f. right Q. *Geve,* t. 31. f. 368 to 390.
Variety C. With the whirls reversed.
  Helix nemoralis contraria. *Chemnitz,* ix. part 1. p. 92. t.
    109. f. 924.
Common in woods and hedges throughout Europe.
Shell about three-quarters of an inch long, and the breadth rather exceeds the length ; it is somewhat globular and pellucid, and has five whirls of various colours ; the following varieties have been enumerated by Muller : A. Red, without
  any variegation ; B. Red, with a very narrow pale band ; C.
  Red, with two pale bands ; D. Red, with a very broad brown
  band ; E. Red, with a broad brown band ; F. Red, with a
  narrow brown band ; G. Red, with three brown bands, of
  which the lower and middle are very broad ; H. Chestnut-
  coloured, with a yellowish band ; I. Flesh-coloured, without
  any variegation ; K. Flesh-coloured, with three darker bands ;
  L. Flesh-coloured, with one darker band ; M. Whitish, with
  fine red transverse bands and lines ; N. Yellow, without any
  variegation ; O. Yellow, with a narrow brown band ; P. Yellow, with a very broad brown band ; R. Yellow, with two
  brown equal bands at the base ; S. Yellow, with two brown
  bands, of which the lower is very broad ; T. Yellow, with a
  white band edged on both sides with red ; U. Yellow, with
  two very broad brown bands ; V. Yellow, with three equal
  brown bands ; X. Yellow, with three brown bands, and the
  middle one very narrow ; Y. Yellow, with three brown-bands,
  and the uppermost very narrow ; Z. Yellow, with three brown
  bands, and the lower one very broad ; AA. Yellow, with three
  brown bands, of which the two lower are equal and the uppermost very narrow ; B B. Yellow, with four brown bands ;

C C. Yellow, with five brown bands at equal distances ; DD.
Yellow, with five brown bands at unequal distances ; &c.

CARTUSIANA. 125. Shell imperforate, rather depressed, and white, with six whirls.

Helix Cartusiana. *Muller Verm.* ii. p. 15. *Gmelin,* p.
3664.
Le Chartreuse. *Geoffroy,* p. 33. t. 2.
Inhabits woods about Paris. *Geoffroy.*
Muller has described this shell to be six lines in diameter, pellucid, white, and convex on both sides, but rather depressed above ; with a mark of doubt he has quoted Schroeter
Erdconch. f. 27, which is very like the variety B. of *H. nemoralis,* and it probably differs principally in having six
whirls.

LUCORUM. 126. Shell imperforate, roundish, smooth,
and banded ; aperture oblong and brown.

Helix Lucorum. *Linnæus Syst. Nat.* p. 1247. *Muller
Verm.* ii. p. 46. *Gmelin,* p. 3649.
Lister Conch. t. 1058. f. 1 and 2. ? *Gualter,* t. 1. f. C.
Favanne, t. 64. f. K 3. ?
Inhabits trees in Europe. *Linnæus.* Italy. *Muller.*
Linnæus, without any addition to the above description, has
given only a reference to Gualter, t. 1. f. C, so that this will
probably always remain a doubtful species ; Muller's *H. Lucorum* he has described to be nineteen lines in diameter,
globose, white, and striated transversely, with four or five
circular red transverse stripes ; aperture lunated, with the
inside white, and the lip brown and reflected over the umbilicus.

GRISEA. 127. Shell imperforate, sub-ovate, and
obtuse, with four whirls, and the aperture
rather oblong.

Helix grisea. *Linnæus Syst. Nat.* p. 1247. *Gmelin,* p.
3649.
Helix aspersa. *Muller Verm.* ii. p. 59. *Chemnitz,* ix.
part 2. p. 125. t. 130. f. 1156 to 1158. *Gmelin,* p. 3631.
Montagu Test. p. 407.
Helix hortensis. *Pennant Zool.* iv. p. 136. t. 84. f. 129.
Donovan, iv. t. 131. *Maton and Racket, in Lin. Trans.*
viii. p. 208. *Dorset Cat.* p. 55. t. 20. f. 1.

Helix Lucorum. *Pulteney Dorset. Cat.* p. 48.
Helix variegata. *Gmelin,* p. 3650.
Helix, No. 195. *Schroeter Einl.* ii. p. 230. t. 4. f. 7.
Cochlea vulgaris. *Da Costa Brit. Conch.* p. 72. t. 4. f. 1.
Lister Anim. Ang. t. 2. f. 2, and *Conch.* t. 49. f. 47.
  Gualter, t. 1. f. B. ? *Argenville,* t. 28. f. 3 and 11.
  Knorr, iv. t. 27. f. 3. *Geve,* t. 30. f. 344. *Favanne,* t.
  62. f. D 2, D 3, and D 4.
Inhabits Southern Europe, and is a land shell ; and also Sweden, but there attains only a much smaller size. *Linnæus.*
Shell about an inch and a quarter long, and the breadth rather
exceeds the length ; it has four irregularly creased and wrinkled whirls, and a semilunar rather elongated aperture, with
a white reflected margin ; it varies much in its markings, but
is most commonly of a pale dull yellowish brown or ash-colour, with darker bands, which are usually more or less
broken, and appear mottled all over, or sometimes the colour is greyish with two paler bands.

HÆMASTOMA. 128. Shell imperforate, roundish,
  of a dark colour transversely banded with
  white ; aperture oblong, with the outer lip
  broadly marginated and purplish.

Helix hæmastoma. *Linnæus Syst. Nat.* p. 1247. *Muller
  Verm.* ii. p. 78. *Born Mus.* p. 387. *Schroeter Einl.* ii.
  p. 160. t. 4. f. 5 and 6. *Chemnitz,* ix. part 2. p. 122.
  t. 130. f. 1150 and 1151. *Gmelin,* p. 3650.
Helix hæmatragus. *Born Index.* p. 400.
Lister Conch. t. 1055. f. 2. *Seba,* iii. t. 40. f. 6 and 7.
  Geve, t. 28. f. 328. *Favanne,* t. 64. f. A 4.
Variety. With the outer lip blackish.
Helix melanotragus. *Born Mus.* p. 388.
Lister Conch. t. 45. f. 43. *Geve,* t. 28. f. 329. *Chemnitz,*
  ix. t. 130. f. 1152 to 1154.
Inhabits the Meridional parts of America. *Seba.* Ceylon, and
the Nicobar Islands ; and is a land shell. *Chemnitz.*
Shell about an inch and a half long, and the breadth rather exceeds the length, thick, rather opake, striated, shining, and
of a chestnut-colour, with one or sometimes two broad transverse white bands, and coated with a yellowish brown epidermis ; it has five whirls, of which the body-whirl is almost
twice as large as the spire ; the outer lip has a broad reflected lip, which varies from rose-colour to blackish purple.

LACTEA. **129.** Shell imperforate, depressed, grey with white scattered dots ; aperture blackish, and the outer lip reflected and one-toothed.

Helix lactea. *Muller Verm.* ii. p. 19. *Chemnitz*, ix. part 2. p. 127. t. 130. f. 1161. *Gmelin*, p. 3629.
Lister Conch. t. 95. f. 96.
Variety. With the tooth on the lip obsolete.
Lister Conch. t. 51. f. 49. *Petiver Gaz.* t. 153. f. 8.
Inhabits Jamaica. *Lister.* Portugal. *Chemnitz.*
Shell from ten lines to an inch and a half in diameter, with five whirls, and Muller says it resembles *H. nemoralis,* but is always covered with minute greyish white confluent dots ; the aperture, and both the lips are of a shining blackish or reddish brown colour, and the outer lip has generally a solitary tooth on its inner margin. Muller has described the following varieties : A. Grey, without spots ; B. Greyish yellow, with a white band in the middle ; C. White, with four brown bands ; D. White, with three obsolete red bands.

PICTA. **130.** Shell imperforate, sub-globular, glabrous, obtuse, with the body-whirl ventricose, and a coloured stripe on the margin ; aperture lunated, and the outer lip reflected.

Helix picta. *Born Mus.* p. 386. t. 15. f. 17 and 18. *Gmelin*, p. 3650.
Helix, No. 102. *Schroeter Einl.* ii. p. 206.
Cortex Mali Citrei. *Chemnitz*, ix. part 2. p. 128. t. 130. f. 1162 to 1165.
Bonanni Rec. and *Kirch.* 3. f. 5.? *Rumphius*, t. 22. f. No. 1. *Petiver Amb.* t. 11. f. 19. *Knorr*, i. t. 10. f. 2.
Inhabits the shores of Italy. *Bonanni?* Amboyna. *Rumphius.* China, and is a land shell. *Chemnitz.* West Indies. *Humphreys.*
Shell about eight lines long, and an inch broad, with four convex whirls, of which the body-whirl is ventricose, and the spire depressed; it varies in colour as follows : A. Grey, with a brown stripe at the suture ; B. Rose-coloured, with the stripe at the suture and the dorsal line brown ; C. Sulphur-coloured, with the stripe at the suture and the dorsal stripe red ; D. Flesh-coloured, with the stripe at the suture and the dorsal line dark purple. These stripes are sometimes edged with white, and the pillar is generally of the same colour as the stripes.

VERSICOLOR. **131.** Shell imperforate, sub-globular, white, with transverse bands and intermediate lines; pillar somewhat twisted and rose-coloured.

Helix versicolor. *Born Mus.* p. 386. t. 16. f. 9 and 10. *Gmelin,* p. 3651.
Helix, No. 207. *Schroeter Einl.* ii. p. 235.
Inhabits ———
Born has described this shell to be ten lines long and eleven broad, sub-globular, smooth, with the body-whirl ventricose, and the other whirls forming a convex obtuse spire; colour white, with alternate broader and narrower, brown, rose, and sulphur-coloured transverse stripes.

APERTA. **132.** Shell imperforate, sub-globular, longitudinally striated, with three whirls, and the body-whirl and aperture very large; pillar slightly twisted.

Helix aperta. *Born Mus.* p. 387. t. 15. f. 19 and 20. *Gmelin,* p. 3651.
Helix neritoides. *Chemnitz,* ix. part 2. p. 150. t. 133. f. 1204 and 1205.
Helix, No. 205. *Schroeter Einl.* ii. p. 234.
Inhabits the island of St. Croix. *Chemnitz.*
Shell about an inch long, and the length and breadth are nearly equal; it has three or three and a half whirls, of which the body-whirl is very large, and the spire small and depressed; it is obsoletely striated longitudinally, and coated with a brownish olive epidermis.

FUSCA. **133.** Shell imperforate, thin, pellucid, smooth, with five or six whirls, which are obsoletely wrinkled longitudinally; aperture lunated.

Helix fusca. *Montagu Test.* p. 424. t. 13. f. 1. *Maton and Racket, in Lin. Trans.* viii. p. 209.
Inhabits woods in Devonshire; but is a scarce shell. *Montagu.*
Shell hardly a quarter of an inch long, and three-eighths of an inch in diameter, of a reddish horn-colour; Montagu says it has much the habit of *H. rufescens,* but is more pellucid, not so much wrinkled, and is at once distinguished by not having any umbilicus.

PELLUCIDA. **134.** Shell imperforate, rather depressed, transparent, and greenish, with three whirls.

Helix pellucida. *Muller Verm.* ii. p. 15.
Helix fuscescens. *Gmelin*, p. 3639.
Le Transparente. *Geoffroy*, p. 38. No. 8. t. 2.
Schroeter Erdconch. t. 1. f. 11.
Inhabits wet moss on the edges of pools about Paris. *Geoffroy.* Saxony. *Schroeter.* Denmark. *Muller.*
Muller says this shell varies from one-eighth of a line to a line and a half in diameter, and that it is brittle, glossy, transparent, highly polished, convex on both sides, and without any striæ or umbilicus; empty shells are greenish white, but otherwise they appear of the colour of the animal, which is yellowish.

***** *Turreted.*

CONSOLIDATA. **135.** Shell turreted, thick, transversely striated, brown, and the summit truncated; aperture oval.

Helix decollata nigra. *Chemnitz*, ix. part 2. p. 188. t. 136. f. 1258.
Helix decollata, Var. *J. Gmelin*, p. 3652.
Bulimus consolidatus. *Bruguiere Enc. Meth.* p. 325.
Inhabits Surinam, and is a land shell. *Chemnitz.*
Shell about fourteen lines long and ten broad, with three transversely striated whirls, and it appears as if four or five others had been broken off; it is of a brownish colour, coated with a black epidermis.

DECOLLATA. **136.** Shell turreted, longitudinally striated, white, and truncated at the summit; aperture oval.

Helix decollata. *Linnæus Syst. Nat.* p. 1247. *Muller Verm.* ii. p. 114. *Born Mus.* p. 388. *Schroeter Einl.* ii. p. 161. *Chemnitz*, ix. part 2. p. 182. t. 136. f. 1254 and 1255. *Gmelin*, p. 3651.
Bulimus decollatus. *Bruguiere Enc. Meth.* p. 326.
Lister Conch. t. 17. f. 12, and t. 18. f. 13. *Petiver Gaz.* t. 66. f. 1. *Gualter*, t. 4. f. O, P, and Q. *Klein Ost.* t. 2, f. 44. *Knorr*, vi. t. 32. f. 3. *Favanne*, t. 65. f. B 8.
Inhabits the South of Europe and the East Indies; and is a

land shell. *Linnæus.* South of France. *Lister.* Barbary. *Petiver.* Spain. *D'Avila.* Italy. *Brisson.* Naples. *Ulysses.*

Shell about sixteen lines long, and six broad in the middle of the body-whirl, and rather more than half as broad at the truncated summit; it has four or five white longitudinally striated shining whirls, and appears as if eight or nine more had been broken off. Bruguiere has given a long account of this species, and endeavoured to account for its strange formation.

TRUNCATA. **137.** Shell turreted, smooth, white, with transverse coloured bands, and the summit truncated; aperture roundish.

Helix decollata et fasciata. *Chemnitz,* ix. p. 187. t. 136. f. 1256 and 1257.

Helix decollata, Var. γ. *Gmelin,* p. 3652.

L'Enfant au Maillot rubanné. *Favanne Cat. Rais.* p. 101. *Favanne,* t. 65. f. B 10.

Inhabits St. Domingo. *Favanne.*

Shell about fourteen lines long, and nearly half as broad; it has seven whirls, and also differs from *H. decollata* in their being smooth, in the upper ones being marked with a broad transverse brown or reddish band, and in having the aperture much rounder.

CALCARIA. **138.** Shell turreted, thick, longitudinally striated, white, and the spire entire, but rounded at the summit.

Helix calcaria. *Born Mus.* p. 389. t. 16. f. 13.

Helix calcarea. *Chemnitz,* ix. part 2. p. 162. t. 135. f. 1226.

Helix decollata, Var. β. *Gmelin,* p. 3652.

Helix obtusata. *Gmelin,* p. 3655.

Helix, No. 6. *Schroeter Einl.* ii. t. 179.

Bulimus calcareus. *Bruguiere Enc. Meth.* p. 328.

Turbo terebra lævis, testa ponderosa. *Schroeter Fluss.* p. 362. t. 10 A. f. 1.

Lister Conch. t. 14. f. 9. *Gualter,* t. 6. f. I. *Kæmmerer Cab. Rudolst.* t. 11. f. 3.

Inhabits the East Indies.? *Bruguiere.*

Shell three or four inches long, and about one-third as broad, with ten longitudinally striated white shining convex whirls; the aperture is rather broader and shorter than in *H. decol-*

lata, and some authors have supposed it to be the more perfect growth of that species.

CUSPIDATA. 139. Shell turreted, of a horn-colour, finely striated transversely and longitudinally plaited; aperture oval, and the outer lip acute.

Helix cuspidata. *Chemnitz*, ix. part 2. p. 163. t. 135. f. 1228. *Gmelin*, p. 3655.
Buccinum flumineum. *Gmelin*, p. 3503. ?
Buccinum, No. 183. *Schroeter Einl.* ii. p. 405. ?
Lister Conch. t. 118. f. 13. *Gualter*, t. 6. f. H. *Martini Berl. Mag.* iv. t. 10. f. 52. ?
Inhabits the East Indies, in fresh water. *Chemnitz.*
Shell two and a half or three inches long, and about one-fourth as broad, transparent, and of a pale horn-colour; it has ten or twelve rather convex whirls, with longitudinal plaits and minute transverse striæ. I think there can be no doubt that Gmelin's *Buccinum flumineum* belongs either to this or one of the two following species.

PLICARIA. 140. Shell turreted, pellucid, yellowish with white spots, and the whirls longitudinally plaited above.

Helix plicaria. *Born Mus.* p. 389. t. 16. f. 14. *Gmelin*, p. 3654.
Helix, No. 209. *Schroeter Einl.* ii. p. 235.
Bulimus plicarius. *Bruguiere Enc. Meth.* p. 328.
Inhabits ——
Born has described this shell to be seventeen lines long and six broad, with ten whirls, which towards their upper extremities are longitudinally plaited; it is of a yellowish colour, with scattered white spots below the sutures.

UNDULATA. 141. Shell turreted, smooth, with undulated chestnut longitudinal stripes, and twelve minutely striated whirls; pillar straight.

Helix undulata. *Gmelin*, p. 3654.
Helix maculata. *Born Mus.* p. 390. t. 16. f. 15.
Bulimus pictus. *Bruguiere Enc. Meth.* p. 329.
Shell two inches and four lines long, and half an inch broad, of a whitish or pale fawn-colour, with longitudinal chestnut

or yellowish red somewhat undulated stripes; it has twelve
slightly convex whirls, which are minutely striated transversely; the aperture is ovate, and its breadth is only half
the length.

ASPERA.　　142. Shell ovate-oblong, greyish, with
　　　transverse muricated striæ, and longitudinal
　　　red stripes; pillar-lip brown.

Helix aspera.　*Gmelin,* p. 3656.
Helix scabra.　*Chemnitz,* ix. part 2. p. 188. t. 136. f. 1259
　and 1260.
Helix lugubris.　*Gmelin,* p. 3665.
Helix, No. 243.　*Schroeter Einl.* ii. p. 246.
Buccinum scabrum.　*Muller Verm.* ii. p. 136.　*Schroeter
　Fluss.* p. 299. t. 6. f. 13.
Bulimus scaber.　*Bruguiere Enc. Meth.* p. 350.
Inhabits marshes near the sea-side in Coromandel.　*Muller.*
Shell from five and a half to eight and a half lines long, and
　from two to three and a half lines broad, of a pale greyish
　brown or dirty white colour, with longitudinal red stripes;
　it has seven or eight transversely striated whirls, and each
　armed with about twelve teeth; aperture ovate, with the
　outer lip acute, and the pillar brown and glossy; throat
　pale, with red stripes and pellucid dots.

VIBEX.　　143. Shell turreted, greyish, transversely
　　　striated, with the whirls nodulous and striped
　　　with red; pillar-lip white.

Helix, No. 79.　*Schroeter Einl.* ii. p. 199.
Nerita tuberculata.　*Muller Verm.* ii. p. 191.
Strombus Vibex.　*Gmelin,* p. 3522.
Bulimus tuberculatus.　*Bruguiere Enc. Meth.* p. 330.
Martini Berl. Mag. iv. t. 10. f. 51.
Inhabits marshes near the sea-side in Coromandel.　*Muller.*
Shell varying from four to fourteen lines in length, and from
　one and a quarter to four lines in breadth, and is very nearly
　allied to *H. aspera;* from the descriptions, the most obvious
　mark of distinction appears to be in the pillar-lip, which is
　brown in one and white in the other species.

CRENATA.　　144. Shell turreted, white, obsoletely
　　　striated transversely, with a crenated belt at
　　　the upper extremity of the whirls.

Helix crenata.　*Gmelin,* p. 3655.

Helix turrita crenulata. *Chemnitz*, ix. part 2. p. 165. t. 135. f. 1230.

Bulimus torulosus. *Bruguiere Enc. Meth.* p. 332.

Inhabits fresh-water pools in Madagascar. *Bruguiere.*

Shell two and a half inches long, and about eight lines broad, covered with a blackish brown epidermis, beneath which it is white without any coloured markings; it has ten or eleven slightly convex whirls, which are obsoletely striated transversely, particularly on the lower whirls, and each has a crenated belt at its upper extremity.

FUSCATA. 145. Shell turreted, shining, brown, with the whirls minutely striated transversely, and the aperture bluish.

Helix fuscata. *Born Mus.* p. 390. t. 16. f. 17. *Gmelin*, p. 3654.

Helix ater. *Chemnitz*, ix. part 2. p. 164. t. 135. f. 1229.

Helix, No. 211. *Schroeter Einl.* ii. p. 326.

Bulimus fuscatus. *Bruguiere Enc. Meth.* p. 332.

Lister Conch. t. 116. f. 11.

Inhabits the East Indies, in fresh water. *Chemnitz.*

Shell about two inches long, and seven and a half lines in diameter, of a shining blackish brown colour, with ten or eleven minutely striated whirls. I rather doubt whether this is more than a young shell of *Strombus ater*; and Bruguiere has quoted the same figure of Lister's for his *Bulimus terebralis* and *Bulimus fuscatus.*

CONTORTA-PLICATA. 146. Shell turbinated, grey, perforated, sub-truncated, with five whirls, of which the two uppermost are depressed.

Helix contorta-plicata. *Gmelin*, p. 3661.

Helix, No. 171. *Schroeter Einl.* ii. p. 119.

Nerita contorta. *Muller Verm.* ii. p. 187. *Schroeter Fluss.* p. 354.

Argenville, t. 28. third fig. 5.? *Martini Berl. Mag.* iv. t. 9. f. 44.

Inhabits muddy ditches in Denmark; very rare. *Muller.*

Muller says that he never found more than one shell of this species, and that it was without an animal; he has described it to be two lines and two-thirds long, and a line and a half broad, pyramidal, squalid, opake, with five whirls, of which the two uppermost are so depressed as to give the shell a truncated appearance; aperture roundish. It is probably a

2 B 2

Inhabits a moist wood at Lackham, and Bow Wood, Wiltshire. *Montagu.*

Shell nearly three-quarters of an inch long, and two-fifths as broad, of a rusty brown colour, with seven whirls, of which the lower are somewhat cylindrical, and the upper taper more suddenly to an obtuse point; it is nearly allied to *H. obscura*, but is twice as large, and rather broader in proportion, and the lip is more reflected over the umbilicus.

PEREGRINA. 151. Shell imperforate, turreted, with eight whirls, which are slightly wrinkled longitudinally, and the aperture ovate.

Helix peregrina. *Gmelin*, p. 3668.
Helix octanfracta. *Montagu Test.* p. 396, and p. 588. t. 11. f. 8. *Maton and Racket, in Lin. Trans.* viii. p. 211. *Dorset Cat.* p. 55. t. 18. f. 11.
Helix octona. *Pennant Zool.* iv. p. 138. t. 86. f. 135.
Helix, No. 267. *Schroeter Einl.* ii. p. 254. t. 4. f. 11.

Inhabits stagnant water in Great Britain. *Pennant, &c.* West Indies. *Schroeter.*

Shell about five-eighths of an inch long, and two-tenths broad, of a horn-colour, and pellucid, but covered with a dusky epidermis which it is difficult to remove; it has eight whirls tapering to a fine point, and well defined by a depressed suture.

OCTONA. 152. Shell turreted, sub-cylindrical, diaphanous, with eight rounded whirls, and the summit obtuse; aperture roundish.

Helix octona. *Linnæus Syst. Nat.* p. 1248.? *Chemnitz,* ix. part 2. p. 190. t. 136. f. 1624. *Maton and Racket, in Lin. Trans.* viii. p. 211. t. 5. f. 10. *Montagu Test. Supp.* p. 144.
Helix octona, Var. β. *Gmelin*, p. 3653.
Bulimus octonus. *Bruguiere Enc. Meth.* p. 325.
Lister Conch. t. 20. f. 15. *Klein Ost.* t. 2. f. 45.

Inhabits the West Indies. *Chemnitz.* Guadaloupe and St. Domingo; and is a land shell. *Bruguiere.*

Linnæus has described his *H. octona* to be sub-perforated, in which respect it differs from the present shell, and has cited Gualter t. 6. f. BB, which is *Buccinum Acicula*, and badly accords with his description. The present shell has been described by Chemnitz under the name of *H. octona*, without any reference to Linnæus, and notwithstanding is probably the same species. It is about three-quarters of an inch long, and hardly one-fourth as broad, of a pale horn-colour, with eight or nine rounded whirls, and remarkably obtuse at the summit.

COLUMNA. 153. Shell turreted, obtuse, with eight reversed whirls, and decussated striæ; aperture oblong, and the pillar twisted.

Helix Columna. *Chemnitz,* ix. part 1. p. 112. t. 112. f. 954 and 955, and xi. p. 309. t. 213. f. 3020 and 3021. *Gmelin,* p. 3653. *Burrow's Elements,* p. 171. t. 20. f. 4.

Helix Pyrum. *Gmelin,* p. 3665.

Buccinum Columna. *Muller Verm.* ii. p. 151. *Schroeter Fluss.* p. 291.

Bulimus Columna. *Bruguiere Enc. Meth.* p. 332.

Lister Conch. t. 39. f. 37 b. *Martini Mannig.* iv. t. 2. f. 15 and 16. *Favanne,* t. 61. f. H 13.

Junior. Shorter, and the whirls less numerous.

Buccinum mucronatum. *Gmelin,* p. 3504.

Bonanni Mus. Kirch. 3. f. 400. *Lister Conch.* t. 38. f. 37.

Inhabits Guinea. *Martini.* Jamaica. *Chemnitz.*

Shell about two and a half, or sometimes three and a half inches long, and hardly one-fifth as broad, with seven or eight whirls, of which the three lower are remarkably long, and contracted in the middle; it is white, with the summit brown or slightly spotted, or sometimes variegated throughout with irregular longitudinal chestnut stripes.

INCUMBENS. 154. Shell turreted, white, with longitudinal elevated striæ, and remote tawny stripes; pillar sinuated and inflected.

Buccinum striatum. *Muller Verm.* ii. p. 149. *Schroeter Fluss.* p. 346.

Strombus striatus. *Gmelin,* p. 3524.

Inhabits ——

Shell about two and a half inches long, and one third as broad, thin, pellucid, with seven or eight whirls, which somewhat overlap each other; it is marked with very distinct crowded elevated longitudinal striæ, and ornamented with tawny linear streaks; aperture ovate-oblong. It appears from Muller's description to be nearly allied to *H. Columna,* and I cannot imagine what could have induced Gmelin to place it among the Strombi.

PELLA. **155. Shell imperforate, ovate-oblong, acu-minated, transversely striated, and brown with yellow bands.**

Helix Pella. *Linnæus Syst. Nat.* p. 1248. *Gmelin,* p. 3654.

Inhabits Iceland. *Linnæus.*

Linnæus, on the authority of Zoega, has given the following description, and this species has not been noticed by any subsequent author : " Shell ovate-oblong, reddish brown, decumbent, of the size of a seed of a Lithospermum, and the whirls transversely striated ; the body-whirl has two bands, and the others only a single one at their bases ; aperture semi-ovate."

ACUTA. **156. Shell turreted, oblong, coarse, with one or two transverse bands near the base, and the aperture roundish-ovate.**

Helix acuta. *Muller Verm.* ii. p. 100.

Helix bifasciata. *Pulteney Dorset Cat.* p. 49, and New Edit. p. 55. t. 18. f. 8 and 10. *Maton and Racket, in Lin. Trans.* viii. p. 210.

Turbo fasciatus. *Pennant Zool.* iv. p. 131. t. 82. f. 119. *Da Costa Brit. Conch.* p. 90. *Donovan,* i. t. 18. *Montagu Test.* p. 346. t. 22. f. 1.

Bulimus acutus. *Bruguiere Enc. Meth.* p. 323.

Lister Conch. t. 19. f. 14. *Gualter,* t. 4. f. I.

Inhabits sandy places in the neighbourhood of the sea. Great Britain. *Lister.* Italy. *Muller.* Barbary. *Poiret.* Dauphiny. *Bruguiere.*

Shell about half an inch long, and rather more than one-third as broad, of a greyish white colour, slightly striated longitudinally, and marked with one or two transverse brownish bands on the base of the body-whirl ; it has seven or eight slightly convex whirls, which are somewhat wrinkled, and have rather a coarse appearance. There does not appear to me to be any doubt that this is Muller's *H. acuta,* and that Chemnitz's fig. 1224 is a different species.

CARINULA. **157. Shell conical-turreted, shining, with five transverse and other longitudinal lines on the whirls ; body-whirl slightly keeled at the base, and the aperture roundish.**

Helix Carinula. *Gmelin,* p. 3655.

Helix cretacea. *Chemnitz,* ix. part 2. p. 190. t. 136. f. 1263, No. 1 to 4.

Bulimus lineatus. *Bruguiere Enc. Meth.* p. 323.

Inhabits Guadaloupe ; and is a land shell. *Bruguiere.*

Shell about eight lines long and three broad, white, with five transverse interrupted by other longitudinal brown lines on all but the four uppermost of its twelve whirls ; it has a small umbilicus, over which the lip is reflected in the same manner as in *H. acuta,* and this sufficiently proves that neither of these species, as some authors have supposed, can be the Linnæan *H. Barbara,* which is described to be imperforate.

DETRITA. 158. Shell oblong-ovate, perforated, white with longitudinal oblique grey or brown streaks, and the aperture ovate.

Helix detrita. *Muller Verm.* ii. p. 101. *Chemnitz,* ix. part 2. p. 161. t. 134. f. 1225, *a* to *d. Gmelin,* p. 3660.

Helix Sepium. *Gmelin,* p. 3654.

Helix, No. 216. *Schroeter Einl.* ii. p. 238.

Buccinum Leucozonias. *Gmelin,* p. 3489.

Bulimus radiatus. *Bruguiere Enc. Meth.* p. 312.

Lister Conch. t. 8. f. 2. *Seba,* iii. t. 39. f. 37. *Martini Berlin. Mag.* iii. t. 5. f. 53. *Schroeter Erdconch.* t. 1. f. 1.

Inhabits Italy and Saxony. *Muller.* Dauphiny ; and is a land shell. *Bruguiere.*

Shell about an inch long, and nearly half as broad, smooth, somewhat shining, white, and marked with longitudinal slightly oblique brown or grey lines ; it has six or seven whirls, and the outer lip is not reflected, but rather thickened at the margin.

GUADALOUPENSIS. 159. Shell oblong, perforated, whitish, with transverse brown bands, and the outer lip thickened within.

Helix acuta. *Chemnitz,* ix. part 2. p. 161. t. 134. f. 1224. *Gmelin,* p. 3660.

Bulimus Guadaloupensis. *Bruguiere Enc. Meth.* p. 313.

Lister Conch. t. 8. f. 1.

Inhabits Guadaloupe ; and is a land shell. *Bruguiere.*

Shell about an inch long, and five lines broad, smooth, with six slightly convex whirls ; the colour is brownish white, and most commonly marked with three dark or yellowish brown, or purplish transverse bands on the body-whirl, and one on each of the other whirls, but sometimes there is only one on the body-whirl; the outer lip in full grown shells is thickened on its inner margin, especially towards the middle, where there is some appearance of an elevated belt. Chemnitz has figured a young shell, and mistaken it for Muller's *H. acuta*, which, as Bruguiere observes, is the *Turbo fasciatus* of Pennant.

SUBSTRIATA. 160. Shell ovate-conical, perforated, minutely wrinkled longitudinally, and white without bands; aperture ovate, and the lip slightly reflected and marginated.

Helix substriata. *Gmelin*, p. 3667.
Helix detrita. *Pulteney's Dorset Cat.* p. 49, and new edit. p. 56. t. 19. f. 26. *Montagu Test.* p. 384. t. 11. f. 1. *Maton and Racket, in Lin. Trans.* viii. p. 217.
Buccinum minutissime striatum candidum. *Schroeter Fluss.* p. 345. t. 10 A. f. 6.
Lister Conch. t. 108. f. 1. *Gualter*, t. 5. f. SS.
Inhabits the Rhone at Vienne. *Lister.* Fresh water near Weymouth. *Pulteney.*
Shell about eight lines long, and rather more than half as broad, white without any coloured markings, and rather glossy: it has six slightly convex whirls well defined by the suture; the breadth is proportionably greater, and the form more conical than in either *H. detrita* or *H. Guadaloupensis*, and the aperture is rather more effuse, but has the lip thickened in the same manner. Bruguiere has quoted the above figures of Lister and Gualter for *H. detrita*, but they are both said to be fresh-water shells, and are similar to some which I brought from the neighbourhood of Bantry or Killarney, and which Mr. Montagu assured me were like those which Mr. Bryer had sent him from the neighbourhood of Weymouth. At the time when I found them I attended very little to this branch of Natural History, and cannot recollect whether it was in or out of water.

UNDATA. 161. Shell turreted-ovate, sub-imperforate, ventricose, whitish, with somewhat undulated dark chestnut stripes, and the aperture ovate.

Helix undata. *Gmelin*, p. 3667.
Helix, No. 143. *Schroeter Einl.* ii. p. 216.

Buccinum ex rufo undatum, sex spirarum. *Schroeter Fluss.* p. 324. t. 10 A. f. 4.

Gualter, t. 5. f. N.

Inhabits New Holland. *Dr. Leach.* And is a fresh-water shell. *Schroeter.*

Shell an inch and a quarter long, and three-quarters of an inch broad, of a somewhat glossy waxy white colour, with rather broad more or less undulated dark chestnut longitudinal stripes, of which some are interrupted; it has six or seven rather ventricose whirls, of which the body-whirl is thrice as large as the next, and the lip by the pillar is reflected so as almost entirely to conceal a small obsolete umbilicus. My friend Dr. Leach favoured me with a specimen of this shell from New Holland, under the name of *H. striatus*, and it is very nearly allied to *Bulla Zebra*, which belongs to the same natural family; but my present attempt is only to define the species, and I have therefore left that shell in the Genus to which it has been referred by other authors.

FLUVIATILIS. 162. Shell sub-turreted, very smooth, and brown with darker spots; aperture nearly round, and the throat whitish.

Helix fluviatilis Tanschauriensis. *Chemnitz*, ix. part 2. p. 174. t. 135. f. 1243, *a* and *b*.

Helix Lanschaurica. *Gmelin*, p. 3655.

Inhabits fresh water in Coromandel. *Chemnitz.*

This shell appears by Chemnitz's figure to be ten lines long and five broad, of a brown colour with a few transverse rows of darker or blackish spots; it has seven whirls; the aperture is roundish, and the inside bluish white.

LYONETIANA. 163. Shell conical, obtuse, longitudinally wrinkled, distorted, and the side opposite the aperture gibbous; aperture compressed.

Helix Lyonetiana. *Pallas Spic. Zool.* x. p. 33. t. 3. f. 7 and 8.

Trochus monstrosus Lyonetanus. *Chemnitz*, v. p. 21. t. 160. f. 1513.

Trochus distortus. *Gmelin*, p. 3580.

Trochus, No. 2. *Schroeter Einl.* i. p. 679.

Bulimus Lyonetianus. *Bruguiere Enc. Meth.* p. 299.

Inhabits the East Indies. *Bruguiere.*

Shell about an inch long, and the breadth rather exceeds the length, and is thick, white, and somewhat glossy; it has six slightly convex longitudinally wrinkled whirls, of which the body-whirl is largest, and is strangely gibbous on the side opposite the aperture, which gives the shell a singularly distorted appearance. If I recollect right, Col. Hardwicke once favoured me with the sight of two specimens which he had collected in the woods of the Isle of France; but it was before I paid much attention to any other than British shells, and am therefore fearful of quoting his authority for the Habitat.

****** *Ovate, and imperforate.*

PUPA.   164.   Shell sub-imperforate, ovate-oblong, coarse, with six whirls, and the aperture lunated-oblong.

Helix Pupa. *Linnæus Syst. Nat.* p. 1248. *Gmelin,* p. 3656.
Bulimus Pupa. *Bruguiere Enc. Meth.* p. 349.?
Inhabits Mauritania. *Linnæus.*
Linnæus has not made any reference or addition to the above specific character. Bruguiere suspects it is a species which he has received from Algiers, and which he has described to be sub-cylindrical, striated, and white, with a large ovate aperture, and a tooth on the upper part of the pillar-lip; he says it is half an inch long, and two lines and one third broad.

BARBARA.   165.   Shell imperforate, oblong, coarse, with eight whirls, and the aperture roundish-lunated.

Helix Barbara. *Linnæus Syst. Nat.* p. 1248. *Gmelin,* p. 3656.
Inhabits Algiers. *Linnæus.*
Linnæus says this shell resembles, but is only half as large as *H. Pupa,* or rather larger than a barley-corn, and has often a grey band round the base of the body-whirl. Chemnitz has supposed that *H. Carinula* may be this species; but the description answers better to a diminutive specimen of *H. acuta,* and they both differ in being umbilicated.

AMARULA.   166.   Shell ovate-oblong, with the whirls transversely keeled above, and the keel spinous.

Helix Amarula. *Linnæus Syst. Nat.* p. 1248. *Born Mus.*
p. 391. t. 16. f. 21. *Schroeter Fluss.* p. 297. t. 9. f. 8
and 11 ; *Einl.* ii. p. 166, and *Inn. Bau Conch.* p. 48. t.
1. f. 4. *Chemnitz,* ix. part 2. p. 157. t. 134. f. 1218 and
1219. *Gmelin,* p. 3656. *Brookes's Introd.* p. 129. t.
8. f. 117.

Buccinum Amarula. *Muller Verm.* ii. p. 137.

Helix Mitra. *Mus. Gronovianum,* p. 128.? *Schroeter
Fluss.* p. 300. t. 9. f. 10.? *Gmelin,* p. 3655.?

Bulimus Amarula. *Bruguiere Enc. Meth.* p. 309.

Melania Amarula. *Lamarck Syst. des Anim.* p. 91.
Lister Conch. t. 133. f. 33. *Rumphius,* t. 33. f. FF. *Petiver Amb.* t. 4. f. 3. *Argenville,* t. 27. f. 6. *Seba,* iii.
t. 53. f. 24 and 25. *Martini Berlin. Mag.* iv. t. 9. f.
38. *Favanne,* t. 61. f. G 2.

Variety. Ventricose, and the spire shorter.
Lister Conch. t. 1055. f. 8. *Gualter,* t. 6. f. B. *Martini
Berlin. Mag.* iv. t. 11. f. 64 c. *Favanne,* t. 61. f. G 5.
Chemnitz, ix. t. 134. f. 1220 and 1221.

Inhabits rivers in Asia. *Linnæus.* Amboyna. *Rumphius.*
Isles of France and Bourbon, and Madagascar. *Bruguiere.*
River Ganges. *Humphreys.*

Shell about an inch and a half long, and nearly half as broad,
of a pale chestnut-colour, coated with a black epidermis,
and the aperture white ; it has seven whirls, of which the
body-whirl is as long as the spire, and they have a flattish
space at their upper extremities separated by an angulated
keel, which is armed with eight or ten spine-like teeth ; the
Variety is much more strongly plaited longitudinally, and
more ventricose, and has the spire much shorter, so as to
give the shell quite a different appearance ; but no author
has arranged it separately.

TURBINATA. 167. Shell oblong, imperforate, smooth,
acuminated, and the whirls ventricose ; aperture sub-oval and marginated.

Helix turbinata. *Gmelin,* p. 3668.
Helix, No. 328. *Schroeter Einl.* ii. p. 271.
Marsigli Danub. iv. p. 89. t. 31. f. 2. *Schroeter Fluss.* p.
360. t. 10 B. f. 1.

Variety B. Shorter, and the whirls more ventricose.
Helix Danubialis. *Gmelin,* p. 3668.
Helix, No. 327. *Schroeter Einl.* ii. p. 271.
Marsigli Danub. iv. p. 89. t. 31. f. 1. *Schroeter Fluss.* p.
360. t. 10 B. f. 5.

Variety C. With the spire curved.

Helix curvata. *Gmelin,* p. 3668.

Helix, No. 329. *Schroeter Einl.* ii. p. 272.

Marsigli Danub. iv. p. 89. t. 31. f. 4. *Schroeter Fluss.* p. 360. t. 10 B. f. 3.

Inhabits the Danube. *Marsigli.*

Shell about three inches and a quarter long, and an inch and a half broad, with nine whirls, of which the body-whirl is much larger than the others, and the aperture roundish-oval. This species rests entirely on the authority of Marsigli, whose figures have been copied by Schroeter; and *H. Danubialis,* which is described with seven whirls, may probably be a younger shell, or of a more stunted growth; and *H. curvata* has every appearance of an accidentally distorted variety.

STAGNALIS. 168. Shell imperforate, oblong, ventricose, pellucid, with the spire produced and subulate; aperture ovate.

Helix stagnalis. *Linnæus Syst. Nat.* p. 1249. *Pennant Zool.* iv. p. 138. t. 86. f. 136. *Da Costa Brit. Conch.* p. 93. t. 5. f. 11. *Born Mus.* p. 391. t. 16. f. 16. *Schroeter Fluss.* p. 304. t. 7. f. 1 and 2, and *Einl.* ii. p. 167. *Chemnitz,* ix. part 2. p. 168. t. 135. f. 1237 and 1238. *Gmelin,* p. 3657. *Donovan,* ii. t. 51. f. 2. *Montagu Test.* p. 367. t. 16. f. 8. *Maton and Racket, in Lin. Trans.* viii. p. 214. *Dorset Cat.* p. 55. t. 21. f. 11. *Brookes's Introd.* p. 129. t. 8. f. 109. *Burrow's Elements,* p. 171. t. 20. f. 5.

Helix Corvus. *Gmelin,* p. 3665.

Buccinum stagnale. *Muller Verm.* ii. p. 132.

Bulimus stagnalis. *Bruguiere Enc. Meth.* p. 303.

Lymnæa stagnalis. *Lamarck Syst. des Anim.* p. 91.

Le grand Buccin. *Geoffroy,* p. 72. No. 22. t. 2.

Bonanni Rec. and *Kirch.* 3. f. 55. *Lister Anim. Ang.* t. 2. f. 21, and *Conch.* t. 123. f. 21. *Gualter,* t. 5. f. I. *Argenville,* t. 27. No. 6. f. 1. *Klein Ost.* t. 3. f. 69. *Seba,* iii. t. 39. f. 52 and 53. *Martini Berl. Mag.* iv. t. 9. f. 33 and 36. *Favanne,* t. 61. f. F 16, and F 23.

Inhabits rivers, ditches, and pools; common in most parts of Europe.

Shell often two inches long, and about half as broad, thin, brittle, and pellucid, of a whitish, dusky, or greyish colour, sometimes covered with a greenish epidermis; it has six or seven whirls, of which the body-whirl is very large, and con—

stitutes half the length of the shell. I hardly know what
Linnæus means by ' sub-angulated' in his specific character,
unless he alludes to a gibbosity, which in old shells is not
unfrequent on the body-whirl. *H. Corvus* was first figured
separately by Martini in the Berlin Magazine, and appears
to differ in nothing from some of the most common appearances of *H. stagnalis.*

**FRAGILIS. 169. Shell imperforate, ovate-subulate,
taper, pellucid, and the aperture ovate-oblong.**

Helix fragilis. *Linnæus Syst. Nat.* p. 1249. *Schroeter
Fluss.* p. 309. t. 7. f. 8, and *Einl.* ii. p. 168. *Gmelin,*
p. 3658.
Martini Berl. Mag. iv. t. 9. f. 35.
Inhabits fresh water in Sweden and Denmark. *Linnæus.*
About Berlin. *Martini.*
It appears to me, that what has been called *H. fragilis* by
English authors, is certainly not distinct from *H. stagnalis;*
but the shell figured by Martini and Schroeter has a different appearance, and may possibly belong to another
species; it tapers more regularly from the base, and the
body-whirl is less, though the smaller whirls are more distended than in the shell which Montagu has figured; the
aperture also is more oblong, and not so wide in proportion
to the length. Linnæus says it is so brittle, that it will
hardly bear to be touched; but Bruguiere, notwithstanding,
suspects that it is a variety of *H. palustris.*

**PALUSTRIS. 170. Shell oblong-acuminated, brown,
with the whirls striated and slightly convex;
aperture ovate.**

Helix palustris. *Gmelin,* p. 3658. *Montagu Test.* p. 373.
t. 16. f. 10. *Maton and Racket, in Lin. Trans.* viii.
p. 216. t. 5. f. 8. *Dorset Cat.* p. 55. t. 18. f. 18.
Helix stagnalis, Var. *Pennant Zool.* iv. p. 138. t. 86.
f. 136 B. *Chemnitz,* ix. part 2. p. 170. t. 135. f. 1239
and 1240.
Helix fragilis. *Donovan,* v. t. 175. f. 1.
Helix fontinalis. *Donovan,* v. t. 175. f. 2.
Helix, No. 101. *Schroeter Einl.* ii. p. 205.
Buccinum palustre. *Muller Verm.* ii. p. 131. *Schroeter
Fluss.* p. 308. t. 7. f. 9 to 11.
Bulimus palustris. *Bruguiere Enc. Meth.* p. 302.

Le petit Buccin. *Geoffroy*, p. 75. No. 2. t. 2.
Lister Anim. Ang. t. 2. f. 22.   *Gualter*, t. 5. f. E.   *Seba*,
   iii. t. 39. f. 39 and 40.   *Martini Berl. Mag.* iv. t. 9.
   f. 37.   *Favanne*, t. 61. f. F 22.
Variety.   With a narrow white band round the suture.
Buccinum fuscum fascia alba anfractus transeunte.   *Schroeter Fluss.* p. 310. t. 7. f. 7.
Lister Conch. t. 124. f. 24.
Inhabits marshes and about ponds and ditches, and appears to
   be common in most parts of Europe.
Shell three-quarters of an inch or an inch long, and about one-
   third as broad, brown, sub-pellucid, generally covered with a
   dark epidermis; it has six whirls, which taper to a fine point,
   and are obsoletely marked with both longitudinal and transverse striæ; it is stronger in its texture, and has the body-
   whirl much less distended than *H. stagnalis*, and Schroeter's
   figure of *H. fragilis* differs in having the aperture much
   more oblong.   Schroeter considered his t. 7. f. 7. to be a
   variety of the present species, and Schreibers has placed it
   separately; but this part of Schreibers' work is so badly ex-
   ecuted, as not to be worth notice.   Schroeter, in the *Fluss-
   conchylien*, p. 319. t. 7. f. 14, has described and figured a
   minute species, one line and a quarter long, and has quoted
   the *Nerita minuta* of Muller's Verm. ii. p. 179.   It has
   been kept separate in the Einleitung (Helix, No. 254), but
   is not noticed by Gmelin or any other author, and is probably the young of some neighbouring species.

FOSSARIA.   171. Shell imperforate, sub-ovate, with
         five or six rounded whirls, and the suture
         conspicuous; aperture ovate.

Helix fossaria.   *Montagu Test.* p. 372. t. 16. f. 9.   *Maton
   and Racket, in Lin. Trans.* viii. p. 217. t. 5. f. 9.   *Dorset
   Cat.* p. 56. t. 18. f. 17.
Helix glabra.   *Gmelin*, p. 3658.?
Helix, No. 255.   *Schroeter Einl.* ii. p. 249.?
Buccinum glabrum.   *Muller Verm.* ii. p. 135.?   *Schroeter
   Fluss.* p. 320. t. 7. f. 15.?
Bulimus glaber.   *Bruguiere Enc. Meth.* p. 312.?
Inhabits muddy ditches, drains, and water-courses, and is more
   often found at the edge of the water than in it; not uncom-
   mon in England.   *Montagu.*
Shell about three-eighths of an inch long, and somewhat less
   than half as broad, rather thin, sub-pellucid, and horn-
   coloured; it has much the habit of *H. palustris*, but is not

half so large, and the whirls are more rounded, and the suture more conspicuous. The *Buccinum glabrum* of Muller is probably at most only a variety of this species, and differs principally in having a reticulated band on the body-whirl.

ALBICANS. 172. Shell sub-ovate, opake, elongated, acuminated, white, and the aperture oval.

Helix albicans. *Gmelin,* p. 3666.
Helix, No. 249. *Schroeter Einl.* ii. p. 247.
Buccinum mucrone valde elongato, colore albo. *Schroeter Fluss.* p. 311. t. 7. f. 6.
Inhabits fresh water about Hamburgh. *Schroeter.*
Shell half an inch long, and about one-third as broad, with five whirls, and appears from Schroeter's figure to have a more elongated spire, and the body-whirl less distended, than any of its congeners.

PUTRIS. 173. Shell imperforate, ovate-oblong, rather obtuse, yellowish, with four whirls, and the aperture ovate.

Helix putris. *Linnæus Syst. Nat.* p. 1249. *Pennant Zool* iv. p. 139. t. 86. f. 137. *Maton and Racket, in Lin. Trans.* viii. p. 219. *Dorset Cat.* p. 56. t. 21. f. 13.
Helix peregra. *Montagu Test.* p. 373. t. 16. f. 3.
Helix auricularia. *Schroeter Fluss.* p. 272. t. 6. t. 3 to 6, and t. 11 C. f. 2.
Bonanni Rec. 3. f. 54. *Lister Anim. Ang.* t. 2. f. 24, and *Conch.* t. 123. f. 23. *Klein Ost.* t. 3. f. 70. *Favanne,* t. 61. f. E 4.
Variety. With a white somewhat expanded lip.
Lin. Trans. viii. t. 5. f. 8. *Dorset Cat.* t. 19. f. 30. *Montagu Supp.* p. 139.
Inhabits stagnant water in Europe. *Linnæus.* England. *Lister, &c.*
Shell sometimes an inch long, and five-eighths broad, but is usually smaller; it is thin, rather pellucid, of a yellowish horn-colour, covered with a blackish epidermis, and has four longitudinally wrinkled whirls, of which the body-whirl is very large, and the spire very small.

PEREGRA. 174. Shell sub-conical, horny, with the summit acuminated, and the aperture ovate.

Helix peregra. *Gmelin,* p. 3659.

Helix atrata. *Chemnitz,* ix. part 2. p. 174. t. 135. f. 1244,
No. 1 and 2.
Helix auricularia, Var. β. *Gmelin,* p. 3663.
Buccinum peregrum. *Muller Verm.* ii. p. 130. *Schroeter
Fluss.* p. 275. t. 6. f. 7, and t. 11 C. f. 3.
Bulimus peregrus. *Bruguiere Enc. Meth.* p. 301.
Inhabits the garden at Fredericksburg; and is amphibious.
Muller. In the Seine, and about the banks of rivers near
Paris. *Bruguiere.*
It is the more difficult to ascertain this species, because Muller
has referred to only one figure, which is Gualter's, of *Bulla
fontinalis,* and from this the description entirely differs.
Muller has described the length to vary from two to eight
lines, and the breadth from one to five, but Bruguiere says
he never found a shell larger than three lines long, and one
line and two-thirds broad; it differs from *H. auricularia* in
being less ventricose, and in having the spire more produced,
and from *H. palustris* in being more conical; it appears
also by the figures to have the spire more produced, and to
be less distended than *H. putris,* which moreover is not an
amphibious species.

LIMOSA. **175. Shell oblong-ovate, reddish yellow,
diaphanous, with three whirls, and the aperture ovate.**

Helix limosa. *Linnæus Syst. Nat.* p. 1249.
Helix succinea. *Muller Verm.* ii. p. 97. *Chemnitz,* ix.
part 2. p. 178. t. 135. f. 1248. *Maton and Racket, in
Lin. Trans.* viii. p. 218. *Dorset Cat.* p. 56. t. 18. f. 19.
Helix putris. *Schroeter Einl.* ii. p. 169. *Gmelin,* p. 3659.
Montagu Test. p. 376, t. 16. f. 4, and *Supp.* p. 139.
Donovan, v. t. 168. f. 1.
Bulimus succineus. *Bruguiere Enc. Meth.* p. 308.
L'Amphibie. *Geoffroy,* p. 60. No. 22. t. 2.
Gualter, t. 5. f. H. *Schroeter Erdconch.* t. 1. f. 2.
Inhabits wet places on aquatic plants, in most parts of Europe.
Shell half or three-quarters of an inch long, and about half as
broad, and its form is more oblong than that of *H. putris,*
from which it also differs in having only three whirls.
Schroeter in the Einleitung, Muller, Chemnitz, and Gmelin,
have confounded several references which belong to this
species with those of *H. putris;* and the error probably ori-
ginated with Geoffroy, who has referred to Lister, though
his own description as well as Duchesne's figure seem to
have been made from *H. succinea.* It is said in the Fauna

Suecica that *H. limosa* has five whirls, but the number is omitted in the Systema Naturæ, and this shell agrees in other respects with the description, as well as with the reference to Gualter.

TRUNCATULA. 176. Shell ovate-oblong, with the whirls detruncated, and the aperture ovate.

Helix truncatula. *Gmelin*, p. 3659.
Buccinum truncatulum. *Muller Verm.* ii. p. 130. *Schroeter Fluss.* p. 318. t. 7. f. 13.
Bulimus truncatus. *Bruguiere Enc. Meth.* p. 310.
Inhabits stagnant waters about Thangelstadt, in Saxony. *Schroeter.*
Shell five lines long and three broad, thin, pellucid, and of a dark brown or blackish transparent colour; it has five whirls, which are truncated above; the aperture is ovate, and the inner lip spread on the pillar; the colour is dirty white.

INFLATA. 177. Shell ovate-oblong, thickish, opake, ventricose, with a short spire, and the aperture large and marginated.

Helix inflata. *Gmelin*, p. 3666.
Helix, No. 248. *Schroeter Einl.* ii. p. 247.
Buccinum prima spira ventricosa, mucrone brevi. *Schroeter Fluss.* p. 311. t. 7. f. 5.
Inhabits the river Unstrut, in Saxony. *Schroeter.*
This shell appears to be nearly allied to the before-mentioned variety of *H. putris,* but has the aperture less expanded; and in this respect it also differs from *H. lutea,* to which it bears some resemblance; it is white, rather strong, and has the body-whirl twice as large as the spire.

OPACA. 178. Shell ovate, acuminated, thickish, opake, and of a dirty white colour; aperture ovate-oblong.

Helix opaca. *Gmelin*, p. 3667.
Helix, No. 257. *Schroeter Einl.* ii. p. 249.
Buccinum album, opacum, ore angustiore. *Schroeter Fluss.* t. 7. f. 17.
Inhabits fresh water about Hamburg. *Schroeter.*
Shell nearly half an inch long, and about half as broad, thickish, tolerably strong, opake, and of a dirty white colour; it has five whirls, of which the body-whirl is larger than the spire.

2 c 2

TENTACULATA.   179. Shell imperforate, ovate-conical, obtuse, with five or six convex whirls,
and the aperture sub-ovate.

Helix tentaculata. *Linnæus Syst. Nat.* p. 1249. *Pennant
Zool.* iv. p. 140. t. 86. f. 140. *Schroeter Fluss.* p. 321.
t. 7. f. 19 to 22, and *Einl.* ii. p. 171. *Chemnitz,* ix.
part 2. p. 175. t. 135. f. 1245. *Gmelin,* p. 3662. *Do-
novan,* iii. t. 93. *Montagu Test.* p. 389. *Maton and
Racket, in Lin. Trans.* viii. p. 220. *Dorset Cat.* p. 56.
t. 21. f. 12.
Nerita Jaculator. *Muller Verm.* ii. p. 185.
Turbo Nucleus. *Da Costa Brit. Conch.* p. 91. t. 5. f. 12.
La petite Operculée. *Geoffroy,* p. 114. t. 3.
Lister Anim. Ang. t. 2. f. 19, and *Conch.* t. 132. f. 32.
Petiver Gaz. t. 18. f. 8. *Gualter,* t. 5. f. B. *Martini
Berlin. Mag.* iv. t. 7. f. 12 to 14.
Junior. More ventricose, with only four whirls.
Helix sphærica. *Gmelin,* p. 3627.
Nerita sphærica. *Muller Verm.* ii. p. 170.
Inhabits rivers and ponds; common throughout Europe.
Shell about half an inch long, and half as broad, of a dirty
white or horn-colour, frequently covered with a dusky epi-
dermis, and sometimes with a calcareous concretion; it has
five or six convex whirls, divided by a deep suture; the
aperture is nearly orbicular, and contracted at the upper
end; and it is one of the few British fresh-water shells
which are furnished with an operculum.

REPANDA.   180. Shell sub-imperforate, ovate, acuminated, and ventricose; aperture oval-oblong.

Helix repanda. *Gmelin,* p. 3666.
Helix, No. 256. *Schroeter Einl.* ii. p. 249.
Buccinum pellucidum, aufractuum quatuor, ore amplo.
Schroeter Fluss. p. 320. t. 7. f. 16.
Inhabits stagnant waters in Thangelstadt. *Schroeter.*
Shell six or seven lines long, and about three broad, with three
or four whirls, of which the body-whirl is ventricose and
larger than the spire. Schroeter says the aperture is longish
oval, and Gmelin's having described it to be semicircular is
obviously an error.

CANALIS.   181. Shell sub-conical, with five rounded
smooth whirls, and the pillar grooved.

Helix canalis. *Maton and Racket, in Lin. Trans.* viii. p. 220.

Turbo canalis. *Montagu Test.* p. 309. t. 12. f. 11.

Inhabits the sea near Southampton. *Montagu.*

Shell about three-eighths of an inch long, and a quarter of an inch broad, pellucid, horn-coloured, with five smooth rounded whirls, of which the body-whirl is proportionably large; the pillar-lip is broad, white, and furnished with a channel or groove terminating in an umbilicus. Mr. Montagu says it has the habit of *H. tentaculata,* but is at once discriminated by the aperture and groove on the pillar-lip; it is also a marine, and the other a fresh-water shell.

LUTEA. 182. Shell imperforate, sub-oval, thickish, and of a yellowish orange colour; aperture spreading and oval.

Helix lutea. *Montagu Test.* p. 380. t. 16. f. 6. *Maton and Racket, in Lin. Trans.* viii. p. 222.

Inhabits the sea-shore in Devonshire. *Montagu.*

Mr. Montagu says that this species has somewhat the habit of a young shell of *H. auricularia,* but that it is less tumid, and though of inferior size, it is always much thicker and stronger; he has thus described it: "H. with a sub-oval, sub-pellucid, moderately strong, smooth shell, of a dull orange yellow colour; volutions scarcely three, the first extremely large, the others very small; apex obtuse, not prominent; aperture patulous, oval; outer lip not attenuated; inner lip pretty strong, and a little spreading on the columella: length nearly half an inch; breadth rather more than a quarter."

AURICULARIA. 183. Shell imperforate, ovate, obtuse, with a very short pointed spire; aperture much expanded.

Helix auricularia. *Linnæus Syst. Nat.* p. 1250. *Pennant Zool.* iv. p. 139. t. 86. f. 138. *Born Mus.* p. 392. t. 16. f. 20.? *Schroeter Einl.* ii. p. 172. *Chemnitz,* ix. part 2. p. 171. t. 135. f. 1241 and 1242. *Gmelin,* p. 3662. *Donovan,* ii. t. 51. f. 1. *Montagu Test.* p. 375. t. 16. f. 2. *Maton and Racket, in Lin. Trans.* viii. p. 221. *Dorset Cat.* p. 56. t. 21. f. 17.

Buccinum Auricula. *Muller Verm.* ii. p. 126.

Turbo patulus. *Da Costa Brit. Conch.* p. 95. t. 5. f. 17.

Lister Anim. Ang. t. 2. f. 23, and *Conch.* t. 123. f. 22. *Gualter,* t. 5. f. F. *Schroeter Fluss.* t. 6. f. 4 and 5. *Favanne,* t. 61. f. E 3, and E 11.

Junior. With the body-whirl less ventricose, and the aperture not so patulous.

Helix limosa. *Chemnitz,* ix. part 2. p. 177. t. 135. f. 1246 and 1247. *Gmelin,* p. 3661. *Montagu Test.* p. 381. t. 16. f. 1.

Schroeter Erdconch. t. 1. f. 3.

Inhabits rivers, ponds, and ditches; not uncommon in any part of Europe.

Shell about an inch long, and three-quarters of an inch broad, rather pellucid, thin, brittle, and of a pale yellow or horn-colour; it has four whirls, of which the body-whirl constitutes almost the whole shell, and ends abruptly in a small and depressed, but well-defined spire; the aperture is oval, and in a large specimen is three-quarters of an inch long, and half an inch wide.

SICULA. 184. Shell ovate, smooth, whitish, with the apex acute, and the pillar one-plaited.

Helix teres. *Gmelin,* p. 3667. ?
Bulimus Siculus. *Bruguiere Enc. Meth.* p. 335.
Gualter, t. 5. f. NN. *Schroeter Flussconch.* t. 10 A. f. 7.
Inhabits fresh water in Sicily. *Bruguiere.*

Gmelin's *H. teres* is derived wholly from Gualter, of whose figure Schroeter's is a copy, and it is quoted with a mark of doubt by Bruguiere for his *B. Siculus,* who says he received it as a fresh-water shell from Sicily; Bruguiere has described it to be eight lines and a half long, by four and two thirds broad, with five whirls, of which the body-whirl is proportionably very large, and the last so small as to be almost imperceptible; to the naked eye it is smooth, but when magnified appears to be minutely striated transversely; it is yellowish white, and is thicker than *H. auricularia,* which Bruguiere says it somewhat resembles.

GLUTINOSA. 185. Shell ventricose, extremely fragile and pellucid, with a short obtuse spire; aperture very large.

Helix glutinosa. *Gmelin,* p. 3659. *Montagu Test.* p. 379. t. 16. f. 5, and *Supp.* p. 139. *Maton and Racket, in Lin. Trans.* viii. p. 222.

Buccinum glutinosum. *Muller Verm.* ii. p. 129. *Schroeter Fluss.* p. 271. No. 79.
Bulimus glutinosus. *Bruguiere Enc. Meth.* p. 306.
Inhabits fresh water on the leaves of Nymphæa lutea, in Denmark, but not common. *Muller.* England. *Montagu.*
Shell rather more than half an inch long, and three-eighths
broad, of a glossy yellowish horn-colour, and nearly smooth
or only obsoletely wrinkled; it has two or three whirls, of
which the body-whirl occupies almost the whole shell, and
the spire, as Mr. Montagu says, is so extremely small and
so little produced as scarce to be seen, when the shell is ly-
ing with its aperture upwards; it may be readily distinguished from the young of *H. auricularia* by its almost membranaceous texture, and by its obtuse and depressed summit.
Mr. Montagu has remarked that the sub-umbilicus formed
by the repand lip of *H. auricularia* is wanting in the present
species, and that the animals also are essentially different.

LÆVIGATA.   186.   Shell imperforate, ob-ovate, extremely obtuse, pellucid, and very smooth.

Helix lævigata. *Linnæus Syst. Nat.* p. 1250. *Pennant
Zool.* iv. p. 140. t. 86. f. 139. *Gmelin,* p. 3663. *Dono-
van,* iii. t. 105. *Montagu Test.* p. 382, and *Supp.* p.
140. *Maton and Racket, in Lin. Trans.* viii. p. 222.
Dorset Cat. p. 56. t. 18. f. 9.
Helix haliotoidea. *Fabricius Fauna Groenlandica,* No.
387.
Helix haliotoidea, Var. β. *Gmelin,* p. 3664.
Bulla velutina. *Muller Zool. Dan.* iii. p. 32. t. 101. f. 1
to 4.
Inhabits the Northern Ocean. *Muller.* Coasts of Great Britain. *Pennant, &c.*
This shell is often not larger than a pea, and rarely attains half
an inch, but is sometimes three-quarters of an inch long,
and five-eighths broad; it is thin, brittle, pellucid, flesh-coloured, slightly striated longitudinally and transversely wrinkled; when alive it is always covered with a dark brown
wrinkled epidermis; there are three whirls, of which the
body-whirl is very large, and ends suddenly in a lateral depressed and very small, but well defined spire; the aperture
is vastly large, and the inside is white or of a pale purplish
brown colour. It is not in the least like the Linnæan *H.
haliotoidea.*

**BALTHICA. 187. Shell imperforate, ovate, acumi-
nated, with elevated wrinkles, and the aperture large and ovate.**

Helix Balthica. *Linnæus Syst. Nat.* p. 1250. *Gmelin,*
p. 3663.
Inhabits the Baltic. *Linnæus.*
Linnæus has not mentioned the size, or given any reference
to a figure, and this species has not been noticed by any
subsequent author. In the Fauna Suecica it is stated to be
pellucid, with four whirls, and that it has a black inhabitant
with two tentacula.

**NERITOIDEA. 188. Shell imperforate, convex, lon-
gitudinally striated, with the lip reflected
over the umbilicus, and the aperture roundish.**

Helix neritoidea. *Linnæus Syst. Nat.* p. 1250. *Gmelin,*
p. 3663.
Gualter, t. 64. f. I.
Inhabits the Baltic. *Linnæus.*
This species is said to be of a livid colour, with more than
forty striæ. Gualter, t. 64. f. I, to which Linnæus has referred, is a magnified figure of a minute shell, which is granulated by cancellated striæ, and is said to have a large white
aperture,

**PERSPICUA. 189. Shell imperforate, convex-ovate,
with crenulated striæ, and the spire slightly
produced ; aperture roundish and very large,
so as to expose the whole inside.**

Helix perspicua. *Linnæus Syst. Nat.* p. 1250. *Gmelin,*
p. 3663.
Inhabits the Mediterranean. *Linnæus.*
Shell about an inch in diameter, and much depressed, but convex on both sides, and of a pale brownish white colour ; it
has three and a half whirls, which are strongly marked with
crenulated spiral striæ, and the spire is slightly elevated ;
the inside is striated, and somewhat iridescent ; though more
nearly allied to Haliotis than *Helix haliotoidea,* it is very
different from the imperforated species of that Genus, and
in a Linnæan arrangement will hardly stand any where better
than in the place which Linnæus has assigned it.

HALIOTOIDEA. 190. Shell imperforate, depressed, with undulated striæ, and the spire flat; aperture oval, very large, and exposing the whole inside.

Helix haliotoidea. *Linnæus Syst. Nat.* p. 1250. *Martini*, i. p. 197. t. 16. f. 151 to 153. *Schroeter Einl.* ii. p. 176. *Born Mus.* p. 393. *Gmelin*, p. 3663. *Brookes's Introd.* p. 130. t. 8. f. 111.

Sigaretus haliotoideus. *Lamarck Syst. des Anim.* p. 65.

Le Sigaret. *Adanson Senegal*, p. 24. t. 2. f. 2.

L'Oreille de Venus. *Favanne*, i. p. 590. t. 5. f. C.

Bonanni Rec. Supp. f. 14, and *Kirch.* 3. f. 404. *Lister Conch.* t. 570. f. 21. *Rumphius*, t. 40. f. R. *Petiver Gaz.* t. 12. f. 4. *Gualter*, t. 69. f. F. *Argenville*, t. 3. f. C. *Klein Ost.* t. 7. f. 114. *Knorr*, vi. t. 39. f. 5.

Variety? Shell extremely thin.

Bulla haliotoidea. *Montagu Test.* p. 211. t. 7. f. 6, and Vign. 2. f. 6. *Maton and Racket, in Lin. Trans.* viii. p. 123. *Dorset Cat.* p. 43. t. 22. f. 5*.

Inhabits the Mediterranean, American, and Asiatic Seas. *Linnæus.* Coasts of Amboyna. *Rumphius.* Mouths of the Niger. *Adanson.* St. Domingo, Martinique, and Isle of France. *Favanne.*

This shell is somewhat ear-shaped, and the base measures about fifteen by thirteen lines; the base is strongly wrinkled, and the upper surface marked with concentrical slightly undulated striæ, which are crossed and interrupted by irregular wrinkles. The shell is enveloped by the animal, and on this account Lamarck has placed it in a separate Genus among the naked Mollusca. I have no doubt from his description, that Montagu's *Bulla haliotoidea* belongs to the Genus Sigaretus, and not to the Bullæa of Lamarck; and it is probably the present species in a younger, or perhaps in a somewhat depauperated state, owing to the colder latitudes of the British coasts on which it was taken. Gmelin says, the animal is small and white, with two short tentacula; but this is a mistake, originating in Muller's having mistaken his *Bulla velutina* for this species.

AMBIGUA. 191. Shell sub-imperforate, convex, with remote compressed transverse ribs, and the aperture semi-orbicular.

Helix ambigua. *Linnæus Syst. Nat.* p. 1251. *Gmelin*, p. 3665.

Le Fossar. *Adanson Senegal.* p. 173. t. 13. f. 1. *Favanne,*
    ii. p. 253. t. 70. f. G 1, and G 2.
Inhabits the Mediterranean. *Linnæus.*
Linnæus, with a reference to Adanson, has given the following
    description of this species: " Shell small, resembling a
    Nerite, white, convex, with a lateral obtuse spire, and six or
    eight remote compressed elevated transverse lines ; aperture
    semi-orbicular like that of a Nerite, but the inner lip is not
    reflected ; the umbilicus is open in young, and nearly closed
    in old shells." Adanson says it is not more than two lines or
    two lines and a half in diameter, and uniformly white.

𝔊enus XXX.

NERITA:

SHELL UNIVALVE, SPIRAL, GIBBOUS, AND FLATTISH
UNDERNEATH; APERTURE SEMI-ORBICULAR, WITH
THE PILLAR-LIP TRANSVERSE, TRUNCATED, AND
FLATTISH.

Subdivisions.†

* Umbilicated.
** Imperforate, with the pillar-lip toothless.
*** Imperforate, with the pillar-lip toothed.

---

* Umbilicated.

CANRENA. 1. Shell umbilicated, smooth, and the
spire slightly mucronated, umbilicus gibbous
and bifid.

Nerita Canrena. *Linnæus Syst. Nat.* p. 1251. *Born Mus.*
p. 396. *Schroeter Einl.* ii. p. 275. *Gmelin,* p. 3669.
Schreibers Conch. i. p. 313. *Maton and Racket, in
Lin. Trans.* viii. p. 223. *Montagu Supp.* p. 148.
Variety A. Chestnut brown, with paler transverse bands, and
darker interrupted longitudinal zic-zac stripes.
Nerita Canrena. *Chemnitz,* v. p. 250. t. 186. f. 1860 and
1861.
Natica Canrena. *Lamarck Syst. des Anim.* p. 95.
L'Aile de Papillon, ou La Perdrix. *Favanne,* ii. p. 276.
t. 11. f. D 4.

---

† The following of Gmelin's species at p. 3675 appear to me to be undeserving of notice: *N. affinis, N australis,* and *N. Islandica.*——*N. clathrata,* p.
3675, and *N. perversa,* p. 3686, are Fossils.

Bonanni Rec. 3. f. 372, and *Kirch*, f. 365. *Lister Conch.*
t. 560. f. 4. *Petiver Gaz.* t. 156. f. 4. *Gualter*, t. 67.
f. V. *Argenville*, t. 7. f. A. *Knorr*, iii. t. 15. f. 4, and
t. 20. f. 4. *Seba*, iii. t. 38. f. 27, 51, and 52 ; and t. 40.
f. 65 and 66. *Geve*, t. 27. f. 290.
Inhabits the coasts of the Isle of France, the Moluccas, Phi-
lippines, St. Domingo, and Martinique. *Favanne.*
Variety B. White, with very numerous ferruginous dots.
Nerita Canrena, Var. *Linnæus Mus. Lud. Ulr.* p. 674.
No. 383.
Nerita punctata. *Mus. Leskianum*, p. 288. *Ulysses's Tra-
vels*, p. 473.
Le Fanel. *Adanson Senegal.* p. 176. t. 13. f. 3.
Le Mille Points. *Favanne*, ii. p. 270. t. 11. f. D 9.
Bonanni Rec. and *Kirch.* 3. f. 228. *Lister Conch.* t. 564.
f. 11. *Petiver Gaz.* t. 101. f. 10. *Gualter*, t. 67. f. S.
Argenville, t. 7. f. C. *Seba*, iii. t. 38. f. 60 and 61.
Chemnitz, v. t. 186. f. 1862 and 1863.
Inhabits the coasts of Sicily. *Bonanni.* Senegal. *Adanson.*
Madagascar. *Favanne.*
Variety C. Whitish, with crowded chestnut or ferruginous
spots.
Nerita Stercus-muscarum. *Chemnitz*, v. p. 265. t. 187.
f. 1894. *Gmelin*, p. 3673. *Schreibers Conch.* i. p. 318.
La Chiure de Puces. *Favanne*, ii. p. 273.
Inhabits the coasts of America and Provence. *Favanne.*
Variety D. White, with crowded narrow longitudinal stripes.
La Natice Siamoise. *Favanne*, ii. p. 274. t. 11. f. D 5.
Lister Conch. t. 559. f. 1. *Petiver Gaz.* t. 101. f. 9. *Seba*,
iii. t. 38. f. 57. *Geve*, t. 27. f. 302. *Born Mus.* t. 17.
f. 1 and 2. *Chemnitz*, v. t. 186. f. 1864 and 1865.
Inhabits the coasts of Amboyna and the Philippine Islands.
Favanne.
Variety E. With transverse articulated black or red and white
bands.
Nerita Ala-Papilionis. *Ulysses's Travels*, p. 473.
L'Aile de Papillon doré. *Favanne*, ii. p. 278.
Lister Conch. t. 560. f. 3. *Knorr*, i. t. 10. f. 5. *Seba*, iii.
t. 38. f. 29, 64, and 65. *Geve*, t. 28. f. 294 and 301.
Chemnitz, v. t. 186. f. 1868 to 1871.
Inhabits the coasts of Naples. *Ulysses.*
Variety F. Pale brown, marbled with darker spots and dots.
Le Mille Points marbré. *Favanne*, ii. p. 271.
Bonanni Rec. and *Kirch.* 3. f. 224. *Gualter*, t. 67. f. Q.
Chemnitz, v. p. 260. t. 187. f. 1876 and 1877.

Inhabits the coasts of Provence, Martinique, and St. Domingo. *Favanne.*

Variety G.   Brownish white with darker dots, and two transverse rows of large spots.

Nerita maculata.   *Ulysses's Travels*, p. 473.

Le Mille Points à bandes.   *Favanne*, ii. p. 272.

Gualter, t. 67. f. R.   *Chemnitz*, v. t. 187. f. 1878 to 1880.

Inhabits the coasts of Naples.   *Ulysses.*

Variety H.   Reddish, with three transverse rows of crescent-shaped spots.

Nerita pennata.   *Chemnitz*, v. p. 275. t. 188. f. 1921.

Nerita, No. 19.   *Schroeter Einl.* ii. p. 311.

Inhabits the West Indian Seas.   *Chemnitz.*

Variety I.   White, with longitudinal undulated chestnut stripes.

Nerita Zebra.   *Chemnitz*, v. p. 263. t. 187. f. 1885 and 1886.

Nerita, No. 4.   *Schroeter Einl.* ii. p. 305.

Nerita, No. 9.   *Schreibers Conch.* i. p. 317.

Le Zebre.   *Favanne*, ii. p. 275. t. 11. f. D.

Lister Conch. t. 561. f. 7.   *Rumphius*, t. 22. f. G.   *Petiver Amb.* t. 4. f. 4.   *Gualter*, t. 67. f. O.   *Seba*, iii. t. 38. f. 26.   *Geve*, t. 27. f. 295.

Inhabits the coasts of the Molucca Islands.   *Favanne.*

Variety K.   White, with about five transverse rows of chestnut spots.

Nerita, No. 5.   *Schroeter Einl.* ii. p. 306.

Nerita, No. 10.   *Schreibers Conch.* i. p. 317.

Le Pavé Chinois.   *Favanne*, ii. p. 280. t. 11. f. E.

Rumphius, t. 22. f. C.   *Petiver Amb.* t. 10. f. 11.   *Seba*, iii. t. 38. f. 62.   *Geve*, t. 27. f. 303 and 304.   *Chemnitz*, v. t. 187. f. 1887 to 1891.

Inhabits the coasts of the Molucca Islands.   *Chemnitz.*

Variety L.   Clouded with blue, and spotted with reddish brown.

Nerita Pellis Tigrina.   *Museum Leskianum*, p. 289.

Nerita, No. 6.   *Schroeter Einl.* ii. p. 306.

Nerita, No. 11.   *Schreibers Conch.* i. p. 318.

La Peau de Tigre.   *Favanne*, ii. p. 266.

Lister Conch. t. 560. f. 5.   *Gualter*, t. 67. f. N.   *Argenville*, t. 7. f. 4.   *Knorr*, i. t. 10. f. 3.   *Seba*, iii. t. 38. f. 70.   *Chemnitz*, v. t. 187. f. 1892 and 1893.

Inhabits the coasts of the Isle of France, the West India Islands, and the Mediterranean.   *Favanne.*

Variety M.   Purplish flesh-coloured, with transverse rows of angulated ferruginous spots.

Nerita intricata. *Donovan*, v. t. 167.

Inhabits the coasts of England. *Donovan, &c.*

Shells from half an inch to two inches long, and the length and breadth are nearly equal ; the colours and markings are very variable, and the above are only a few of the leading Varieties of. the twenty-five which Gmelin has enumerated ; *N. Stercus-muscarum* has a similar form, and does not appear to me to have a greater claim than the others to be considered a distinct species.

CANCELLATA.   2. Shell with decussated striæ and impressed dots; spire somewhat club-shaped; umbilicus gibbous and bifid.

Nerita cancellata. *Herman Naturf.* xvi. p. 55. t. 2. f. 8. and 9.   *Gmelin*, p. 3670.   *Schreibers Conch.* i. p. 319.

Nerita rugosa granosa. *Chemnitz*, v. p. 270. t. 188. f. 1911 to 1914.

Nerita, No. 14.   *Schroeter Einl.* ii. p. 309.

Nerita Canrena, Var. x.   *Gmelin*, p. 3670.

Lister Conch. t. 566. f. 16.

Inhabits the coasts of the West India Islands.   *Chemnitz.*

Shell about three-quarters of an inch long, and equally broad, of a whitish colour with obsolete spots, and sometimes with a fulvous band and irregular rays.   Chemnitz considered this and the *N. sulcata* of Born to be the same species, but Schroeter, Gmelin, and Schreibers, have placed them separate.

SULCATA.   3. Shell sub-globular, with oblique-longitudinal grooves, and the umbilicus bifid.

Nerita sulcata. *Born Mus.* p. 400. t. 17. f. 5 and 6.   *Gmelin*, p. 3673.   *Schreibers Conch.* i. p. 319.

Nerita, No. 15.   *Schroeter Einl.* ii. p. 309.

Inhabits ——

Born has described this shell to be eight lines long, and equally broad, and of a milk-white colour, with four obliquely grooved whirls.

GLAUCINA.   4. Shell smooth, with a rather obtuse spire; umbilicus partly closed by the lip, which is gibbous and of two colours.

Nerita glaucina. *Linnæus Syst. Nat.* p. 1251.   *Pennant Zool.* iv. p. 140. t. 87. f. 141.   *Born Mus.* p. 397.

Chemnitz, v. p. 246. t. 186. f. 1856 to 1859. *Schrœter Einl.* ii. p. 279. *Gmelin*, p. 3671. *Schreibers Conch.* i. p. 315. *Donovan*, i. t. 20. f. 1. *Montagu Test.* p. 469. *Maton and Racket, in Lin. Trans.* viii. p. 224. *Dorset Cat.* p. 57. t. 21. f. 7. *Burrow's Elements*, p. 171. t. 20. f. 6.

Cochlea Catena. *Da Costa Brit. Conch.* p. 83. t. 5. f. 7.

La Diorchite. *Favanne*, ii. p. 281. t. 10. f. K and L.

Bonanni Rec. and *Kirch.* 3. f. 225. *Lister Anim. Ang.* t. 3. f. 10, and *Conch.* t. 568. f. 19. *Petiver Gaz.* t. 93. f. 7. *Gualter*, t. 67. f. A and B. *Argenville*, t. 7. f. V. *Seba*, iii. t. 39. f. 16. *Knorr*, ii. t. 11. f. 1, and vi. t. 13. f. 7.

Inhabits the coasts of Europe and Africa. *Linnæus.* Britain. *Lister.* Holland. *Petiver.* Moluccas, and the Isle of France. *Favanne.* Tranquebar. *Chemnitz.*

Shell about an inch, or an inch and a half long, and the length rather exceeds the breadth ; it has six whirls of a glaucous, yellowish-brown, or purplish flesh-colour, and is most commonly marked with a row of darker spots or short streaks on the shoulders ; the outer lip is thin, and the pillar-lip forms a strong callus, which partly closes the umbilicus.

VITELLUS. 5. Shell sub-globular, with a perforated equal umbilicus.

Nerita Vitellus. *Linnæus Syst. Nat.* p. 1252. *Born Mus.* p. 398. *Chemnitz*, v. p. 255. t. 186. f. 1866 and 1867. *Schroeter Einl.* ii. p. 280. *Gmelin*, p. 3671. *Schreibers Conch.* i. p. 316.

Le Jaune d'Œuf. *Favanne*, ii. p. 257. t. 11. f. D 3.

Lister Conch. t. 565. f. 12. *Rumphius*, t. 22. f. A. *Petiver Amb.* t. 10. f. 13. *Gualter*, t. 67. f. L. *Seba*, iii. t. 38. f. 30. *Knorr*, i. t. 7. f. 2, and ii. t. 8. f. 5. *Geve*, t. 27. f. 292.

Inhabits the Asiatic Ocean. *Linnæus.* Red Sea. *Forskael.* Coasts of Amboyna. *Rumphius.* Bantam and the Isle of France. *Favanne.*

Shell about an inch and a quarter, or an inch and a half long, and the length and breadth are very nearly equal ; it has five whirls, and the spire is only slightly produced ; the colour is orange clouded with yellow, and variegated with indistinct white spots in transverse rows ; the inside and pillar-lip are white. Favanne erroneously considered this shell to be the Linnæan *N. glaucina*.

SPADICEA.   6. Shell sub-globular, solid, of a chestnut-colour, with a white base, and a white
band round the middle of the body-whirl and
at the suture.

Nerita spadicea.   *Gmelin*, p. 3672.   *Schreibers Conch.* i.
p. 317.
Nerita dilute rufescens.   *Chemnitz*, v. p. 258. t. 187. f.
1872 and 1873.
Nerita leucozonias.   *Gmelin*, p. 3672.
Nerita, No. 1.   *Schroeter Einl.* ii. p. 304.
Nerita rufa, Var.? *Montagu Supp.* p. 150.
Kæmmerer Cab. Rudolst. p. 187. t. 12. f. 5 and 6.
Variety B.   With white cancellated stripes.
Nerita Forskalii.   *Chemnitz*, xi. p. 172. t. 197. f. 1901 and
1902.
Variety C.   Minutely striated longitudinally.
Nerita globosa.   *Chemnitz*, v. p. 267. t. 188. f. 1896 and
1897.   *Mus. Leskianum*, p. 290.
Nerita, No. 8.   *Schroeter Einl.* ii. p. 307.
Inhabits the coasts of the Mauritius. *Chemnitz.*
Shell about an inch and a half long, and equally broad; it is of
a pale chestnut-colour, with a white band across the middle
of the body-whirl, and also at the suture and along the outside of the inner lip; the throat has two reddish stripes;
the umbilicus is said to be deep and spiral, but no mention
is made of its being bifid, and in this respect it appears
principally to differ from *N. rufa.*   *N. leucozonias* is not
well defined, but from the description and Kæmmerer's
figure, it appears to be at most only a Variety of this species.

RUFA.   7. Shell sub-globular, smooth, of a reddish
brown colour, with a white band at the suture; umbilicus bifid.

Nerita rufa.   *Born Mus.* p. 398. t. 17. f. 3 and 4.   *Chemnitz*, v. p. 259. t. 187. f. 1874 and 1875.   *Mus. Leskianum*, p. 289.   *Gmelin*, p. 3672.   *Schreibers Conch.* i.
p. 317.   *Montagu Supp.* p. 150. t. 30. f. 3.
Nerita, No. 2.   *Schroeter Einl.* ii. p. 304.
Cochlea parva.   *Da Costa Brit. Conch.* p. 85.
Lister Conch. t. 606. f. 34.   *Rumphius*, t. 22. f. D.   *Petiver Amb.* t. 11. f. 3.   *Geve*, t. 27. f. 296.
Inhabits the coasts of Barbadoes. *Lister.*   Amboyna. *Rumphius.*   Mauritius. *Chemnitz.*   West Indies and Great
Britain. *Montagu.*

Shell half an inch or an inch long, and equally broad, of a livid purple, reddish, or dark brown colour, with a white band at the suture, and sometimes in the middle of the body-whirl; Mr. Montagu says it has somewhat the contour of *N. glaucina*, but differs in having a large projection which narrows the umbilicus, and occasions an indenture on each side of it.

PUNCTATA. 8. Shell sub-globular, whitish, with numerous scattered yellowish dots, and three transverse rows of reddish irregular spots.

Nerita punctata. *Chemnitz*, xi. p. 173. t. 197. f. 1903 and 1904.
Inhabits the Mediterranean. *Chemnitz.*
Shell about an inch and a quarter long, and nearly equally broad, and in form resembles *N. spadicea*, but differs in the markings; the colour is whitish, with numerous scattered yellowish dots, and three transverse rows of irregular jagged large reddish spots; the umbilicus is pervious, the inner lip callous, and the aperture semi-orbicular.

FULMINEA. 9. Shell sub-globular, with longitudinal zic-zac chestnut stripes; umbilicus toothed and nearly pervious.

Nerita fulminea. *Gmelin*, p. 3672. *Schreibers Conch.* i. p. 317.
Nerita zik-zak. *Museum Leskeanum*, p. 289.
Nerita umbilicata, &c. *Chemnitz*, v. p. 261. t. 187. f. 1881 to 1884.
Nerita, No. 3. *Schroeter Einl.* ii. p. 305.
Le Gochet. *Adanson Senegal*, p. 177. t. 13. f. 4.
Le Point d'Hongrie. *Favanne*, ii. p. 261. t. 10, lower fig. Z.
Lister Conch. t. 567. f. 17. *Gualter*, t. 67. f. M. *Seba*, iii. t. 38. lower fig. 33. *Knorr*, i. t. 10. f. 4.
Inhabits the coasts of Senegal. *Adanson.* Molucca Islands. *Favanne.*
Shell about an inch long, and nearly equally broad, of a whitish or yellowish colour, with dark or pale chestnut longitudinal zic-zac stripes; I rather doubt whether it has any greater claim to be arranged separately than many of the Varieties of *N. Canrena.*

ORIENTALIS.   10. Shell sub-globular, polished, with
    the spire slightly wrinkled at its base, and
    the umbilicus nearly closed by a broad
    callus.

Variety A.   Yellowish brown, with a white transverse nar-
    row band in the middle of the body-whirl, and the pillar
    white.
    Nerita orientalis.   *Gmelin*, p. 3673.   *Schreibers Conch.* i.
       p. 318.
    Nerita subfulva.   *Chemnitz*, v. p. 268. t. 188. f. 1898 and
       1899.
    Nerita, No. 9.   *Schroeter Einl.* ii. p. 307.
Variety B.   White, without any markings.
    Nerita eburnea.   *Chemnitz*, v. p. 268. t. 188. f. 1904.
    Nerita candidissima.   *Ulysses's Travels*, p. 474.
    Nerita, No. 10.   *Schroeter Einl.* ii. p. 308.
    Nerita, No. 14.   *Schreibers Conch.* i. p. 318.
      Geve, t. 28. f. 308.
    Inhabits the East Indian Seas. *Chemnitz.*   Bay of Naples.
       Ulysses.
    Shell about an inch long, and nearly equally broad, and the
       two Varieties appear to differ in nothing but colour.

CRUENTATA.   11. Shell sub-globular, obtuse, with the
    whirls rounded, and the suture channelled;
    umbilicus spiral and pervious.

    Nerita cruentata.   *Gmelin*, p. 3673.   *Schreibers Conch.* i.
       p. 319.
    Nerita, No. 11.   *Schroeter Einl.* ii. p. 308.
    Nerita maculis rufescentibus, &c.   *Chemnitz*, v. p. 269. t.
       188. f. 1900 and 1901.
    Nerita rufescente varia.   *Mus. Lesksanum*, p. 290.
    Inhabits the coasts of Tranquebar. *Chemnitz.*
    Shell about an inch long, and nearly equally broad; it is white,
       covered with reddish spots, and the summit and inside are
       bluish.

RUGOSA.   12. Shell sub-globular, with the whirls
    longitudinally wrinkled, and the umbilicus
    bordered with white.

    Nerita rugosa.   *Chemnitz*, v. p. 269. t. 188. f. 1902 and
       1903.   *Gmelin*, p. 3673.   *Schreibers Conch.* i. p. 319.
    Nerita, No. 12.   *Schroeter Einl.* ii. p. 308.

Inhabits the coasts of the West India Islands. *Chemnitz.*
Shell about ten lines long, and equally broad, of a pale nut-
brown colour, with darker crowded longitudinal wrinkles;
the inside is glabrous, and pale testaceous.

MAROCCANA. 13. Shell sub-globular, slightly wrinkled at the angle of the whirls, greenish, and
the inside brown; umbilicus somewhat compressed.

Nerita Maroccana. *Chemnitz,* v. p. 270. t. 188. f. 1905 to
1910. *Ulysses's Travels,* p. 474.
Nerita Marochiensis. *Gmelin,* p. 3673. *Schreibers Conch.*
i. p. 318.
Nerita, No. 13. *Schroeter Einl.* ii. p. 309.
Inhabits the coasts of Africa, and the West Indies. *Chemnitz.*
Bay of Naples. *Ulysses.*
Shell about three-quarters of an inch or an inch long, and the
length rather exceeds the breadth; the colour is greenish or
brownish olive, with darker longitudinal more or less undulated stripes, and the summit livid or violet; the stripes are
sometimes interrupted so as to form transverse feathered
bands; the spire is somewhat produced, and the summit
acute.

ARACHNOIDEA. 14. Shell sub-globular, white with
reddish reticulated veins, and the summit
blackish; whirls convex, and the umbilicus
nearly closed.

Nerita arachnoidea. *Gmelin,* p. 3674. *Schreibers Conch.*
i. p. 319.
Nerita, No. 16. *Schroeter Einl.* ii. p. 310.
Chemnitz, v. p. 271. t. 188. f. 1915 and 1916.
Inhabits ——
This shell appears by the figures to be ten lines long and nine
broad; and Gmelin doubts whether it is more than a Va-
riety of some one of the neighbouring species.

VITTATA. 15. Shell sub-globular, brown with two
paler reticulated transverse bands, and their
margins sinuated.

Nerita vittata. *Gmelin,* p. 3674. *Schreibers Conch.* i. p.
319.

Nerita parva. *Chemnitz,* v. p. 271. t. 188. f. 1917 and 1918.

Nerita, No. 17. *Schroeter Einl.* ii. p. 310.

Inhabits the coasts of Morocco. *Chemnitz.*

This shell appears, by the figures, to be about seven lines long and six broad, and as well as many other of the sub-globular umbilicated species, is principally distinguished by its colours.

ALBUMEN. 16. Shell depressed, with the umbilicus somewhat heart-shaped, and nearly filled by a flattened callus.

Nerita Albumen. *Linnæus Syst. Nat.* p. 1252. *Born Mus.* p. 399. *Chemnitz,* v. p. 276. t. 189. f. 1924 and 1925. *Schroeter Einl.* ii. p. 281. t. 4. f. 13. *Gmelin,* p. 3671. *Schreibers Conch.* i. p. 316.

Le Jaune d'Œuf aplati, ou le Pain d'Epice. *Favanne,* ii. p. 283. t. 11. f. H 1.

Rumphius, t. 22. f. B. *Petiver Amb.* t. 10. f. 14. *Seba,* iii. t. 41. f. 9 to 11.

Inhabits the coasts of Amboyna. *Rumphius.* Isle of France. *Favanne.* Nicobar and Molucca Islands. *Chemnitz.* Barbary, and the Cape of Good Hope. *Gmelin.*

Shell much depressed, and the base measures about an inch and three-quarters, by an inch and a half; the upper side is convex, and of a buff or sometimes of a liver colour, and the under side is flat and white; the umbilicus is semi-heart-shaped and very large, and is almost closed by a broad flattened callus.

MAMMILLA. 17. Shell ovate, glabrous, with the umbilicus closed, and the aperture ovate.

Nerita Mammilla. *Linnæus Syst. Nat.* p. 1252. *Born Mus.* p. 399. *Schroeter Einl.* ii. p. 282. *Gmelin,* p. 3672. *Schreibers Conch.* i. p. 316.

Variety A. White with the umbilicus closed, and the outer lip acute.

Nerita Mammilla. *Chemnitz,* v. p. 280. t. 189. f. 1928 to 1931.

Le vrai Mammelon, ou Teton blanc de Venus. *Favanne,* ii. p. 288. t. 11. f. H 2.

Lister Conch. t. 571. f. 22. *Rumphius,* t. 22. f. F. *Petiver Gaz.* t. 152. f. 12, and *Amb.* t. 11. f. 15. *Gualter,* t. 67. f. C. *Argenville,* t. 7. f. X. *Seba,* iii. t. 41. f. 22. *Knorr,* i. t. 6. f. 6 and 7.

Variety B. Of a yellowish brown colour, with the aperture white, the outer lip thickish, and the umbilicus open.

Mamma Veneris fuscata. *Chemnitz*, v. p. 282. t. 189. f. 1932 and 1933.

Le Mammelon, ou Teton brun de Venus. *Favanne*, ii. p. 285. t. 11. f. H 4.

Lister Conch. t. 566. f. 14. *Knorr*, iv. t. 8. f. 4. *Seba*, iii. t. 38. f. 32 and 33. *Geve*, t. 28. f. 306.

Variety C. Smaller, white, and the umbilicus partly open.

Nerita nitida. *Donovan*, iv. t. 144. *Montagu Supp.* p. 149.

Nerita Mammilla. *Maton and Racket, in Lin. Trans.* viii. p. 225.

Inhabits the coasts of Alexandria. *Linnæus.* Barbadoes and Jamaica. *Lister.* Amboyna. *Rumphius.* Red Sea. *Forskael.* Moluccas, Isle of France, Madagascar, and Cape of Good Hope. *Favanne.* Tranquebar. *Chemnitz.*

Shell varying from an inch to an inch and three-quarters in length, and more than two-thirds as broad, thick, very glossy, most commonly of a snow-white colour, or sometimes yellowish brown; the spire is small, but somewhat produced and pointed; a large vitreous mass extends from the upper end of the pillar-lip, and generally wholly fills up the umbilicus, especially in oriental specimens, while in those from the West Indies, the umbilicus is commonly open, or at most only partially closed. Both Chemnitz and Favanne considered the Variety B to be a distinct species; but other authors have entertained a contrary opinion, and the two shells approach each other so intimately, that I have not ventured on separating them.

PAPILLA. 18. Shell oblong, thin, pellucid, of a dirty yellow-colour with decussated striæ, and the aperture sub-oval; pillar white, and the umbilicus partially closed.

Nerita Papilla. *Gmelin*, p. 3675. *Schreibers Conch.* i. p. 321.

Nerita, No. 25. *Schroeter Einl.* ii. p. 313.

Papilla seu Ruma Felis. *Chemnitz*, v. p. 285. t. 189. f. 1939.

Le Teton de Chat. *Favanne*, ii. p. 177, and p. 292.

Plentiful on the coasts of Tranquebar. *Chemnitz.* New Zealand. *Favanne.*

Shell about thirteen lines long and ten broad, with four whirls, and of a dirty yellowish colour; Favanne considered it to be a Variety of his *Teton de Negresse*, which is the Linnæan *Helix mammillaris*, and it may be doubted whether either this or any of the shells figured by Chemnitz from 1936 to 1941 ought not to be arranged with that species; but 1934 and 1935, which have also been quoted by Gmelin for a Variety of his N. *melanostoma*, have more appearance of belonging to N. *Mammilla*.

PALLIDULA. 19. Shell smooth, sub-pellucid, with a semi-lunated patulous aperture, and the umbilicus large.

Nerita pallidula. *Montagu Test.* p. 468. *Maton and Racket, in Lin. Trans.* viii. p. 226. *Dorset Cat.* p. 57. t. 20. f. 4 and 5.
Nerita pallidulus. *Da Costa Brit. Conch.* p. 51. t. 4. f. 4 and 5. *Donovan,* i. t. 16. f. 1.
Inhabits the coasts of England. *Da Costa, &c.*
Shell about half an inch long, and three-eighths of an inch broad, of an uniform pale brown or horn-colour, coated with a darker rough epidermis, beneath which it is smooth, or only very slightly wrinkled; it has three whirls, with a flattish spire, and the pillar-lip is thick, white and concave.

** *Imperforate, and the Pillar-lip toothless.*

CORONA. 20. Shell with a transverse row of spines on the body-whirl.

Nerita Corona. *Linnæus Syst. Nat.* p. 1252. *Schroeter Fluss.* p. 217, and *Einl.* ii. p. 283. *Chemnitz,* ix. part 2. p. 68. t. 124. f. 1083 and 1084. *Gmelin,* p. 3675. *Schreibers Conch.* i. p. 322.
Clithon coronata. *Leach Zool. Misc.* ii. p. 122. t. 104.?
Rumphius, t. 22. f. O. *Petiver Amb.* t. 3. f. 4. *Argenville,* t. 7. f. No. 2. *Martini Berl. Mag.* iv. t. 8. f. 30. *Favanne,* t. 61. f. D 7.
Variety B. Smaller, and the pillar-lip minutely toothed.
Nerita Corona Australis. *Chemnitz,* xi. p. 175. t. 197. f. 1909 and 1910.
Variety C. Yellowish brown, and glabrous.
Nerita Corona Bengalensis. *Chemnitz,* xi. p. 176. t. 197. f. 1911.
Inhabits rivers in Asia. *Linnæus.* The Ganges, and China. *Humphreys.*

Shell half, or sometimes three-quarters of an inch long, and about two-thirds as broad, with a row of long spines on the shoulder of the body-whirl, and generally coated with a black epidermis; Dr. Leach thinks his *Clithon coronata* is a distinct species, but has not stated in what the difference consists.

RADULA.   21. Shell with transverse equally tuber-culated ribs, and the aperture roundish.

Nerita Radula.   *Linnæus Syst. Nat.* p. 1252.   *Born Mus.* p. 400. t. 17. f. 7 and 8.   *Chemnitz*, v. p. 289. t. 190. f. 1946 and 1947.   *Schroeter Einl.* ii. p. 284.   *Gmelin*, p. 3676.   *Schreibers Conch.* i. p. 323.
La Pelote de Neige.   *Favanne*, ii. p. 250. t. 11. f. N.
Rumphius, t. 22. f. M.   *Petiver Amb.* t. 11. f. 8.   *Seba*, iii. t. 41. f. 18 and 19.   *Geve*, t. 23. f. 241.
Inhabits the coast of Amboyna. *Rumphius.*   Bantam and Java. *Favanne.* Tranquebar. *Chemnitz.*
Shell about an inch or an inch and a quarter long, and nearly equally broad, thin, pellucid, and of a white or greyish or brownish white colour; it has four whirls, with the spire somewhat produced and pointed, and the interstices of the tuberculated ribs are striated.

MAGDALENÆ.   22. Shell broadly ribbed transverse-ly, and black, with the inside white, and the pillar yellowish; operculum smooth and two-toothed.

Nerita Magdalenæ.   *Gmelin*, p. 3677.
Nerita, No. 225.   *Schroeter Einl.* ii. p. 370.
Le Kiset. *Adanson Senegal*, p. 192. t. 13. f. 5.   *Favanne*, ii. p. 206.
Inhabits the coasts of the Magdalen Islands. *Adanson.*
Adanson has described this shell to be half an inch long, with three whirls, and the summit much depressed; the body-whirl has twenty broad flattish ribs, and the operculum two broad approximated teeth; Favanne considered it to be a Variety of *N. grossa;* but this is impossible, for Adanson says that both the lips are toothless.

CORNEA.   23. Shell roundish, obsoletely striated, and the aperture toothless.

Nerita cornea.   *Linnæus Syst. Nat.* p. 1253.   *Forskæl*

Desc. Anim. p. 128.   *Born Mus.* p. 401.   *Gmelin,* p.
3676.   *Schreibers Conch.* i. p. 323.
Argenville, t. 7. f. M.   *Geve,* t. 22. f. 226.
Inhabits the Red Sea. *Forskael.*
Born says this species is six lines long and eight broad, and he
has given the following description : " Shell sub-globular,
brittle, with an obtuse short spire ; whirls contiguous and
obsoletely truncated transversely ; aperture toothless, with
the outer lip acute, and the inner lip flattened with its margin entire ; colour black, with two dorsal transverse bands,
and scattered yellow dots."

FLUVIATILIS.   24. Shell minutely wrinkled, and
the lips toothless.

Nerita fluviatilis.   *Linnæus Syst. Nat.* p. 1253.   *Muller*
Verm. ii. p. 194.   *Pennant Zool.* iv. p. 141, t. 87. f. 142.
Da Costa Brit. Conch. p. 48. t. 3. f. 17 and 18.   *Schroe-*
ter Fluss. p. 210. t. 5. f. 5 to 10, and t. 11 C. f. 8 ; and
Einl. ii. p. 286.   *Chemnitz,* ix. part 2. p. 72. t. 124.
f. 1088.   *Gmelin,* p. 3676.   *Schreibers Conch.* i. p. 323.
Donovan, i. t. 16. f. 2.   *Montagu Test.* p. 470.   *Maton*
and Racket, in Lin. Trans. viii. p. 225.   *Dorset Cat.*
p. 57. t. 16. f. 17 and 18.
Nerita lacustris.   *Born Mus.* p. 402.
Le Nerite des Rivières.   *Geoffroy,* p. 118. No. 5. t. 3.
Lister Anim. Ang. t. 2. f. 20, and *Conch.* t. 141. f. 38, and
t. 607. f. 43 and 44.   *Petiver Gaz.* t. 91. f. 3.   *Gualter,*
t. 4. f. LL.   *Argenville,* t. 27. f. 3.   *Geve,* t. 24. f. 258
to 265.   *Martini Berl. Mag.* iv. t. 8. f. 27 and 28.   *Fa-*
vanne, t. 61. f. D 3, D 4, D 17, and D 20.
Variety. Shell larger, and yellowish, with parallel black longitudinal stripes.
Nerita cornea.   *Mus. Leskeanum,* p. 291. ?
Zebra Neritarum fluviatilium.   *Chemnitz,* ix. part 2. p. 67.
t. 124. f. 1080 and 1081.
Inhabits rivers in Europe. *Linnæus.* Great Britain. *Lister,*
&c. France. *Geoffroy.* Saxony. *Schroeter.* The Variety
is oriental. *Chemnitz.*
Shell commonly about three-eighths of an inch long, and two-
eighths broad, frequently covered with a brownish green
epidermis, beneath which it is elegantly spotted, streaked,
mottled or reticulated with white and purplish brown, black,
red, or green, in such an infinite variety of ways, that it is
difficult to procure two specimens alike ; it has three whirls,
of which the body-whirl is very large, and the others very

small; at first sight it appears to be quite smooth, but on closer examination it is found to be minutely wrinkled. The oriental shell which Chemnitz has figured differs in being almost an inch long, and is probably a distinct species.

LITTORALIS. 25. Shell smooth, with the summit rather obtuse, and the lips toothless.

Nerita littoralis. *Linnæus Syst. Nat.* p. 1253. *Pennant Zool.* iv. p. 141. t. 87. f. 143. *Da Costa Brit. Conch.* p. 50. t. 3. f. 13 to 16, and t. 4. f. 2 and 3. *Born Mus.* p. 401. *Schroeter Einl.* ii. p. 287. *Gmelin*, p. 3677. *Schreibers Conch.* i. p. 323. *Donovan*, i. t. 20. f. 2. *Montagu Test.* p. 467. *Maton and Racket, in Lin. Trans.* viii. p. 226. t. 5. f. 15. *Dorset Cat.* p. 57. t. 16. f. 13 to 16, and t. 20. f. 2 and 3.

Turbo Neritoides. *Chemnitz*, v. p. 234. t. 185. f. 1854. *Schroeter Einl.* ii. p. 4. *Gmelin*, p. 3588. *Pulteney Dorset Cat.* p. 44.

Le Grain de Maiz. *Favanne*, ii. p. 251. t. 10. lower fig. H. *Lister Anim. Ang.* t. 3. f. 11 to 13, and *Conch.* t. 607. f. 39 to 42. *Petiver Gaz.* t. 34. f. 4 to 6. *Gualter*, t. 64. f. N. *Klein Ost.* t. i. f. 25 and 26.

Inhabits the coasts of Europe. *Linnæus.* Great Britain. *Lister, &c.* France, Italy, Spain, St. Domingo, Martinique, and Barbadoes. *Favanne.*

Shell about three-quarters of an inch long, and the breadth generally rather exceeds the length, but the shape varies considerably; it is thick and strong, and has four whirls, with the spire small, and generally flattish or only slightly produced; the colour is orange, reddish, brown, chestnut, olive or yellow, and is most commonly uniform, but sometimes mottled or marked with one or two paler transverse bands; the summit is said in the Systema Naturæ to be carious; the aperture is roundish oval, and many authors are of opinion that Linnæus has described this species both as a Turbo and a Nerite.

LACUSTRIS. 26. Shell rather smooth, of a horn-colour, and terminating in a very fine point.

Nerita lacustris. *Linnæus Syst. Nat.* p. 1253. *Gmelin*, p. 3677.

Gualter, t. 4. f. MM.

Inhabits stagnant waters and warm springs, in Europe. *Linnæus.*

Linnæus says that this shell is like *N. littoralis*, but is of a horn-colour, and the summit not carious; Gualter's figure, which Linnæus has quoted, is much more like *N. fluviatilis*, and Chemnitz has quoted the Linnæan *N. lacustris* for a Variety of that species.

DUBIA.   27. Shell ovate, thin, pellucid, shining, of a dull yellow colour, variegated with black; outer lip acute, and the inner lip glabrous.

Nerita dubia.   *Chemnitz*, v. p. 324. t. 193. f. 2019 and 2020. *Gmelin*, p. 3676. *Schreibers Conch.* i. p. 332.
Nerita, No. 120.   *Schroeter Einl.* ii. p. 340.
Geve, t. 24. f. 248.
Inhabits ――――
This shell appears by Chemnitz's figures to be about an inch long, and three-quarters of an inch broad, and is said to be very rare; the body-whirl is large, and the spire very small, with the summit slightly produced; the colour is brownish yellow, with black spots and longitudinal streaks in transverse rows. Schreibers has placed this shell in the third division.

MARGINATA.   28. Shell sub-globular, thin, tuberculated, with decussated striæ, and black with ochraceous spots; aperture marginated.

Nerita marginata.   *Gmelin*, p. 3677.
Nerita, No. 226.   *Schroeter Einl.* ii. p. 371. t. 4. f. 16.
Inhabits ――――
This shell appears by Schroeter's figure to be about ten lines long, and is distinguished by having the aperture marginated outwardly; the inside is said to be glabrous, and the pillar-lip flat, wrinkled and tuberculated.

*** *Imperforate, with the pillar-lip toothed.*

PULLIGERA.   29. Shell smooth, unpolished, with the spire excavated at the summit, and the margin of the inner lip crenulated.

Nerita pulligera. *Linnæus Syst. Nat.* p. 1253. *Born Mus.* p. 402. t. 17. f. 9 and 10. *Schroeter Fluss.* p. 215, and *Einl.* ii. p. 289. *Chemnitz*, ix. part 2. p. 64. t. 124. f. 1078 and 1079. *Gmelin*, p. 3678. *Schreibers Conch.* i. p. 326.
Nerita rubella.   *Muller Verm.* ii. p. 195.

Lister Conch. t. 143. f. 37. *Rumphius*, t. 22. f. H. *Petiver Amb.* t. 11. f. 4. *Gualter*, t. 4. f. H H. *Seba*, iii. t. 41. f. 24 to 26. *Knorr*, vi. t. 13. f. 3. *Geve*, t. 23. f. 242. *Martini Berl. Mag.* iv. t. 8. f. 31. *Favanne*, t. 61. f. D 18.

Inhabits rivers in India. *Linnæus.* Amboyna. *Rumphius.* Otaheite. *Chemnitz.*

Shell about sixteen lines long and thirteen broad, of a blackish, reddish, or brownish olive-colour; it has two whirls, of which the body-whirl occupies almost the whole shell, and ends in an acute tooth; the margins of both the lips are tinged with red or brownish yellow.

ACULEATA. 30. Shell with transverse spinous striæ; inner lip flattish, smooth, and slightly toothed at the margin.

Nerita aculeata. *Chemnitz*, x. p. 305. t. 169. f. 1642. *Gmelin*, p. 3686.

Inhabits rivers in the East Indies. *Chemnitz.*

This shell appears by the figure to be about ten lines long and seven broad; and is said to be of a blackish colour.

PUPA. 31. Shell roundish, smooth, of a milk-white colour, with transverse parallel black striæ on the whirls.

Nerita Pupa. *Linnæus Syst. Nat.* p. 1253. *Gmelin*, p. 3679.
Nerita, No. 142. *Schroeter Einl.* ii. p. 345.
Lister Conch. t. 605. f. 31. *Petiver Gaz.* t. 15. f. 8.

Inhabits the coasts of Jamaica. *Lister.*

Linnæus, without any reference, has described this shell to be of the size of *N. littoralis*, and intermediate between that species and *N. Virginea*, with the lip very flat, and very minutely toothed at the margin; colour milk-white, with distant transverse oblique black striæ on the whirls. A shell which Lister and Petiver have figured, and which answers this description, is frequently received from the West Indies, and has a yellow aperture.

BIDENS. 32. Shell smooth, with two teeth on the inner lip.

Nerita bidens. *Linnæus Syst. Nat.* p. 1254. *Gmelin*, p. 3679. *Schreibers Conch.* i. p. 326.

La Nerite verte. *Favanne*, ii. p. 244. t. 10. lower fig. R.?
Inhabits the coasts of New Zealand. *Favanne?*
Linnæus has described his *N. bidens* to be as large as a pea, of
a black or reddish colour, and obsoletely striated, and Favanne considered his *Nerite verte* to be the same; the latter
is described to be about four or five lines long, and eight or
nine broad, of a yellowish green colour, or sometimes white,
and has a spire consisting of one and a half or two whirls,
which form an obtuse summit.

FLAVESCENS. 33. Shell smooth, yellowish variegated with white, and three black bands;
inner lip two-toothed, and the outer lip
slightly striated.

Nerita flavescens. *Chemnitz*, x. p. 304. t. 165. f. 1594 and
1595.
Nerita bidens, Var. β. *Gmelin*, p. 3679.
Nerita, No. 69. *Schreibers Conch.* i. p. 329.
Inhabits the coasts of the Nicobar Islands. *Chemnitz.*
Shell about ten lines long, and the breadth by Chemnitz's
figures appears to exceed the length; the colour is yellowish
mottled with white, and marked with three black bands;
the aperture is semi-circular, and the throat yellowish.
Gmelin considered this shell to be a Variety of *N. bidens*,
but neither Chemnitz or Schreibers have entertained this
opinion.

VIRIDIS. 34. Shell smooth and green, with the inner lip crenulated in the middle.

Nerita viridis. *Linnæus Syst. Nat.* p. 1254. *Born Mus.*
p. 403. *Schroeter Einl.* ii. p. 291. *Chemnitz*, ix. part 2.
p. 73. t. 124. f. 1089. *Gmelin*, p. 3679. *Schreibers
Conch.* i. p. 326.
Nerita fluviatilis subviridis dentata. *Schroeter Fluss.* p. 211.
t. 5. f. 11.?
Le petit Pois vert. *Favanne*, ii. p. 245. t. 10. lower fig. &.
Lister Conch. t. 601. f. 18.
Inhabits Minorca and Jamaica. *Linnæus.* St. Domingo,
Martinique, Barbadoes, Jamaica, and the coasts of Virginia.
Favanne. Fresh waters in St. Croix and St. Thomas.
Chemnitz.
Shell about two lines long, and not much more than half as
broad, generally of an uniform pale green colour, or marked
with only a few small white spots, and sometimes with mi-

nute longitudinal undulated brown lines; the aperture is large, and has a broad pillar-lip, which is generally of a paler colour and somewhat gibbous. Favanne says it comes from the coasts of Virginia, but I believe it is a fresh-water species, and boxes of the shells are often brought from the West Indies, under the name of *Green Pease*.

VIRGINEA. 35. Shell smooth, ovate, with the inner lip gibbous and denticulated on the margin.

Nerita Virginea. *Linnæus Syst. Nat.* p. 1254. *Born Mus.* p. 404. *Schroeter Einl.* ii. p. 292. t. 4. f. 14. *Gmelin,* p. 3679. *Schreibers Conch.* p. 326.
Nerita fluviatilis, Var. γ. *Gmelin,* p. 3677.
Neritæ fluviatiles Indiæ occidentales. *Chemnitz,* ix. part 2. p. 71. t. 124. f. 1086, 1087, and *a* to *l.*
La Nerite à zic-zacs. *Favanne,* ii. p. 246. t. 10. f. B.?
Bonanni Rec. and *Kirch.* 3. f. 204 and 205. *Lister Conch.* t. 604. f. 24 to 27, and t. 606. f. 35 to 37. *Argenville* t. 7. f. P.? *Klein Ost.* t. 2. f. 32. *Geve,* t. 24. f. 250 to 257.
Inhabits the Mediterranean. *Linnæus.* Barbadoes. *Lister.* Rivers in the West India Islands. *Chemnitz.*
Shell about a quarter, or sometimes three-quarters of an inch long, and nearly equally broad; it is allied to *N. fluviatilis,* but is larger, proportionably broader, and more glossy, and has the margin of the inner lip denticulated; the colour is white or yellowish, elegantly marked longitudinally with fine crowded black, brown, or purplish undulated or somewhat reticulated lines, and spotted or banded in an infinite variety of ways. It is generally considered to be the Linnæan *N. Virginea,* though in the Systema Naturæ that species is described to be marine, and this is an inhabitant of fresh water.

TURRITA. 36. Shell with alternate black and white longitudinal parallel stripes, and the spire much produced; outer lip acute, and the inside white.

Nerita turrita. *Chemnitz,* ix. part 2. p. 71. t. 124. f. 1085. *Gmelin,* p. 3686.
Inhabits fresh water, in the West India Islands. *Chemnitz.*
This shell appears by the figure to be rather more than an inch long, and two-thirds as broad, with four whirls, and the spire much produced; the outer lip is said to be acute, and

the inner lip gibbous, but it is not stated whether the latter is toothed or entire.

POLITA.   37. Shell ovate, solid, slightly striated longitudinally, with the spire flat and somewhat obliterated; inner lip toothed, and the outer lip crenulated on its inner margin.

Nerita polita.   *Linnæus Syst. Nat.* p. 1254.   *Born Mus.* p. 405. *Chemnitz,* v. p. 315.   *Schroeter Einl.* ii. p. 293. *Gmelin,* p. 3680.   *Schreibers Conch.* i. p. 327.   *Shaw Nat. Misc.* xvi. t. 678. upper fig.

Variety A.   Pale ash-coloured, or whitish, undulated with grey.

Rumphius, t. 22. f. I.   *Petiver Amb.* t. 11. f. 5.   *Gualter,* t. 66. f. D, F, and G.   *Geve,* t. 22. f. 219 to 221. *Seba,* iii. t. 38. f. 56.   *Chemnitz,* v. t. 193. f. 2001.

Variety B.   With two or three transverse red bands, and the interstices mottled with brown and white.

Le Jaspe sanguin.   *Favanne,* ii. p. 235. t. 10. f. upper S. *Lister Conch.* t. 602. f. 20.   *Rumphius,* t. 22. f. K. *Petiver Amb.* t. 11. f. 6.   *Gualter,* t. 66. f. G.   *Argenville,* t. 7. f. K.   *Klein Ost.* t. 1. f. 29.   *Knorr,* iii. t. 1. f. 4.   *Geve,* t. 22. f. 223 and 224.   *Regenfuss,* i. t. 4. f. 43.   *Chemnitz,* v. t. 193. f. 2002 and 2003.

Variety C.   With transverse red and white alternate bands.

Born Mus. t. 17. f. 13 to 16.   *Chemnitz,* v. t. 193. f. 2004. and 2005.

Variety D.   Elegantly variegated with dark brown and white, and three darker transverse bands.

Chemnitz, v. t. 193. f. 2006 and 2007.

Variety E.   Blackish with white interrupted longitudinal stripes, and paler transverse bands.

Chemnitz, v. t. 193. f. 2008 and 2009.

Variety F.   Olive variegated with white, and three transverse rows of angulated or feather-shaped black spots.

Nerita pennata.   *Born Mus.* p. 404. t. 17. f. 11 and 12.   *Shaw Nat. Misc.* xvi. t. 678 lower fig.

Chemnitz, v. t. 193. f. 2011 and 2012.

Variety G.   Minutely striated transversely, with a yellowish red line round the margin of both lips.

Nerita polita Oceani Australis.   *Chemnitz,* v. t. 193. f. 2013 and 2014.

Lister Conch. t. 600. f. 17.

Variety H.   Black, with two greyish white transverse bands, and the summit white.

Nerita nigra. *Chemnitz,* v. p. 321. t. 193. f. 2015.
Nerita bifasciata. *Gmelin,* p. 3685. *Schreibers Conch.* i. p. 332.
Nerita, No. 116. *Schroeter Einl.* ii. p. 340.
Rumphius, t. 22. f. No. 7. *Petiver Amb.* t. 11. f. 23.
Inhabits the Asiatic Ocean. *Linnæus.* Coasts of the Isle of France. *Lister.* Amboyna. *Rumphius.* Moluccas. *Favanne.* Madagascar, China, and the Sandwich and Society Islands. *Humphreys.*
Shell about fourteen lines long and eleven broad, solid, polished, of different colours, and variously ornamented with stripes, spots, or clouds, of which Varieties only a few are above enumerated; the aperture towards the margin is white, and the throat generally yellow; if the length is taken from the lateral summit of the spire to the end of the pillar, it measures less than the breadth, but the length and breadth of the Nerites have been generally taken obliquely, and in fact are the greatest and smallest diameters.

HIEROGLYPHICA. 38. Shell sub-globular, white, with blackish undulated streaks; inner lip toothed and crenulated.

Nerita hieroglyphica. *Chemnitz,* v. p. 322. t. 193. f. 2016 and 2018.
Nerita litterata. *Gmelin,* p. 3685.
Nerita, No. 117, and No. 118. *Schroeter Einl.* ii. p. 340.
Nerita, No. 83. *Schreibers Conch.* i. p. 332.
Inhabits the East Indian Seas. *Chemnitz.*
This shell is nearly of the same size, but rather shorter and more globular than *N. polita;* from which it also differs in being less polished.

PIPERINA. 39. Shell sub-globular, smooth, thin, of a dull yellowish colour, with triangular black spots, and the inner lip denticulated.

Nerita Piperina. *Chemnitz,* xi. p. 173. t. 197. f. 1905 and 1906.
Inhabits the coasts of Malabar. *Chemnitz.*
The diameter appears by the figures to be about ten or eleven lines, and the colour a dull yellowish olive, with numerous large black triangular spots; the inner lip has four small teeth, and the outer lip is acute.

LARVA.   40. Shell sub-globular, smooth, with the summit very obtuse, and the inner lip very slightly denticulated.

Nerita Larva. *Chemnitz*, v. p. 323. t. 193. f. 2017. *Gmelin*, p. 3679.   *Schreibers Conch*. i. p. 332.

Nerita, No. 119.   *Schroeter Einl*. ii. p. 340.

Rumphius, t. 22. f. No. 6.   *Petiver Amb*. t. 11. f. 22.

Inhabits the coasts of Amboyna.   *Rumphius*.

N. *Larva* appears by the figures to be about three-quarters of an inch long, and nearly equally broad, of a whitish colour, with two or three broad yellowish brown transverse bands. Chemnitz has quoted *La Nuancée* of Favanne, ii. p. 229, but that shell appears by the description to be somewhat ribbed transversely.

ATRATA.   41. Shell smooth, black, and the aperture white; outer lip striated, and the inner margin slightly toothed; inner lip concave, wrinkled, and the margin denticulated.

Nerita atrata. *Chemnitz*, v. p. 296. t. 190. f. 1954 and 1955. *Gmelin*, p. 3683.

Nerita Senegalensis.   *Gmelin*, p. 3686.

Nerita, No. 100.   *Schroeter Einl*. ii. p. 334.

Le Dunar. *Adanson Senegal*, p. 188. t. 13. f. 1.   *Favanne*, ii. p. 209. t. 70. f. F 1, and F 2.

Inhabits the coasts of the West India Islands. *Chemnitz*. Goree. *Adanson*.

The largest and smallest diameters of this shell, when measured obliquely, are about thirteen and ten lines; it is minutely striated transversely, and of an uniform very black colour, except the aperture, which is white; the summit when decorticated, as it often occurs, is also white. *Le Dunar* of Adanson appears to be this shell; and it has also been figured by Martyn in the Universal Conchology, but I cannot ascertain the number of his plate.

NIGERRIMA.   42. Shell sub-globular, smooth, thick, opake, and minutely striated transversely; outer lip entire, and the inner lip slightly wrinkled.

Nerita nigerrima. *Chemnitz*, v. p. 309. t. 192. f. 1985 and 1986.

Nerita aterrima. *Gmelin*, p. 3679. *Schreibers Conch.* i. p. 331.

Nerita, No. 109. *Schroeter Einl.* ii. p. 337.

Inhabits ——

This shell appears by the figures to be rather larger than *N. atrata*, and to have the spire rather more produced; it is of a blackish colour, and the aperture white.

ANTILLARUM. 43. Shell sub-globular, slightly ribbed transversely, and the summit obtuse; lips wrinkled and denticulated.

Nerita Antillarum. *Gmelin*, p. 3685. *Schreibers Conch.* i. p. 331.

Nerita, No. 110. *Schroeter Einl.* ii. p. 338.

Nerita nigerrima, Var. *Chemnitz*, v. p. 309. t. 192. f. 1987.

Inhabits the coasts of the West India Islands. *Chemnitz*.

This shell appears by the figures to be seven or eight lines in diameter, and very nearly globular; it is black, but the ribs are painted darker than their interstices, and the aperture is white.

PELORONTA. 44. Shell somewhat ribbed transversely, and the spire pointed; inner lip slightly and the outer lip strongly toothed.

Nerita Peloronta. *Linnæus Syst. Nat.* p. 1254. *Born Mus.* p. 406. *Chemnitz*, v. p. 305. t. 192. f. 1977 to 1984. *Schroeter Einl.* ii. p. 295. *Gmelin*, p. 3680. *Schreibers Conch.* i. p. 327.

La Quenotte saignante. *Favanne*, ii. p. 215 to 217. t. 10. f. L 1, and L 2.

Bonanni Rec. and *Kirch.* 3. f. 214. *Lister Conch.* t. 595. f. 1. *Gualter*, t. 66. f. Z. *Argenville*, t. 7. f. G. *Knorr*, v. t. 3. f. 2.

Inhabits the coasts of Banda. *Linnæus*. Barbadoes and Jamaica. *Lister*. Red Sea. *Forskael*. Martinique, St. Domingo, and the Molucca Islands. *Favanne*.

Shell an inch or an inch and a quarter long, and about equally broad, of a whitish or pale grey colour, with irregular black and red or purplish longitudinal zic-zac stripes; the inner lip is rather concave, with two or three large teeth, and an irregular saffron or blood-coloured spot in the middle; the aperture is white, and the throat of a pale saffron-colour.

ALBICILLA. 45. Shell ovate, somewhat ribbed transversely, with the lips slightly toothed, and
the inner one tuberculated.

Nerita Albicilla. *Linnæus Syst. Nat.* p. 1254. *Born
Mus.* p. 406. *Chemnitz,* v. p. 313. t. 193. f. 2000, *a* to
h. Schroeter Einl. ii. p. 296. *Gmelin,* p. 3681. *Schrei-
bers Conch.* i. p. 327.
La Nerite à bec. *Favanne,* ii. p. 207. t. 10. upper fig. E.
Le Palais de Bœuf. *Favanne,* ii. p. 208. t. 11. f. F.
Lister Conch. t. 600. f. 16. *Rumphius,* t. 22. f. 8. *Petiver Amb.* t. 21. f. 10. *Argenville,* t. 7. f. F. *Knorr,* vi.
t. 13. f. 4.
Inhabits the coasts of Hitoe. *Rumphius.* Cape of Good
Hope, Isle of France and Manilla. *Favanne.* China.
Humphreys.
Shell about half an inch or more commonly an inch long, and
three-fourths as broad, thick, and of a pale dull yellow,
white, or pale red, with irregular black or dark brown spots,
clouds, or bands, or sometimes blackish with white spots;
the spire is lateral, obsolete and flat, and the body-whirl towards the summit becomes narrower and compressed; the
pillar-lip is tuberculated, and has about three small teeth on
the middle of its margin.

MAXIMA. 46. Shell thick, glabrous, with black and
yellow undulated longitudinal stripes; inner
margin of the outer lip crenulated, and the
inner lip a little concave and four-toothed.

Nerita maxima. *Chemnitz,* v. p. 287. t. 190. f. 1942 and
1943. *Gmelin,* p. 3683.
Nerita, No. 98. *Schroeter Einl.* ii. p. 333.
Inhabits ———
The length, or rather the larger diameter of this shell is an inch
and a half, and the breadth an inch and a quarter, and it
appears in the figures to be slightly striated longitudinally;
the outer margin of the outer lip is entire, but its inner
margin is crenulated, and that of the inner lip has four broad
teeth.

PLEXA. 47. Shell roundish, with transverse crenulated ribs which are alternately larger; inner
lip plaited above, and tuberculated below.

Nerita Plexa. *Chemnitz,* v. p. 288. t. 190. f. 1944 and
1945.

Nerita textilis. *Gmelin*, p. 3683. *Schreibers Conch.* i. p. 330.

Nerita, No. 99. *Schroeter Einl.* ii. p. 334.

Nerita Exuvia, Var. orientalis. *Lamarck Syst. des Anim.* p. 95.

La Grive à vives-arrêtes. *Favanne*, ii. p. 203. t. 11. f. M.

Lister Conch. t. 598. f. 14. *Rumphius*, t. 22. f. No. 3. *Petiver Amb.* t. 21. f. 5. *Da Costa Brit. Conch.* t. 3. f. 14 and 15.

Inhabits the coasts of Tranquebar, Ceylon, and Nicobar Islands. *Chemnitz.*

Shell three-quarters of an inch or an inch long, and about equally broad, glossy, and white, tessellated with distant black spots; the transverse ribs are alternately large and small, and crenulated by longitudinal striæ which pass over them; the outer margin of the outer lip is crenulated, and the inner margin toothed; the inner lip is plaited above and tuberculated below. Chemnitz has quoted the above-mentioned shell of Favanne's for *N. Exuvia*, but it undoubtedly belongs to the present species, and *La Grande Grive orientale* (Favanne, ii. p. 201.), may probably be a large Variety of the same; but the figure, which is a copy from Argenville's t. 7. f. B, badly accords with the description.

HISTRIO. 48. Shell with unequal transverse ribs; outer lip crenated on the inner margin, and the inner lip four-toothed.

Nerita Histrio. *Linnæus Syst. Nat.* p. 1254. ? *Chemnitz*, v. p. 291. t. 190. f. 1948 and 1949, and t. 191. f. 1960 and 1961. *Schroeter Einl.* ii. p. 297. *Museum Lesk.* p. 294. *Gmelin*, p. 3681. *Schreibers Conch.* i. p. 327.

La Jonquille. *Favanne*, ii. p. 232. t. 10. f. R. ?

Lister Conch. t. 598. f. 11. *Geve*, t. 22. f. 218. *Knorr*, vi. t. 13. f. 2. ?

Inhabits the East Indian Seas. *Chemnitz.*

Linnæus for *N. Histrio* has only quoted Rumphius, t. 22. f. 6, which badly accords with his description, and is much more like *N. Larva.* The shell which Chemnitz has figured appears to be about an inch and a quarter long, and nearly equally broad, and variously mottled or somewhat tessellated with black and white; the pillar-lip is nearly smooth, yellowish, and the margin has four strong teeth. Chemnitz has quoted *La Jonquille* of Favanne, but Argenville, t. 7. f. I, of which Favanne's figure is a copy, has been quoted by Linnæus for *N. Exuvia,* and appears to be different from the description.

PLICATA. **49. Shell transversely ribbed, and the spire pointed ; outer lip crenated, and toothed** on the inner margin ; inner lip convex, wrinkled, and strongly toothed.

Nerita plicata. *Linnæus Syst. Nat.* p. 1255. *Born Mus.* p. 407. t. 17. f. 17 and 18. *Chemnitz*, v. p. 293. t. 190. f. 1952 and 1953. *Schroeter Einl.* ii. p. 298. *Gmelin,* p. 3681. *Schreibers Conch.* i. p. 328. *Museum Lesk.* p. 295.

Nerita tricolor. *Gmelin,* p. 3686.

Le Selot. *Adanson Senegal,* p. 191. t. 13. f. 4.

La Nerite à dents de Cheval. *Favanne,* ii. p. 226. t. 10. f. Q 3.

Bonanni Rec. 3. f. 386, and *Kirch,* f. 371. *Lister Conch.* t. 595. f. 3. *Gualter,* t. 66. f. V. *Klein Ost.* t. 5. f. 100. *Seba,* iii. t. 59. f. 18.

Variety. White without any spots.

Nerita lactaria. *Linnæus Mantissa,* p. 551. *Schroeter Inn. Bau Conch.* p. 70. t. 1. f. 5.

Inhabits the coasts of the Isle of France and Madagascar. *Favanne.* Tranquebar. *Chemnitz.*

Shell three-quarters of an inch or an inch long, and about equally broad, of a white or reddish white colour, with brown or black spots on the ribs, or sometimes marked with undulated dark longitudinal stripes ; the spire is sometimes yellowish ; the inner margin of the outer lip has about eight teeth ; the inner lip is convex and wrinkled, and has about four strong teeth on the margin, which are bent somewhat inwards ; the aperture is narrow and white, and the throat yellow.

ASCENSIONIS. **50. Shell strongly ribbed transversely, with the spire produced, and the outer lip entire ; inner lip slightly concave, glabrous, and the margin denticulated.**

Nerita Ascensionis. *Gmelin,* p. 3683.

Nerita, No. 101. *Schroeter Einl.* ii. p. 334.

Nerita diversicolor. *Martyn Univ. Conch.* t. 108. ?

Nerita, &c. *Chemnitz*, v. p. 297. t. 191. f. 1956 and 1957.

Inhabits the coasts of the island of Ascension. *Chemnitz.*

Shell an inch and a half long, and about equally broad, of a white or reddish colour, with about sixteen strong transverse ribs, which are spotted with black; the inner lip is yellowish, slightly concave, and the margin denticulated; the outer lip has both its inner and outer margin entire.

LINEATA. 51. Shell sub-globular, with black transverse striæ, and the interstices violet; outer lip striated within, and the margin of the inner lip denticulated.

Nerita lineata. *Chemnitz,* v. p. 297. t. 191. f. 1958 and 1959. *Gmelin,* p. 3684. *Schreibers Conch.* i. p. 329.
Nerita, No. 102. *Schroeter Einl.* ii. p. 335.
Geve, t. 22. f. 228.
Inhabits the Straights of Malacca. *Chemnitz.*
This shell appears by the figures to be about an inch long, and nearly equally broad, and I have seen specimens still larger; it is of a pale violet colour, with transverse elevated black striæ, and the lips and summit are of a pale brownish yellow.

VERSICOLOR. 52. Shell transversely ribbed, with the spire produced, and tessellated with red and dark spots in transverse rows; inner and outer lip toothed, and the latter striated within.

Nerita versicolor. *Gmelin,* p. 3684. *Schreibers Conch.* i. p. 329.
Nerita variegata. *Mus. Leskeanum,* p. 296.
Nerita, &c. *Chemnitz,* v. p. 298. t. 191. f. 1962 and 1963.
Nerita, No. 103. *Schroeter Einl.* ii. p. 335.
La petite Livrée. *Favanne,* ii. p. 221. t. 10. f. S.
Argenville, t. 7. f. &c.
Inhabits the coasts of the West India Islands. *Chemnitz.* St. Domingo, Martinique, Barbadoes, Jamaica, and Africa. *Favanne.*
Shell about three-quarters of an inch or an inch long, and nearly equally broad, white, marked more or less regularly with alternate tessellated bands of black or purplish and red spots; the upper part of the inner lip is wrinkled, and the margin toothed, with the middle teeth largest.

PICA. 53. Shell transversely ribbed, and the spire produced; outer lip with twelve and the inner with four strong teeth.

Nerita Pica. *Chemnitz*, v. p. 298. t. 191. f. 1964 and 1965. *Gmelin*, p. 3684. *Mus. Lesk.* p. 295. *Schreibers Conch.* i. p. 330.
Nerita, No. 104. *Schroeter Einl.* ii. p. 335.
Rumphius, t. 22. f. No. 4. *Petiver Amb.* t. 11. f. 20.
Inhabits the coasts of Amboyna. *Rumphius.*
This shell is rather smaller but of the same form as *N. versicolor*, of which Karsten, in the Museum Leskeanum, has queried whether it is more than a Variety; it is white, variegated with black tessellated spots, or interrupted longitudinal streaks.

FLAMMEA. **54.** Shell sub-globular, with equal transverse crowded ribs, and the spire slightly produced; lips toothed; outer lip grooved within, and the inner one wrinkled.

Nerita flammea. *Gmelin*, p. 3685.
Nerita striata. *Chemnitz*, v. p. 311.
Variety A. White, tessellated with black spots.
Nerita, No. 111. *Schroeter Einl.* ii. p. 338.
Chemnitz, v. t. 192. f. 1992 and 1993.
Variety B. Tinged with red, and the spots blackish, forming longitudinal rays.
Nerita, No. 112. *Schroeter Einl.* ii. p. 338.
Chemnitz, v. t. 192. f. 1994.
Variety C. With bands of blackish and reddish spots.
Nerita, No. 113. *Schroeter Einl.* ii. p. 339.
Chemnitz, v. t. 192. f. 1995.
Inhabits the coasts of the West India islands. *Chemnitz.*
Chemnitz, both for this shell and for *N. versicolor*, has quoted *La petite Livrée* of Favanne, and the present is not a well defined species; by the figures it appears to be eight or nine lines in diameter, and nearly globular.

COSTATA. **55.** Shell sub-globular, with black transverse ribs and white interstices, and the spire very obtuse; lips toothed, and the inner one slightly convex, wrinkled and tuberculated.

Nerita costata. *Chemnitz*, v. p. 299. t. 191. f. 1966 and 1967. *Gmelin*, p. 3684. *Schreibers Conch.* i. p. 330.
Nerita grossa. *Born Mus.* p. 407. t 17. f. 19 and 20. ?
Inhabits the coasts of the Nicobar Islands. *Chemnitz.*
Shell an inch long and about equally broad, with fifteen black or dark brown transverse ribs, and the interstices white;

outer lip with the outer margin crenated, and the inner margin has eight teeth, of which those at each extremity are largest; the inner lip is somewhat convex, wrinkled and tuberculated; and has three or four strong teeth on the margin. It appears to be nearly allied to *N. grossa,* but has the spire very obtuse.

GROSSA. 56. **Shell transversely ribbed, and the spire acute; lips toothed, and the inner one convex and wrinkled.**

Nerita grossa. *Linnæus Syst. Nat.* p. 1255. *Chemnitz,* v. p. 299. t. 191. f. 1968 and 1969. *Schroeter Einl.* ii. p. 299. *Gmelin,* p. 3682. *Schreibers Conch.* i. p. 328.

La Grive rousse. *Favanne,* ii. p. 204.

Rumphius, t. 22. f. N. *Petiver Amb.* t. 5. f. 8, and t. 11. f. 9. *Geve,* t. 23. f. 239.

Inhabits the Asiatic Ocean. *Linnæus.* Coasts of the Molucca Islands. *Chemnitz.*

Shell an inch long and about equally broad, with thirteen or fourteen somewhat spotted brown transverse ribs, and the interstices white; the outer lip has its outer margin crenated, and on the inner margin there is a strong tooth at each extremity, with intermediate crenatures; the inner lip is wrinkled, tinged with yellow, and three-toothed. Linnæus has described this species 'spira acuta,' and the shell which Born has figured under the name of *N. grossa* differs in being obtuse, and therefore answers better to the foregoing species.

CHAMÆLEON. 57. **Shell sub-globular, with about twenty transverse ribs, and the spire obtuse; lips toothed, and the inner one wrinkled and tuberculated.**

Nerita Chamæleon. *Linnæus Syst. Nat.* p. 1255. *Born Mus.* p. 408. *Chemnitz,* v. p. 310. t. 192. f. 1988 to 1991. *Schroeter Einl.* ii. p. 300. *Gmelin,* p. 3682. *Schreibers Conch.* i. p. 328.

La grande Livrée. *Favanne,* ii. p. 219. t. 10. lower fig. C. *Rumphius,* t. 22. f. L. *Petiver Amb.* t. 11. f. 7. *Gualter,* t. 66. f. X. *Argenville,* t. 7. f. Q. *Knorr,* v. t. 15. f. 4.

Inhabits the coasts of Banda. *Linnæus.* Molucca Islands. *Chemnitz.*

Shell about three-quarters of an inch long, and very nearly

equally broad, with a lateral depressed spire, and nearly globular; it has about twenty unequal ribs, which are crossed
with minute striæ; the outer lip has the outer margin very
slightly crenated, and is striated within; the inner lip is flattish, wrinkled, and tuberculated, and has two or three teeth
on the margin; the colour is white or yellow, with black
spots, and undulated longitudinal stripes.

STELLA. 58. Shell sub-globular, transversely ribbed, with the summit obtuse and radiated;
lips denticulated.

Nerita Stella. *Chemnitz*, xi. p. 174. t. 197. f. 1907 and
1908.
Inhabits the East Indian Seas. *Chemnitz*.
This shell appears by the figures to be about three-quarters of
an inch in diameter, and nearly globular, with the summit
very obtuse; the colour is dark red with black spots, and
the summit radiated with black; the operculum is said to
be covered with elevated dots, which is also the case in several other species.

UNDATA. 59. Shell with about thirty transverse
ribs, and the spire acute; lips toothed, and
the inner one wrinkled and tuberculated.

Nerita undata. *Linnæus Syst. Nat.* p. 1255. *Born Mus.*
p. 408. ? *Chemnitz*, v. p. 292. t. 190. f. 1950 and 1951.
Schroeter Einl. ii. p. 302. *Gmelin*, p. 3682. *Schreibers
Conch.* i. p. 328.
Nerita Promontorii. *Gmelin*, p. 3686.
Le Legar. *Adanson Senegal*, p. 191. t. 13. f. 3.
Bonanni Rec. and *Kirch.* 3. f. 215. *Lister Conch.* t. 596.
f. 7. *Gualter*, t. 66. f. P.
Inhabits the East Indian Seas. *Chemnitz*. Between Cape
Manuel and Cape Verd, on the coast of Africa. *Adanson*.
Linnæus, in the Systema Naturæ, for this species has quoted
Rumphius, t. 22. f. 4, and Gualter, t. 66. f. X, neither of
which answer the description given in his account of the
Museum of the Queen of Sweden, and the former is more
like *N. Pica*, and the latter *N. Chamæleon*. The shell is
about an inch long and nearly equally broad, of a yellowish
grey colour, with black or dark brown broad longitudinal
confluent stripes; the margin of the outer lip is crenated,
and the inner margin is crenulated by striæ, which are formed
into teeth at the extremities; the inner lip is somewhat con-

vex, wrinkled and tuberculated, and has two or three strong teeth on its margin.

UNDULATA. 60. Shell with about thirty transverse ribs, and the summit obtuse; outer lip slightly striated and toothless; inner lip wrinkled and denticulated.

Nerita undulata. *Gmelin*, p. 3678.
Nerita, No. 106. *Schroeter Einl.* ii. p. 336.
Nerita undata subtilior et levior. *Chemnitz*, v. p. 301. t. 191. f. 1970 and 1971.
Inhabits the East Indian Seas. *Chemnitz.*
This shell appears to be nearly of the same size and shape as the foregoing, but is said to be thinner, with the ribs much narrower, and the summit not pointed; it is of a greyish colour, longitudinally waved with black.

EXUVIA. 61. Shell transversely ribbed, with the summit very obtuse; lips toothed, and the inner one slightly concave and tuberculated.

Nerita Exuvia. *Linnæus Syst. Nat.* p. 1255. *Born Mus.* p. 409. *Chemnitz*, v. p. 302. t. 191. f. 1972 and 1973. *Schroeter Einl.* ii. p. 303. *Gmelin*, p. 3683. *Schreibers Conch.* i. p. 329.
Lister Conch. t. 596. f. 5, and t. 599. f. 15. *Gualter*, t. 66. f. CC. ? *Seba*, iii. t. 59. f. 4 to 10. *Geve*, t. 23. f. 240. *Knorr*, iii. t. 1. f. 5.
Inhabits the coasts of America and Asia. *Linnæus.* Jamaica. *Lister.*
Shell about half or sometimes three-quarters of an inch long, and the breadth rather exceeds the length; it has from fifteen to nineteen broad flattened ribs; the inner lip is slightly concave, and studded with small tubercles, and has two or three minute teeth on the margin; the colour is black, with white spots or longitudinal undulated stripes. Some of the figures which are quoted in the Systema Naturæ belong to *N. plexa*, and Linnæus probably considered it to be only a Variety of this species.

QUADRICOLOR. 62. Shell transversely ribbed, with the upper ribs larger, and the summit acute; lips toothed, and the inner one wrinkled.

Nerita quadricolor. *Gmelin*, p. 3684. *Schreibers Conch.* i. p. 330.

Nerita Maris Rubri. *Chemnitz,* v. p. 304. t. 191. f. 1974 and 1975.

Nerita, No. 107. *Schroeter Einl.* ii. p. 337.

Inhabits the Red Sea. *Forskael.*

Shell thirteen or fourteen lines long, and about equally broad, with the ribs becoming smaller and rather obsolete towards the base; the outer lip is grooved and toothed within, and the inner lip is toothed and wrinkled; it is of a dull violet colour, with the ribs black, the summit yellowish, and the inside white.

MALACCENSIS. 63. Shell with broad much elevated ribs, and a produced spire; outer lip crenated on the outer margin, and the inner margin entire; inner lip glabrous, yellowish, and slightly denticulated.

Nerita Malaccensis. *Gmelin,* p. 3684.

Nerita, No. 108. *Schroeter Einl.* ii. p. 337.

Chemnitz, v. p. 305. t. 192. f. 1976.

Inhabits the straights of Malacca. *Chemnitz.*

This shell appears, by the figure, to be rather more than an inch long, and about equally broad, of a yellowish white colour, with regular rows of black spots on the ribs.

FULGURANS. 64. Shell sub-globular, with crowded transverse striæ, and the spire very slightly produced; lips somewhat toothed, and the inner one tuberculated in the middle.

Nerita fulgurans. *Gmelin,* p. 3685.

Nerita, No. 114. *Schroeter Einl.* ii. p. 339.

Chemnitz, v. p. 312. t. 192. f. 1996 and 1997.

Inhabits the coasts of the West India Islands. *Chemnitz.*

Shell about seven or eight lines in diameter, and nearly globular, of a blackish colour, with undulated longitudinal yellowish white stripes; outer lip grooved within, and toothed at each extremity, and its outer margin crenated and variegated with black and white; inner lip somewhat concave, tuberculated in the middle, and the margin slightly toothed.

TESSELLATA. 65. Shell obtuse, transversely striated, and tessellated with black and white; lips denticulated, and the inner lip concave and glabrous.

Nerita tessellata. *Gmelin*, p. 3685. *Mus. Lesk.* p. 295.
Nerita, No. 115. *Schroeter Einl.* ii. p. 339.
Le Tadin. *Adanson Senegal*, p. 190. t. 13. f. 2. *Favanne*,
    ii. p. 222.
Chemnitz, v. p. 313. t. 192. f. 1998 and 1999.
Inhabits the West Indies. *Chemnitz.* Coasts of the Magda-
len Islands. *Adanson.*
Shell about half an inch long, and the breadth rather exceeds
    the length ; the outer lip is somewhat striated within, and
    has two teeth at each extremity ; the inner lip is slightly
    concave, glabrous, with a few elevated dots, and the margin
    denticulated ; it is tessellated between the striæ with nearly
    square black and white spots.

𝕲𝖊𝖓𝖚𝖘 XXXI.

HALIOTIS:†

**SHELL EAR-SHAPED AND DILATED, WITH A LONGI-
TUDINAL ROW OF PERFORATIONS; SPIRE LA-
TERAL AND NEARLY CONCEALED.**

MIDÆ. 1. Shell roundish, and polished on both
sides.

Haliotis Midæ. *Linnæus Syst. Nat.* p. 1255. *Martini,*
i. p. 178. t. 14. f. 136, and t. 15. f. 141. *Born Mus.*
p. 411. *Schroeter Einl.* ii. p. 374. *Gmelin,* p. 3687.
Schreibers Conch. i. p. 330.
L'Oreille de Mer feuilletée. *Favanne,* i. p. 587. t. 5. f. B.
Lister Conch. t. 613. f. 5. *Gualter,* t. 69. f. B. *Knorr,*
v. t. 20. f. B. ?
Inhabits the East Indian Seas. *D'Avila.* Coasts of the Isle
of France. *Favanne.* Cape of Good Hope. *Humphreys.*
The base is about five inches long, and four broad, with a longitudinal row of from eight to ten perforations; the outer
surface, when deprived of its yellowish brown epidermis, is
white more or less tinged with orange and other colours, and
is strongly wrinkled transversely; the inside is smooth and
pearly.

PULCHERRIMA. 2. Shell roundish, with granulated
striæ, and the spire exserted.

Haliotis pulcherrima. *Martyn Univ. Conch.* ii. t. 62.
Chemnitz, x. p. 313. t. 166. f. 1605 and 1606. *Gmelin,*
p. 3690.
Inhabits King George's Sound in the South Sea. *Martyn.*
Shell about three-quarters of an inch long, and nearly equally
broad, variegated with white and rose-colour, and the inside

---

† *H. Guineensis* of Gmelin, p. 3689, appears to me to be too uncertain a
species to be retained; and *H. perversa,* and *H. plicata,* p. 3690, are fossils.

silvery; it has about thirty tubercles, of which six are perforated; the inner margin is said to be very broad, and the outer lip crenated.

VIRGINEA. 3. Shell ovate, with decussated undulated striæ, and the inside iridescent.

Haliotis virginea. *Chemnitz,* x. p. 314. t. 166. f. 1607 and 1608. *Gmelin,* p. 3690.

Inhabits the coasts of New Zealand. *Chemnitz.*

Shell about an inch and a half long, and an inch broad, of a dull green colour, spotted and on the margin striped with white; it has six of the tubercles perforated; the inside is striated, and strikingly iridescent.

TUBERCULATA. 4. Shell sub-ovate, with the outside transversely wrinkled and tuberculated.

Haliotis tuberculata. *Linnæus Syst. Nat.* p. 1256. *Pennant Zool.* iv. p. 141. t. 88. f. 144. *Martini,* i. p. 187. t. 16. f. 146 to 149. *Schroeter Einl.* ii. p. 375. *Gmelin,* p. 3687. *Schreibers Conch.* i. p. 330. *Donovan,* i. t. 5. *Montagu Test.* p. 473. *Maton and Racket, in Lin. Trans.* viii. p. 227. *Dorset Cat.* p. 57. t. 22. f. 1 and 2. *Brookes's Introd.* p. 135. t. 9. f. 121. *Burrow's Elements,* p. 173. t. 21. f. 1.

Haliotis vulgaris. *Da Costa Brit. Conch.* p. 15. t. 2. f. 1 and 2. *Lamarck Syst. des Anim.* p. 97.

L'Ormier. *Adanson Senegal,* p. 19. t. 2. f. 1.

L'Oreille de Mer des Côtes de France. *Favanne,* i. p. 582. t. 5. f. A 2.

Bonanni Rec. and *Kirch.* 1. f. 10 and 11. *Lister Anim. Ang.* t. 3. f. 16, and *Conch.* t. 611. f. 2. *Rumphius,* t. 40. f. G and H. *Gualter,* t. 69. f. I. *Argenville,* t. 3. f. A. *Knorr,* i. t. 17. f. 2 and 3. *Ginanni Adr.* ii. t. 3. f. 27. *Regenfuss,* i. t. 8. f. 20, and t. 10. f. 42.

Inhabits the coasts of Europe. *Linnæus.* Great Britain. *Lister, &c.* Senegal. *Adanson.* Adriatic. *Ginanni.* France. *Favanne.*

Shell about three or four inches long, and about three-fourths as broad, strong, of a reddish brown colour more or less mottled, and is generally covered with Balani and Serpulæ; the outer surface is longitudinally striated, and transversely wrinkled; the marginal ridge which terminates in the spire has about twenty-eight tubercles, which increase in size as they recede from the summit, and from six to eight of the lowermost of these are perforated; the whole inside is beautifully pearly. I much doubt whether several of the shells which Linnæus has arranged separately, are more than Varieties of this species.

STRIATA.   5. Shell ovate, ferruginous, transversely wrinkled, and longitudinally striated.

Haliotis striata.   *Linnæus Syst. Nat.* p. 1256.   *Martini,* i. p. 179. t. 14. f. 138.   *Born Mus.* p. 411.   *Schroeter Einl.* ii. p. 377.   *Gmelin,* p. 3688.   *Schreibers Conch.* i. p. 330.
Haliotis, No. 8.   *Schroeter Einl.* ii. p. 387. t. 4. f. 17.?
Inhabits the Asiatic Ocean. *Linnæus.*   Coasts of Barbary. *Gmelin.*
This shell is said to resemble *H. tuberculata,* but is smaller, and has the wrinkles more regularly cut and not tuberculated, and the marginal ridge is described with thirty-eight tubercles, of which from four to eight are perforated; the colour is ferruginous, more or less spotted with white, red, or green.

BISTRIATA.   6. Shell ovate, greenish, with brown spots, and elevated transverse striæ in pairs.

Haliotis bistriata.   *Gmelin,* p. 3689.   *Schreibers Conch.* i. p. 331.
Haliotis, No. 1.   *Schroeter Einl.* ii. p. 384.
Lister Conch. t. 612. f. 3.   *Martini,* i. t. 15. f. 142.
Inhabits the coasts of Africa. *Lister.*
This shell appears by the figures to be nearly of the same form and size as *H. tuberculata,* but is said to have the margin under the spire more sinuous, and the transverse striæ in pairs; the colour is greenish, and under the spire is a pale purple spot branching out into curved rays which extend to the margin.

VARIA.   7. Shell ovate, with longitudinal striæ, of which the larger ones are tuberculated.

Haliotis varia.   *Linnæus Syst. Nat.* p. 1256.   *Martini,* i. p. 184. t. 15. f. 144. *Schroeter Einl.* ii. p. 378. *Gmelin,* p. 3688.   *Schreibers Conch.* i. p. 330.
Lister Conch. t. 612. f. 4.   *Gualter,* t. 69. f. L and M.
Inhabits the East Indian Seas. *Martini.*
This shell appears by the figures to be about an inch and a half long, and rather more than an inch broad, of a white

or yellowish brown colour clouded with dirty green; the outer surface is described to be covered with elevated curved parallel striæ, which become larger towards the spire; and the number of the tubercles is said to be from twenty to thirty, of which only four or five are perforated; Linnæus says the margin in this species is unequal, in the following species equal, and small in both.

MARMORATA. 8. Shell ovate, with longitudinal and obsolete transverse striæ.

Haliotis marmorata. *Linnæus Syst. Nat.* p. 1256. *Martini,* i. p. 180. t. 14. f. 139. *Born Mus.* p. 412. *Schroeter Einl.* ii. p. 379. *Gmelin,* p. 3688. *Schreibers Conch.* i. p. 331.

La petite Oreille de Mer des Indes. *Favanne,* i. p. 581. t. 5. f. A 1.

Lister Conch. t. 614. f. 6. *Gualter,* t. 69. f. A. *Argenville,* t. 3. f. B. *Knorr,* ii. t. 17. f. 4 and 5.

Inhabits the coasts of Africa. *Linnæus.* East Indies. *Chemnitz.* Guinea. *Humphreys.*

Shell from two and a half to four inches long, and two-thirds as broad, and is prettily mottled with greenish or yellowish brown and white; the longitudinal striæ are fine, and the transverse ones obsolete; it has about thirty tubercles, of which from four to six are perforated.

GLABRA. 9. Shell ovate, rather smooth, green mottled with white, and the spire small.

Haliotis glabra. *Chemnitz,* x. p. 311. t. 166. f. 1602. *Gmelin,* p. 3690. *Ulysses's Travels,* p. 474.

Inhabits the Bay of Naples. *Ulysses.*

Shell about two inches and five lines long, and an inch and three-quarters broad, with only a few transverse striæ near the spire, and a few longitudinal ones behind the tubercles, of which about six are perforated; the inner lip is said to be very broad.

ASININA. 10. Shell oblong, rather smooth, with elevated nerves on the outside, and the margin somewhat sickle-shaped.

Haliotis asinina. *Linnæus Syst. Nat.* p. 1256. *Martini,* i. p. 190. t. 16. f. 150. *Born Mus.* p. 412. *Schroeter*

Einl. ii. p. 381.   *Gmelin,* p. 3688.   *Schreibers Conch.*
i. p. 331.
L'Oreille de Mer de la Chine.   *Favanne,* i. p. 584. t. 5. f.
A 4.
Lister Conch. t. 610. f. 1.   *Rumphius,* t. 40. f. E and F.
Petiver Amb. t. 16. f. 34.   *Gualter,* t. 69. f. D.   *Ar-*
genville, t. 3. f. E.   *Knorr,* iii. t. 15. f. 1.   *Regenfuss,*
i. t. 9. f. 29.
Inhabits the Indian Ocean. *Linnæus.* Coasts of Amboyna.
Rumphius. Batavia. *Martini.* China. *Favanne.*
Shell two and a half, or three inches long, and about half as
broad, variegated with green, white, and brown, and the inside pearly; it is very finely striated longitudinally, and the
striæ become somewhat granulated towards the spire; it has
about thirty tubercles, of which from five to seven are perforated.

AUSTRALIS.   11. Shell oval-oblong, with longitudinal plaits and wrinkles, and the spire prominent and gibbous.

Haliotis australis.   *Gmelin,* p. 3689.
Haliotis rugoso-plicata.   *Chemnitz,* x. p. 311. t. 166. f.
1604 and 1604 *a.*
Haliotis ruber.   *Leach Zool. Misc.* i. p. 54. t. 23.
Haliotis, No. 7.   *Schroeter Einl.* ii. p. 386.
Spengler Naturf. ix. p. 150. t. 5. f. 1, a and b.
Inhabits the coasts of New Zealand. *Chemnitz.*
Shell about two inches and three-quarters long, and an inch
and three-quarters broad, mottled with red or rose-colour,
and grey or white; the inside is pearly, and splendidly tinged
with red and yellow; from seven to nine of the tubercles
are perforated.

GIGANTEA.   12. Shell oval, rugged with undulated
plaits and transverse wrinkled striæ; spire
depressed.

Haliotis gigantea.   *Chemnitz,* x. p. 315. t. 167. f. 1610 and
1611.   *Gmelin,* p. 3691.
Haliotis naevosa.   *Martyn Univ. Conch.* ii. t. 63.
Inhabits the coasts of New Holland and New South Wales.
Chemnitz.
Shell about five or six inches long, and four-fifths as broad, of
a dark red orange colour, and is said to be sometimes tinged
with white; the spire, though depressed and flattish above,

has the first whirl considerably elevated; the tubercles are very prominent and numerous, and about seven are perforated; the inside is uneven, and very pearly.

IRIS. 13. Shell ovate, convex, ventricose, with obsolete longitudinal plaits and transverse wrinkles, and the inside highly iridescent.

Haliotis Iris. *Martyn Universal Conch.* ii. t. 61. *Gmelin,* p. 3691.

Haliotis Iridis. *Chemnitz,* x. p. 317. t. 167. f. 1612 and 1613.

L'Oreille de Mer de la Nouvelle Zelande. *Favanne,* i. p. 585.

Inhabits the coasts of New Zealand. *Martyn.*

Shell four or five inches long, and about two-thirds as broad, of a yellowish brown colour more or less mottled with olive; it is distinguishable by being more convex and ventricose than any of the preceding species, and by the remarkably brilliant and variable colours on its inside; it has about six perforations, and the others which have been closed are less conspicuous than is usual in this Genus.

CRACHERODII. 14. Shell roundish-ovate, nearly smooth, of a bluish black colour, and the inside iridescent.

Haliotis Cracherodii. *Leach Zool. Misc.* i. p. 131. t. 58.

Inhabits the coasts of California. *Leach.*

Shell about three inches and three-quarters long, and three inches broad, and is distinguishable by its general smoothness and dark colour; the specimen now before me has eleven perforations, and the remains of the others are perfectly flat, and not very conspicuous.

OVINA. 15. Shell roundish, depressed, wrinkled and plaited, with three elevated approximated ridges, of which the middle one is perforated; spire large.

Haliotis ovina. *Chemnitz,* x. p. 315. t. 166. f. 1609. *Gmelin,* p. 3691.

Inhabits ——

Shell about an inch and a half long, and nearly equally broad, of a chestnut-colour marbled with white; it has three elevated ridges near the margin, of which the middle one contains the perforated tubercles, and the outer surface is also tuberculated behind the wrinkles; the inside is pearly, and the inner lip is said to be broad.

PARVA. 16. Shell ovate, depressed, with decussated striæ, and three elevated remote ribs, of which the middle one is perforated; spire elevated.

Haliotis parva. *Linnæus Syst. Nat.* p. 1256. *Martini,* i. p. 181. t. 14. f. 140. *Schroeter Einl.* ii. p. 382. *Gmelin,* p. 3689. *Schreibers Conch.* i. p. 331.
Padollus scalaris. *Leach Zool. Misc.* i. p. 66. t. 28.
L'Oreille de Mer à gouttière. *Favanne,* i. p. 588. t. 5. f. D. *Knorr,* i. t. 20. f. 5.
Inhabits the coasts of Africa. *Linnæus.* Isle of France. *Favanne.* China. *Humphreys.*
Shell an inch and a quarter, or an inch and a half long, and about two-thirds as broad, of a dull orange-red more or less marbled with ash-colour; the tubercles on the middle one of the three ribs, which run parallel to the margin, are slightly raised, and five or six are perforated.

IMPERFORATA. 17. Shell ovate-oblong, convex, imperforate, with tuberculated ribs, and the spire exserted.

Haliotis imperforata. *Chemnitz,* x. p. 309. t. 166. f. 1600 and 1601. *Gmelin,* p. 3690. *Brookes's Introd.* p. 135. t. 9. f. 120.
Stomatia Phymotis. *Helbins Abhand. Boehm. Privat.* iv. p. 124. t. 2. f. 34 and 35. *Lamarck Syst. des Anim.* p. 96.
Auris antiqua. *Mus. Geversianum,* p. 250.
L'Oreille de Mer en Buccin. *Favanne,* i. p. 592. t. 5. f. F. *Lister Conch.* t. 1056. f. 6 and 7.
Inhabits the East Indian Seas. *Chemnitz.*
Shell about ten or eleven lines long, and not much more than half as broad, of a whitish dull ash-colour, and is distinguishable by its want of any perforations.

Genus XXXII.

PATELLA:

SHELL UNIVALVE, SOMEWHAT CONICAL, AND WITH-
OUT A SPIRE.

Sub-divisions.†

* With an internal appendage at the summit.
** With an internal transverse partition.
*** With the margin angular or irregularly
toothed.
**** With the summit pointed and recurved.
***** With the summit obtuse, and the margin
entire.
****** With a marginal fissure.
******* With the summit perforated.

---

* *With an internal appendage at the summit.*

EQUESTRIS. 1. Shell sub-orbicular, wrinkled, and
minutely striated longitudinally; summit lateral and obtuse.

Patella equestris. *Linnæus Syst. Nat.* p. 1257. *Martini,*
i. p. 151. t. 13. f. 119 and 120. *Born Mus.* p. 415.
Schroeter Einl. ii. p. 394. *Gmelin,* p. 3691. *Schreibers Conch.* i. p. 337. *Brookes's Introd.* p. 138. t. 9.
f. 122. *Burrow's Elements,* p. 175. t. 21. f. 4.

---

† So many of Gmelin's descriptions appear to me to be wholly undeserving
of notice, that I have principally confined my attention to the Linnæan Species,
and to those which have been figured by Martini and Chemnitz.

Calyptræa equestris. *Lamarck Syst. des Anim.* p. 70.
La Cloche, ou Sonnette. *Favanne,* i. p. 556. t. 4. f. B 4.
Lister Conch. t. 546. f. 38. *Rumphius,* t. 40. f. P and Q.
 Petiver Amb. t. 16. f. 28. *Gualter,* t. 9. f. Z. *Argen-*
 ville, t. 2. f. K. *Knorr,* vi. t. 35. f. 5.
Inhabits the coasts of Barbadoes. *Lister.* Amboyna. *Rum-*
 phius. St. Domingo and the East Indies. *Favanne.*
Shell about half or three-quarters of an inch in diameter, transparent and white, or of a brownish white colour; it is variously wrinkled, and delicately striated longitudinally, but the
 striæ are sometimes so minute as to be hardly discernible;
 the margin is irregularly sinuated.

NEPTUNI. 2. Shell sub-orbicular, with longitudinal
 strong elevated toothed striæ, and the summit rather lateral.

Le Bonnet de Neptune. *Favanne,* i. p. 555. t. 4. f. B 3.
D'Avila, t. 2. f. B. *Martini,* i. p. 150. t. 13. f. 117 and
 118.
Inhabits the coasts of St. Domingo. *D'Avila.* Falkland Islands. *Favanne.*
This shell has been confounded with *P. equestris,* but it appears to me to be perfectly distinct, and I have followed
 Martini in placing it as a separate species; it is twice as
 large and much less wrinkled, and may be at once distinguished by its longitudinal strongly toothed or spinous ribs.

TECTUM. 3. Shell pyramidal, transversely foliaceous, and the summit central and erect.

Patella Tectum Sinense. *Chemnitz,* x. p. 337. t. 168.
 f. 1630 and 1631.
Patella equestris, Var. β. *Gmelin,* p. 3692.
Patella, No. 6. *Schreibers Conch.* i. p. 337.
Le Toit Chinois, ou la Molette. *Favanne,* i. p. 553. t. 4.
 f. B 1.
Argenville, t. 2. f. S. *Knorr,* vi. t. 35. f. 4. *Martini,* i.
 t. 13. f. 125 and 126. *Humphreys Conch.* t. 6. f. 9.
Inhabits the coasts of Batavia. *Martini.* China. *Chemnitz.*
Linnæus, in his History of the Queen of Sweden's Museum,
 and Favanne, as well as Gmelin, considered this to be a
 Variety of *P. equestris;* but, on the other hand, we have the
 authorities of Martini, Chemnitz, and Schreibers, for regard-
 ing it as a separate species. The size is generally considerably smaller than that of *P. equestris,* and its transverse

membranes give it somewhat the appearance of several shells of that species piled on each other.

SINENSIS. 4. Shell conical, depressed, nearly smooth, with the summit erect, and the internal appendage oblique.

Patella Chinensis, or Sinensis. *Linnæus Syst. Nat.* p. 1257. *Martini*, i. p. 154. t. 13. f. 121 and 122. *Schroeter Einl.* ii. p. 398. *Gmelin*, p. 3692. *Schreibers Conch.* i. p. 337. *Montagu Test.* p. 489. t. 13. f. 4. *Maton and Racket, in Lin. Trans.* viii. p. 228.
Patella albida. *Donovan*, iv. t. 129.
Le Bonnet Chinois de la Méditerranée. *Favanne,* i. p. 560. t. 4. f. C 3.
Bonanni Rec. and *Kirch.* 1. f. 12. *Lister Conch.* t. 546. f. 39. *Argenville,* t. 2. f. F. *Born Mus.* Vign. at p. 414. f. e.
Inhabits the Mediterranean. *Linnæus.* Coasts of Batavia and the Canary Islands. *Martini.* Provence. *Favanne.* Great Britain. *Montagu, &c.*
Shell half an inch, or sometimes an inch in diameter, and the height is rarely more than half the diameter; the colour is brownish white, sometimes faintly striped longitudinally with reddish brown, and the inside glossy; the margin of the base is circular, and not at all sinuated.

AURICULATA. 5. Shell longitudinally grooved and striated, with the summit recurved, and the appendage ear-shaped.

Patella auriculata. *Chemnitz,* x. p. 336. t. 168. f. 1628 and 1629.
Patella Auricula. *Gmelin,* p. 3694.
Patella Sinensis, Var. β. *Gmelin,* p. 3692.
Le Bonnet Chinois rayé. *Favanne,* i. p. 559. t. 4. f. C 2.
Petiver Gaz. t. 21. f. 11. *Knorr,* vi. t. 22. f. 1. *Martini,* i. t. 13. f. 123 and 124. *Humphreys Conch.* t. 6. f. 10.
Inhabits the coasts of the island of Borneo. *Petiver.* Java and Batavia. *Martini.* St. Croix and St. Thomas, in the West Indies. *Chemnitz.*
This shell appears to be of about the same size as *P. Sinensis,* and is white, more or less grooved longitudinally, and radiated with yellow or dark brown; the summit is generally brown, and the inside glossy and stained with brown about the appendage. *Patella auricularia* of the Portland Cata-

logue, lot 3983, is different, and the present shell at lot 2526 is called the *Cup and Saucer*.

** *With an internal transverse partition.*

TROCHIFORMIS.   6. Shell conical, somewhat spiral, with the whirls longitudinally plaited, and the margin entire.

Patella trochiformis.   *Chemnitz*, x. p. 335. t. 168. f. 1626 and 1627. *Gmelin*, p. 3693.
Patella, No. 147. *Schroeter Einl.* ii. p. 503.
Le Lépas volute chambré. *Favanne*, i. p. 551. t. 4. f. A 1.
Knorr, iii. t. 29. f. 1 and 2. *Meuschen Naturf.* xviii. t. 2. f. 17. *Humphreys Conch.* t. 6. f. 2.
Inhabits the coasts of the Falkland Islands. *Favanne*.   Tranquebar. *Chemnitz*.
Shell about an inch in diameter, and ten lines high, of a dirty white colour, with yellowish brown longitudinal oblique plaits; it has a spiral obsolete suture, with the appearance of three or four whirls, and the inside has a transverse appendage, which becomes narrower as it turns somewhat spirally upwards.

TROCHOIDES.   7. Shell convex, spiral, with longitudinal undulated ribs, and the margin crenated.

Patella trochiformis, Var. β.   *Gmelin*, p. 3694.
Patella, No. 136. *Schroeter Einl.* ii. p. 498.
Ancilia volutata. *Mus. Geversianum*, p. 248.
Le Bouton de Chapeau. *Favanne*, i. p. 552. t. 4. f. A 2.
Argenville, t. 2. f. L. *Martini*, i. t. 13. f. 135.
Inhabits ———
Shell eleven, or sometimes fifteen lines in diameter, and has the summit more rounded and less elevated than in *P. trochiformis*, and appears from Favanne's description to be a still more remarkable link between the Patella and Trochi; it is of a reddish brown colour, with the inside white, and the transverse somewhat spiral partition tinged with lilac.

NERITOIDEA.   8. Shell ovate, with the summit slightly spiral and recurved, and the margin entire.

Patella neritoidea. *Linnæus Syst. Nat.* p. 1257. ?   *Martini*, i. p. 161. t. 13. f. 133 and 134. *Gmelin*, p. 3692. ?
Nerita violacea. *Gmelin*, p. 3686.

Nerita, No. 218. *Schroeter Einl.* ii. p. 367.

La Nacelle, ou Coquille de Noix. *Favanne,* i. p. 562. t. 4. lower fig. E 1.

Lister Conch. t. 545. f. 36.

Inhabits the Indian Seas. *Martini.* Coasts of the Isle of France. *Favanne.*

Linnæus's description is very short, and is unaccompanied by any reference, so that it is almost impossible with tolerable certainty to ascertain this species, and all authors differ respecting it. Martini, with a mark of doubt, has cited the Linnæan *P. neritoidea* for his figures 133 and 134, and these Gmelin has referred to for his *Nerita violacea.* Neither Schroeter, Gmelin, or Schreibers have made any reference without a mark of doubt; and Favanne, in the Appendix, page 849, has referred to *La Retorte épineuse,* t. 4. f. F 2, which is *Patella aculeata.* Martini's shell is about ten lines long and eight broad, and of a whitish colour with violet spots.

PORCELLANA. 9. Shell oval, depressed, with the margin entire, and the summit recurved; internal partition flat.

Patella porcellana. *Linnæus Syst. Nat.* p. 1257. *Martini,* i. p. 157. t. 13. f. 127 and 128. *Schroeter Einl.* ii. p. 399. *Gmelin,* p. 3692. *Schreibers Conch.* i. p. 338.

Nerita porcellana. *Chemnitz,* ix. part ii. p. 68. t. 124. f. 1082.

Crepidula porcellana. *Lamarck Syst. des Anim.* p. 70.

Le grand Bateau ponté. *Favanne,* i. p. 565. t. 4. f. G 1.

Le Sulin. *Adanson Senegal,* p. 38. t. 2. f. 8.

Lister Conch. t. 545. f. 34. *Rumphius,* t. 40. f. O. *Petiver Amb.* t. 16. f. 25. *Knorr,* vi. t. 11. f. 5. *Da Costa Conch.* t. 2. f. 6.

Inhabits the Indian Ocean. *Linnæus.* Coasts of Goree. *Adanson.*

Shell from ten to fourteen lines long, and nearly three-fourths as broad, of a whitish or brownish colour, variegated with dark brown or red spots, or purplish undulated lines; the inside is generally white and glossy.

FORNICATA. 10. Shell oval, with the margin entire, and the summit obliquely recurved; internal partition concave.

Patella fornicata. *Linnæus Syst. Nat.* p. 1257. *Martini,*

i. p. 160. t. 13. f. 129 and 130. *Born Mus.* p. 416.
Schroeter Einl. ii. p. 400. *Gmelin,* p. 3693. *Schreibers Conch.* i. p. 338. *Brookes's Introd.* p. 138. t. 9. f. 124.
Burrow's Elements, p. 174. t. 21. f. 3.
La Chaloupe de Saint Pierre. *Favanne,* i. p. 563. t. 4. lower fig. E 2.
Lister Conch. t. 545. f. 33 and 35. *Argenville,* t. 2. f. N.
Knorr, vi. t. 21. f. 3.
Variety. With darker broad longitudinal rays.
La Chaloupe pontée. *Favanne,* i. p. 563. t. 4. lower fig. E 3.
Le Garnot. *Adanson Senegal,* p. 40. t. 2. f. 9.
Inhabits the Mediterranean. *Linnæus.* Coasts of Goree.. *Adanson.*
Shell three-quarters of an inch, or sometimes an inch and a quarter long, and the length, depth, and breadth are about in the proportions of eleven, six, and four; the colour is brown, or sometimes chestnut, with a white spot on each side the apex, or white variously mottled or longitudinally rayed with brown; the internal partition is concave, and rounded at the margin.

ACULEATA. 11. Shell oval, depressed, shallow, with longitudinal prickly ribs, and the summit recurved and lateral.

Patella aculeata. *Gmelin,* p. 3693.
Patella fornicata aculeata. *Chemnitz,* x. p. 334. t. 168. f. 1624 and 1625.
Patella fornicata, Var. *Schreibers Conch.* i. p. 338.
La Retorte épineuse. *Favanne,* i. p. 564. t. 4. f. F 2.
Da Costa Elements, t. 2. f. 2. *Humphreys Conch.* t. 6. f. 1.
Inhabits the coasts of the Isle of France. *Favanne.* West India Islands, *Chemnitz.*
Shell about eight or nine lines long, and six or seven broad, of a brownish, greyish, or chestnut-colour both inside and out, and the internal partition white; the spire lies on the side, somewhat in the manner of an Haliotis; and the shell itself, though in this respect subject to considerable variation, is also generally much depressed.

GOREENSIS. 12. Shell oval, shallow, much depressed, transversely wrinkled, with the spire obsolete, and the summit recurved and lateral.

Patella Goreensis. *Gmelin*, p. 3694.

Le Jenac. *Adanson Senegal*, p. 41. t. 2. f. 10.

La Sandale. *Favanne*, i. p. 561. t. 4. lower fig. D. ?

Martini, i. p. 160. t. 13. f. 131 and 132.

Inhabits the coasts of Goree. *Adanson*.

Shell five or six lines long, and about four-fifths as broad, of a
whitish colour, and is coarse, and transversely wrinkled like
a common Oyster-shell; the inside is very white and glossy.
La Sandale, which Favanne considered to be the Linnæan
P. Crepidula, appears from the description to be rather
larger and thinner, and is said to be sometimes marked near
the summit with obsolete longitudinal yellowish brown lines.

CREPIDULA.   13. Shell oval, flattish, and smooth,
with the internal partition flat.

Patella Crepidula. *Linnæus Syst. Nat.* p. 1257. *Gmelin*,
p. 3695.

Patella porcellana, Var. *Martini*, i. p. 157.

Inhabits the Mediterranean, especially on the coasts of Barbary.
Linnæus.

Linnæus for this species has referred to Gualter, t. 69. f. H,
from whence he has derived the name of *Crepidula*, and to
Adanson, t. 1. f. 1, of which the former is probably at most
only a Variety of *P. porcellana*, and the latter is a shield
placed at the tail of its inhabitant, which it only partially
covers.   In the account of the Museum of the Queen of
Sweden, *P. Crepidula* is described to be white, diaphanous,
and flattish, with the internal partition somewhat crescent-
shaped and flat.   Martini considered it to be a Variety of
P. porcellana, and its having been placed by Linnæus in
the following, instead of the present division of the Genus, I
imagine must have been wholly accidental.

*** *With the margin angular, or irregularly toothed.*

LACINIOSA.   14. Shell with unequal longitudinal
ribs, which become thicker and obtuse at
the margin.

Patella laciniosa. *Linnæus Syst. Nat.* p. 1258. *Martini*,
i. p. 127. t. 10. f. 81. *Schroeter Einl.* i. p. 403. *Gmelin*,
p. 3695. *Schreibers Conch.* i. p. 340.

Le Lépas aux deux Yeux. *Favanne*, i. p. 516. t. 2. f. I.

Rumphius, t. 40. f. C. *Petiver Amb.* t. ii. f. 13. *Argenville*, t. 2. f. O.

Inhabits the coasts of Amboyna. *Rumphius*.

Shell about an inch and a half long, and rather more than half
as broad, of a brownish colour, with the ribs white, and
their interstices transversely striated; the summit is obtuse,
and marked with two white eyes. This species appears
to me to rest almost entirely on the authority of Rumphius, from whose figure all the others have probably been
copied.

PLICATA. 15. Shell angulated, with undulated
blunt much elevated ribs, and transversely
wrinkled; summit obtuse.

Patella plicata. *Born Mus.* p. 417. t. 18. f. 1. *Mus. Lesk-
eanum,* p. 301.
Patella plicaria. *Gmelin,* p. 3708.
Patella, No. 84. *Schroeter Einl.* ii. p. 476.
Patella, No. 43. *Schreibers Conch.* i. p. 349.
Inhabits ——

Born says, the length of this shell is two inches and a half, and
the breadth two inches and one line, and that the colour is
white spotted with brown, but the figure is coloured of a
pale brownish white without any spots; it has about thirty
obtuse ribs, which form strong teeth on the margin. The
synonyms which Born has added to his description are very
erroneous, and his having stated that the species inhabits the
Straights of Magellan, arose from his having considered it to
be the same as D'Avila, t. 3. f. D, which is *P. ferruginea.*
Schreibers, for *P. plicaria* of Gmelin, has only quoted
Knorr, iii. t. 30. f. 1, and erroneously arranged Born's shell
separately. *P. plicata* of Gmelin is taken from Meuschen
Naturf. t. 2. f. 12. and is a very uncertain species.

MONOPIS. 16. Shell depressed, with about eleven
strong carinated and intermediate smaller
ribs.

Patella monopis. *Gmelin,* p. 3707. *Schreibers Conch.* i.
p. 347.?
Patella, No. 30. *Schroeter Einl.* ii. p. 453.
Martini, i. p. 126. t. 9. f. 80.
Inhabits the coasts of the West India Islands. *Gmelin.*
Shell about two inches long, and nearly equally broad, thick,
of a whitish brown colour, more or less streaked with dark
chestnut, and the latter sometimes prevails; some of the
ribs project much further than others, which gives the base
a very irregular form, and the inside is white. Gmelin has

referred to Martini's figures 82 and 83, both as a Variety of this species and of his *P. octoradiata*, but they appear to me to be quite different from either; and they answer so imperfectly to the description which has been given of them, that they are hardly worth notice. Schreibers has erroneously quoted Martini, f. 79. for *P. monopis*, and with 82 and 83 has constituted his species No. 37.

SACCHARINA. 17. Shell oblong, angulated, with seven or eight carinated obtuse ribs, and the interstices longitudinally striated.

Patella saccharina.. *Linnæus Syst. Nat.* p. 1258. *Martini*, i. p. 121. t. 9. f. 70, 75? 76 and 77. *Born Mus.* p. 416. *Schroeter Einl.* ii. p. 404. *Mus. Lesk.* p. 300. *Gmelin*, p. 3695. *Schreibers Conch.* i. p. 340.

L'Etoile. *Favanne*, i. p. 513. t. 2. f. F 2, and F 3.

Lister Conch. t. 532. f. 10. *Rumphius*, t. 40. f. B. *Petiver Amb.* t. 3. f. 3. *Gualter*, t. 9. f. I and M. *Argenville*, t. 2. f. M. *Klein Ost.* t. 8. f. 4.

Inhabits the coasts of Amboyna. *Rumphius.* Batavia. *Martini.* Cape of Good Hope. *Favanne.* China. *Humphreys.*

Shell about an inch long, and three-quarters of an inch broad, of a yellowish or greenish white colour, marbled with red or brown in the interstices of the ribs; the inside is white tinged with blue, and the mark left by the animal at the summit is generally milk-white, with the edge angulated in correspondence with the margin of the shell. Linnæus says it inhabits Barbadoes; but this is probably a mistake, owing to a misquotation from Lister.

ANGULOSA. 18. Shell sub-octangular, thick, depressed, with fine somewhat granulated striæ, and the spire acute.

Patella angulosa. *Gmelin*, p. 3707. *Schreibers Conch.* i. p. 346.

Patella, No. 26. *Schroeter Einl.* ii. p. 452.

Le Camelot. *Favanne*, i. p. 509. t. 2. f. C.

Lister Conch. t. 538. lower fig. *Martini*, i. p. 119. t. 8. f. 69.

Inhabits the coasts of Provence. *Favanne.*

Shell an inch and a half long, and appears by the figures to be about three-fourths as broad; the colour is greyish white, rayed with dark brown and spotted with red, and the summit is black, pointed, and somewhat lateral; the muscular impression is milk-white, and the remainder of the inside has a greyish cast.

REPANDA. 19. Shell ovate, thin, much depressed, with longitudinal somewhat scaly striæ, and the summit acute; margin irregularly indented.

Patella repanda. *Gmelin*, p. 3707. *Schreibers Conch.* i. p. 346.
Patella, No. 25. *Schroeter Einl.* ii. p. 451.
Le Petit Soleil de l'Isle de Cythere. *Favanne*, ii. p. 513. t. 2. f. G 1.
Inhabits the coasts of the Island of Cerigo. *Favanne.*
The base measures about an inch and a quarter by one inch, and the margin is much angulated and indented, with an appearance of having been broken; the shell is grooved, and finely striated longitudinally, and made rough by minute scales or granules; the colour is yellow or whitish, and greyish in the grooves, with broad rays of reddish brown, and the inside white and pearly.

TENUIS. 20. Shell ovate, depressed, and very narrow at one end, with ten angulated diverging ribs, and the interstices longitudinally striated.

Patella tenuis. *Gmelin*, p. 3708.
Patella tenuissima. *Turton's Gmelin*, p. 575.
Patella, No. 35. *Schroeter Einl.* ii. p. 455.
L'Ambre jaune. *Favanne*, ii. p. 506. t. 2. f. A.?
Martini, i. p. 130. t. 10. f. 87.
Inhabits ———
Shell about an inch, or sometimes an inch and a half long, and nearly equally broad at one end, of a pale ochre-colour, frequently marked with chestnut towards the margin; the summit is pointed, and is much nearer the narrow than the broad end; it has ten angulated ribs with intermediate striæ, and the margin is strongly angulated, particularly at the broad end. There are two species in Gmelin with the name of *tenuis*, which induced Dr. Turton in his translation to change the name to *tenuissima*; but I much doubt whether it is applicable, for Favanne's *L'Ambre jaune* differs principally in being thick, and I believe that the thickness of the Patellæ is much increased by age.

MARGARITACEA. 21. Shell ovate, depressed, with irregular angulated ribs, and transverse striæ; summit acute, and the margin angulated and crenated.

Patella margaritacea. *Gmelin*, p. 3707.
Patella barbara. *Born Mus.* p. 417.? *Mus. Leskeanum*, p. 300.
Patella, No. 33. *Schroeter Einl.* ii. p. 455.
Le Grand Soleil. *Favanne*, i. p. 513. t. 2. f. G 2.
Gualter, t. 8. f. L. *Martini*, i. p. 129. t. 10. f. 85, A and B.
Inhabits Iceland. *Gmelin?* Coasts of Patagonia. *Favanne.*
The length of the base is rather more, and the breadth rather less than two inches; the shell is white clouded with yellow, and the inside pearly with an orange ring round the muscular impression, and the teeth at the margin on the inside are of a chestnut-colour. Favanne's figure is marked with much stronger concentric striæ than Martini's, but in his Appendix, p. 848, they are said to be the same, and the latter may probably have been made from a worn shell.

BARBARA. 22. Shell toothed, with nineteen elevated vaulted muricated ribs.

Patella barbara. *Linnæus Syst. Nat.* p. 1258. *Schroeter Einl.* ii. p. 405. t. 5. f. 1. *Gmelin*, p. 3696.
Knorr, v. t. 13. f. 5.
Inhabits the coasts of the Falkland Islands. *Gmelin,*
I am wholly unacquainted with the shell which Schroeter has figured for this species, and am inclined to believe that *P. Cypria* would answer better to the Linnæan description; the greatest diameter of the base appears by the figure to be about an inch and a half, and it is said to be convex, white, ornamented with a broad brown band and rays, and to have nineteen longitudinal ribs with intermediate smaller ones.

CYPRIA. 23. Shell oval, thick, rather shallow, with foliaceous elevated ribs, and the margin plaited.

Patella Cypria. *Gmelin*, p. 3698.
Patella, No. 29. *Schroeter Einl.* ii. p. 453.
Le Marbre blanc à côtes. *Favanne*, ii. p. 509. t. 11. f. D 1.
Martini, i. p. 125. t. 9. f. 79.
Inhabits the coasts of the Isle of France, and New Zealand. *Favanne.*

Shell from two to four inches long, and about three-fourths as
broad, of a whitish colour, sometimes marked with a circle
of brown round the summit; the ribs, of which there are
upwards of sixteen, are here and there tuberculated, and
somewhat foliaceous. Favanne says, " Sa coquille est fort
épaisse, de forme ovale, très élevée, et à sommet pointu,
d'où partent en rayonnant un grand nombre de côtes alterna-
tivement plus et moins grosses, souvent aiguës, et à vive-
arrête, et souvent aussi raboteuses par les crues de la coquille. On y apperçoit quelquefois des stries mal formées
et presque imperceptibles."

OCULUS.   24. Shell oval, thick, rather shallow, with
     broad   angulated   ribs,   and   the   margin   un-
     equally serrated.

Patella Oculus.   *Born Mus.* p. 418.
Patella Cypria, Var. *β.   Gmelin,* p. 3698.
Patella, No. 34.   *Schroeter Einl.* ii. p. 455.
Patella, No. 39.   *Schreibers Conch.* i. p. 347.
L'Oeil de Bouc radié.   *Favanne,* i. p. 518. t. 2. f. M.
   Gualter, t. 9. f. H.   *Argenville,* t. 2. f. B.   *Da Costa
   Elements,* t. 2. f. 1.   *Martini,* i. p. 130. t. 10. f. 86.
Inhabits the coasts of the Island of Raclia, in the Archipelago.
Tournefort.   Brazil. D'Avila.   Common in the Mediter-
ranean, on the coasts of Africa, the Isle of Cyprus, and
Provence. *Favanne.*
The length of the base is often three inches and a half, and the
breadth three inches ; the colour is yellowish white, and
the summit surrounded by one or two dark brown rings;
the inside is of a greyish white, and the muscular impression
milk-white. I have followed Martini, Schroeter, and Schreibers, in placing this separate from *P. Cypria,* and Martini's two figures appear very different, but some of the others
are so intermediate that it is difficult to say to which of them
they belong.   It is plain, from Favanne's description and references, that his *Oeil de Bouc* belongs to this species, but
his figure is not a good one.

PENTAGONA.   25. Shell pyramidal, sub-pentangular,
     with the summit obtuse, and the margin ir-
     regularly dilated and crenated.

Patella pentagona.   *Born Mus.* p. 421. t. 18. f. 4 and 5.
   Gmelin, p. 3708.

Patella, No. 85. *Schroeter Einl.* ii. p. 476.
Inhabits ——
The following is Born's description: " Shell pyramidal, with
five obtuse angles, and broader on one side ; base dilated,
pentangular, and crenated ; summit central and obtuse ;
from the base to the summit measures ten lines, and the diameter is one inch."

GRANULARIS. 26. Shell ovate, with narrow granulated ribs, and the margin toothed.

Patella granularis. *Linnæus Syst. Nat.* p. 1258. *Martini,*
i. p. 114. t. 8. f. 61. *Schroeter Einl.* ii. p. 406. *Gmelin,*
p. 3696. *Schreibers Conch.* i. p. 341.
Patella granatina. *Born Mus.* p. 419. *Mus. Leskeanum,*
p. 301.
Le Lépas à Grains de Millet. *Favanne,* i. p. 502. t. 3. f.
D 4.
Lister Conch. t. 536. f. 15. *Petiver Gaz.* t. 85. f. 11.
Argenville, t. 2. f. H. *Regenfuss,* i. t. 2. f. 24.
Inhabits the coasts of Southern Europe. *Linnæus.* Cape of
Good Hope. *Petiver.* Isle of France. *Favanne.*
The base is about an inch and a quarter long, and three-fourths
as broad ; the colour is brown, with the summit and the imbricated granules white ; the inside is whitish and slightly
pearly, and has a yellowish muscular impression at the apex ;
the summit is not central, but placed almost twice as near
to one end as to the other. Linnæus has mixed the references which belong to this species with those of *P. granatina,* and Born in consequence has mistaken both these
species.

GRANATINA. 27. Shell ovate, depressed, with distant alternately smaller ribs, and transverse
undulated rows of dark scales; margin angular.

Patella granatina. *Linnæus Syst. Nat.* p. 1258. *Martini,*
i. p. 123. t. 9. f. 71 and 72. *Schroeter Einl.* ii. p. 408.
Gmelin, p. 3696. *Schreibers Conch.* i. p. 331. *Burrow's Elements,* p. 175. t. 21. f. 5.
Patella granularis. *Born Mus.* p. 418. *Mus. Lesk.* p.
301.
L'Oeil de Rubis radié. *Favanne,* i. p. 508. t. 2. f. B 3,
and B 4.
Bonanni Mus. Kirch. 1. f. 30. *Lister Conch.* f. 534. f.

13.   *Gualter*, t. 9. f. F.   *Argenville*, t. 2. f. G.   *Knorr*, i. t. 30. f. 2.   *Regenfuss*, i. t. 9. f. 31.
Inhabits the Southern coasts of Europe. *Linnæus.* Jamaica. *Martini.* Isle of France. *Favanne.* Cape of Good Hope. *Humphreys.*

The base is two or three inches long, and three-fourths as broad, and the height is less than half the breadth; the shell is white, rather transparent, with the summit more or less bare, and the sides coated with transverse zic-zac rows of dark purplish brown scales; the outside colouring is seen on the inside through a thin coat of pearl, and the muscular impression is generally of a rich reddish brown. Martini's figures 73 and 74 are copied from Lister, t. 733. f. 12, and this again from Columna; on which account I do not think them worth attending to.

CHLOROSTICTA.   28.   Shell ovate, depressed, with eleven broad and intermediate narrower ribs; margin angulated.

Patella chlorosticta.   *Gmelin*, p. 3707.   *Schreibers Conch.* ii. p. 347.
Patella, No. 32.   *Schroeter Einl.* ii. p. 454.
La Gorge de Pigeon.   *Favanne*, i. p. 464.
Martini, i. p. 128. t. 10. f. 84, A and B.
Inhabits the coasts of Jamaica. *Martini.*

The length at the base rather exceeds an inch and a half, and the breadth at the broader end is almost equal; the colour is brownish yellow, with broad chestnut rays and green scattered dots; the inside has longitudinal grooves corresponding with the ribs on the outside; the muscular impression and the inner margin are chestnut-brown, and in the figure there appears a yellowish belt between them. From his appendix, p. 844, it is plain that Favanne considered this shell to be only a Variety of *P. granatina.*

TIGRINA.   29.   Shell oval-oblong, with seven depressed ribs and intermediate longitudinal striæ; margin sub-octangular and crenated.

Patella tigrina.   *Gmelin*, p. 3707.   *Schreibers Conch.* i. p. 347.
Patella, No. 28.   *Schroeter Einl.* ii. p. 453.
Martini, i. p. 124. t. 9. f. 78.
Inhabits ———
Shell about fifteen lines long, and eleven broad at the base,

slender, polished, and pellucid; the summit is acute, and placed much nearer to one end than to the other; the sides are bluish grey spotted with brown, and the summit orange; the inside is pearly, and the muscular impression orange.

ORNATA. 30. Shell oval, longitudinally ribbed and ornamented with rays, which are alternately green and black, dotted with white; margin sub-angular and nearly entire.

Patella margaritaria. *Chemnitz,* xi. t. 197. f. 1914 and 1915.

Inhabits the coasts of New Zealand. *Chemnitz.*

This shell appears by the figures to be eleven lines long and nine broad, and of a dark green colour with eleven black longitudinal rays, which are ornamented with a row of white dots on each; the inside is glossy, and exhibits the external rays, and the muscular impression is blackish. Chemnitz says that a shell, which is in England called the ' black and white beaded Auricula Limpet," belongs to this species.

MELANOGRAMMA. 31. Shell ovate, depressed, with numerous narrow blackish ribs, and the margin crenated.

Patella melanogramma. *Gmelin,* p. 3706. *Schreibers Conch.* i. p. 346.

Patella, No. 24. *Schroeter Einl.* ii. p. 451.

Martini, i. p. 118. t. 8. f. 67.

Inhabits ——

The base, according to Martini's figure, measures rather more than an inch and a half in length, and the breadth is about an inch and a quarter; the colour is ochraceous, with the ribs blackish, and the summit white; the inside is silvery, and the muscular impression straw-coloured.

FERRUGINEA. 32. Shell ovate, with sub-nodulous narrow ribs, and somewhat imbricated transverse wrinkles; margin plaited.

Patella ferruginea. *Gmelin,* p. 3706. *Schreibers Conch.* i. p. 346.

Patella Scutum deauratum. *Chemnitz,* x. p. 327. t. 168. f. 1616.

Patella deaurata. *Gmelin,* p. 3719. *Schreibers Conch.* i. p. 361.

Patella Magellanica. *Mus. Leskeanum*, p. 302.

Patella, No. 23. *Schroeter Einl.* ii. p. 450.

Lepas denticulata. *Martyn Univ. Conch.* ii. t. 65.?

Le Grand Lépas tuilé et nacré. *Favanne*, i. p. 496. t. 3. f. D 2.

D'Avila, t. 3. f. D. *Knorr*, iv. t. 9. f. 1 and 2. *Martini*, i. p. 117. t. 8. f. 66.

Inhabits the Straights of Magellan. *D'Avila*. Coasts of the Falkland Islands. *Favanne*.

The length of the base sometimes exceeds two inches and a half, aud it is then nearly two inches and a quarter broad; Favanne has described the ribs to be of a dark brown, with their interstices of a bluish slate-colour; but it is said that the shell varies much, both in form and colour, and that it is sometimes of a dark mottled brown with white bands, which answers to Martini's figure; the inside is channelled and white, and the muscular impression silvery. In Favanne's appendix, Martini's fig. 66, is referred to, and they have both quoted D'Avila, t. 3. f. D, which is strongly marked with imbricated concentric wrinkles; Born has strangely cited the same figure of D'Avila's for his *P. plicata*, which it is very unlike. It is said by Favanne, that those shells of this species which are of a yellowish brown colour will, by being kept a proper time under hot cinders, become of the colour of gold, and equally brilliant.

CRENATA. 33. Shell ovate, rather depressed, slightly ribbed, and unequally striated longitudinally; margin crenated.

Patella crenata. *Gmelin*, p. 3706. *Schreibers Conch.* i. p. 345.

Patella, No. 22. *Schroeter Einl.* ii. p. 450.

Le Petit Gris. *Favanne*, i. p. 511. t. 2. f. E.

Gualter, t. 9. f. G. *Martini*, i. p. 116. t. 8. f. 64 and 65.?

Inhabits the coasts of Africa, Malaga, and Lisbon. *Martini*. Mediterranean. *Favanne*.

The length of the base is from an inch to an inch and a half, and the breadth is about five-sixths of the length; the colour is pale yellowish brown with grey between the ribs, and marked with fine longitudinal striæ and grey transverse bands. Martini's fig. 65 has the margin angulated, as well as crenated, but Favanne considered this to be distinct from fig. 64, and to be a Variety of *Le Grand Soleil*, which is *P. margaritacea*.

MINIATA. 34. Shell ovate, depressed, with irregular decussated striæ, and narrow longitudinal grooves; margin crenated.

Patella miniata. *Born Mus.* p. 420.
Patella Umbella. *Gmelin,* p. 3706. *Schreibers Conch.* i. p. 345.
Patella sanguinolenta. *Gmelin,* p. 3716.
Patella, No. 14. *Schroeter Einl.* ii. p. 446, and No. 21. p. 449.
Le Bouclier couleur de Rose. *Favanne,* i. p. 488. t. 1. f. H 1, and H 2.
Lister Conch. t. 538. f. 21. *Knorr,* v. t. 8. f. 5, and t. 19. f. 3. *Martini,* i. p. 108, and p. 115. t. 7. f. 52 and 53, and t. 8. f. 63.
Inhabits the coasts of Africa. *Martini.* Isles of France and Bourbon. *Favanne.*

Shell varying in the length of its base from an inch and a half to three inches, and is about four-fifths as broad; the colour is white or yellowish, streaked and sprinkled with dots of pale vermillion or rose-colour, in a variety of ways. Gmelin's *P. sanguinolenta* is probably a young shell of this species, and I have followed Born in referring to three of Martini's figures; fig. 52, on the other hand, has been quoted in the Museum Leskeanum for a species, under the name of *P. Carthaginensis,* which is placed in the division with recurved summits.

ULYSSIPONENSIS. 35. Shell sub-oval, depressed, thin, with nine broad rays, and striated and slightly ribbed longitudinally; margin denticulated.

Patella Ulyssiponensis. *Gmelin,* p. 3706.
Patella, No. 20. *Schroeter Einl.* ii. p. 449.
Le Bouclier la Punaise. *Favanne,* i. p. 487. t. 1. f. B 1.?
Martini, i. p. 115. t. 8. f. 62.
Inhabits the shores of Lisbon. *Martini.*

This shell appears by Martini's figure to have the base about thirteen lines long and eleven broad; the colour is yellowish, with the summit orange, and there are nine broad dark brown rays diverging to the margin.

RADIATA. 36. Shell sub-ovate, slightly striated, and stellated both inside and out with white and blackish rays.

2 G 2

Patella radiata. *Chemnitz*, xi. p. 180. t. 197. f. 1916 and 1917.

Inhabits the coasts of the Nicobar and Molucca Islands. *Chemnitz.*

This shell appears by the figures to be about eleven lines long, and nine broad at the broader end, and to be marked with decussated striæ; the margin is somewhat angulated and crenulated; the inside has a pearly coat, through which the external rays appear, and the muscular impression is brown.

LUGUBRIS. 37. Shell sub-oval, depressed, thick, black, with longitudinal ribs and transverse striæ; margin white and crenated.

Patella lugubris. *Gmelin*, p. 3705. *Schreibers Conch.* i. p. 345.

Patella, No. 19. *Schroeter Einl.* ii. p. 448.

Le petit Deuil. *Favanne*, i. p. 491. t. 1. f. M. *Martini*, i. p. 112. t. 8. f. 60.

Inhabits the coasts of Provence and Cyprus. *Favanne.*

Shell with the base about two inches long, and four-fifths as broad; the sides are blackish with the ribs darker, and the summit as well as a band round the margin are white; *La petit Deuil* is described by Favanne with two irregular rows of spots round the summit, and the margin orange; but I have no doubt of its belonging to the same species. Adanson's description of *Le Libot*, agrees much better with this shell than with *P. miniata*, to which Gmelin has referred it.

VULGATA. 38. Shell sub-oval, with about fourteen obsolete ribs, and intermediate striæ; margin sub-angulated, dilated and acute.

Patella vulgata. *Linnæus Syst. Nat.* p. 1258. *Martini*, i. p. 99. t. 5. f. 38. *Pennant Zool.* iv. p. 142. t. 89. f. 145. *Schroeter Einl.* ii. p. 411. *Gmelin*, p. 3697. *Schreibers Conch.* i. p. 342. *Donovan*, i. t. 14. *Montagu Test.* p. 475. *Maton and Racket, in Lin. Trans.* viii. p. 229. *Dorset Cat.* p. 58. t. 23. f. 1, 2, and 8.

Patella vulgaris. *Da Costa Brit. Conch.* p. 3. t. 1. f. 1, 2, and 8.

Patella citrina. *Gmelin*, p. 3720. ?

Patella Islandica. *Gmelin*, p. 3698. ?

Lister Anim. Ang. t. 5. f. 40, and *Conch.* t. 535. f. 14. *Knorr*, vi. t. 27. f. 8. *Kæmmerer Cab. Rud.* t. 2. f. 6, and t. 3. f. 6. ?

Variety. Shell much depressed.

Patella depressa. *Pennant Zool.* iv. p. 142. t. 89. f. 146.

L'Aile de Chauve-souris. *Favanne,* i. p. 516. t. 2. f. K.

Common on all the coasts of Europe. *Linnæus.*

This species differs considerably in size, shape, substance, and colour, and the conical and depressed Varieties run so much into each other that it is impossible to draw any separating line ; the diameter of a full grown shell varies from an inch to two inches and a half, and the depth in some specimens is more than double that of the others with the same diameter ; it is most commonly of a dark brownish ash-colour, but young shells are striped in a variety of ways with red, white, and yellow, and these colours appear through the transparent coat on the inside in beautiful variegations ; old shells on the inside are of a horn-colour, and the muscular impression is white ; the number of longitudinal ribs is generally about fourteen, but varies considerably, and the surface in old shells is sometimes nearly level. From Kæmmerer's figures, it appears probable that Gmelin's *P. citrina* and *P. Islandica* belong to this species.

CÆRULEA. 39. Shell erose-subangulated with numerous unequal striæ, and the inside blue.

Patella cærulea. *Linnæus Syst. Nat.* p. 1259. *Born Mus.* p. 419. t. 18. f. 2. *Schroeter Einl.* ii. p. 412. *Gmelin,* p. 3697.

La Variété des Côtes d'Aunis. *Favanne,* i. p. 502. t. 3. f. D 5. ?

Inhabits the Mediterranean. *Linnæus.*

Linnæus has not made any reference or addition to the above short character, so that this must probably always remain a doubtful species ; and *P. vulgata,* in the earlier stages of its growth, often answers the description, as well as the shell which Born has figured : the latter is described to be seven lines long, and five broad, oval, convex, with unequal ribs, and the margin unequally toothed ; summit nearly central ; colour blackish, and the inside blue and shining.

TUBERCULATA. 40. Shell conical, tuberculated, with the posterior end retuse, and the margin toothed.

Patella tuberculata. *Linnæus Syst. Nat.* p. 1259. *Gmelin,* p. 3697.

Inhabits ———

No reference to any figure has been made for this species, and the description only says that it is yellowish with white tubercles. Favanne, by a note, vol. i. p. 849, appears to have considered his *Bonnet epineux*, which is *P. intorta*, to be this species ; and on the other hand it is called *La Tete de Meduse* in Callone's Catalogue.

COCHLEAR. **41. Shell longitudinally ribbed and striated, with one end contracted like a scoop, and the summit acute ; margin slightly angular.**

Patella Cochlear. *Born Mus.* p. 420. t. 18. f. 3. *Gmelin,* p. 3721. *Schreibers Conch.* i. p. 350. *Mus. Lesk.* p. 302.
Patella caudata. *Mus. Gevers.* p. 242.
Patella, No. 66. *Schroeter Einl.* ii. p. 467.
La Raquette. *Favanne,* ii. p. 505. t. 79. upper fig. B. *Knorr,* ii. t. 26. f. 3.
Inhabits the coasts of New Zealand. *Favanne.*
Shell about an inch and a half long, and two-thirds as broad at the base, and is sometimes almost twice as large ; the colour is bluish or greyish white, frequently marked with chestnut between the ribs, especially towards the margin ; and the inside is whitish, with a large pale brown or yellowish spot, which nearly surrounds the muscular impression, and is shaped like a horse-shoe.

**** *With the summit pointed and recurved.*

UNGARICA. **42. Shell conical, finely striated longitudinally, and somewhat wrinkled transversely ; summit recurved and spiral.**

Patella ungarica. *Linnæus Syst. Nat.* p. 1259. *Born Mus.* p. 421, and Vign. at p. 414. fig. d. *Gmelin,* p. 3709. *Schreibers Conch.* i. p. 351. *Montagu Test.* p. 486. *Maton and Racket, in Lin. Trans.* viii. p. 230. *Dorset Cat.* p. 58. t. 23. f. 7. *Brookes's Introd.* p. 163. t. 9. f. 125.
Patella Hungarica. *Pennant Zool.* iv. p. 143. t 90. f. 147. *Martini,* i. p. 144. t. 12. f. 107 and 108. *Schroeter Einl.* ii. p. 413. *Donovan,* i. t. 21. f. 1. *Burrow's Elements,* p. 176. t. 21. f. 6.

Patella Pileus Morionis major. *Da Costa Brit, Conch.*
  p. 12. t. 1. f. 7.
Le Bonnet de Dragon. *Favanne*, ii. p. 540. t. 4. f. E 2.
Gualter, t. 9. f. V, and VV. *Knorr*, vi. t. 16. f. 3. *Argenville*, t. 2. f. R.
Inhabits the Mediterranean. *Linnæus.* Coasts of England.
  Pennant, &c.
The base, which is nearly round, is about an inch or sometimes two inches in diameter, and the shell is rather more
  than half as high; it is often coated with a brown shaggy
  epidermis, beneath which the colour is reddish, and the inside is white and smooth; the summit is much reflected, and
  ends in two or three spiral turns; the margin is more or less
  regularly indented, and crenated.

MILITARIS.   43. Shell conical, with fine cancellated
    striæ, and the summit obliquely recurved;
    margin entire.

Patella militaris. *Linnæus Mantissa,* p. 552. *Pulteney
  Dorset Cat.* p. 51. *Montagu Test.* p. 488. t. 13. f. 11.
  Donovan, v. t. 171. *Maton and Racket, in Lin. Trans.*
  viii. p. 231. *Dorset Cat.* p. 58. t. 22. f. 7.
Le Réticulé. *Favanne,* i. p. 538. t. 4. f. B.
Lister Conch. t. 544. f. 32. *Petiver Gaz.* t. 95. f. 12, and
  t. 152. f. 15. *Klein Ost.* t. 8. f. 10.
Inhabits the coasts of Barbadoes. *Lister.* Martinique and St.
  Domingo. *Favanne.* England. *Pulteney, &c.*
West India specimens are often an inch in diameter, but those
  which have been found on the English coasts are only half
  so large; it is white, and has somewhat the habit of *P. ungarica;* but is at once distinguished by its obliquely recurved
  summit, by its cancellated striæ, and much thicker substance.

ANTIQUATA.   44. Shell sub-conical, with concentric
    imbricated wrinkles, and the summit slightly
    recurved.

Patella antiquata. *Linnæus Syst. Nat.* p. 1259. *Gmelin,*
  p. 3709. *Pulteney's Dorset Cat.* p. 51. *Montagu Test.*
  p. 485. t. 13. f. 9.
Patella Mitrula. *Gmelin,* p. 3708. *Maton and Racket,
  in Lin. Trans.* viii. p. 230. *Dorset Cat.* p. 58. t. 22. f.
  7 a.
Patella nivea. *Gmelin,* p. 3727.

Patella, No. 36, and No. 90.   *Schroeter Einl.* i. p. 456, and p. 478.

Le Soron.   *Adanson Senegal*, p. 32. t. 2. f. 3.

Le Feuilleté, ou l'Etage.   *Favanne*, i. p. 541. t. 4. upper fig. F 1.

Le Ridé.   *Favanne*, i. p. 542. t. 4. upper fig. F 2.

Lister Conch. t. 544. f. 30 and 31.   *Petiver Gaz.* t. 152. f. 14.   *Klein Ost.* t. 8. f. 9, 11, and 12.   *Martini*, i. p. 146. t. 12. f. 111 to 113.   *Humphreys Conch.* t. 4. f. 10.

Inhabits the coasts of Barbadoes. *Lister.* Senegal. *Adanson.* Martinique and St. Domingo. *Favanne.* England. *Pulteney, &c.*

Shell a quarter, or sometimes half an inch in diameter, with the summit more or less recurved, or sometimes much extended and nearly erect; it is a thick opake white shell, and in some specimens the transverse imbricated wrinkles are much stronger and more prominent than in others. Favanne, as well as Pulteney and Montagu, considered this to be the Linnæan *P. antiquata*; and *Le Soron* of Adanson, which is *P. nivea* of Gmelin, is obviously the same. Gmelin, besides the present, has also at p. 3735 an obscure species of Schroeter's under the name of *P. antiquata*.

COCHLEATA.   45. Shell with distant strong longitudinal ribs, and the summit recurved; base ovate, and the margin sinuated.

Patella cochleata.   *Chemnitz*, xi. p. 182. t. 197. f. 1919 and 1920.

Inhabits the South Sea. *Chemnitz.*

The base appears, by the figures, to be about five lines long, and four broad, and the summit is incurved and reflected over the posterior end; the shell on both sides is white, and marked with about six distant strong longitudinal ribs.

CALYPTRA.   46. Shell convex, with longitudinal somewhat wrinkled ribs, and the summit recurved; base roundish, and the margin crenated.

Patella Calyptra.   *Martyn Univ. Conch.* i. t. 18.   *Chemnitz*, x. p. 340. t. 169. f. 1643 and 1644.   *Gmelin*, p. 3712.   *Schreibers Conch.* i. p. 354.

Inhabits the North West Coast of America. *Martyn.*

The base is about half, or three-quarters of an inch broad, and the height is nearly equal to two-thirds of the diameter; the colour on the outside is dull orange-brown, and the inside is tinged with bluish or reddish brown, and somewhat glossy; the ribs are slightly wrinkled and undulated towards the margin, which is crenated; the summit is nearly lateral, and recurved over the posterior end.

INTORTA. 47. Shell convex, with longitudinal striæ, which are alternately larger, and armed with vaulted scales; summit recurved and the base ovate.

Patella intorta. *Pennant Zool.* iv. p. 143. t. 90. f. 148. *Donovan*, v. t. 146. *Maton and Racket, in Lin. Trans.* viii. p. 231. *Montagu Supp.* p. 154.
Patella pectinata. *Born Mus.* p. 423. t. 18. f. 7.
Patella Pectunculus. *Gmelin*, p. 3713.
Patella, No. 40, and No. 41. *Schroeter Einl.* ii. p. 458.
Patella, No. 49, and No. 50. *Schreibers Conch.* i. p. 350.
Le Cabochon, ou Bonnet epineux. *Favanne*, i. p. 547. t. 4. f. K.
Martini, i. p. 148. t. 12. f. 115.

Inhabits the coasts of England. *Pennant, &c.* Jamaica. *Martini.* America and the Falkland Islands. *Favanne.*
The length of the base varies from three-quarters of an inch to an inch and a quarter, and the greatest depth is about half the diameter; the shell also varies greatly in colour, and according to Favanne, is either orange with the summit white, ferruginous with the striæ of a blackish brown, or fawn-coloured with yellowish red rays and mottled with white; it has upwards of twenty muricated longitudinal striæ, and intermediate smaller ones.

CASSIDA. 48. Shell convex, longitudinally ribbed, and striated transversely, with the summit recurved; base roundish, and the margin crenated.

Patella Pectunculus, Var. *Gmelin*, p. 3713.
La Lentille. *Favanne*, i. p. 539. t. 4. f. C.
Martini, i. t. 12. f. 116.

Inhabits the coasts of the Isle of France and Otaheite. *Favanne.*
The diameter, according to Favanne, varies from one third to three-quarters of an inch, and the aperture is roundish-oblong; the shell is thick, and the longitudinal ribs are crossed by fine transverse striæ; the colour is reddish or yellowish brown, or sometimes bluish white, and the inside is white

and pearly. Martini and Schroeter have confounded this species with *P. intorta*, and they together constitute the *P. Pectunculus* of Gmelin.

TRANQUEBARICA. **49. Shell depressed, with crowded very fine longitudinal striæ, and white scales; summit slightly recurved, and the base ovate.**

Patella Tranquebarica. *Gmelin*, p. 3714. *Schreibers Conch*. i. p. 348.
Patella, No. 38. *Schroeter Einl*. ii. p. 457.
L'Oiseau-mouche. *Favanne*, i. p. 495. t. 1. f. R.
Lister Conch. t. 530. f. 8. *Martini*, i. p. 147. t. 12. f. 114.
Inhabits the coasts of Tranquebar. *Martini*. Batavia. *Favanne*.

The length of the base is about seven or eight lines, and the shell is thin, pellucid, of a chestnut-colour marked with white scales or spots, and blue or white about the summit, where Favanne says the colours are as brilliant as those of the crest of a humming bird; the muscular impression is brown, and the remainder of the inside white and pearly. Martini says the summit is slightly recurved, but neither Favanne or Schreibers have placed the species in this division.

MAMMILLARIS. **50. Shell conical, sub-pellucid, finely striated longitudinally, and the summit reflected; base oblong, and the margin entire.**

Patella mammillaris. *Linnæus Syst. Nat.* p. 1259. *Martini*, i. p. 111. t. 7. f. 58 and 59. *Born Mus.* p. 422. *Schroeter Einl.* ii. p. 416. *Gmelin*, p. 3709. *Schreibers Conch.* i. p. 352.
Patella grisea. *Gmelin*, p. 3727.
Le Mouret. *Adanson Senegal*, p. 34. t. 2. f. 5.
Le Téton de Venus. *Favanne*, i. p. 522. t. 3. f. F 1.
Lister Conch. t. 537. f. 17. *Klein Ost.* t. 8. f. 1.
Inhabits the Mediterranean. *Linnæus.* Coasts of Africa. *Lister.* Island of Goree. *Adanson.* St. Domingo, Martinique, and the Falkland Islands. *Favanne.*

Favanne has described a specimen from the Falkland Islands, of which the base measured about seventeen lines long and fourteen broad, but the usual length is from half to three-quarters of an inch; the shells vary considerably, both in their thickness and markings; the summit is generally white, and the sides either whitish grey or brown, or mottled with both, and longitudinally striped with numerous black elevated capillary lines, and there are sometimes one or two transverse white zones near the summit.

LEUCOPLEURA. 51. Shell conical, with broad slightly elevated longitudinal striæ, and the summit obtusely reflected ; base ovate.

Patella leucopleura. *Gmelin*, p. 3699.
Patella melanoleuca. *Gmelin*, p. 3713.
Patella, No. 18. *Schroeter Einl.* ii. p. 448.
La Punaise mouchetée. *Favanne*, i. p. 487. t. 1. f. B 2.
Lister Conch. t. 539. f. 22. *Martini*, i. p. 110. t. 7. f. 56 and 57.
Inhabits the coasts of the West India Islands.
This shell is white with brown or blackish rays, and is of the same size, and very nearly allied to *P. mammillaris*, of which Martini considered it to be only a Variety.

TRICARINATA. 52. Shell slightly striated, with three keels in front, and the summit revolute.

Patella tricarinata. *Linnæus Syst. Nat.* p. 1259. *Born Mus.* t. 18. f. 6. ? *Schroeter Einl.* ii. p. 417. t. 5. f. 2. ? *Gmelin*, p. 3710. *Portland Catalogue*, p. 165, lot 3601.
Patella tricostata. *Chemnitz*, x. p. 333. t. 168. f. 1622 and 1623. *Gmelin*, p. 3698. *Schreibers Conch.* i. p. 354.
Le Petit Concho-lepas. *Favanne*, i. p. 545. t. 4. f. H 1.? *Humphreys Conch.* t. 4. f. 9.
Inhabits the coasts of New Zealand ? *Humphreys*.
Linnæus has not given any reference, but to the above character has added the following description : " Shell white, ovate, of the size of a hazel-nut, with three distinct keels on the fore side, which spread out and form an angulated margin ; spire recurved, and placed on the hinder side." I am entirely unacquainted with this shell, but the *P. tricostata* of Chemnitz appears by the figure and description to be the same species. Born's *P. tricarinata* has a different appearance, and has been very strangely quoted by Gmelin for his *P. octoradiata*, p. 3699.

PECTINATA. 53. Shell ovate, entire, with wrinkled somewhat branched striæ ; summit nearly central, reflected, and pointed.

Patella pectinata. *Linnæus Syst. Nat.* p. 1259. *Schroeter*
Einl. ii. p. 418. t. 5. f. 3. *Gmelin*, p. 3710.
Inhabits the Mediterranean. *Linnæus.*
Linnæus, without any reference, has described this species to
be "opake, with the inside smooth and like tortoise-shell;
outside with numerous unequal longitudinal striæ." The
shell which Schroeter has figured is about two inches long,
of a greyish colour, and has the summit mottled with chestnut and white; the inside is mottled like tortoise-shell with
chestnut and lead-colour. Born erroneously considered *P.*
intorta to be this species.

FUSCO-LUTEA. 54. Shell convex, oval-oblong, sub-
membranaceous, with longitudinal striæ,
which are alternately larger, and wrinkled
transversely; summit reflected.

Patella lutea. *Born Mus.* p. 424. t. 18. f. 8.
Inhabits ———
The length of the base is about an inch, and the breadth rather
more than seven lines; the colour, excepting a white muscular impression, is yellowish brown both inside and out;
the shell is thin and brittle, somewhat depressed towards the
base, and has the summit placed about three times as near
to one end as to the other.

LUTEA. 55. Shell convex, oval-oblong, striated, with
the summit pointed, incurved, and nearly
marginal, and the inside iridescent.

Patella lutea. *Linnæus Syst. Nat.* p. 1260. *Martini*, i.
p. 191. t. 17. f. 154 and 155. *Schroeter Eint.* ii. p. 419.
Gmelin, p. 3710. *Schreibers Conch.* i. p. 352.
L'Oreille de Mer piquetée. *Favanne*, i. p. 591. t. 5. f. E.
Rumphius, t. 40. f. I. *Petiver Amb.* t. 16. f. 30.
Inhabits the coasts of Amboyna. *Rumphius.*
Martini, as well as Favanne, considered this shell to be an imperforated species of Haliotis, and it may probably belong to
the same natural family as *Helix perspicua;* the colour is
described to be fulvous, but Martini's figure has a greenish
tinge with spots and transverse stripes of yellow, and from
Favanne's description it appears to vary considerably in the
markings, as well as colour; the length is about three-quarters of an inch. His reference to Rumphius, t. 40. f. I, in
some measure proves this shell to be the *P. lutea* of Linnæus; but it is not at all like the upper valve of *Anomia*

patelliformis, to which he has compared it, in describing that species in the Fauna Suecica.

PERVERSA. 56. Shell membranaceous, brittle, and very smooth, with the summit pointed and recurved ; base roundish-oblong.

Patella perversa. *Gmelin*, p. 3714.
Patella, No. 39. *Schroeter Einl.* ii. p. 457.
Le Liri. *Adanson Senegal*, p. 32. t. 2. f. 2.
Martini, i. p. 148. t. 12. f. 114, *A* and *B*.
Inhabits the coasts of the Cape de Verd, Goree, and Magdalen Islands, on rocks. *Adanson*.
This, though a marine species, appears from Adanson's account to be very nearly allied to those which inhabit fresh water : he says, that the summit is placed nearest to that end of the shell which covers the tail of the animal, and that it is recurved in the same direction ; it is coated with an epidermis, which gives the shell a rust-colour ; the base is about four lines long and three broad.

LACUSTRIS. 57. Shell membranaceous, brittle, with the summit pointed and reflected ; base oval.

Patella lacustris. *Linnæus Syst. Nat.* p. 1260. *Schroeter Fluss.* p. 203. t. 5. f. 1, 2, and 3 ; and *Einl.* ii. p. 421. *Lightfoot in Phil. Trans.* lxxvi. p. 168. t. 3. f. 4. *Pennant Zool.* iv. p. 143. *Donovan*, v. t. 147. *Maton and Racket, in Lin. Trans.* viii. p. 232. *Dorset Cat.* p. 58. t. 22. f. 8.
Patella fluviatilis. *Da Costa Brit. Conch.* p. 1. t. 2. f. 8. *Gmelin*, p. 3711. *Montagu Test.* p. 482.
Ancylus fluviatilis. *Muller Verm.* ii. p. 201.
L'Ancille. *Geoffroy*, p. 124. t. 3.
Lister Anim. Ang. t. 2. f. 32, and *Conch.* t. 141. f. 39. *Gualter*, t. 4. f. BB. *Argenville*, t. 27. f. 1.
Inhabits most of the rivers and lakes in Europe. *Da Costa*.
The base is about three-eighths of an inch long, and two-eighths broad ; the shell is thin, brittle, and transparent, of a whitish horn-colour, usually coated with a dark brown or greenish epidermis, and slightly wrinkled transversely. Da Costa, though he changed the name, was aware that this is the Linnæan *P. lacustris*. Montagu says, " a Variety is sometimes met with, finely striated longitudinally from the beak, and we received some specimens, from a fresh-water

stream near Folkestone, in Kent, more strongly striated than usual, which were considered as a distinct species."

OBLONGA. 58. Shell membranaceous, brittle, slightly contracted in the middle, with the summit pointed and obliquely reflected; base oblong.

Patella oblonga. *Lightfoot in Phil. Trans.* lxxvi. p. 168. t. 3. f. 1, 2, 3, and 5. *Donovan*, v. t. 150. *Maton and Racket, in Lin. Trans.* viii. p. 233. *Dorset Cat.* p. 58. t. 18. f. 20, and t. 22. f. 8 a.
Patella piccolissima. *Ginanni, Op. Post.* ii. p. 50. t. 2. f. 11.
Patella lacustris. *Gmelin*, p. 3710. *Montagu Test.* p. 484.
Patella, No. 101. *Schroeter Einl.* ii. p. 483.
Ancylus lacustris. *Muller Verm.* ii. p. 199.
Ancylus fluviatilis. *Schroeter Fluss.* p. 205. t. 5. f. 4, a and b.
Gualter, t. 4. f. AA.
Inhabits rivers on aquatic plants, in most parts of Europe.
The base is about a quarter of an inch long, and hardly half as broad; it differs from *P. lacustris* in being much narrower in proportion to the length, and in having the summit smaller, more central, and obliquely reflected.

***** *With the summit obtuse, and the margin entire.*

PELLUCIDA. 59. Shell convex, membranaceous, pellucid, with about four longitudinal azure rays; summit obsolete and nearly marginal; base ovate.

Patella pellucida. *Linnæus Syst. Nat.* p. 1260. *Pennant Zool.* iv. p. 143. t. 90. f. 150. *Born Mus.* p. 424. t. 18. f. 9. *Schroeter Einl.* ii. p. 423. *Chemnitz*, x. p. 331. t. 168. f. 1620. *Gmelin*, p. 3717. *Schreibers Conch.* i. p. 355. *Montagu Test.* p. 477 and *Supp.* p. 153. *Dorset Cat.* p. 58. t. 23. f. 5.
Patella cœruleata. *Da Costa Brit. Conch.* p. 7. t. 1. f. 5 and 6.
Patella pellucida, junior. *Maton and Racket, in Lin. Trans.* viii. p. 233.
Lister Conch. t. 543. f. 27.
Inhabits the coasts of Norway, and the Mediterranean. *Linnæus.* Britain. *Lister, &c.*

Shell with the base generally six or eight lines long, and about two-thirds as broad; it is extremely thin, brittle, and pellucid, of a yellowish horn-colour, with from three to seven more or less dotted azure lines, extending longitudinally from the summit to the margin.

LÆVIS.   60. Shell sub-conical, smooth, with diverging longitudinal lines, and the summit nearly marginal ; base sub-ovate.

Patella lævis.   *Pennant Zool.* iv. p. 144. t. 90. lower fig.
Patella pellucida.   *Muller Zool. Dan.* iii. p. 37. t. 104. f. 1
to 4.   *Donovan,* i. t. 3. f. 1.
Patella pellucida, senior.   *Maton and Racket, in Lin.
Trans.* viii. p. 234.
Patella pellucida, Var.   *Chemnitz,* x. p. 332. t. 168. f. 1621.
Dorset Cat. p. 58. t. 23. f. 6.
Patella cærulea.   *Montagu Supp.* p. 152.
Le Cabochon Ventre de biche. *Favanne,* i. p. 550. t. 4. f. N.?
Lister Anim. Ang. App. t. 2. f. 10, and *Conch.* t. 542. f. 26.
   Petiver Gaz. t. 75. f. 3.   *Klein Ost.* t. 8. f. 6 and 7.
Variety.   With two black spots on the summit.
Patella bimaculata.   *Montagu Test.* p. 482. t. 13. f. 8.
Maton and Racket, in Lin. Trans. viii. p. 235.
Inhabits the coasts of England. *Lister, &c.*   Northern Ocean.
Muller.
Many Conchologists have supposed that this is *P. pellucida* at a more advanced stage of growth, and this opinion was at first held by Mr. Montagu ; but in the Supplement to his Testacea Britannica, he says that further observations have induced him to think otherwise : in the following passage he has pointed out the difference between them, but neither the one or the other at all answers the Linnæan description of *P. cærulea:* " The principal distinction of the *pellucida* is the regular ovate and convex appearance, with scarcely any obvious beak ; but what little it has, is always close to the margin ; besides, it is always pellucid, and seldom has more than four or five blue lines.   The *cærulea,* on the contrary, is extremely various in its shape at all ages, some being much depressed and others greatly elevated, and the beak is never so low as to be destitute of margin.   It is also usually rayed from the vertex on all sides, sometimes with a few blue lines, and the rest brown ; the beak is generally decorticated even in the smallest specimens, and often stands abrupt as if a small shell was placed upon a larger."   *P. lævis* of Gmelin is different, and is a very doubtful species.

RADIANS. 61. Shell oval, entire, pellucid, much depressed, striated longitudinally, and radiated with blackish spots.

Patella radians. *Gmelin*, p. 3720.
Patella radiata Novæ Zelandiæ. *Chemnitz*, x. p. 329. t. 168. f. 1618.
Inhabits the coasts of New Zealand and Terra del Fuego. *Chemnitz.*
The base appears by the figure to be eighteen lines long, and fourteen broad ; the shell is said to be of a horn-colour, with longitudinal radiated rows of rather irregular blackish spots, or short oblique streaks, and the summit yellowish ; the inside is somewhat silvery. Chemnitz says, this shell was bought in London with the name of 'The grey mottled Shield Limpet from New Zealand.'

ROTA. 62. Shell roundish, with longitudinal reddish stripes radiated in pairs, and the margin yellowish.

Patella Rota. *Chemnitz*, x. p. 330. t. 168. f. 1619. *Gmelin*, p. 3720. *Schreibers Conch.* i. p. 362.
Inhabits the East and West Indian Seas. *Chemnitz.*
This shell appears by the figure to be about eighteen lines long, and sixteen broad, of a pale tawny colour, with eighteen reddish brown longitudinal stripes, radiated in pairs ; the inside is said to be somewhat silvery.

TESTUDINARIA. 63. Shell ovate, convex-conical, very entire, smooth, and glabrous ; summit obtuse and somewhat lateral.

Patella testudinaria. *Linnæus Syst. Nat.* p. 1260. *Martini*, i. p. 104. t. 6. f. 45 to 48. *Muller Zool. Dan. Prod.* p. 237.? *Born Mus.* p. 425. *Schroeter Einl.* ii. p. 425. *Gmelin*, p. 3717. *Schreibers Conch.* i. p. 356. *Lamarck Syst. des Anim.* p. 68. *Brookes's Introd.* p. 138. t. 9. f. 126.
Le Bouclier l'Ecaille de Tortue. *Favanne*, i. p. 493. t. 1. f. Q 1.
Lister Conch. t. 531. f. 9. *Rumphius*, t. 40. f. A.? *Gualter*, t. 8. f. B. *Argenville*, t. 2. f. P. *Knorr*, i. t. 21. f. 1, and iii. t. 30. f. 2 to 5.
Inhabits the coasts of Norway. *Muller?* East Indies. *Favanne.*

The base is sometimes three inches long, and nearly two inches
and three-quarters broad, but the shell is usually smaller,
and often not half so large; perfect shells I believe are always slightly striated, and I rather doubt whether the Linnæan character was not taken from an artificially polished
specimen; the colour is whitish or pale ash-colour, with
chestnut-brown spots, disposed either in radiated bands or
over the whole surface, and the inside has a pearly coat,
through which the external markings are seen; the muscular
impression is generally brown.

TESTUDINALIS.   64. Shell ovate, entire, with rather
obsolete decussated striæ; summit slightly
elevated, and nearly central.

Patella testudinalis. *Muller Zool. Dan. Prod.* p. 237.
Fabricius Fauna Groenl. p. 385. *Gmelin,* p. 3718.
Patella testudinaria, Var. β.   *Kæmmerer Cab. Rudolst.*
p. 12. t. 2. f. 4 and 5.
Patella testudinaria Gröenlandica. *Chemnitz,* x. p. 325.
t. 168. f. 1614 and 1615.
Patella radiata. *Born Mus.* p. 425. t. 18. f. 10.
Patella virgata. *Gmelin,* p. 3727.
Patella, No. 88. *Schroeter Einl.* ii. p. 477.
La petite Ecaille de Tortue. *Favanne,* i. p. 495. t. 1. f. Q 2.
Inhabits the coasts of Norway. *Muller.*   Among Fuci on the
shores of Greenland; very common. *Fabricius.*   St. Domingo. *Favanne.*
Fabricius says that the largest specimen which he has seen,
measured fourteen lines long, eleven broad, and seven high:
he says the largest shells are greenish, the smaller tinged
with violet, and that the smallest have radiated undulated
stripes or tessellated spots of brown; inside white, glabrous,
and the muscular impression coloured like tortoise-shell; the
shell figured by Chemnitz has the inner margin elegantly
spotted with brown.

COMPRESSA.   65. Shell oblong-conical, compressed
at the sides, and longitudinally striated.

Patella compressa. *Linnæus Syst. Nat.* p. 1261.   *Born
Mus.* p. 426. *Schroeter Einl.* ii. p. 427. *Gmelin,* p. 3718.
Schreibers Conch. i. p. 356.
Le grand Comprimé, ou le Bateau.   *Favanne.* i. p. 521.
t. 3. f. B 3.
Lister Conch. t. 541. f. 25. *Knorr,* ii. t. 26. f. 4. *Martini,*
i. p. 142. t. 12. f. 106.

Inhabits the coasts of the Isles of France and Otaheite. *Favanne*. Cape of Good Hope. *Humphreys*.

The length of the base varies from two to four inches, and the breadth and height are each about half the length; the shells which I have seen are of a pale nearly uniform horn-colour, both inside and out, but according to Martini it is sometimes longitudinally striped with yellowish brown, and Favanne mentions some Varieties differently mottled with reddish fawn-colour and white.

AFRA.   66. Shell conical, elevated, with about one hundred narrow longitudinal ribs; summit obtuse, glabrous, and central.

Patella Afra. *Gmelin*, p. 3715.
Patella, No. 1. *Schroeter Einl.* ii. p. 441.
Le Gadin. *Adanson Senegal*, p. 33. t. 2. f. 4.
Gualter, t. 9. f. C. *Martini*, i. p. 95. t. 5. f. 34.
Inhabits the coasts of Cape Manuel, and the island of Goree. *Adanson*.
Shell about ten lines or an inch in diameter at the base, and more than half as high, of a pale greenish brown colour, and the inside white; it has about one hundred narrow longitudinal ribs, and the base is nearly circular. Favanne considered *Le Gadin* to be the same as Martini, t. 10. f. 84, which is *P. chlorosticta;* but it appears to me to be perfectly different.

RUSTICA.   67.   Shell conical, very entire, with fifty narrow obtuse longitudinal ribs.

Patella rustica. *Linnæus Syst. Nat.* p. 1261. *Born Mus.* p. 426. t. 18. f. 11. *Favanne App.* p. 845. *Schroeter Einl.* ii. p. 428. t. 5. f. 4.? *Gmelin*, p. 3718.? *Mus. Lesk.* p. 304.
Patella Lusitanica. *Gmelin*, p. 3715.
Patella. No. 2. *Schroeter Einl.* ii. p. 441; No. 3. p. 442; and No. 9. p. 445.
Le Bouclier moucheté. *Favanne*, i. p. 489. t. 1. f. I.
Lister Conch. t. 537. f. 20.? *Gualter*, t. 8. f. N, M and P. *Martini*, i. t. 5. f. 35 and 36, and t. 6. f. 43.
Inhabits the coasts of Portugal. *Martini*. Jamaica, Provence, and China. *Favanne*.
The base is about nineteen lines long and seventeen broad, and the height according to Favanne varies considerably, some shells being much elevated and others much depressed; the colour is whitish, variegated with brown or pale chestnut longitudinal rays, and the summit surrounded with one or more yellowish rings; sometimes the rays are dotted with

black, and the summit spotted with yellow; the inside is bluish white, more or less glossy, and the muscular impression ferruginous or tawny.

JAMAICENSIS. 68. Shell conical, convex towards the base, with longitudinal striæ and distant transverse ridges; base roundish.

Patella Jamaicensis. *Gmelin*, p. 3715.
Patella cancellata, Var. β. *Gmelin*, p. 3704.
Patella radiata. *Turton's Gmelin*, p. 581.
Patella, No. 4. *Schroeter Einl.* ii. p. 442.
Martini, i. p. 97. t. 5. f. 37.
Inhabits the coasts of Jamaica. *Martini*.

The base is roundish, about an inch and a quarter in diameter, and the height of the shell is about three-quarters of an inch; the outside is greyish yellow, with twelve longitudinal orange lines, and the inside is of a horn-colour. Dr. Turton has changed the name of this species on account of Gmelin's having also given the name of *Jamaicensis* to another species; but as the latter had before received the name of *nodosa* from Born, the change is unnecessary; and both Born and Chemnitz have moreover given the name of *radiata* to other shells.

STELLIFERA. 69. Shell oval, entire, of a blackish brown colour, with narrow longitudinal ribs, and the summit stellated with white.

Patella stellata, seu stellifera. *Chemnitz*, x. p. 329. t. 168. f. 1617.
Patella stellifera. *Gmelin*, p. 3719. *Schreibers Conch.* i. p. 361.
Inhabits the coasts of the Friendly Islands and New Zealand. *Chemnitz.*

This shell appears by the figure to be about two inches long, and an inch and a half broad, and is marked with short radiated white stripes at the summit; the outer surface has narrow longitudinal ribs, crossed with transverse wrinkled striæ, and the inside is silvery. Chemnitz says it has been received from London, with the name of "the brown and white starred Shield Limpet;" but I rather doubt whether it is more than a Variety of some neighbouring species, or perhaps of *P. ferruginea.*

FUSCA. 70. Shell conical, obtuse, with about thirty-nine narrow longitudinal ribs; base ovate, and the margin slightly crenated.

2 H 2

Patella fusca. *Linnæus Syst. Nat.* p. 1261. *Born Mus.*
    p. 427. *Gmelin,* p. 3719. *Shaw Nat. Misc.* xv. t. 606.
Patella Magellanica. *Gmelin,* p. 3703.
Patella, No. 6. *Schroeter Einl.* ii. p. 443.
Le Pain de Sucre. *Favanne,* i. p. 486. t. 1. f. A 2.
Martini, i. p. 100. t. 5. f. 40.
Inhabits the Straights of Magellan. *D'Avila.* Coasts of the
    Falkland Islands. *Favanne.*
The base is about two inches and a half long, and two inches
    broad, and the height of the shell is sometimes nearly equal
    to the breadth ; in the natural state it is of a dull ash-colour,
    but when polished it becomes of a chestnut-brown with the
    ribs nearly white, and in this state it has been figured by
    Martini ; the inside is white and silvery, and the muscular
    impression richly bronzed.   It is perhaps impossible to place
    it beyond a doubt what the *P. fusca* of Linnæus is, for he
    has given only a short description, without any reference to a
    figure.   Born and Shaw have considered it to be the present
    shell, although its margin is distinctly crenated, whereas in
    the Linnæan species it is said to be very entire ; but in other
    respects it answers the description, and Martini, notwith-
    standing the crenatures, has described it ' limbo integro.'

AREOLATA.   71,   Shell pyramidal, depressed, with
        remote decussated striæ ; base round, and
        the margin nearly entire.

Patella areolata. *Gmelin,* p. 3716.
Patella, No. 7. *Schroeter Einl.* ii. p. 444.
Le Pyramidal. *Favanne,* i. p. 485. t. 1. f. A 1.
Argenville, t. ii. f. A.   *Martini,* i. p. 101. t. 5. f. 41.
Inhabits the Straights of Magellan. *Argenville.*
Favanne considered this to be a Variety of the preceding spe-
    cies, from which it differs in having the base rounder, the
    margin less distinctly crenated, and the height not much
    more than half the diameter ; it has about twenty-one longi-
    tudinal striæ, and these are crossed by others transversely, so
    as to form quadrangular departments ; the colour is yellowish
    red, with a purple summit, and the inside white and pearly,
    except the muscular impression, which is richly bronzed.

FLAMMEA.   72.   Shell convex, with fine transverse
        elevated striæ, and waved longitudinal stripes ;
        summit nearly central, and the base ovate.

Patella flammea. *Gmelin,* p. 3716.

Patella, No. 8. *Schroeter Einl.* ii. p. 444.
L'Agate flambée. *Favanne*, i. p. 493. t. 1. f. P 2.
Argenville, t. 2. f. Q. *Martini*, i. p. 102. t. 5. f. 42.
Inhabits the coasts of the Falkland Islands. *Favanne.*
The base appears by the figures to be about two inches and a
quarter long, and an inch and a half broad ; the shell is reddish white with waved radiated brown stripes, and when
held to the light the summit is said to appear of the colour
of a ruby. It is rather a doubtful species, and both Favanne's and Martini's figures are copies from Argenville.

INDICA. 73. Shell depressed, longitudinally striated, and decussated towards the base ; summit pointed, central, and the base ovate.

Patella Indica. *Gmelin*, p. 3716.
Patella, No. 11. *Schroeter Einl.* ii. p. 445.
Le Papier brouillard. *Favanne*, i. p. 488. t. 1. f. G.
Gualter, t. 8. f. E. *Martini*, i. p. 106. t. 7. f. 49.
Inhabits the East Indian Seas. *Martini.*
Shell about three inches and a half long, and two inches and
three-quarters broad at the base, of a yellowish white colour,
with darker tawny longitudinal rays, and there is sometimes
a reddish ring round the summit, which is smooth and point-
ed ; the inside is white.

VITELLINA. 74. Shell depressed, yellow, striated
and plaited towards the base, with the summit smooth and convex ; base sub-ovate.

Patella vitellina. *Gmelin*, p. 3716.
Patella, No. 13. *Schroeter Einl.* ii. p. 446.
Knorr, i. t. 20. f. 2. *Martini*, i. p. 107. t. 7. f. 51.
Inhabits ——
The base appears by the figure to be twenty-three lines long,
and nineteen broad ; the colour is bright yellow, and the ribs
towards the base, in Martini's figure, are shaded with reddish
brown, and Knorr says that the inside is yellow ; the margin
has the appearance of being slightly crenated, but all the
above-mentioned authors have described it to be entire.

LÆVIGATA. 75. Shell ovate, rather depressed, with
flattened alternately thicker narrow ribs and
oblique striæ.

Patella lævigata. *Gmelin*, p. 3717.
Patella, No. 16. *Schroeter Einl.* ii. p. 447.

Martini, i. p. 109. t. 7. f. 54.

Inhabits ——

The base, according to Martini's figure, is fourteen lines long, and eleven broad at the broader end; the colour is yellowish, with numerous narrow brown rays, and the summit is white, polished, and placed rather nearer to one end than the other.

SURINAMENSIS. **76.** Shell depressed, thick, with remote decussated striæ towards the base, and the summit smooth and central; base nearly oval, and the inside yellow.

Patella Surinamensis. *Gmelin,* p. 3716.
Patella, No. 12. *Schroeter Einl.* ii. p. 446.
Le Cadran. *Favanne,* i. p. 518. t. 2. f. N.?
Martini, i. p. 107. t. 7. f. 50.
Common on the coasts of Surinam. *Martini.*

The base measures seventeen by fourteen lines, and Martini's figure is of a whitish colour, becoming browner and irregularly striped with brown towards the base; the summit is smooth and slightly elevated, and the inside yellowish. Favanne, in the Appendix, has quoted Martini's fig. 50 for *Le Cadran,* but the descriptions do not well agree, and the figures are quite unlike each other.

PUNCTULATA. **77.** Shell roundish-oblong, with radiated striæ towards the base, and scattered dots; summit central and smooth; margin slightly sinuated.

Patella punctulata. *Gmelin,* p. 3705, and p. 3717.
Patella, No. 17. *Schroeter Einl.* ii. p. 448.
Bonanni Rec. and *Kirch.* i. f. 7. *Martini,* i. p. 109. t. 7. f. 55.
Inhabits ——

The base is about fourteen lines long, and an inch broad; the colour is white, with scattered red, black, and yellow dots, and two brown transverse lines near the base, which, though otherwise described, appears in the figure to be slightly sinuated. Gmelin has inadvertently described this shell with the same name in two places, and Dr. Turton has been thereby misled to change the name in one place to *punctata.*

NOTATA. **78.** Shell depressed, with longitudinal dark striæ, and a pointed erect summit; inside white, with the muscular impression spatula-shaped.

Patella notata. *Linnæus Syst. Nat.* p. 1261. *Schroeter*
Einl. ii. p. 431. t. 5. f. 5. *Chemnitz*, x. p. 324, and Vign.
25 at p. 320. f. *C* and *D*. *Gmelin*, p. 3719. *Schreibers*
Conch. i. p. 357.
La Gerbe de Bled. *Humphreys Conch.* p. 26. t. 5. f. 1.
Kæmmerer Cab. Rudolst. p. 13. t. 2. f. 3 and 7.
Inhabits the Mediterranean. *Linnæus.* Coasts of Guinea and
the West Indies. *Chemnitz.*
Shell nine, ten, or sometimes eleven lines long, and rather more
than two-thirds as broad, of a dullish ash-colour, with blackish or brown longitudinal crowded fine elevated striæ; the
summit is placed twice as near to one end as to the other,
and is sometimes surrounded with a blackish ring; the inside
is white, with a chestnut spotted margin, and the muscular
impression of a blackish or brownish colour, is shaped like a
spatula, or, according to Da Costa, like a wheat-sheaf, and
the shell is thence called in the Portland Catalogue the *Wheat-
sheaf Patella.*

CRUCIATA. 79. Shell oval, sub-convex, very entire,
and brown with a white cross.

Patella cruciata. *Linnæus Syst. Nat.* p. 1261. *Schroeter*
Einl. ii. p. 432. t. 5. f. 6. *Gmelin*, p. 3719. *Schreibers*
Conch. i. p. 357.
Inhabits ———
Linnæus, without any reference to a figure, has given only a
short description of this species, which is said by Schroeter
to be an inch long, and three-quarters of an inch broad, of a
brown colour, with four broad white rays, and a paler summit, the inside white, and the muscular impression brown.

RETICULATA. 80. Shell conical, compressed, entire,
and reticulated with veins.

Patella reticulata. *Linnæus Syst. Nat.* p. 1261. *Schroeter*
Einl. ii. p. 433. t. 5. f. 7. *Gmelin*, p. 3719. *Schreibers*
Conch. i. p. 357.
Inhabits the Mediterranean. *Linnæus.*
Linnæus has given only a short description, without any reference to a figure, and I much doubt whether the shell
which Schroeter has figured belongs to the same species;
by the figure it appears to be roundish, and about eight lines
in diameter, and it is said to be milk-white, much narrower
behind, with elevated crowded decussated striæ, and the
summit nearly central and obtuse.

CAECA.　81. Shell oval, with crowded longitudinal elevated dotted striæ, and the summit acute and erect.

Patella caeca. *Gmelin*, p. 3711.
Patella coeca. *Muller Zool. Dan.* i. p. 12. t. 12. f. 1 to 3.
Inhabits the bay of Dröbach, in Norway, on stones. *Muller*.
Shell about four lines long, and two-thirds as broad, pellucid, of a dark colour with a white border, and marked with from sixty to eighty striæ, which extend from the summit to the margin, and are furnished with minute elevated dots; the summit is mucronated but not recurved.

VIRGINEA.　82. Shell oval, very entire, with pale purplish longitudinal rays, and the summit nearly marginal.

Patella virginea. *Muller Zool. Dan.* i. p. 13. t. 12. f. 4 and 5. *Gmelin*, p. 3711. *Maton and Racket, in Lin. Trans.* viii. p. 235. *Dorset Cat.* p. 59. t. 14. f. 11.
Patella parva. *Da Costa Brit. Conch.* p. 7. t. 8. f. 11. *Donovan*, i. t. 21. f. 2. *Montagu Test.* p. 480.
Inhabits the coasts of Norway on Fuci. *Muller*. West of England. *Da Costa, &c.* Langlan Bay, near Swansea. *L. W. D.*
Shell about three-eighths or sometimes half an inch long, and two-thirds as broad, minutely striated, and of a reddish or greyish white colour, with about eighteen pale purplish rays extending from the summit to the margin; the inside is glossy, and appears radiated like the outside from the transparency of the shell.

TESSELLATA.　83. Shell oval-oblong, very entire, finely striated longitudinally, and whitish, tessellated with red spots.

Patella tessellata. *Muller Zool. Dan.* i. p. 13. t. 12. f. 6 and 7. *Gmelin*, p. 3711.
Olaffsen Islandica. t. 11. f. 11.?
Inhabits Fuci on the coasts of Norway. *Muller*.
Shell about nine or ten lines long, and not much more than half as broad, and has an obsoletely mucronated summit, which is placed more than twice as near to one end as to the other; the inside is coloured like the outside, and has the muscular impression red.

FULVA. 84. Shell oval-oblong, very entire, with minute somewhat decussated striæ, and the spire nearly marginal.

Patella fulva. *Muller Zool. Dan.* i. p. 24. t. 24. f. 1 to 3. *Gmelin*, p. 3712.

Inhabits the Bay of Dröbach, in Norway. *Muller.*

Shell about two lines long, and rather more than half as broad, of an uniform orange or tawny yellow colour, without any spots, and when examined with a glass appears to be slightly marked with both transverse and longitudinal striæ.

AMBIGUA. 85. Shell oblong, depressed, with the posterior margin rounded, and the anterior truncated; summit nearly marginal.

Patella ambigua. *Chemnitz*, xi. p. 181. t. 197. f. 1918. *Humphreys Conch.* t. 5. f. 11.

Inhabits the coasts of New Holland. *Humphreys.*

Shell two inches or two inches and a half long, and rather less than half as broad, thick, and rather strongly marked with concentrical wrinkles; the outside is dirty white or pale brown, and the inside is white. In shape it somewhat resembles *Mytilus Lingua*, and in the Portland Catalogue (lot 3565) it is called the *White Duck's-Bill Patella*.

UMBELLATA. 86. Shell roundish, nearly flat, with concentric wrinkles, and slightly undulated longitudinally; margin very acute.

Patella umbellata. *Gmelin*, p. 3720.
Patella Sinica. *Gmelin*, p. 3705.
Patella Umbraculum. Portland Cat. p. 178. lot 3830.
Patella, No. 10. *Schroeter Einl.* ii. p. 445.
Operculatum læve. *Linnæus Mus. Tessinianum*, p. 116. t. 6. f. 5.
Umbella Chinensis. *Chemnitz*, x. p. 341. t. 169. f. 1645 and 1646.
Acardo Umbella. *Lamarck Syst. des Anim.* p. 130.
Le Parasol Chinois. *Favanne*, i. p. 524. t. 3. f. H.
D'Avila, t. 2. f. A. *Martini*, i. p. 103. t. 6. f. 44.

Inhabits the coasts of China. *D'Avila.* Isle of France and St. Domingo. *Favanne.* Nicobar Islands. *Chemnitz.*

Shell roundish or slightly oval, and from two to five inches in diameter; it is thickest towards the summit, which is placed nearer to one end than to the other, and the margin is more or less slightly sinuated; the colour is generally whitish, with the summit pale yellow, and the inside either yellow or brownish, and finely striated with raised dots. Linnæus has figured this shell in the Museum Tessinianum, and appears to have been at a loss whether it should be considered as a shell or an operculum. Lamarck supposes that it is a single valve of a species of his Genus Acardo, which he defines to be composed of two flat nearly equal valves, without either a hinge or cartilage, and with a muscular impression in the center of each.

****** *With a marginal fissure.*

FISSURA. 87. Shell conical, cancellated, with a deep and narrow marginal fissure, and the summit recurved.

Patella Fissura. *Linnæus Syst. Nat.* p. 1261. *Martini,* i. p. 145. t. 12. f. 109 and 110. *Pennant Zool.* iv. p. 144. t. 90. f. 152. *Da Costa Brit. Conch.* p. 11. t. 1. f. 4. *Born Mus.* p. 427. t. 18. f. 12. *Schroeter Einl.* ii. p. 434. *Muller Zool. Dan.* i. p. 25. t. 24. f. 7 to 9. *Gmelin,* p. 3728. *Schreibers Conch.* i. p. 362. *Donovan,* i. t. 3. f. 2. *Montagu Test.* p. 490. *Maton and Racket, in Lin. Trans.* viii. p. 235. *Dorset Cat.* p. 59. t. 23. f. 4. *Brookes's Introd.* p. 138. t. 9. f. 127.
Emarginula conica. *Lamarck Syst. des Anim.* p. 69.
Le petit Cabochon l'Entaille. *Favanne,* i. p. 548. t. 4. f. M 1.
Lister Conch. t. 543. f. 28. *Petiver Gaz.* t. 75. f. 2.
Inhabits the coasts of Great Britain. *Lister, &c.* Algiers. *Linnæus.*
Shell five, or sometimes six lines long, and the height and breadth are each about equal to three-fourths of the length; it is white or brownish white, and elegantly cancellated with longitudinal ribs and transverse striæ; the fissure extends from the margin about one-third of the way to the summit, and is very narrow. D'Avila calls this a fresh-water shell, and says it inhabits the Lake of Geneva, but this is most probably a mistake.

INCISA. 88. Shell oval, rather depressed, cancellated, with a broad marginal fissure, and the summit recurved.

Patella Fissura reticulata. *Chemnitz*, xi. p. 185. t. 197. f. 1925 and 1926.
Le grand Cabochon l'Entaille. *Favanne*, i. p. 549. t. 4. f. M 2.
The Cracked Limpet. *Humphreys Conch.* t. 4. f. 2.
Inhabits the coasts of the Falkland Islands. *Chemnitz.*
Shell about three-quarters of an inch long, and half an inch broad, and the height is only about half the length; it is white and cancellated, and nearly allied to *P. Fissura*, from which it differs in being larger and more depressed, and in having the marginal fissure broader and shorter; but Da Costa, when he wrote his British Conchology, appears to have considered it a Variety of that species.

FISSURATA. 89. Shell oval, depressed, with crowded longitudinal striæ, and a short marginal fissure.

Patella fissurata. *Humphreys Conch.* p. 20. t. 4. f. 3. *Chemnitz*, xi. p. 188. t. 197. f. 1929 and 1930.
L'Echancré. *Favanne*, i. p. 549.
Inhabits the coasts of New Zealand. *Favanne.* Ceylon. *Chemnitz.*
Shell about seven lines long, five broad, and three high, of a rose-colour, with the summit and muscular impression white, and the remainder of the inside is flesh-colour; the marginal fissure is scarcely half a line deep, and almost equally broad.

******* *With a perforated summit.*

NOACHINA. 90. Shell conical, longitudinally striated, and the summit recurved, with a rhomboidal perforation extending towards the anterior margin.

Patella Noachina. *Linnæus Mantissa.* p. 551. *Chemnitz*, xi. p. 186. t. 197. f. 1927 and 1928.
Patella Fissurella. *Muller Zool. Dan. Prod.* p. 237, and *Zool. Dan.* i. p. 24. t. 24. f. 4 to 6. *Fabricius Fauna Groenl.* p. 384. *Gmelin*, p. 3728.
Patella Apertura. *Montagu Test.* p. 491. t. 13. f. 10, and *Supp.* p. 155. *Maton and Racket, in Lin. Trans.* viii. p. 236.
Humphreys Conch. t. 7. f. 8.
Inhabits the coasts of Norway. *Muller.* Greenland. *Fabri-*

cius. Ferroe Islands and Iceland. *Chemnitz.* England. *Montagu.*

Shell about three lines long, rather more than two broad, and one and a half high, white, with strong longitudinal tuber- culated ribs, and somewhat wrinkled transversely; it has a fissure or perforation, extending longitudinally from the re- curved summit towards the anterior margin, and appears to be a link which connects the Emarginulæ and the Fissurellæ of Lamarck.

PUSTULA. 91. Shell oval, flattish, with radiated ribs, and sub-truncated at one end; margin thick and crenated.

Patella Pustula. *Linnæus Syst. Nat.* p. 1262. ? *Schroeter Einl.* ii. p. 436. t. 5. f. 8. ? *Chemnitz, x.* p. 338. t. 168. f. 1632 and 1633. *Gmelin,* p. 3728. *Schreibers Conch.* i. p. 362.

Patella Unguis. *Martini Besch. Berl. Naturf.* ii. t. 12. f. 4 and 5.

Patella, No. 157. *Schroeter Einl.* ii. p. 508.

Lister Conch. t. 528. f. 3. *Petiver Gaz.* t. 3. f. 12. *Hum- phreys Conch.* t. 7. f. 12.

Inhabits the Mediterranean. *Linnæus.* Jamaica. *Sloane.* West India Islands. *Chemnitz.*

Linnæus has given of *P. Pustula* the following specific charac- ter: " Shell oval, gibbous-convex, with reticulated striæ, and a crenated margin;" to this he has added a remark that it is very like *P. Fissura,* and has quoted the above-men- tioned figure of Petiver's, which is the shell now described, and also Klein Ost. t. 8. f. 3, which is *P. Græca;* so that with such contradictions the Linnæan *P. Pustula* must al- ways continue a doubtful species. The length is most com- monly about ten, and the breadth eight lines, and the shell is shaped somewhat like a finger-nail, but a Variety is less truncated, and measured ten by only seven lines; the outer surface is of a dull white, sometimes variegated with dark red in irregular blotches, and the inside is white and glossy; the perforation is placed nearer to the truncated than to the other end, and excepting a small smooth space which sur- rounds the perforation, the surface is longitudinally ribbed and slightly striated transversely. Favanne has erroneously described this shell for a Variety of *P. Scutellum.*

GRÆCA. 92. Shell ovate, cancellated, with the sum- mit somewhat lateral, and the inner margin crenated.

Patella Græca. *Linnæus Syst. Nat.* p. 1262. *Martini,* i. p. 139. t. 11. f. 98 and 99. *Pennant Zool.* iv. p. 144. t. 89. f. 153. *Born Mus.* p. 423. *Schroeter Einl.* ii. p. 437. *Gmelin,* p. 3728. *Schreibers Conch.* ii. p. 362. *Montagu Test.* p. 492. *Maton and Racket, in Lin. Trans.* viii. p. 236. *Dorset Cat.* p. 59. t. 23. f. 3. *Brookes's Introd.* p. 138. t. 9. f. 123.? *Burrow's Elements,* p. 176. t. 21. f. 8.

Patella Larva reticulata. *Da Costa Brit. Conch.* p. 14. t. 1. f. 3.

Patella reticulata. *Donovan,* i. t. 21. f. 3.

Le Gival. *Adanson Senegal,* p. 37. t. 2. f. 7.

Le Lépas a réseau, ou le Treillis. *Favanne,* i. p. 532. t. 3. f. B.

Bonanni Rec. and *Kirch.* 1. f. 6. *Lister Conch.* t. 527. f. 2. *Gualter,* t. 9. f. N. *Argenville,* t. 2. f. 1. *Humphreys Conch.* t. 7. f. 15.

Inhabits the Mediterranean. *Linnæus.* Coasts of Barbadoes and Jamaica. *Lister.* Island of Goree. *Adanson.* England. *Pennant, &c.*

Shell an inch, or sometimes two inches long, and the length, breadth, and height are about in the proportions of three, two, and one; it is elegantly cancellated with unequal longitudinal narrow ribs and transverse striæ crossing over them; the margin is not even, but somewhat arcuated, and strongly crenated on the inside; the colour is pale dull brown, or greenish or yellowish white, sometimes spotted or marked with one or two brown concentrical rings.

ATRICAPILLA. 93. Shell ovate, cancellated, with the longitudinal ribs nodulous and alternately smaller; summit somewhat lateral, and the inner margin entire.

Patella Græca, Var. *Martini,* i. p. 141. t. 12. f. 104.
Patella, No. 158. *Schroeter Einl.* ii. p. 508.
Lister Conch. t. 528. f. 5.

Inhabits the coasts of Barbadoes. *Lister.*

This shell in shape resembles *P. Græca*, with which it has been confounded by Gmelin, but the length rarely exceeds an inch, and it differs in having the larger and smaller ribs regularly alternate, and the inner margin very nearly entire; the colour is dull green or olive, with the larger ribs white, and the perforation at the summit is, I believe, always surrounded both inside and out with a black ring.

NODOSA.   94. Shell oval, sub-conical, with elevated strongly tuberculated ribs, and the summit nearly central ; margin crenated.

Patella nodosa.   *Born Mus.* p. 429.
Patella Jamaicensis.   *Gmelin*, p. 3730.
Patella spinosa.   *Gmelin*, p. 3731.
Patella, No. 153.   *Schroeter Einl.* ii. p. 506. and No. 168.
    p. 513. t. 6. f. 12.
Le Lépas ergoté.   *Favanne*, i. p. 535. t. 3. f. D.
Lister Conch. t. 528. f. 6.   *Petiver Gaz.* t. 153. f. 5.
Inhabits the coasts of Barbadoes. *Lister.* St. Domingo and Martinique. *Favanne.*
Shell about fifteen lines long, eleven broad, and eight high, and is white both inside and out ; it has about forty-eight alternately much elevated and small longitudinal ribs, crossed by four or five convex concentrical rings, which form strong pointed tubercles at the intersections ; the perforation at the summit is oblong, contracted at the middle like a key-hole, and surrounded on the inside with a milk-white callus.

PERFORATA.   95. Shell ovate, with unequal longitudinal nodulous ribs, and the margin toothed ; perforation circular.

Patella perforata.   *Gmelin*, p. 3730.
Patella Barbadensis.   *Gmelin*, p. 3729.
Patella, No. 152.   *Schroeter Einl.* p. 506.
La Perdrix.   *Favanne*, i. p. 536. t. 3. f. F.
Lister Conch. t. 528. f. 7.   *Petiver Gaz.* t. 80. f. 12.
    Martini, i. p. 135. t. 11. f. 93, 96, and 97.
Inhabits the coasts of Barbadoes. *Lister.* St. Domingo and Martinique. *Favanne.*
Shell about an inch and a half long, one inch broad, and half an inch high, of a pale grey, yellowish green, or whitish colour, with pale or dark purplish brown or reddish rays, and sometimes spotted in concentric circles ; the upper half of the inside is white, and the part towards the margin is green ; it is slightly contracted towards one end, and the summit is placed not far from the center ; the circular perforation is often bordered with a pale chestnut, reddish, or straw-coloured ring, and the callus, which surrounds it on the inside, is oval.   Martini considered his figures 93, 96, and 97, to be the same species ; and they are all quoted by Gmelin for *P. Barbadensis*, though he has afterwards also quoted the latter for *P. perforata*.

CAFFRA. **96.** Shell ovate, with the sides compressed towards one end, and striated and rayed with black ; perforation nearly central and sub-ovate.

Patella Caffra. *Gmelin,* p. 3730.
Patella, No. 154. *Schroeter Einl.* ii. p. 506.
Martini, i. p. 137. t. 11. f. 95.
Inhabits the coasts of the Cape of Good Hope. *Martini.*
Shell about an inch and a half long, seven lines high, and one inch broad in the broadest part, but the sides are somewhat compressed, and become gradually narrower towards one end; it is slightly ribbed longitudinally, and marked with minute transverse striæ ; the colour is greyish or dirty white, with numerous narrow black rays in pairs or fours, and the inside is white tinged with green ; the inner margin is slightly crenated.

PILEOLUS. **97.** Shell oblong, with longitudinal striæ, and the sides compressed ; perforation circular and sub-marginal.

Patella pileata, seu Pileolus. *Chemnitz,* xi. p. 183. t. 197. f. 1922.
Humphreys Conch. t. 7. f. 2. ?
Inhabits ———
This shell appears by Chemnitz's figure to be about an inch long, and half as high, and of a brownish yellow colour, with a few darker transverse stripes ; the summit is much nearer to one end than to the other, and has a circular perforation on the posterior side.

SCUTELLUM. **98.** Shell sub-oval, with longitudinal striæ, and transverse elevated belts ; sides compressed near the middle, and the perforation large.

Patella Scutellum. *Gmelin,* p. 3731. *Schreibers Conch.* i. p. 363.
Patella nimbosa, Var. *Martini,* i. p. 134.
Patella, No. 166. *Schroeter Einl.* ii. p. 512. t. 6. f. 11.
Le Comprimé à trou de serrure. *Favanne,* i. p. 527. t. 3. f. A 1.
Petiver Gaz. t. 3. f. 11. *Meuschen Naturf.* xviii. p. 11. t. 2. f. 2 and 3. ?
Inhabits the coast of the Falkland Islands. *Favanne.*
Shell about an inch and a half long, eleven lines broad, and

the height is about half the breadth ; the colour is yellowish white, or sometimes pale lilac, irregularly radiated with violet, grey, or reddish stripes, or sometimes almost wholly grey or ash-coloured, and the inside is white ; the sides are strikingly compressed in the middle, and one end is rather broader than the other ; the perforation is oblong-oval, about four lines long, and half as broad.

PICTA. 99. Shell ovate, with concentrical elevated belts, and alternately white and violet longitudinal rays ; perforation oval.

Patella picta. *Gmelin*, p. 3729.
Patella, No. 151. *Schroeter Einl.* ii. p. 505.
Fissurella radiata. *Lamarck Syst. des Anim.* p. 69.
Le grand Lépas ovale à trou de serrure. *Favanne*, ii. p. 530. t. 3. f. A 4.
Argenville, t. 2. f. E. *D'Avila*, t. 3. f. C. *Martini*, i. p. 131. t. 11. f. 90.
Inhabits the Straights of Magellan. *D'Avila.* Coasts of the Falkland Islands. *Favanne.*
Shell about three inches and a quarter long, two inches and a quarter broad, and rather more than one inch high ; Gmelin has described the perforation to be round, but it is oval, though not contracted in the middle like that of *P. nimbosa*, to which this species is very nearly allied, and from which Favanne says it differs in nothing but what may be attributed to the effects of age or climate.

NIMBOSA. 100. Shell sub-oval, with longitudinal distant slightly elevated ribs, and transverse striæ ; perforation oblong and contracted in the middle.

Patella nimbosa. *Linnæus Syst. Nat.* p. 1262. *Martini*, i. p. 134. t. 11. f. 91 and 92. *Born Mus.* p. 429. *Schroeter Einl.* ii. p. 439. *Gmelin*, p. 3729. *Schreibers Conch.* i. p. 362.
Le Dasan. *Adanson Senegal*, p. 35. t. 2. f. 6.
Le petit Lépas ovale à trou de serrure. *Favanne*, i. p. 528. t. 3. f. A 3.
Le Lépas ovale tuilé à trou de serrure. *Favanne*, i. p. 528. t. 3. f. A 2.
Gualter, t. 9. f. R and S.
Inhabits the coasts of Southern Europe and America. *Lin-*

næus. Raclia in the Archipelago. *Tournefort.* Goree. *Adanson.* St. Domingo and Martinique. *Favanne.*

Shell varying in length from one inch to two inches and a half, and the breadth is about two-thirds of the length; the co.lour is white or greyish white, with broadish brown or violet rays; the longitudinal ribs are often alternately larger, but not regularly so, and they are generally more or less furnished with small vaulted scales; the inside is white with a border, or otherwise tinged with green, and the margin is slightly crenated.

NUBECULA. 101. Shell ovate, with alternate rose-coloured and white rays, and very slightly striated; perforation oblong.

Patella Nubecula. *Linnæus Syst. Nat.* p. 1262. *Gmelin,* p. 3729.
Patella rosea. *Gmelin,* p. 3730.
Patella, No. 156. *Schroeter Einl.* ii. p. 507.
Lister Conch. t. 529. f. 22.? *Martini,* i. p. 141. t. 12. f. 105.
Inhabits the Mediterranean. *Linnæus.* Coasts of Jamaica. *Martini.*

Shell about three-quarters of an inch long, and half an inch broad, and is elegantly variegated with red and white rays extending from the summit to the margin; the inside is white more or less tinged with green, and has a brownish or rose-coloured ring surrounding the perforation; Linnæus has described the perforation of *P. Nubecula* to be ovate, whereas that of Gmelin's *P. rosea* is said to be oval, and in my specimens, which otherwise answer both these descriptions, it is oblong. Born has erroneously quoted Martini's figure 105, together with 91 and 92, for *P. nimbosa.*

PORPHYROZONIAS. 102. Shell oblong, with unequal longitudinal sub-nodulous ribs, and the perforation small and circular.

Patella Porphyrozonias. *Gmelin,* p. 3730.
Patella, No. 155. *Schroeter Einl.* ii. p. 507.
Martini, i. p. 140. t. 12. f. 102 and 103.
Inhabits the coasts of North America. *Martini.*

This shell is of about the same size as *P. Nubecula,* but has the summit more elevated, and is irregularly ribbed longitudinally; the colour is dirty white, variegated with purplish or reddish rays, which are sometimes interrupted by four or five narrow transverse stripes of the same colour ; the inside is greenish white, and has a brown or red line surrounding the perforation, which is small and circular.

MACROSCHISMA. 103. Shell ovate-oblong, slightly striated, and the sides somewhat compressed; perforation very long, and widened at the posterior end.

Patella Macroschisma. *Portland Cat.* p. 71, lot 1601. *Callone's Cat.* p. 4, No. 62. *Chemnitz,* xi. p. 184. t. 197. f. 1923 and 1924.

Humphreys Conch. t. 7. f. 3.

Inhabits the coasts of Japan. *Chemnitz.*

This scarce and valuable shell is about thirteen lines long, and six broad, with the outside of a reddish brown colour, and whitish on the inside ; the perforation is half an inch long, extending from the summit nearly to the posterior margin, where its breadth becomes rather abruptly doubled.

𝕲𝖊𝖓𝖚𝖘 XXXIII.

DENTALIUM :†

SHELL UNIVALVE, TUBULAR, NEARLY STRAIGHT, WITHOUT ANY INTERNAL PARTITION, AND OPEN AT BOTH ENDS.

RECTUM. 1. Shell nearly straight, longitudinally ribbed, and slightly striated transversely; and somewhat annulated.

Dentalium rectum. *Gmelin*, p. 3738.
Dentalium, No. 2. *Schroeter Einl.* ii. p. 527.
Le grand Dentale à stries longitudinales. *Favanne*, i. p. 638. t. 5. f. E 6.
Petiver Gaz. t. 95. f. 11. *Gualter*, t. 10. f. H. *Martini*, i. p. 30. t. 1. f. 4 *A*.
Inhabits the East Indian Seas. *Favanne.*
This shell is of about the same size as *D. elephantinum*, and principally differs from it in being nearly straight, and in having the ribs smaller, and the growth marked by irregular elevated rings; it is green with the annulations of a darker colour, and is rather a doubtful species.

ELEPHANTINUM. 2. Shell slightly curved, with about ten longitudinal ribs, and intermediate striæ.

Dentalium elephantinum. *Linnæus Syst. Nat.* p. 1263. *Martini*, i. p. 33. t. 1. f. 5 *A*. *Born Mus.* p. 431, and

---

† *D. nebulosum* of Gmelin, p. 3738, appears to me to be too uncertain a species to be retained; and the following are fossils, *D. annulatum*, p. 3738, *D. fossile*, p. 3738, *D. interruptum*, p. 3739, *D. Radula*, p. 3738, *D. sexangulum*, p. 3739, and *D. vitreum*, p. 3739.

2 I 2

Vign. at p. 430. *Schroeter Einl.* ii. p. 520. *Gmelin,* p. 3736. *Lamarck Syst. des Anim.* p. 526. *Shaw's Nat. Misc.* vii. t. 226. *Burrow's Elements,* p. 177. t. 22. f. 1.

Le grand Dentale à cannelures. *Favanne,* i. p. 636. t. 5. f. E 5.

Lister Conch. t. 547, upper fig. 1. *Rumphius,* t. 41. f. I. *Petiver Amb.* t. 16. f. 33. *Gualter,* t. 10. f. I. *Argenville,* t. 3. f. H. *Knorr,* i. t. 29. f. 3.

Variety. More curved, and the summit less truncated.

Dentalium arcuatum. *Gmelin,* p. 3738.

Dentalium, No. 6. *Schroeter Einl.* ii. p. 529.

Gualter, t. 10. f. G.

Inhabits the Indian and European Seas. *Linnæus.* Coasts of Amboyna. *Rumphius.* Sicily. *Martini.* Isle of France. *Favanne.*

Shell from two and a half to four inches long, and the diameter at the broad end is four or five lines ; it is strongly marked longitudinally with from eight to twelve ribs and a few intermediate striæ; the colour is green becoming paler towards the summit and is sometimes stained transversely with darker bands. Gmelin for a Variety has cited Argenville, t. 3. f. I, which is very like a fragment of this species with the upper end accidentally broken off. Gualter, t. 10. f. G, from which *D. arcuatum* has been constituted, differs from the common appearances of *D. elephantinum* only in being rather more curved and less truncated at the summit.

APRINUM. 3. Shell slightly curved, with ten longitudinal ribs, and the interstices smooth.

Dentalium aprinum. *Linnæus Syst. Nat.* p. 1263. *Martini,* i. p. 31. t. 1. f. 4 B. *Schroeter Einl.* ii. p. 521. *Gmelin,* p. 3736.

Bonanni Rec. and *Kirch.* 1. f. 8.

Inhabits the Indian Seas. *Linnæus.*

This shell is nearly allied to *D. elephantinum,* but differs in being white and strongly ribbed without any intermediate striæ.

STRIATUM. 4. Shell pointed at the summit, with eight longitudinal ribs, and the interstices striated.

Dentalium striatum. *Born Mus.* p. 431. *Mus. Leskeanum,* p. 306.

Dentalium striatulum. *Gmelin*, p. 3738. *Maton and Racket, in Lin. Trans.* viii. p. 238. *Montagu Supp.* p. 155. *Brookes's Introd.* p. 140. t. 9. f. 129.

Dentalium octangulatum. *Donovan*, v. t. 162.

Dentalium, No. 3. *Schroeter Einl.* ii. p. 528.

Lister Conch. t. 547, lower fig. *Martini*, i. p. 33. t. 1. f. 5 B.

Inhabits the coasts of Sicily. *Martini.* Cornwall. *Donovan.* Shell about two inches long, and appears, from Mr. Donovan's description, to differ principally from *D. aprinum* in having only eight ribs, and the interstices striated; much reliance cannot, however, be placed on the number of the ribs, for *D. elephantinum* is often found with only eight, and is so represented in most of the figures to which Linnæus has referred; if *D. striatum* and *D. octangulatum* are the same, of which there can be very little doubt, the colour of this species is sometimes green and sometimes white, or reddish brown, if the colouring of Mr. Brookes's figure is correct.

DENTALIS. 5. Shell slightly curved, with about twenty longitudinal striæ, and the summit pointed.

Dentalium Dentalis. *Linnæus Syst. Nat.* p. 1263. *Born Mus.* p. 432. t. 18. f. 13. *Schroeter Einl.* ii. p. 522. *Gmelin*, p. 3736. *Maton and Racket, in Lin. Trans.* viii. p. 237.

Dentalium striatum. *Montagu Test.* p. 495.

L'Epingle courbe. *Favanne*, i. p. 634. t. 5. f. E 4.

Rumphius, t. 41. f. 6.

Inhabits the Mediterranean. *Linnæus.* Coasts of Amboyna, *Rumphius.* West of England. *Montagu.*

Shell twelve or fourteen lines long, and at the base only one line in diameter; it is sharp pointed, and closely and regularly striated longitudinally throughout, and sometimes a few faint annulations are observable towards the base; foreign specimens are reddish, but those which Mr. Montagu found on the coasts of Cornwall and Devon were white.

ENTALIS. 6. Shell slightly curved, taper, continuous, and smooth.

Dentalium Entalis. *Linnæus Syst. Nat.* p. 1263. *Martini*, i. p. 26. t. 1. f. 1 and 2. *Pennant Zool.* iv. p. 145. t. 90. f. 154. *Born Mus.* p. 432. *Schroeter Einl.* ii. p. 523. *Gmelin*, p. 3736. *Donovan*, ii. p. 48. *Mont-*

agu Test. p. 494.   *Maton and Racket, in Lin. Trans.*
viii. p. 237.   *Dorset Cat.* p. 59. t. 22. f. 10.
Dentale vulgare.   *Da Costa Brit. Conch.* p. 24. t. 2. f. 10.
Le petit Dentale.   *Favanne,* i. p. 633. t. 5. f. E 1.
Bonanni Rec. and *Kirch.* 1. f. 9.   *Lister Conch.* t. 547.
    f. 2, and t. 1056. f. 4.   *Petiver Gaz.* t. 65. f. 9.   *Gual-*
    ter, t. 10. f. E.   *Argenville,* t. 3. f. K.   *Knorr,* i. t. 29.
    f. 4.
Inhabits the Indian and European Seas. *Linnæus.* Coasts of
    Britain. *Lister.*   Persia. *Martini.*   Norway, Spain, and
    France. *Favanne.*
Shell about an inch and a half long, and two lines in diameter
    at the broader end, smooth, glossy, pervious, and tapering to
    a small point; Gmelin's *D. arietinum,* which is said to dif-
    fer in being only one-eighth part as large and more curved,
    is most probably the young of this species.

CORNEUM.   7. Shell slightly curved, taper, inter-
     rupted, and opake.

Dentalium corneum. *Linnæus Syst. Nat.* p. 1263.   *Schroe-*
    ter Einl. ii. p. 523. t. 6. f. 16.   *Gmelin,* p. 3737.
Inhabits the African Ocean. *Linnæus.*
Linnæus has not given any reference, and to the above specific
    character only adds that this species is very like *D. Entalis,*
    but is of a dark horn-colour, and often interrupted; the shell
    figured by Schroeter is an inch and a quarter long, and differs
    from *D. Entalis* in being cylindrical, and in having a round-
    ed obtuse summit.

POLITUM.   8. Shell slightly curved, taper, and con-
     tinuous, with much crowded annular striæ.

Dentalium politum.   *Linnæus Syst. Nat.* p. 1264.   *Mar-*
    tini, i. p. 29. t. 1. f. 3 *A.*   *Born Mus.* p. 433.   *Schroe-*
    ter Einl. ii. p. 524.   *Gmelin,* p. 3737.
Le Dentale à sillons circulaires.   *Favanne,* i. p. 635.
Rumphius, t. 41. f. 5.   *Gualter,* t. 10. f. F.
Inhabits India. *Linnæus.*   Coasts of Sicily. *Martini.*
This shell is of the same size, and has nearly the appearance of
    D. Entalis, from which it is distinguished by its crowded
    transverse annular striæ; it is generally milk-white, but some-
    times of a pale rose-colour, and the whole surface has a po-
    lished appearance.

EBURNEUM.   9. Shell slightly curved, taper, and
     continuous, with remote annular striæ.

Dentalium eburneum. *Linnæus Syst. Nat.* p. 1264. *Gmelin,* p. 3737.

Inhabits India. *Linnæus.*

Linnæus has not given any reference, and I cannot find that any subsequent naturalist has ascertained this species ; it is described to be as white as ivory, extremely smooth and glossy, and to be very like *D. Entalis,* but to differ in having numerous convex annular striæ at equal distances apart.

FASCIATUM. 10. Shell slightly curved, very finely striated longitudinally, and pale grey with darker transverse bands.

Dentalium fasciatum. *Gmelin,* p. 3737.
Dentalium, No. 1. *Schroeter Einl.* ii. p. 526.
Le petit Dentale à stries longitudinales. *Favanne,* i. p. 635. t. 5. f. E 2.
Martini, i. p. 29. t. 1. f. 3 *B.*
Inhabits the coasts of Sicily. *Martini.*

Shell about an inch or an inch and a quarter long, and two lines in diameter at the broader end ; the longitudinal striæ sometimes terminate about the middle, leaving the upper half smooth, and Favanne says they rarely extend through the whole length of the shell ; the colour is whitish or pale grey, tinged with orange towards the summit, and marked transversely with dusky bands.

GADUS. 11. Shell slightly curved, very smooth, with a pointed summit, and the base contracted.

Dentalium Gadus. *Montagu Test.* p. 496. t. 14. f. 7. *Maton and Racket, in Lin. Trans.* viii. p. 238.
Inhabits many parts of the British Channel. *Montagu.*

Shell about three-eighths of an inch long, and in the broadest part, which is towards the middle, about one-sixteenth of an inch in diameter ; it tapers to a small point at the summit, and is also slightly contracted towards the larger end ; it is white, glossy, and perfectly smooth, and Mr. Montagu says it is known to mariners by the name of *Hake's Tooth.*

IMPERFORATUM. 12. Shell slightly curved, cylindrical, minute, transversely striated, and the summit truncated and imperforate.

Dentalium imperforatum. *Walker's Minute Shells,* f. 15.

Adams's Micro. p. 635. t. 14. f. 8.   *Montagu Test.* p. 496.   *Maton and Racket, in Lin. Trans.* viii. p. 238.

Inhabits the sea near Sandwich. *Boys.*   Among the sand in Falmouth harbour. *Montagu.*

Shell about one-eighth of an inch long, and one-third as broad, of a greyish white colour, and striated transversely; aperture round, a little contracted at the margin, and Montagu says the opposite end is closed, truncated, and furnished with a small round protuberance.

TRACHEA.   13. Shell curved, sub-cylindrical, and minute, with crowded annular striæ; summit truncated and imperforate.

Dentalium Trachea.   *Montagu Test.* p. 497. t. 14. f. 10. *Maton and Racket, in Lin. Trans.* viii. p. 239.

Found in sand from Milton in Devonshire, but is extremely rare. *Montagu.*

Shell rather more than one-eighth of an inch long, and the diameter is about one-fifth of the length; Montagu says it differs from *D. imperforatum* " by being more arcuated, and a little tapering; is longer in proportion to its breadth, the margin of the aperture even, and not contracted, and the annulations stronger, giving it the appearance of the windpipe or Trachea of an animal;" the colour is ferruginous brown, becoming paler towards the summit.

MINUTUM.   14. Shell slightly curved, cylindrical, minute, and smooth.

Dentalium minutum. *Linnæus Syst. Nat.* p. 1264. *Schroeter Einl.* ii. p. 526.   *Gmelin,* p. 3737.

Dentalium glabrum.   *Montagu Test.* p. 497.   *Maton and Racket in Lin. Trans.* viii. p. 239.

Inhabits the Mediterranean. *Linnæus.*   North coast of Devonshire. *Montagu.*

In addition to a very short specific character, Linnæus only says that this species is so very minute as not to be discernible with the naked eye, which does not at all accord with his reference to Plancus's t. 2. f. 2, and Gmelin has conjectured that Linnæus by mistake described the spire of some minute Echinus: it is, however, by far more probable that *D. glabrum* is the species intended, and of this Mr. Montagu has given the following description : " Shell cylindrical, arcuated, smooth, glossy, white, devoid of either striæ or wrinkles, and equal in size throughout; aperture orbicular; the other end closed, rounded, and sub-marginated. Length scarce one line, and the diameter one-fifth of the length."

PELLUCIDUM. 15. Shell taper, nearly straight, horny, flexible and smooth.

Dentalium pellucidum. *Gmelin,* p. 3738.
Dentalium, No. 9. *Schroeter Einl.* ii. p. 529. t. 6. f. 17.
Inhabits the North Sea. *Schroeter.*
Shell two inches and a quarter long, very narrow and thin, and of a pale honey-colour; Schroeter says it will not effervesce with acids, and it may therefore be doubted whether it belongs properly to the Testacea.

𝕲𝖊𝖓𝖚𝖘 XXXIV.

—

SERPULA:†

SHELL UNIVALVE, TUBULAR, ADHERING, AND SOME-
TIMES DIVIDED BY IMPERFORATED DISSEPIMENTS
AT UNEQUAL DISTANCES.

———

STELLARIS.   1. Shell sub-orbicular, umbilicated, convex, with radiated wrinkles.

Serpula stellaris.   *Fabricius Fauna Groenl.* p. 383. No.
380.   *Gmelin*, p. 3747.

Inhabits the shores of Greenland, on Sertulariæ, and sometimes
on stones and shells. *Fabricius.*

Shell very small, not larger than a small pin's head, of a reddish brown, yellow, or violet-colour, and rayed with white.
It has one whirl or bend, is flat beneath, and has an extremely minute aperture.

SEMINULUM.   2. Shell regular, oval, detached, and
glabrous.

Serpula Seminulum.   *Linnæus Syst. Nat.* p. 1264.   *Martini,* i. p. 61. t. S. f. 22, a and b.   *Schroeter Einl.* ii.
p. 535.   *Fabricius Fauna Groenl.* p. 376.   *Gmelin*, p.
3739.   *Maton and Racket, in Lin. Trans.* viii. p. 245.
Dorset Cat. p. 60. t. 19. f. 31.

Serpula ovalis.   *Adams in Lin. Trans.* v. t. 1. f. 28 to 30.

———

† I have omitted the following of Gmelin's species, which appear to me to be
very doubtful and obscure.   *S. cinerea*, p. 3747, *S. Infundibulum*, p. 3745, *S.
Norwegica*, p. 3746, and *S. pyramidalis*, p. 3746.———*S. Melitensis*, p. 3746, is a
fossil.   The *S. filograna* of Linnæus and the *Tubipora ramosa* of Gmelin are the
same ; and *S. intestinalis* of Gmelin, p. 3745, appears also to be a Zoophyte.

Vermiculum intortum. *Montagu Test.* p. 520.
Plancus, t. 2. f. 1. *Gualter,* t. 10. f. S.
Inhabits sandy shores in all parts of Europe. Red Sea. *Gmelin.*
Shell one-tenth of an inch in diameter, white, opake, glabrous, and variable in its formation; aperture compressed and semi-lunar. Messrs. Maton and Racket express their doubts whether the three species, described by Mr. Montagu under the names of *S. subrotunda, S. oblonga,* and *S. ovalis,* are more than Varieties of this species; and there are also several bottle-shaped species, which, on account of their extreme minuteness and doubtful place in the system, I have not thought it necessary at present to notice.

INCURVATA. 3. Shell regular, detached, with three close involutions at the smaller end.

Serpula incurvata. *Adams's Microscope,* p. 634. t. 14. f. 7. *Maton and Racket, in Lin. Trans.* viii. p. 246.
Vermiculum incurvatum. *Montagu Test.* p. 518.
Walker's Minute Shells. f. 11.
Variety. Vermiculum pervium. *Montagu Test.* p. 518.
Walker's Minute Shells, f. 12.
Inhabits the sandy shore at Sandwich. *Boys.* And the Variety has been found on the Devonshire coast by *Mr. Montagu.*
Shell scarcely one line in diameter, white, and semi-transparent. The Variety is rather smaller, and makes only one turn at the upper end, which is open or pervious.

PLANORBIS. 4. Shell regular, orbicular, flat, and equal.

Serpula planorbis. *Linnæus Syst. Nat.* p. 1264. *Gmelin,* p. 3740.
Inhabits the sea on shells. *Linnæus.*
No author has noticed this species besides Linnæus, who says it is very minute, resembling a round scale firmly fixed to other shells, without any external appearance of whirls, but when broken horizontally exhibiting the appearance of a spire in minute concentric circles.

CEREOLUS. 5. Shell taper, smooth, and sub-orbicular, with many involutions.

Serpula Cereolus. *Gmelin,* p. 3745.
Serpula, No. 12. *Schroeter Einl.* ii. p. 560.
Le Pain de Bougie. *Favanne,* i. p. 665. t. 6. f. D.

D'Avila, t. 4. f. F. *Martini*, i. p. 58. t. 3. f. 20 E.
Inhabits the coasts of America, St. Domingo, Martinique, and
St. Helena. *Favanne.*
The numerous volutions are often turned laterally inwards, like
the whirls of *Helix Vortex*, and form a flat disk an inch or
more in diameter, but they sometimes rise spirally with less
regularity upon each other. The tube scarcely exceeds a
line in diameter, and the colour on the outside is dull white,
but within it is glossy.

SPIRILLUM. 6. Shell regular, spiral, orbicular, and
pellucid; whirls taper.

Serpula Spirillum. *Linnæus Syst. Nat.* p. 1264. *Martini,*
i. p. 57. t. 3. f. 20, C and D. *Schroeter Einl.* ii. p. 537.
Fabricius Fauna Groenl. p. 376. *Gmelin,* p. 3740.
Montagu, p. 499. *Maton and Racket, in Lin. Trans.*
viii. p. 240. *Dorset Cat.* p. 59. t. 19. f. 27.*
Plancus, t. 1. f. 8. *Ginanni Adr.* ii. t. 1. f. 7.
Inhabits Sertulariæ and other Zoophites, on most of the European shores.
Shell white, glossy, slightly wrinkled, with four or five rounded
and longitudinally striated volutions, which together are not
more than one line in diameter, and by this it may be known
from *S. spirorbis*, which has only two volutions. Mr.
Montagu's *S. sinistrorsa* appears to be nothing more than
a reversed Variety of this species; and Gmelin's *S. porrecta*
appears also to be only a Variety, for the principal, if not the
only difference, consists in its not having any wrinkles, and in
being of a more snowy white.

MINUTA. 7. Shell regular, spiral, and orbicular,
with the whirls reversed, taper, and transversely wrinkled.

Serpula minuta. *Montagu Test.* p. 505. *Maton and Racket, in Lin. Trans.* viii. p. 241.
Inhabits the sea on Corallina officinalis; frequent on the English coasts. *Montagu.*
Shell white, opake, with two or three volutions, and not half
the size of the reversed Variety of *S. Spirillum*, from which
it also differs in being transversely wrinkled, and not longitudinally striated. Mr. Montagu says it has sometimes a slight
longitudinal furrow on each side, forming a ridge or carina
along the back.

SPIRORBIS. 8. Shell regular, spiral, and orbicular; whirls slightly grooved above and within, and gradually tapering to a point.

Serpula spirorbis. *Linnæus Syst. Nat.* p. 1265. *Martini,* i. p. 59. t. 3. f. 21, *A* and *B.* *Pennant Zool.* iv. p. 145. t. 91. f. 155. *Da Costa Brit. Conch.* p. 22. t. 2. f. 11. *Schroeter Einl.* ii. p. 538. *Fabricius Fauna Groenl.* p. 377. *Muller Zool. Dan.* iii. p. 8. t. 86. f. 1 to 6. *Gmelin,* p. 3740. *Donovan,* i. t. 9. *Montagu Test.* p. 498. *Maton and Racket, in Lin. Trans.* viii. p. 241. *Dorset Cat.* p. 59. t. 22. f. 11. *Brookes's Introd.* p. 142. t. 9. f. 134.

Spirorbis nautiloides. *Lamarck Syst. des Anim.* p. 326. *Lister Conch.* t. 553. f. 5. *Petiver Gaz.* t. 35. f. 8. *Gualter,* t. 10. f. O. *Ginanni Adr.* ii. t. 1. f. 8.

Variety. With the whirls reversed.

Chemnitz, ix. part 1. p. 151. t. 116. f. 999.

Inhabits the sea on Fuci and Zoophytes, on all the coasts of Europe.

Shell about a line in diameter, round, white, smooth, and opake, consisting of two whirls gradually and regularly coiling to a point, like some of the depressed Helices.

TRIQUETRA. 9. Shell creeping, flexuose, and three-sided.

Serpula triquetra. *Linnæus Syst. Nat.* p. 1265. *Martini,* i. p. 68. t. 3. f. 25. *Pennant Zool.* iv. p. 146. t. 91. f. 157. *Da Costa Brit. Conch.* p. 20. t. 2. f. 9. *Born Mus.* p. 436. t. 18. f. 14. *Schroeter Einl.* ii. p. 540. *Gmelin,* p. 3740. *Dorset Cat.* p. 59. t. 22. f. 9. *Donovan,* iii. t. 95. *Montagu Test.* p. 511. *Maton and Racket, in Lin. Trans.* viii. p. 244.

Gualter, t. 10. f. P. *Ellis's Corallines,* t. 38. f. 2.

Common on the shores of Europe and America, on shells, stones, wood, and Algæ.

Shell strong, opake, wrinkled, variously twisted, and sometimes nearly straight; it has a dorsal elevated ridge, and the base spreads so as to give it a triangular appearance. Mr. Montagu has remarked, that the funnel-shaped proboscis of the inhabiting Terebella is single, and therein differs from that of *S. vermicularis,* which is double.

INTRICATA. 10. Shell filiform, rough, taper, and flexuose.

Serpula intricata. *Linnæus Syst. Nat.* p. 1265. *Pennant Zool.* iv. p. 146.

Inhabits the shores of the Mediterranean, on Pinnæ and the rubbish thrown up by the sea. *Linnæus.* Common on the coasts of Britain.

In the opinion of Da Costa and Montagu, this is not distinct from *S. vermicularis,* and it differs chiefly in being more grouped, and much more slender.

CARINATA. 11. Shell regular, spiral, with the outer whirl rising into a carinated ridge on the top.

Serpula carinata. *Montagu Test.* p. 502. *Maton and Racket, in Lin. Trans.* viii. p. 242.

Inhabits the sea on other shells in Salcombe Bay, Devonshire. *Mr. Montagu.*

Shell about half the size of *S. spirorbis,* and a little spreading at the base; besides the carinated ridge on the outer whirl, it also differs from *S. spirorbis* in having the middle whirl rather concave, and sometimes pervious.

GRANULATA. 12. Shell taper, spiral, and glomerated, with three elevated ribs on the upper side.

Serpula granulata. *Linnæus Syst. Nat.* p. 1266. *Fabricius Fauna Groenl.* p. 380. *Gmelin,* p. 3741. *Donovan,* iii. t. 100. *Montagu Test.* p. 500. *Maton and Racket, in Lin. Trans.* viii. p. 241.

Serpula sulcata. *Adams in Lin. Trans.* iii. p. 254.

Inhabits the Northern Ocean, on stones and shells; and is not uncommon on the shores of Britain.

Shell opake, white, with two volutions, strongly ribbed longitudinally and transversely wrinkled, especially in the furrows; aperture round; it is of the same size as *S. spirorbis.*

CANCELLATA. 13. Shell spiral, glomerated, with three grooves, the lower of which is interrupted by transverse ribs; aperture bidentated.

Serpula cancellata. *Fabricius Fauna Groenl.* p. 383. *Gmelin,* p. 3746.

Inhabits the shores of Greenland; common with Serpula granulata. *Fabricius.*

Shell white, greyish, or greenish, and nearly allied to *S. granu-*

lata, from which it may readily be distinguished by the bidentated aperture.

HETEROSTROPHA. 14. Shell taper, spiral, with three elevated ribs on the upper side, and the whirls reversed.

Serpula heterostropha. *Montagu Test.* p. 503. *Maton and Racket, in Lin. Trans.* viii. p. 242.

Inhabits the sea on oysters and other shells, and marine Algæ, in Kingsbridge Bay, Devonshire. *Montagu.*

Shell thick, brownish or dirty white, and not half the size of *S. granulata,* from which it also differs in having the whirls reversed.

CORRUGATA. 15. Shell regular, spiral, transversely wrinkled, and umbilicated.

Serpula corrugata. *Montagu Test.* p. 502. *Maton and Racket, in Lin. Trans.* viii. p. 242.

Inhabits the sea on slate rocks, at Milton, Devonshire. *Montagu.*

Mr. Montagu says that this is a much stronger shell than *S. Spirillum,* and never exposes so much of the second volution to view, and that, though more wrinkled, it possesses a superior gloss when cleared from extraneous matter.

LUCIDA. 16. Shell taper, spiral, very smooth and glossy; whirls reversed.

Serpula lucida. *Montagu Test.* p. 507. *Maton and Racket, in Lin. Trans.* viii. p. 243.

Serpula reflexa. *Adams in Lin. Trans.* v. p. 4. t. 1. f. 31 and 32.

Inhabits the sea, on Sertularia abietina very frequent, and sometimes on Sertularia argentea. *Montagu.*

Shell half a line in diameter, and distinguishable from its congeners by its extremely smooth, glossy, and vitreous appearance; whirls two or three, irregular, sometimes lateral, and sometimes turned over each other, with the aperture projecting upwards, and not unfrequently unconnected. *S. cornea* of Adams appears to be a very doubtful species, and not worth notice.

VITREA. 17. Shell taper, regularly spiral, orbicular, and wrinkled; aperture thickened.

Serpula vitrea. *Fabricius Fauna Groenl.* p. 382. *Gmelin,* p. 3746.

Inhabits the shores of Greenland on Sertulariæ, shells, stones, and Fuci. *Fabricius.*

Shell not a line in diameter, thick, white, or sometimes reddish, glossy, and pellucid. Fabricius says it is allied to *S. glomerata;* and Messrs. Montagu, Maton, and Racket have supposed it to be the same as *S. lucida,* but in that species the edge of the aperture is not thickened, and the whirls are reversed.

CONTORTUPLICATA. 18. Shell semi-round, wrinkled, glomerated, and carinated.

Serpula contortuplicata. *Linnæus Syst. Nat.* p. 1266. *Martini,* i. p. 67. t. 3. f. 24 *A. Born Mus.* p. 437. *Schroeter Einl.* ii. p. 545. *Gmelin,* p. 3741. *Lamarck Syst. des Anim.* p. 326.

Le Boyau de Mer. *Favanne,* i. p. 650. t. 6. f. E 1.

Bonanni Rec. and *Kirch.* i. t. 20. f. G. *Argenville,* t. 4. f. D.

Inhabits the European Ocean. *Linnæus.* Mediterranean. *D'Avila.* Coasts of Provence. *Favanne.*

It appears to me that this species was formed by Linnæus from those shells of *S. triquetra* which are not spread out at the base so as to give them a triangular appearance, and, as is frequently the case, they then answer tolerably well to the most striking part of his description, 'semitereti carinata.' Da Costa, Montagu, and Racket have also considered *S. contortuplicata* and *S. triquetra* to be the same; but the figures referred to by Martini, Born, Schroeter, Favanne, and Gmelin, appear to be different, and probably may belong to a separate species; though never having seen any shell which answers to them, I have not ventured to make the alteration.

NEBULOSA. 19. Shell thick, wrinkled, and much twisted; aperture large and indented.

Les Intestins. *Favanne,* i. p. 652. t. 6. f. I.

D'Avila, t. 4. f. H.

Inhabits the American seas. *Favanne.*

Shell thick, extremely wrinkled, white clouded with dark fawn-colour, and much twisted and plaited together, so as generally to conceal the lower extremity, which terminates in a point; the diameter of the largest tube is said by Favanne not to exceed three or four lines, and it expands at the aperture, which widens at the edge. From the appearance of the figures, it might be supposed that this is a Variety of the species which Gmelin has considered to be the Linnæan *S. contortuplicata*; but from Favanne's description it appears to be perfectly distinct.

GOREENSIS. 20. Shell taper, twisted, and cancellated with minute elevated striæ.

Serpula Goreensis. *Gmelin*, p. 3745.
Serpula, No. 27. *Schroeter Einleitung*, ii. p. 566.
Les Boyaux de Mer d'Afrique. *Favanne*, i. p. 651. t. 6. f. E 2.
Le Dofan. *Adanson Senegal*, p. 164. t. 11. f. 3.
Inhabits the shore of the Island of Goree, on shells and pieces of timber. *Adanson.*
Tube eight or nine inches long, and three or four lines broad, irregularly twisted, and forming masses a foot in diameter; the surface is cancellated with minute elevated striæ, and is white on the outside, and horn-coloured within.

GLOMERATA. 21. Shell taper, glomerated, and marked with decussated wrinkles.

Serpula glomerata. *Linnæus Syst. Nat.* p. 1266. *Martini,* i. p. 65. t. 3. f. 23. *Born Mus.* p. 437. *Schroeter Einleitung,* ii. p. 546. *Gmelin*, p. 3742. *Brookes's Introd.* p. 142. t. 9. f. 133.
Le Gateau de Vermisseaux. *Favanne*, i. p. 654. t. 6. f. Q.
Bonanni Rec. and *Kirch.* 3. f. 20 E. *Gualter*, t. 10. f. T. *Argenville*, t. 4. f. G.
Inhabits the European Seas. *Linnæus.* Mediterranean. *Martini.* Coasts of Provence. *Favanne.*
Shell an inch or two long, and about two lines broad, but so interwoven that it is impossible to separate them. Gualter mentions a mass nine inches in diameter, which weighed twenty-three pounds, and Favanne says that the diameter often exceeds a foot, but that the height is not more than two or three inches. The surface of the shell is longitudinally grooved, and transversely striated, shining, particularly on the inside, and of a white, brown, or reddish colour.

LUMBRICALIS. 22. Shell taper, flexuous, spiral and acute at one end.

Serpula lumbricalis. *Linnæus Syst. Nat.* p. 1266. *Martini.* i. p. 48. t. 2. f. 12 B. *Born Mus.* p. 438. *Schroeter Einl.* ii. p. 547. *Gmelin*, p. 3742 *Brookes's Introd.*

p. 142. t. 9. f. 132. *Burrow's Elements*, p. 177. t. 22. f. 2.
Vermicularia lumbricalis. *Lamarck Syst. des Anim.* p. 97.
Le Vilebrequin. *Favanne*, i. p. 659. t. 5. f. G.
Bonanni Rec. 1. f. 20 D. *Lister Conch.* t. 548. f. 1. *Rumphius*, t. 41. f. No. 1. *Gualter*, t. 10. f. Q and V. *Petiver*, t. 22. f. 1. *Argenville*, t. 4. f. I. *Knorr*, ii. t. 13.
f. 1. *D'Avila*, t. 4. f. G.
Inhabits the Indian Ocean. *Linnæus.* Coasts of Amboyna.
Rumphius. Moluccas. *Favanne.*
The length varies from two to four inches, and the diameter
from three to five lines; the colour is brown or yellowish,
becoming darker towards the summit, and nearly white in the
inside; the tube is more or less curved throughout its whole
length, and terminates at one end in a rather compact pointed
coil, or spire, of from six to ten whirls. Linnæus describes
the spiral end to be the summit of the shell, of which the
correctness is rather doubtful, and it appears far more likely
that the spiral end is the part by which it adheres to the rock
or other substance to which it is naturally attached; if this
is the case, *Le Vermet* of Adanson, t. 11. f. 1, Martini,
t. 3. f. 24 B, and Favanne, t. 6. f. H, and t. 68. f. G, may
probably belong to this species, but not as the Linnæan character now stands, ' *apice* spirali acuto.'

CONICA. **23.** Shell sub-cylindrical, flexuose, and spi-
　　ral at the base.

Serpula lumbricalis, Var. β. *Gmelin*, p. 3742.
Serpula, No. 6. *Schroeter Einl.* ii. p. 557.
La Trompe d'Eléphant. *Favanne*, i. p. 664. t. 5. f. C.
Rumphius, t. 41. f. 4. *Martini*, i. p. 52. t. 2. f. 15.
Inhabits the coasts of America, adhering to rocks. *Favanne.*
Amboyna. *Rumphius.*
Tube rather thick, brownish white, and coiled into a conical
spire of about eight whirls at the base; the summit rises
about nine inches from the rock to which the shell adheres,
but the length is probably thrice as great, owing to the coils
at the base; and the diameter of the tube is about two lines
and a half.

ARENARIA. **24.** Shell jointed, entire, distinct, and
　　rather flat beneath.

Serpula arenaria. *Linnæus Syst. Nat.* p. 1266. *Martini*,
　　i. p. 55. t. 3. f. 19, *A*, *B* and *C.* *Born Mus.* p. 439.
　　Schroeter Einl. ii. p. 550. *Gmelin*, p. 3743.
Serpula intiotina. *Ulysses's Travels*, p. 448.

Le Solitaire couleur de Rose. *Favanne,* i. p. 668. t. 6. f. P, and f. R. ?

Le Masier. *Adanson Senegal,* p. 165. t. 11. f. 5.

Bonanni Rec. and *Kirch.* i. f. 20 B. *Gualter,* t. 10. f. L and T. *Argenville,* t. 4. f. H.

Inhabits the coasts of India. *Linnæus.* St. Domingo. *Favanne.* Bay of Naples. *Ulysses.*

Shell solitary, rose-coloured, with obvious joints, and the upper surface marked somewhat like a Cornu Ammonis; the length is three inches, and the diameter at the aperture about five lines, and forms a flat coil about an inch and a half in diameter. Rumphius, t. 41. f. L, and Martini, t. 1. f. 10, which are quoted by Gmelin for a Variety, are described to be 'claboniformis,' and are more allied to *S. Ocrea* than the present species, which they are not in the least like. *S. arenaria* of the Mus. Lud. Ulrica is *Teredo gigantea.*

AFRA. 25. Shell taper, spiral, solitary, with three somewhat compressed volutions.

Serpula Afra. *Gmelin,* p. 3745.

Serpula, No. 11. *Schroeter Einl.* ii. p. 559.

Le Datin. *Adanson Senegal,* p. 165. t. 11. f. 4, A and B. *Martini,* i. p. 57. t. 3. f. 20, *A* and *B.*

Inhabits the shores of Goree on rocks and shells. *Adanson.*

Shell about two inches long, and two lines in diameter, solitary, yellowish or dark brown, and generally smooth, but sometimes marked with five or six longitudinal striæ; the volutions are sometimes flat, and turned regularly like those of a Nautilus, but they are frequently a little raised irregularly one above the other.

VOLVOX. 26. Shell sub-octangular, cancellated, and irregularly spiral.

La Chenille. *Favanne,* i. p. 653. t. 6. f. M.

Rumphius, t. 41. f. H. *Petiver Amb.* t. 16. f. 32. *Martini,* i. t. 2. f. 14.

Inhabits the East Indies. *Favanne.*

Most other authors have followed Linnæus in referring to Rumphius, t. 41. f. H, for *S. anguina,* but Favanne says it is a distinct species, and has thus described it: " Shell either solitary or in groups, somewhat quadrilateral, and coiled into an irregular spire of rather distant whirls; it is moderately thick, and cancellated with fine striæ, and has not any longitudinal cleft; the diameter of the tube is nearly four lines, and the length, measured along the course of the whirls, is

2 K 2

more than five inches, but the summit is elevated only two inches above the base."

ANGUINA.　27. Shell rather taper, somewhat spiral, with a longitudinal sub-articulated fissure.

Serpula anguina.　*Linnæus Syst. Nat.* p. 1267.　*Martini,* i. p. 51. t. 2. f. 13, *B* and *C.*　*Born Mus.* p. 440. t. 18. f. 15.　*Schroeter Einl.* ii. p. 552.　*Gmelin,* p. 3743. *Shaw Nat. Misc.* xiv. t. 571.

Siliquaria anguina.　*Lamarck Syst. des Anim.* p. 98.

Le Tire-bourre.　*Favanne,* i. p. 660. t. 6. f. G 1.

Lister Conch. t. 548. f. 2.　*Rumphius,* t. 41. f. 2.　*Petiver Amb.* t. 22. f. 11.　*D'Avila.* t. 4. f. E.

Inhabits the Indian Ocean.　*Pallas.*　Coasts of Amboyna. *Rumphius.*　China and Sicily.　*Humphreys.*

Shell whitish, clouded with brown or yellow, variously curved, and particularly towards one end is often spiral; the length varies from two to eight inches, and the diameter of the tube from five to eight lines, and it is marked throughout the whole length by a fissure formed of a continued series of oblong perforations.　The shell figured by Lister and Rumphius, and Martini, f. A, is much shorter, and consists wholly of a spire of five whirls, and is the Variety β of Linnæus; but it appears to me to be only the spiral end broken off from a larger tube.　Gmelin's Var. β is not in the least like the species, and is much more nearly allied to *Teredo gigantea.*

MURICATA.　28. Shell angular, muricated, with a longitudinal sub-articulated fissure.

Serpula muricata.　*Born Mus.* p. 440. t. 18. f. 16.　*Shaw Nat. Misc.* xiv. t. 575.

Serpula echinata.　*Gmelin,* p. 3744.

Serpula anguina.　*Brookes's Introd.* p. 142. t. 9. f. 131.

Serpula anguina, Var. γ.　*Gmelin,* p. 3743.

Serpula, No. 1.　*Schroeter Einl.* ii. p. 556; and No. 26. p. 565.

Gualter, t. 10. f. R.　*Martini,* i. t. 2. f. 8.

Inhabits the Indian Ocean.　*Born.*

This species has the longitudinal fissure and oblong perforations of *S. anguina,* and differs only in being of a rose-colour and armed with short spines.　Gualter has omitted to notice the longitudinal fissure, but his description and figure in all other respects answer so strikingly to this species, that there can remain no doubt of its being the same.　Martini's t. 2.

f. 8. is copied from Gualter's figure, and from thence Gmelin's *S. echinata* has been obviously taken.

ANNULARIS. 29. Shell sub-cylindrical, with annular contractions, and an obsolete longitudinal fissure.

Serpula lumbricalis, Var. γ. *Gmelin*, p. 3742.
Serpula, No. 7. *Schroeter Einl.* ii. p. 557.
Le Tire-bourre annulaire. *Favanne*, i. p. 662. t. 6. f. G 2.
Bonanni Rec. 1. f. 20 C. *Martini*, i. p. 53. t. 2. f. 16.
Inhabits ——
Shell thick, variously curved and twisted, and marked with annular contractions, which in Favanne's figure appear to be about a quarter of an inch apart; the colour is brownish yellow on the outside, and pale fawn-colour within; the longitudinal fissure is so very narrow as to be hardly observable, and is placed in a groove between two small elevated striæ. Favanne has not mentioned the length; but as it may be doubted whether the specimens were perfect from which the figures above referred to were taken, it probably extends to six or eight inches, and the diameter is near half an inch.

CORNU-COPIÆ. 30. Shell sub-conical, taper, rather obtuse, and twice twisted spirally; aperture orbicular.

Serpula Cornu-copiæ. *Gmelin*, p. 3745.
Serpula helicina. *Portland Cat.* p. 190. lot 4040.
Serpula, No. 25. *Schroeter Einl.* ii. p. 564.
Cornu Copiæ. *Born Mus.* p. 362. t. 13. f. 10 and 11, and Vign. at p. 361.
Cornu Copiæ monstrosum. *Chemnitz*, xi. p. 292. t. 211. f. 2092 and 2093.
Cornucopia helicina. *Shaw Nat. Misc.* xiv. t. 518.
Inhabits the Mauritius, where it burrows into stone and coral. *Humphreys.*
This rare and elegant shell is about two inches long, and eight lines in diameter at the broader end, from which it tapers gradually to an obtuse point; the colour is yellowish, with three variegated longitudinal brown bands; it has the aperture of a Turbo, with the habit of a Helix allied to *Pomatia* or *scalaris*, but its volutions are entirely unconnected and distant from each other, and the habitat which Mr. Humphreys has given precludes the possibility of its being a distorted land shell.

DECUSSATA. **31.** Shell taper, flexuose, irregularly contracted and marked with much elevated striæ.

Serpula decussata. *Gmelin,* p. 3745.
Serpula, No. 8. *Schroeter Einl.* ii. p. 558.
Le Bois de Charme. *Favanne,* i. p. 652. t. 6. f. L.
Lister Conch. t. 547. f. 4. *Martini,* i. p. 53. t. 2. f. 17.
Inhabits Barbadoes. *Lister.* Coast of America. *Favanne.*
Martini's figure is copied, and his description (as also are those of Schroeter and Gmelin) is taken wholly from Lister's, but there can be no doubt that *Le Bois de Charme* belongs to the same species, and of this shell Favanne has given the following account : " Le Bois de Charme est un Tuyau d'Amérique, à stries longitudinales, fines et serrées, médiocrement épais, de figure conique fort alongée et sinueuse, se repliant sur lui-même, avec des nodosités ou renflemens de distance en distance. Il est en dehors d'un fauve rougeâtre foncé, et blanchâtre en dedans ; son ouverture antérieure est mince et cylindrique, mais sa pointe est conique et fermée. Il est représenté ici solitaire et de grandeur ordinaire, quoiqu'il se trouve souvent en groupes, et d'un volume plus considérable. Il est peu commun." In Lister's figure a few irregular transverse striæ are represented, which are not noticed in the above description ; but they appear too irregular to justify Gmelin's introduction of ' decussated striæ' in the specific character.

VERMICULARIS. **32.** Shell taper, subulate, curved, and transversely wrinkled.

Serpula vermicularis. *Linnæus Syst. Nat.* p. 1267, *Pennant Zool.* iv. p. 146. t. 91. f. 158. *Martini,* i. p. 61. *Da Costa Brit. Conch.* p. 18. t. 2. f. 5. *Schroeter Einl.* ii. p. 553. *Muller Zool. Danica,* iii. p. 9. t. 86. f. 7 to 9. *Gmelin,* p. 3743. *Donovan,* iii. t. 95. *Montagu Test.* p. 509. *Maton and Racket, in Lin. Trans.* viii. p. 243. *Dorset Cat.* p. 60. t. 22. f. 5.
Tubus vermicularis. *Ellis Corallines,* t. 38. f. 2.
Common on the coasts of Europe on stones and shells.
S. vermicularis differs from *S. triquetra* in being more slender and cylindrical, without any carinated edge, more strikingly wrinkled transversely, and in size, which is generally much larger : from the circumstance of only one trumpet-shaped proboscis being represented in the figures of Ellis and Donovan, Mr. Montagu has referred to them for *S. triquetra,*

but they want the carinated edge, and have every other appearance of belonging to this species. It has been already remarked, that *S. intricata* is not distinct, and *S. reversa* of Montagu appears to be only a reversed Variety of this species.

TUBULARIA. 33. Shell taper, subulate, and flexuose, with the larger end detached and ascending.

Serpula tubularia. *Montagu Test.* p. 513. *Maton and Racket, in Lin. Trans.* viii. p. 244.
Inhabits the coasts of Devonshire on Shells. *Montagu.*
The larger end of the shell is frequently detached for half its length, and ascends at a considerable angle, wherein it principally differs from *S. vermicularis*, which is usually attached to some other substance throughout its whole length. According to the observations of Mr. Montagu, the animal of this species is an Amphitrite, and in *S. vermicularis* it is a Terebella.

DENTICULATA. 34. Shell taper, subulate, nearly straight, with the sides toothed, and a longitudinal glabrous rib in the middle.

Serpula denticulata. *Gmelin,* p. 3746.
Serpula, No. 33. *Schroeter Einl.* ii. p. 570. t. 6. f. 18.
Inhabits the sea on Lepas Tintinnabulum. *Schroeter.*
Shell about three-quarters of an inch long, white, and straight, except at the narrow end, where it is slightly curved; aperture orbicular.

AQUARIA. 35. Shell taper, straight, with a radiated border and perforated disk at the summit.

Serpula Aquaria. *Burrow's Elements,* p. 178. t. 22. f. 3.
Serpula Penis. *Linnæus Syst. Nat.* p. 1267. *Martini,* i. p. 43. t. 1. f. 7. *Born Mus.* p. 441. *Schroeter Einl.* ii. p. 554. *Gmelin,* p. 3744. *Brookes's Introd.* p. 141. t. 9. f. 136.
Serpula perforata. *Shaw's Nat. Misc.* vi. t. 188.
Penicillus Javanus. *Bruguiere Enc. Meth.* p. 128. *Lamarck Syst. des Anim.* p. 98.
L'Arrosoir. *Favanne,* i. p. 640. t. 5. f. B.
Bonanni Kirch. i. f. 38, and *Rec. Supp.* f. 45. *Lister Conch.* t. 548. f. 3. *Rumphius,* t. 41. f. 7. *Petiver Amb.* t. 21. f. 17. *Gualter,* t. 10. f. M. *Argenville,*

t. 3. f. G.   *Knorr,* iv. t. 28. f. 1, and vi. t. 40. f. 1.   *Da Costa Elements,* t. 2. f. 8.

Variety.   Sub-fusiform, with the summit only slightly bordered, and the perforated disk small.

Penicillus Novæ Zelandiæ. *Bruguiere Enc. Meth.* p. 129.

L'Arrosoir de la Nouvelle Zelande.   *Favanne,* i. p. 642. t. 79. f. E.

Inhabits the coasts of Java. *Linnæus.*   Amboyna. *Rumphius.* Coromandel. *D'Avila.*   Moluccas. *Favanne.*   Madagascar, and the Nicobar Islands. *Humphreys.*

This shell when perfect is nearly a foot long, and more than an inch in diameter at the dilated summit, which is convex, and perforated so as to resemble the spout of a watering-pot; the summit has a radiated border, and the perforations are circular, excepting the one in the center which is linear; the colour is whitish with a slight tinge of pale red or grey, and the whole of the shell very brittle; according to Favanne, the lower part by which it adheres to the rocks is flexuose and twisted, and the upper part, which is straight, proceeds from it nearly at a right angle.   The Variety appears to have been wholly described by Favanne from a drawing which was sent him, and without any additional authority it has been arranged by Bruguiere as a separate species; it differs in being thicker, and much more distended below the summit, towards which it is rounded off; the perforated disk is also much smaller, and the radiated border hardly projects at all, so that the general outline of the shell is fusiform.

OCREA.   36. Shell sub-cylindrical, widening at the base, and shaped like a boot.

Serpula Ocrea.   *Gmelin,* p. 3744.

Serpula, No. 2.   *Schroeter Einl.* ii. p. 556.

Le Tuyau en forme de Botte. *Favanne,* i. p. 644. t. 5. f. D. *Rumphius,* t. 41. f. K. *Petiver Amb.* t. 20. f. 12. *Martini,* i. p. 44. t. 1. f. 9.

Inhabits the East Indian Seas on coral. *Rumphius.*

This shell is represented to be nearly two inches high, and eight lines broad, of a dark brown colour, and widened at the base so as almost exactly to resemble a man's boot; all the other figures have been copied from Rumphius, on whose sole authority it rests; and it is probably nothing more than an accidentally distorted Variety of some other species.

PROBOSCIDEA.   37. Shell taper, smooth, with the broad end straight, and transversely plaited.

Serpula proboscidea. *Gmelin*, p. 3745.
Serpula, No. 9, and No. 10. *Schroeter Einl.* ii. p. 559.
Proboscis Elephantis minor, et major. *Martini*, i. p. 54.
   t. 2. f. 18, *A* and *B*.
Inhabits ——
Shell from two to four inches long, and the diameter at the
   larger end is nearly half an inch; the colour is white clouded
   with black or yellow, and in the figures the whole length is
   represented straight, except that there is one bend at the
   distance of about one-third of the length from the broader
   end. It appears to be a badly defined and very uncertain
   species.

PROTENSA. 38. Shell sub-cylindrical, with transverse plaits and wrinkles, and divided by
   internal transverse partitions.

Serpula protensa. *Gmelin*, p. 3744.
Serpula erecta. *Ulysses's Travels*, p. 448.?
Serpula, No. 5. *Schroeter Einl.* ii. p. 557.
Le Tuyau cordé. *Favanne*, i. p. 676. t. 5. f. F.
Rumphius, t. 41. f. 3. *Petiver Amb.* t. 21. f. 18. *Martini*,
   i. p. 46. t. 2. f. 12 *A*.
Inhabits the coasts of Amboyna. *Rumphius*.
Shell about four inches long, and three lines in diameter,
   nearly equal at both ends, slightly flexuous, and of a bluish
   white, or reddish or brownish colour; the outer surface is
   strongly plaited and wrinkled transversely, so as to give it
   somewhat of an annulated appearance, and the inside is divided by transverse partitions into numerous chambers without any connecting siphon. From the appearance of the
   figures, it may be suspected that *S. protensa* and *Teredo
   Utriculus*, are the same species. *Le Serpenteau* of Favanne, i. p. 676. t. 68. f. N, which is taken from Gualter,
   t. 10. f. LL, is very nearly allied, but is much more flexuous.

GIGANTEA. 39. Shell taper, somewhat three-sided,
   and the aperture at the narrow end furnished
   with a conical tooth.

Serpula gigantea. *Pallas Misc. Zool.* p. 139. t. 10. f. 2 to
   10. *Born Mus.* p. 441. *Gmelin*, p. 3747.
Seba, i. t. 29. f. 1 and 2.
Inhabits the coasts of Curaçoa, and other Caribbee Islands,

generally in coral beds, and incrusted with Millepora alcicornis. *Pallas.*

Shell about six inches long, and as broad as a little finger, tubular, thick, and flexuose; the tooth at the mouth is produced from an obsolete rib which runs along the whole shell; the surface is smooth, of a red, white, yellow, or violet colour on the outside, and the aperture is yellow and striated within. Pallas has given several figures, and a long description of the animal.

Genus XXXV.

TEREDO:

SHELL TAPER, FLEXUOUS, AND TUBULAR, WITH
TWO HEMISPHERICAL VALVES AT ONE END OF
THE ANIMAL, AND TWO LANCEOLATE ONES AT
THE OTHER.

GIGANTEA. 1. Shell taper, sub-cylindrical, nearly
straight, thick, and pellucid, with two internal tubes at the summit.

Teredo gigantea. *Sir E. Home, in Phil. Trans. for* 1806,
p. 276, and t. 10 and 11. f. 1 to 7.
Serpula Polythalamia. *Linnæus Syst. Nat.* p. 1266. *Martini,* i. p. 40. t. 1. f. 6 and 11. *Schroeter Einl.* ii. p.
549. *Gmelin,* p. 3742.
Serpula arenaria. *Linnæus Mus. Lud. Ulr.* p. 700.
Serpula gigantea. *Portland Cat.* p. 6, lot 97.
Serpula, No. 4. *Schroeter Einl.* ii. p. 557.
Fistulana cornicula. *Lamarck Syst. des Anim.* p. 129.
Le grand Tuyau de Mer. *D'Avila,* i. p. 97, No. 52.
Le Cierge. *Favanne,* i. p. 673. t. 5. f. N.
Rumphius, t. 41. f. D and E. *Klein Tubul.* t. 1. f. 3.
Seba, iii. t. 94, largest fig.
Inhabits the sandy shores of Ceram, one of the Molucca
Islands, in shallow water. *Rumphius.* Coasts of the Island
of Battoo near Sumatra. *J. Griffiths, Esq.*
It appears to me to be almost beyond a doubt that the *Serpula
Polythalamia* of Linnæus, and the *Teredo gigantea* of the
Philosophical Transactions are the same, for the two smaller
tubes appear in Rumphius's figure to be still partially en-

closed by the outer cylinder, and the latter, of which the summit is said by Mr. Griffiths to be very brittle, might probably, either to shew the internal structure, or by accident, have been more broken away than usual; the Linnæan specific name is moreover rather peculiarly applicable from the circumstance of the base being divided into separate chambers, as has been observed by Sir E. Home, who after relating the manner in which the lower end of a full-grown shell is closed, adds, that in some of Mr. Griffiths's specimens " the animal has receded from its first enclosure, and has formed a second three inches up the tube, and afterwards a third two inches further on." The discovery of the two boring shells and two flattened opercula prove the necessity for its removal to Teredo, and the divided summit gives it a place in Lamarck's Genus Fistulana. Pallas, in his Miscellanea Zoologica, p. 140, asserts that Rumphius's shell differs only from *Teredo navalis* in being an inhabitant of sand instead of wood; but Favanne, in describing *Le Cierge,* maintains a contrary opinion, and among other marks of distinction points out the bifurcated summit. Mr. Griffiths could not obtain a perfect specimen, and the following remarks are extracted from a memoir with which he favoured the Royal Society, and which in their Transactions immediately precedes the observations of Sir E. Home: " The length of the longest of these shells that came into my possession was five feet four inches, and the circumference at the base nine inches, tapering upwards to two inches and a half; the colour on the outside milk-white, the inner surface rather of a yellow tinge. This specimen was nearly perfect, having a small part of the lower extremity entire. I have others of various dimensions, a very good one about three feet long, and four inches round, tapering to an inch and a half at the point." The outer surface is uneven, and strongly wrinkled transversely, but the inner surface is perfectly smooth; the internal tubes at the summit are about eight or nine inches long; in a fragment rather more than an inch and a half in diameter, which Sir Joseph Banks gave me, the thickness of the tube is a quarter of an inch, and the structure is so singularly radiated, that I at first mistook it for a mineral stalactite. The *Serpula arenaria* of the 12th edition of the Systema Naturæ is an entirely different species; but the following description in the Mus. Lud. Ulr. is more likely to have been intended for the present shell: " Testa crassa, formâ intestini, curvata, irregularis, cinerea, inamoena, extrorsum sensim latior apice angustiore, sæpe in duos ramos bifida."

NAVALIS. **2. Shell sub-cylindrical, very thin and smooth.**

Teredo navalis. *Linnæus Syst. Nat.* p. 1267. *Pennant Zool.* iv. p. 147. *Schroeter Einl.* ii. p. 572. *Gmelin,* p. 3747. *Montagu Test.* p. 527, and *Supp.* p. 7. *Home in Phil. Trans.* 1806. p. 276. t. 12 and 13. *Donovan,* v. t. 145. *Maton and Racket, in Lin. Trans.* viii. p. 249. *Dorset Cat.* p. 60. t. 18. f. 21. *Burrow's Elements,* p. 178. t. 22. f. 4.

Teredo vulgaris. *Lamarck Syst. des Anim.* p. 128.

Serpula Teredo. *Da Costa Brit. Conch.* p. 21.

Le Taret. *Adanson Senegal,* p. 263. t. 19. f. 1.

Favanne, t. 60. f. C 1, C 2, and C 3. *Enc. Meth.* t. 167. f. 1 to 3.

Inhabits the sea in the sides and bottoms of ships and other timber, which it perforates.

Shell sometimes a foot long, and about three-quarters of an inch in diameter at the lower extremity, from which it tapers slightly to the summit, but the shell in these climates is rarely more than half so large ; it is thin, brittle, slightly flexuous, and of a whitish colour. Linnæus, considering the valves which are placed at the two ends of the animal to be of the nature of opercula, has arranged this shell among the Univalves ; whilst others, who consider them to be similar to the accessory valves of a Pholas, think it ought to be placed with the Multivalves. Both the shell and the animal have been described at great length in the works above quoted, by Adanson, Home, and Montagu. Another species in Lamarck's Système des Animaux sans Vertèbres is mentioned under the name of *Teredo bipalmulata,* as follows : " Ce Taret, qui vit aussi dans l'intérieur des bois plongés sous les eaux marines, est plus grand que le Taret commun. Il est remarquable par deux bras ou palettes articulées, subpinnées, situées à son extrémité inférieure. On en voit un individu dans la Collection anatomique du Muséum, qui est devenue si intéressante par les belles dissections et préparations du citoyen Cuvier."

UTRICULUS. **3. Shell sub-cylindrical, thick, flexuous, and the aperture at the summit divided by a partition.**

Teredo Utriculus. *Gmelin,* p. 3748.

Kæmmerer Cab. Rudolst. p. 7. t. 1.

Inhabits wood in the sea. *Kæmmerer.*

This shell is stated to be seven inches long, and appears by Kæmmerer's figure to be four or five lines in diameter at one end, which is nearly twice as broad as the other ; it is white, sub-pellucid, taper, transversely wrinkled, and considerably curved irregularly ; the aperture is said to be oval and divided by a partition in the middle.

CLAVA.   4. Shell with one end club-shaped, and the other narrower, incurved, obtuse, and perforated in the middle.

Teredo Clava.   *Gmelin*, p. 3748.
Teredo, No. 1.   *Schroeter Einl.* ii. p. 574. t. 6. f. 20.
Fistulana gregata.   *Lamarck Syst. des Anim.* p. 129.
Walch Naturf. x. t. 1. f. 9 and 10.   *Spengler Naturf.* xiii. t. 1. f. 1 to 11, and t. 2. f. 12 to 14.   *Guettard Mem.* t. 70. f. 6 to 9.   *Enc. Meth.* t. 167. f. 6 to 16.
Inhabits the coasts of Coromandel. *Schroeter.*   In the seed-vessels of Xylocarpus Granatum. *Gmelin.*
Shell nearly two inches long, and about half an inch broad at the lower extremity, from which it tapers upwards, and the other end is said to be narrower, obtuse, somewhat incurved, and perforated ; it is generally found in clusters, and is more or less flexuous, of a brownish colour, with the outside rough and the inside smooth. I have in this instance quoted Walch and Spengler on the authority of Schroeter, and Guettard on that of Lamarck. Besides *F. cornicula* and *F. gregata,* which are above noticed, Lamarck in his Système des Animaux, has given the following notices of two other Fistulanæ, and *Le Pillon* of Favanne, i. p. 672. t. 5. f. K, may probably also belong to the same natural family :

Fistulana Clava.   *Enc. Meth.* t. 167. f. 17 to 22.
Fistulana lagenula.   *Enc. Meth.* t. 167. f. 23.

ADDENDA ET CORRIGENDA.

P. 2. l. 6. read *Chemnitz*, viii. p. 271. t. 94, &c.

p. 8. l. 13.   The reference to *Enc. Meth.* should begin the next line.   The same remark applies to the references at p. 13. l. 16, and p. 17. l. 31, both beginning *Enc. Meth.*, at p. 28. l. 7, beginning with *Phil. Trans.*, and at p. 29. l. 11, beginning *Rumphius.*

p. 11. l. 29. read *Montagu Test.* p. 3, and *Supp.* p. 1.

p. 15. l. 17 and 19.   PUNCTATA., *and* Lepas punctata.

    l. 35.   are punctated like a thimble.

p. 16. l. 7.   STRIATA. 4. Shell conical, truncated, &c.

    l. 22.   referred to for *L. balanoides,*

p. 19. l. 8.   *Pennant's Zool.* iv. p. 73. t. 38. f. 7.

p. 23. l. 12.   *Lister Conch.* t. 443. f. 285.

    l. 14.   *Argenville,* t. 26. f. A.

    l. 25 et passim. *Ulysses* instead of *Sir C. Ulysses.*

p. 24. l. 5.   Balanus Tintinnabulum, Var. *Bruguiere, &c.*

    l. 14.   differ from *L. Tintinnabulum,*

p. 25. l. antep.   *Petiver Gaz. &c.*   Also at p. 29. l. 12; p. 40. l. 18; p. 42. l. 30; and p. 44. l. 3.

p. 27. l. 12.   SPONGEOSA.

p. 28. l. 12 and 18.   *L. spongeosa.*

p. 31. l. 13 f. b.   Lepas dilata.   *Donovan, &c.*

p. 32. l. 24.   Balanus anatiferus.   *Da Costa, &c.*

p. 35. l. 17.   Pholas muricatus.   *Da Costa, &c.*

p. 36. l. 18 f. b.   Pholas costatus.   *Linnæus, &c.*

p. 37. l. 9.   *Gualter,* t. 105. left hand fig. E.

    l. ult.   *P. pusillus.*

p. 38. l. 24 and 25.   the Linnæan *P. striatus,* and *P. pusillus,*

p. 42. l. 27.   *Da Costa, Brit. Conch.* p. 233. t. 16. f. 1.

p. 43. l. 6.   *Da Costa, Brit. Conch.* p. 232.

p. 45. l. antep.   *Montagu Test.* p. 44.

p. 52. l. 15 f. b.   *Lister Anim. Ang. App.* t. 1. f. 1, &c.

p. 53.   The *Mya rhomboidea* of Schroeter here mentioned is the *Unio littoralis* of Lamarck, Syst. des Anim. p. 114.

p. 59. l. 8 f. b.   *Chemnitz,* vi. p. 46. t. 4. f. 29 and 30.

p. 60. l. 15 f. b.   *Pennant Zool.* iv. p. 84. t. 46. f. 24.

p. 77. l. 8.   *Linnæus Syst. Nat.* p. 1117.

p. 82. l. 16 f. b.   *Linnæus Syst. Nat.* p. 1117.

p. 86. l. 21.   *Gronovius Zooph.* p. 287. t. 1. f. 3.

p. 93. l. 5 f. b.   Tellina calcarea.   *Chemnitz, &c.*

p. 95. l. ult.   Tellina, No. 84.   *Schroeter, &c.*

p. 104. l. 16 f. b.   *Argenville,* t. 27. f. 9.

p. 107. l. 12. read *Chemnitz*, vi. p. 321. t. 30. f. 322 and 323.

l. 27. *Chemnitz*, vi. p. 320. t. 30. f. 321.

p. 140. l. 16 f. b. *Chemnitz*, x. p. 350. t. 170. f. 1656.

p. 141. l. 10 f. b. *Gmelin*, p. 3221.

p. 155. l. 11, et passim, read *Schreibers Conch.*

p. 170. l. 9 f. b. *Chemnitz*, xi. p. 225. t. 201. f. 1974.

p. 177. l. 4. Venus, No. 135. *Schroeter Einl.* iii. p. 195.

p. 185. l. 24. without any references, has given nothing, &c.

p. 187. l. 14. Venus purpurata, *Gmelin, &c.*

l. 15. Venus, No. 80. *Schroeter, &c.*

p. 194. between l. 1 and 2. Lucina Jamaicensis. *Lamarck, Syst. des Anim.* p. 124.

p. 200. l. 1 and 5. ÆQUIVOCA., *and* Venus æquivoca.

p. 202. l. 7 f. b. *Chemnitz*, x. p. 354. t. 171. f. 1661.

p. 228. l. 10. *Chemnitz*, vii. p. 195. t. 55. f. 542.

p. 231. l. 22. but Chemnitz's figure represents, &c.

p. 252. l. 9 and 6. f. b. PES-LUTRÆ., *and* Ostrea Pes Lutræ.

p. 335. l. 3 f. b. L'Ecope de Batelier. *Favanne, &c.*

p. 368. l. 1. CÆRULESCENS.

p. 417. l. 12. Conus Terebellum, Var. γ. *Gmelin, &c.*

p. 436. l. 4 f. b. *C. leucogaster*, p. 3413.

p. 458. l. 17 and 18. CRIBRARIA., *and* Cypræa cribraria.

p. 469. l. 26 and 29. STAPHYLÆA., *and* Cypræa staphylæa.

p. 519. l. 17 f. b. *Favanne*, ii. p. 836. t. 19. f. I 2.

p. 520. l. 1. *Favanne*, ii. p. 835. t. 19. f. I 1.

p. 532. l. 17. Voluta Ziervoyelii. *Gmelin, &c.*

p. 559. l. 17 f. b. Mitra millepora. *Lamarck, &c.*

l. 16 f. b. Voluta Cribrum. *Solander's MSS.*

p. 590. l. 1. *Born Mus.* p. 243.

p. 606. l. 9 f. b. *Schreibers Conch.* i. p. 259.

p. 611. between l. 10 and 11. Concholepas Peruviana. *Lamarck Syst. des Anim.* p. 70.

p. 621. between l. 24 and 25. Eburna flavida. *Lamarck Syst. des Anim.* p. 78.

p. 625. l. 11. Buccinum glaberrima. *Gmelin*, &c.

p. 654. l. 8 f. b. *Schroeter Einl.* i. p. 416.

p. 655. l. 23. *Schroeter Einl.* i. p. 417.

p. 678. l. 21. *Buccinum torridum.*

p. 772. l. 14. *Linnæus Syst. Nat.* p. 1225.

p. 902. l. 19 f. b. *Scopoli Del Insub.* i. p. 66. t. 25. f. A.

p. 925. l. 6. *Lister Anim. Ang.* t. 2. f. 4, &c.

p. 961. l. 1. *Linnæus Syst. Nat.* p. 1249.

p. 1036. l. 9. *Humphreys Conch.* t. 4. f. 5.

FINIS.

INDEX,

WITH THE SYNONYMA IN ITALICS.

The Second Volume being paged in continuation of the First, commences with page 581.

2 г.

INDEX.

INDEX.

2 L 2

INDEX.

INDEX.

INDEX.

2 M 2

INDEX.

INDEX.

INDEX.

INDEX.

INDEX.

2 N 2

INDEX.

INDEX.

/

INDEX.

INDEX.

J. M'Creery, Printer,
Black-Horse-Court, London.

ae

HN

Check Out More Titles From HardPress Classics Series In this collection we are offering thousands of classic and hard to find books. This series spans a vast array of subjects – so you are bound to find something of interest to enjoy reading and learning about.

Subjects:
Architecture
Art
Biography & Autobiography
Body, Mind &Spirit
Children & Young Adult
Dramas
Education
Fiction
History
Language Arts & Disciplines
Law
Literary Collections
Music
Poetry
Psychology
Science
…and many more.

Visit us at www.hardpress.net

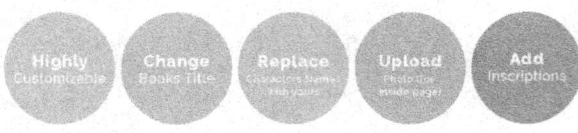